Population Dynamics for Conservation

Population Dynamics for Conservation

LOUIS W. BOTSFORD, J. WILSON WHITE, AND ALAN HASTINGS

OXFORD
UNIVERSITY PRESS

OXFORD
UNIVERSITY PRESS

Great Clarendon Street, Oxford, OX2 6DP,
United Kingdom

Oxford University Press is a department of the University of Oxford.
It furthers the University's objective of excellence in research, scholarship,
and education by publishing worldwide. Oxford is a registered trade mark of
Oxford University Press in the UK and in certain other countries

First Edition published in 2019
Impression: 1

Published in the United States of America by Oxford University Press
198 Madison Avenue, New York, NY 10016, United States of America

British Library Cataloguing in Publication Data
Data available

Library of Congress Control Number: 2019945433

ISBN 978–0–19–875836–5 (hbk.)
ISBN 978–0–19–875837–2 (pbk.)

DOI: 10.1093/oso/9780198758365.001.0001

Printed and bound by
CPI Group (UK) Ltd, Croydon, CR0 4YY

Preface

Over the past several decades, a useful body of knowledge about how ecological populations function has accumulated from theoretical analyses and empirical observations. Knowledge of these general characteristics of population behavior can improve our ability to conserve threatened populations. Our hope is that a formal presentation of this knowledge will benefit ecologists, managers, and students by reducing the need to reinvent the wheel for every new problem or question. In this book we describe the lessons we have learned about population dynamics in the form of a consistent mathematical theory tying together a range of specific topics. For example, readers will see how analysis of insect life histories from the 1970s provides valuable information regarding both the adaptive management of marine protected areas and the way that age-structured populations respond to randomly varying environments.

From our experience in several areas of applied ecology there appears to be a gap between theoreticians and practitioners. In the decision-making processes associated with population resource management, decision-makers and associated scientific staff are often troubled by their lack of understanding of the workings of mathematical models whose results they must use. Moreover, the necessity of using population dynamic models in resource decision-making has even been construed as causing a lack of transparency in the process, thus it is perceived as interfering with a fundamental guarantee of most decision-making processes.

A better understanding of population dynamics will also allow empiricists to take advantage of the fact that their observations are (or are not) consistent with population dynamics theory. The resulting synergism between empirical observations and theory will lead to more rapid development of our understanding of population dynamics. Observations can be specific tests of theory, while theory is a way of filling in the gaps between observations. Also, theory can provide a check on conclusions drawn from noisy data; in some cases, the mechanism proposed to underlie empirical results may not be possible.

The knowledge gap between theoreticians and empiricists motivates a special aspect of this book: we attempt to make the underlying principles of population dynamics available to those without strong mathematical backgrounds. We do this by stressing intuitive principles, in addition to their underlying mathematical basis. For example, most ecologists readily grasp the concept that a population will persist only if each individual reproduces more than enough in their lifetime to replace themselves, however they rarely consider what that concept implies for the robustness of estimating probabilities of extinction, or the persistence of marine fish in a specified pattern of marine protected areas. Some of the complicated equations arising from conditions for population persistence in this book become much more "user friendly" when interpreted in the light of this concept. We propose that the intuition gained from this book can enable valuable application by the reader later, regardless of whether the reader retains detailed knowledge of the underlying mathematics. This book is a unified, comprehensive survey of the field of population dynamics for the more mathematically trained modeler, but it will also serve the user of modeling results (i.e. resource ecologists and managers), enabling them to understand and apply the principles of population dynamics.

Ideally, the examples in a book about population dynamics would be uniformly distributed across terrestrial, freshwater, and marine plant and animal species, but that is seldom the case. Like most ecologists, we have been attracted to specific organisms, specific scientific questions, and specific problems. As a consequence the reader will find more examples about marine animals than other taxa. However, this book is not solely about marine plants and animals; we have attempted to point to a broad range of taxa as examples. Also, each question addressed here in a marine context will likely have some practical relevance in other realms. More specifically, many will address population persistence, a central goal common to most conservation problems regardless of the focal species.

Among the practical ecological problems mentioned in this book, one will find a preponderance of applications to fisheries management. This stems simply from our backgrounds and experience, but on the other hand fisheries management has distinct advantages as an example. Fisheries management involves an essential conflict between the use of a natural resource on the one hand, and avoiding so much "use" that the population fails to persist, on the other. Thus it serves as a caricature of many problems where human activities that benefit society also diminish population viability, and decisions have to be made regarding how much use is possible without losing the sustainability of the species (one can easily think of examples in wildlife or forestry management). Moreover, fisheries is a unique example of such a problem in that there are probably more data available for fisheries than any other practical ecological problem. For example, few problems involving endangered species have 30 or more years worth of abundance data and removal mortality, as well as fairly good estimates of the dependence of reproduction, growth, and mortality on age. Thus, fisheries are an example of the difficulties of wise management of renewable resources, even with relatively large amounts of data.

We begin this book with a philosophical discussion that leads to the definition of a state variable, the characteristics of descriptive variables that allow mathematical models to function properly. In population dynamics, the various state variables are those characteristics of individual organisms that we think will tell us how individuals will contribute to population dynamics (i.e. age, size, stage, space). We have organized this book in terms of different individual state variables, as chapters about models that describe populations in terms of how their distribution of individuals over that state variable varies with time (i.e. age, size, stage, spatial distributions), at least in the early chapters. This remains a key concept throughout the book, allowing us to preserve the realism of our dynamic depictions of various kinds of populations. The second chapter describes simple models, those that do not track distributions, rather track only total abundance or total biomass. These models are widely used in ecology and its applications. While this chapter identifies the shortcomings of these models, a number of basic concepts of population dynamics that appear in later chapters are first illustrated in this chapter. Chapters 3 and 4 introduce age-structured models in their most primitive, idealized form, i.e. without density dependence and random environmental variability. In that chapter we introduce the idea that population persistence depends on *replacement*, a concept that plays a fundamental role in later chapters throughout the book.

Chapters 5 and 6 move beyond age structure to size and stage structure. Ecologists use models with size and stage structure because size and stage are commonly observed characteristics of individuals, and vital rates often depend on them. We will see there that the way that stage structure is represented in some of the stage-structured matrix models commonly used in ecology leads to only a rough approximation to the true dynamics, and other options are currently becoming available.

Chapters 7 and 8 add the real-world characteristics often necessary to mimic real age-structured populations, density dependence, and random environmental variability. Adding these essential elements of real population dynamics changes model results substantially, confirming the necessity for their use in real applications. These chapters complete our description of single populations of individuals all presumed to be at one location.

In Chapter 9 we move to descriptions of populations distributed over space, which leads to completely new results. This is primarily because of the many ways of moving from state to state, and the multi-dimensionality of the state (as compared, for example, to growth rate from one size to the next). However, some general concepts, such as the requirement for a replacement for persistence, still hold, and remain conceptually useful points of view.

Chapters 10 and 11 focus on applications of the theory developed in the first nine chapters. We separate these two into conservation biology and marine fisheries, although we stress the general similarities between these two areas of application, which stem from their common concern for population persistence.

Finally, in Chapter 12, we reflect upon what we have learned. That chapter ends up summarizing a somewhat fresh and different view of the science of population dynamics, being based on what we thought before writing this book, and as new insights that became obvious while writing it. We hope that whether the reader is a population modeler, an empirical population biologist, a resource manager, or a non-governmental policy advocate, they will come away with a different, enhanced view of population dynamics.

We end this preface with a brief description of the book's genesis, in part to give the reader some sense of the arc of time over which it has developed. Two of the authors were planning to do the bulk of the writing during the first year of life of the daughter of one of them. This daughter is now almost as old as the third author; yet it is clear that the book is even more timely now than when originally planned. Adding the third, younger author provided both the energy and impetus to bring this long dormant project to completion. Both the techniques and understanding available have grown during the intervening decades and the need to use these ideas in managing populations has only grown and become more obvious. In contrast to the pessimistic view that every population is unique, it has become clear that the ideas we present have broad applications; we believe that this book will both make clear the utility and universality of population thinking and provide many with the tools needed to apply these ideas to their particular problems. To that end, the model code (in the MATLAB language) used to perform the model analyses and create the figures in this book is available online, at https://github.com/jwilsowhite/PDC_book.

We close by thanking the many funding sources that have supported the development of this work, particularly the National Science Foundation (including U.S. GLOBEC) and California Sea Grant. We are also grateful to our families for their support: LWB's wife, Marylynn Barkley, JWW's wife, Susanne Brander, and their children, and AH's wife, Elaine Fingerett, and their children; AH also appreciates his family's amusement at the length of time it has taken for this book to reach fruition.

Contents

1 Philosophical approach to population modeling 1

1.1 Simplicity versus complexity, and four characteristics of models 2
1.2 Logical basis for population modeling 6
1.2.1 Deductive reasoning and the scientific uses of modeling 6
1.2.2 Inductive reasoning and practical applications of modeling 9
1.2.3 Consequences of deductive and inductive logic for population dynamics 10
1.3 The state of a system 11
1.3.1 Models of *i*-states and *p*-states 12
1.3.2 Individual based models (IBM) 14
1.4 Uncertainty and population models 15
1.5 Levels of integration in ecology 17
1.6 State of the field 18

2 Simple population models 21

2.1 The first population model—the rabbit problem 22
2.2 Simple linear models (exponential or geometric growth) 26
2.3 Simple nonlinear models (logistic-type models) 30
2.3.1 Continuous-time logistic models 30
2.3.2 Discrete-time logistic models 34
2.4 Illustrating population concepts with simple models 37
2.4.1 Illustrating dynamic stability with simple, linear, discrete-time models 37
2.4.2 Dynamic stability of simple nonlinear models 42
2.4.3 Quasi-extinction in random environments with a discrete-time linear simple model 45
2.4.4 What does the simple logistic model tell us about managing for sustainable fisheries? 51
2.5 What have we learned in Chapter 2? 52

3 Linear, age-structured models and their long-term dynamics 54

3.1 The continuity equation and the M'Kendrick/von Foerster model 55
3.1.1 Solving the M'Kendrick/von Foerster model 61
3.2 The renewal equation—Lotka's model 64
3.3 The Leslie matrix 66
3.3.1 Solving the Leslie model 69
3.3.2 The stable age distribution 71
3.4 Mathematical theory underlying the Leslie matrix 73
3.4.1 The Perron–Frobenius theorem 77
3.5 Sensitivity and elasticity of eigenvalues: the *Totoaba* example 77

3.6 Handling the oldest age classes: age-lumping, terminal age classes, and post-reproductive ages 82
3.7 What have we learned in Chapter 3? 84

4 Age-structured models: Short-term transient dynamics 87

4.1 The other eigenvalues 88
4.1.1 An example of cyclic transient dynamics 90
4.2 How the dependence of reproduction on age influences these cycles 93
4.2.1 Semelparous species and imprimitive Leslie matrices 93
4.2.2 Cycle period: the mean age of reproduction and the echo effect 96
4.2.3 How age structure influences the occurrence of cycles 98
4.2.4 Convergence to the asymptotic dynamics 100
4.2.4.1 Rate of convergence to the stable age distribution: the damping ratio 100
4.2.4.2 The distance to the stable age distribution 101
4.2.4.3 Example: adaptive management of marine protected areas 104
4.3 Transient responses to ongoing environmental variability 105
4.3.1 Determining the equilibrium of a nonlinear age-structured population 106
4.3.2 The frequency response of a population 108
4.3.3 Cohort resonance 112
4.3.3.1 Analysis of cohort resonance 113
4.3.3.2 Cohort resonance: effects of life history, fishing, and eigenvalues 116
4.3.4 Extreme period-T cycles: cyclic dominance in sockeye salmon 118
4.4 What have we learned in Chapter 4? 120

5 Size-structured models 122

5.1 The size-structured M'Kendrick/von Foerster model 124
5.1.1 The solution to the size-structured M'Kendrick/von Foerster model 126
5.1.2 Adding reproduction to obtain a complete population model 127
5.2 Stand distributions 128
5.3 Cohort distributions 134
5.4 Numerical methods 138
5.4.1 Grid-based method 138
5.4.2 The escalator–boxcar train 138
5.4.3 Integral projection models 139
5.5 What have we learned in Chapter 5? 143

6 Stage-structured models 145

6.1 Biological processes 145
6.2 History of development of stage-structured matrix models 147
6.2.1 Early development of stage models 147
6.2.2 Early successes in stage-structured modeling 149
6.2.3 Early applications 152
6.2.4 Stochastic stage-structured models 154
6.3 Problems with stage-structured models 155
6.4 Possible better alternatives to stage-structured models 159

6.5 Replacement in stage-structured models 160
6.6 Delay equations 163
6.7 What have we learned in Chapter 6? 165

7 Age-structured models with density-dependent recruitment 166

7.1 Local stability and $2T$ cycles 167
7.1.1 Local stability analysis 167
7.1.2 An example: $2T$ cycles in Dungeness crab 172
7.2 The simplest general model of age-structured density dependence 177
7.3 Cycles in Dungeness crab: models and data 180
7.4 An intertidal barnacle, *Balanus glandula* 183
7.5 Cannibalism and the flour beetle, *Tribolium* 187
7.6 Effects of equilibrium conditions 189
7.6.1 Single-sex harvest 189
7.6.2 Multiple equilibria 190
7.7 What have we learned in Chapter 7? 191

8 Age-structured models in a random environment 194

8.1 The small fluctuation approximation (SFA) 196
8.2 The first crossing solution 197
8.3 A more general version of the growth of variability 199
8.4 Does the SFA/diffusion approximation work? *Totoaba* as an example 201
8.5 Color of the random environmental variability 203
8.6 Application of SFA to population data 205
8.7 State of the science quantifying extinction risk at the turn of the century 207
8.8 Perils of using stage models to characterize extinction risk 210
8.9 What have we learned in Chapter 8? 212

9 Spatial population dynamics 214

9.1 Modeling the spread of a population 216
9.1.1 The reaction–diffusion model 217
9.1.2 The asymptotic rate of spread 219
9.1.3 Leptokurtic dispersal 223
9.1.4 When diffusion is not a good representation of movement 224
9.2 Population persistence in aquatic habitats 225
9.2.1 The KISS model: persistence of a patch of plankton 225
9.2.2 The drift paradox 226
9.3 Metapopulations 230
9.3.1 The Levins model 230
9.3.2 Incidence function models 231
9.3.3 Patch value in the incidence function model 232
9.4 Models with internal patch dynamics: structure in space and age 235
9.4.1 Metapopulation persistence: replacement over space 236

9.4.2 Population persistence in heterogeneous space 238
9.5 Spatial variability across populations 242
9.6 What have we learned in Chapter 9? 244

10 Applications to conservation biology 247

10.1 Lessons from earlier chapters 248
10.2 Probabilities of extinction: the problem of measurement uncertainty 250
10.3 Probabilities of extinction: the importance of environmental spectra 252
10.4 Replacement as an extinction metric 253
10.5 An example with abundance, replacement, and measurement error 254
10.6 Comparative studies: Pacific salmon 255
10.7 Addressing exogenous variability: drivers and errors 259
10.8 Population diversity 261
10.9 What have we learned in Chapter 10? 264

11 Population dynamics in marine conservation 266

11.1 Three models from the 1950s 267
11.1.1 The logistic fishery model 267
11.1.2 The single cohort model (also known as the dynamic pool model, yield-per-recruit model) 270
11.1.3 The stock and recruitment model 275
11.1.4 Complete age-structured models: linking cohorts with a stock–recruit curve 276
11.2 Replacement in fully age-structured fishery models 276
11.2.1 Stock–recruit curves, lifetime egg production (LEP), and spawning per recruit (SPR) 277
11.2.2 Replacement and optimal fishery yield 280
11.3 The precautionary approach and modern fishery management 281
11.3.1 Precautionary management and reference points 281
11.3.2 Managing to avoid overfishing 282
11.4 Spatial management: marine protected areas 286
11.4.1 Strategic models of MPAs 286
11.4.2 Tactical models of marine protected area design 292
11.4.3 Other types of models used in MPA design 296
11.4.4 Adaptive management of MPAs 297
11.5 What have we learned in Chapter 11? 301

12 Thinking about populations 303

12.1 Modeling philosophy and approach 304
12.2 Replacement, an organizing principle 305
12.3 Population responses to time scales of environmental variability 307

12.4 Applying the lessons of population dynamics 308
12.5 What next? 309

Glossary 311
References 315
Index 337

CHAPTER 1

Philosophical approach to population modeling

Over the past century, population models have evolved from being simply clever ways of plotting a reasonable curve of abundance versus time to being a tool both to improve understanding of population dynamics and to make predictions for the purpose of management. That understanding has been gained by examining how the mechanisms embodied in the equations lead to the particular pattern of abundance over time and space. This focus on the connection between the model structure (the form of the equations) and emergent population properties contained in the model solutions (what kind of population dynamics do the equations produce?) has allowed broader and more useful employment of population models, in both the scientific endeavor of understanding population dynamics and in the application of population dynamics to practical problems. This chapter describes how the approach to population modeling has evolved over the past several decades and the nature of its current state (see Box 1.1 for a vocabulary of conceptual terms that we will use throughout the book).

We begin our description of modeling philosophy in this chapter with a somewhat historical digression to give the reader a sense of the way that ecologists began to approach population modeling in the mid twentieth century, and how that approach has evolved (Section 1.1). One remnant of that meandering has endured as a useful concept: the difference between *strategic* and *tactical* population models.

Next we describe the formal logic underlying both the scientific and the practical use of models (Section 1.2). This logic provides useful ways of structuring how one uses results from population models in the context of one's goals and available data. This section also serves as a reminder that using models to achieve greater scientific understanding and using them for predictions in practical application are actually based on different logical foundations. The principal outcome of this section is the fact that for both scientific progress and practical applications, we need models that provide the opportunity to compare explicitly the structure of the model to the real world, either empirically or, at least, conceptually. We refer to that characteristic as *realism*, and it connotes a quality of testability or observability in models. This characteristic guides the organization of model descriptions throughout the book.

Next we turn to the basic research on system theory during the 1970s for a formal definition of the *state* of a system to provide us with that realism (Section 1.3). While this topic may sound a bit esoteric, it provides the key rationale for development of models to achieve scientific understanding and effective management. The basic idea underpinning this book is that a description of the state of a system at any time instant needs to include all of the information necessary to project uniquely the state of that system at the next

Population Dynamics for Conservation. Louis W. Botsford, J. Wilson White, and Alan Hastings, Oxford University Press (2019).
DOI: 10.1093/oso/9780198758365.001.0001

time instant. This concept forms the basis for understanding the different roles of models with age structure, size structure, stage structure, and spatial structure, the backbone of this book.

The next section (Section 1.4) explores the limitations in our ability to detect the actual structure and state of an ecological system in the real world. These limitations are cast in terms of the different kinds of uncertainty we encounter. Although we do not include estimation of population parameters in this book, our development of population models, our analysis of them, and our examples of applications, are heavily influenced by the relative uncertainty in various components. Basically, we seek ways to formulate models that allow our conclusions to depend as closely as possible on the parameters and concepts we can know well, and we identify the important aspects we would like to know better, i.e. the most critical remaining uncertainties. The role of uncertainties, of stochastic versus deterministic approaches, will be another important theme of this book.

Next we turn to a description of how the analysis and study of populations is related to other disciplines that focus on different *levels of integration* in biology (Section 1.5). Different disciplines focus on different levels of integration. Research at the different levels of integration employs different approaches, and they lead to different emergent properties, with each level tending to explain the next higher level. For example, population dynamics depends on the survival and reproductive rates of individual organisms. Also, lower levels of integration are usually more sensitive to smaller temporal and spatial scales of environmental variability. These differences are important because population models can serve to scale-up individual-level observations to the population scale.

We end this chapter with a synthesis of current thinking on these topics as reflected in recent publications, as of 2019. These illustrate how the conversation in the ecological literature regarding how best to formulate and use population models continues to be active.

1.1 Simplicity versus complexity, and four characteristics of models

An obvious question addressed in any modeling activity is how complex should the model be? Simpler models are preferred because they are easier to analyze and to understand. On the other hand, simple models are not useful if they omit key aspects that are essential to the dynamics. The brute force method of making sure to avoid this error of omission is to develop a model of sufficient complexity to completely replicate the actual system being modeled. However, this elusive goal is not necessarily desirable, especially when the operation of some mechanisms or components included are poorly understood.

Two different approaches for reaching a model of the appropriate complexity are possible. One could start with the simplest possible model, such as deterministic exponential growth, and add features to the model that would allow better understanding. Alternatively, one could start with a complex description that includes all features one could envision, and then make simplifying assumptions. Each approach has its advantages and disadvantages, as will emerge later.

The topic of model complexity for ecological models is currently an area of active discussion in population science and management. For example, national and international panels (e.g. the Pew Oceans Commission (Pew Oceans Commission, 2003) and the US Commission on Ocean Policy (US Commission on Ocean Policy, 2004)) have recommended that resource management transition from using single-species population models to models that include all components of the ecosystem, including food web interactions, the variable physical environment, and socioeconomic factors (i.e. ecosystem based management, EBM; Pikitch et al., 2004). However, these additional factors are often poorly understood. A challenging question, therefore, is whether it is better to add poorly understood structure to the management model; i.e. will it actually improve management (Botsford et al., 1997)? This remains a vexing problem that influences how population dynamics will be described in these models (Collie et al., 2014).

An important step in the history of population modeling was mathematical ecologist Richard Levins' (1966) analysis of the question of simplicity versus complexity. He pointed out the problems with models of high complexity: (a) there are too many parameters to estimate; (b) the equations are not solvable analytically and would exceed the capacity of the fastest computers (in 1966), and (c) the resulting expressions are so complex as to be meaningless. These problems are still present forty years later, except that computational limitations continue to shrink with time.

Levins (1966) declared that it was "of course desirable to work with manageable models which maximized *generality*, *realism*, and *precision* toward the overlapping but not identical goals of understanding, predicting, and modifying nature" [emphasis ours]. He further proposed that we could not achieve all three qualities in a single model, but rather that only two out of three could be achieved in any specific case. We do not quibble with why these three characteristics should be the ones chosen, nor with the basis for the statement that one can achieve two, but not three. However, Levins did not define the three characteristics. We will. The simplest definition of *generality* is the characteristic of applying to all possible examples. *Realism* is open to several different definitions, but it is reasonably clear that what Levins meant is the characteristic introduced in the introduction to this chapter, that of allowing for direct comparison to the real world. For example, a model that included the mortality rates and reproductive rates of individuals at all ages would be considered realistic, but a model that included a (presumed) simple summary of their effects, such as population carrying capacity, would be less realistic. We can observe the amount of reproduction and the rates of death of individuals, but carrying capacity is an emergent property of a model solution and would require longer-term observations to determine. Note that this definition of realism includes no relationship to truth, as the more common, everyday definition of realism implies. In the context of the truth, Levins' definition of realism can be interpreted as meaning that an element's truth in the real world is testable, not that the element is a true depiction of reality. Levins' definition of *precision* is the standard statistical definition, though he probably actually meant accuracy, i.e. not just consistently similar answers, but consistently the correct answer (see Box 1.1).

Box 1.1 DEFINITIONS RELATED TO MODELING PHILOSOPHY

Generality—strictly speaking, the quality of a statement applying to all cases.

Realism—having the same structural form as a real object.

Precision—the quality of a statistical estimate having a narrow distribution of error about a point, but not necessarily about the true value.

Accuracy—the quality of a statistical estimate having a narrow distribution of error about the true value.

Holism—including all of the relevant factors

Strategic models—models devised to answer very general questions about population behavior, with little attention to accurately portraying a specific situation.

Tactical models—models devised to answer specific questions about real situations for the purpose of making projections on which management will be based.

According to Levins' two-out-of-three rule, some models sacrifice generality for the sake of achieving realism and precision. Levins offered the models used in fisheries management as an example. Those models include individual growth rates, reproductive rates, and mortality rates, which are comparable to the real world, and values of catch can be projected precisely, but the modelers do not intend for a single model to describe all fished populations.

Levins offered Volterra's (1926) predator–prey system as an example of the second category: models that sacrifice realism for generality and precision. This model omits the time lags often involved in predator–prey cycles and the effects of a species' population density on its birth and death rates. Instead, the model uses phenomenological model parameters such as carrying capacity. Although this model parameter cannot be directly linked to an observable ecological interaction in the field (hence the lack of realism), the model nonetheless represents the essential characteristics of general predatory–prey dynamics, and the results can be stated very precisely.

The remaining class of models in the two-out-three scheme sacrifices precision for the sake of realism and generality. Levins' example for this case was the set of simple biogeographical models describing broad latitudinal patterns. For example, Bergmann's rule states that animal body size will be larger in colder climates (Bergmann, 1847; Meiri and Dayan, 2003). These models are general in that they apply to many species, and they possess the quality of realism since they can easily be compared to observations, but they do not yield precise values, rather they are merely qualitative predictions in the form of inequalities.

The value of Levins' contribution was that he acknowledged that models can be quite different in function, and that different kinds of models may be best for different uses. This topic has continued to evolve up to the present.

The noted systems ecologist C. S. ("Buzz") Holling (Holling, 1968) assessed Levins' (1966) view and added a fourth characteristic of models: *holism*, the quality of including all relevant factors. As noted in the second paragraph of this section, requiring that all relevant factors be included in a model is a questionable step when there is only a poor understanding of the dynamic behavior of the proposed additions. The modern advice in favor of including all species of an ecosystem in ecosystem models is an example of the quest for greater holism. However, as noted previously, there is not a clear advantage to adding a prey species to a model when the consumption rates and the consequences of consumption are not well known (Collie et al., 2014), and a predator that consumes many different prey species can be represented just as well by a model without prey explicitly included (Murdoch et al., 2002).

An assessment of Levin's (1966) views by Robert May (May, 1973), a physicist-turned-mathematical ecologist, concluded that most ecological models were neither general, realistic, nor precise, and that the central issue did not involve those three qualities, but rather should focus on the relative advantages of simple, general models versus complex, specific models. May held that both simple and complex models had their place, with the former being useful for descriptions of the general trends in large complex systems, while the latter were useful in dealing with specific aspects of parts of such a system, as would occur in a management situation. Additional factors that would influence model complexity include the amount of data available: in data-poor situations there is less justification for complex models (as well as a need for more data, to permit the use of more complex models). These two notions were similar to what Holling (1973) had referred to as *strategic* and *tactical* approaches to modeling, respectively. These two terms are now commonly used to differentiate between these two different kinds of modeling activities (Box 1.1).

An illustrative example of the difference between strategic and tactical approaches is the development of population models to design marine reserves, following their increase in popularity as a management tool in the 1990s. As the idea of managing marine resources by creating areas of no fishing (i.e. marine reserves) began to gain prominence in the 1990s (Botsford et al., 1997; Murray et al., 1999), it became necessary to develop population models to evaluate how marine metapopulations would persist when distributed over heterogeneous habitats with fishing permitted in some locations but not in others. A marine metapopulation is a number of separate subpopulations distributed over space, linked by a dispersing larval stage (Roughgarden et al., 1988; Botsford et al., 1994; Kritzer and Sale, 2004; Chapter 9). Initially, ecologists used simplified, strategic models to understand the effects of marine reserves on fish populations. They addressed broad questions such as how does management with reserves compare to conventional management in terms of fishery yield (e.g. Holland and Brazee, 1996; Mangel, 1998; Hastings and Botsford, 1999; Hart, 2006; White and Kendall, 2007)? The answers all indicated the dual nature of conventional management and management by reserves: a certain catch was possible, and it could be achieved by a range of pairs of values of fishing mortality rates and fractions of the coastline in reserves. Another important question was which spatial configurations of reserves connected by a dispersing larval stage would support persistent fish populations (e.g. Botsford et al., 2001; Gaines et al., 2003; Kaplan, 2006, White et al., 2010a)? The general answers were that (a) single reserves would support persistent populations of species with dispersal distances shorter than the width of the reserve, and (b) combinations of many reserves that covered a certain fraction of the coastline could support species with a range of dispersal distances. These strategic models typically described populations along idealized, linear coastlines with logistic dynamics or simple age structure. They provided early guidance when decisions were needed to be made in the face of limited data and knowledge, and they also later provided a check on whether the eventual, more detailed tactical models made sense.

When management agencies actually began to implement marine reserves early in the twenty-first century, the decision-making process required the ability to compare the costs and benefits to specific fish species of specific proposed MPAs at specific locations, hence tactical modeling was required. The models became much more realistic, including age structured models with density-dependent recruitment, linked to life history data from several species, and using data on habitat distribution and ocean currents along real coastlines. These models projected the relative amounts of biomass and fishery catch expected at specific locations after reserve implementations (e.g. Kaplan et al., 2006; Pelletier et al., 2008; Kaplan et al., 2009; White et al., 2010c, 2013b; Hopf et al., 2015; reviewed by Pelletier and Mahevas, 2005; White et al., 2011).

Fortunately, the results of simulations of these realistic, tactical models turned out to be consistent with the characteristics predicted by the earlier strategic models. For example, when the fishing rate already managed a species at the maximum sustained yield, adding a reserve would only cause the yield to decline, a characteristic of the dual nature of reserve and conventional fishery management identified earlier in the strategic modeling. Moreover, the strategic modeling often provided valuable interpretations of the results of tactical modeling, that may not have been appreciated otherwise. The characteristics of persistence and yield in reserves mentioned here are developed more thoroughly in Chapters 9 and 11.

What useful information can we take from these early musings by Levins, Holling, and May? Ideally we would want all three of Levins' characteristics, but realism seems most important. For scientific progress and useful practical applications, we need to be able to connect our results to the real world, hence we need to use model components that can be observed and measured. It will be advantageous to be general, but we need not always be; sometimes our interest will be focused on specific situations. In some situations we will need to be precise, e.g. when making projections of the consequences of proposed management, but in other situations a comparative description of results (e.g. this parameter value leads to greater population stability than a lower value) will still be valuable.

Obviously we will sometimes be concerned with a broad range of instances and issues, and will be following a strategic approach, while in others we will be concerned with specific cases, and will be following a tactical approach. An essential question concerns the true degree of generality involved in strategic models. Models that yield more general results by applying to a greater number of specific instances often do so because they apply less well to any specific instance. This tradeoff is illustrated by the old joke about the drunk seen looking for his lost ring outside of Sam's Bar. When asked what he is looking for, he replies that he is looking for the ring he lost while coming out of Joe's Bar. When it is pointed out that Joe's Bar is several blocks down the street, the drunk replies, "I know, but the light is better here near Sam's Bar."

The key to making good use of both strategic and tactical approaches is to choose simple strategic models that possess enough realism in their critical aspects that their general results are apparent in the specific results of tactical model implementations. The results described above for strategic and tactical models of reserves are an example (discussed further in Chapter 11).

1.2 Logical basis for population modeling

We can also gain insights into the appropriate use of population models from a formal description of the logical bases for their use. In this book the planned uses of models can be viewed as twofold: (1) *scientific* use to improve understanding of the mechanisms of population dynamics, and (2) *practical* use to provide projections for management. These depend on different logical bases. We review those here with the caveat that these are only sketches of the basic ideas; adequate to discern the nature of the type of model needed for a particular usage, but not necessarily expert views of the current status of the philosophy of science. Also note that these two uses of population models do not cover all of their uses. For example, another use of models is simply as a *pedagogical tool* to illustrate the behavior of populations without great concern for actual mechanisms. We will not be concerned with such use of models in this book, although some of the models in Chapter 2 are often used in that way.

1.2.1 *Deductive reasoning and the scientific uses of modeling*

We characterize the scientific use of models in terms of how we might answer specific questions regarding the causes of observed population phenomena. For example, there is often concern over why certain populations have declined over time, or have increased to dramatically high abundance, or have exhibited cyclic behavior. In almost all instances, there are multiple proposed causes for the observed behavior. Taking cyclic behavior as an example, hypothetical factors potentially responsible for the observed cycles would include (a) an over-compensatory stock-recruitment relationship (i.e. a case where recruitment actually decreases as stock increases, for large enough stock levels), (b) a cyclic environmental variable, (c) a predator/prey relationship, and so forth. We refer to these as hypotheses, and describe a program for using population models to evaluate these hypotheses as causes of the observed cycling, decline, or increase.

The basic deductive approach to testing hypotheses in population modeling is to incorporate each hypothesis in a population model of the species of interest, and run a simulation or solve the model to see whether its output (e.g. the pattern of abundance over time) produces the observed behavior. We then evaluate the model and its derived consequences as a conditional argument from deductive logic (Box 1.2).

Box 1.2 CONDITIONAL ARGUMENTS

Conditional arguments from deductive logic are arguments of the following form, with two premises (P1 and P2) and a conclusion (C):

P1: If *a* then *b*

P2: *a* true (or *b* true, or *a* false, or *b* false)

C: *b* true (or *a* true, or *b* false, or *a* false, respectively).

Here "*a*" is called the *antecedent* and "*b*" is called the *consequent*. Of the four possible outcomes of P2 and C, two are valid arguments and two are invalid. The valid arguments are (1):

P2: *a* true

C: *b* true,

which is called confirming the antecedent, and (2):

P2: *b* not true

C: *a* not true,

which is called denying the consequent.

The logically invalid arguments are (1):

P2: *b* true

C: *a* true,

which is called confirming the consequent, and (2):

P2: *a* not true

C: *b* not true,

which is called denying the antecedent. Note that the arguments are named after the statement in P2, which contains the information available on which we can base the conclusion.

In using models and these conditional arguments from deductive logic, we obviously want to make use of the valid arguments and avoid using the invalid arguments (Box 1.2). We will describe this approach formally using the terminology in Box 1.2, and then show an example of its application to cyclic population dynamics in Dungeness crabs. The basic argument begins with the initial premise P1,

If [MODEL] then [MODEL OUTPUT],

where the MODEL contains one of the hypothetical explanations of the observed phenomenon whose cause is in question; it is "true" if it is the correct explanation. MODEL OUTPUT is the pattern of population dynamics produced by the model; it is "true" if it matches the observed natural phenomenon. If the model output does not match the behavior in question, then the second premise P2 will be "[MODEL OUTPUT] not true" (which corresponds to "P2: *b* not true" in Box 1.2), whereas if the model output does match the behavior in question, the second premise will be "[MODEL OUTPUT] true" (which corresponds to "P2: *b* true"). We can see from Box 1.2 that if *b* is not true (i.e. the model output does not match the observed population behavior), we can draw the conclusion (C) that *a* is not true by applying the valid argument of denying the consequent. This would mean that MODEL is not true, which we take to mean that the hypothesized mechanism in that model is not the cause of the observed behavior (presuming the other assumptions of the model are true). If, on the other hand, *b* is true, the corresponding argument would be confirming the consequent, an invalid argument. If *b* is true, we do not have a valid argument available, hence we cannot draw a conclusion about the truth of the MODEL.

These results from deductive logic have been used to formulate a program for research that depends on strong results that reject hypotheses, while avoiding weak results that confirm hypotheses (Popper, 1959; Platt, 1964). It consists of sequentially testing all proposed hypothetical causes. If the hypothesis is rejected (i.e. model output does not match the empirical population observation), the next step is to test the sub-hypotheses, i.e. the assumptions of the model other than the hypothesis being tested. This step is taken (and repeated) because of the importance of the step of rejecting a hypothesis. If, on the other hand the hypothesis is not rejected, the program simply moves on to the next hypothetical cause of the phenomenon at issue.

Research that used age-structured models to study the cause of the cycles in Dungeness crab populations in northern California in the late 1970s (Fig. 1.1) provide an example of the use of this approach in interpretation of population models. At that time, research on whether the cycles could be caused by over-compensatory density-dependent recruitment

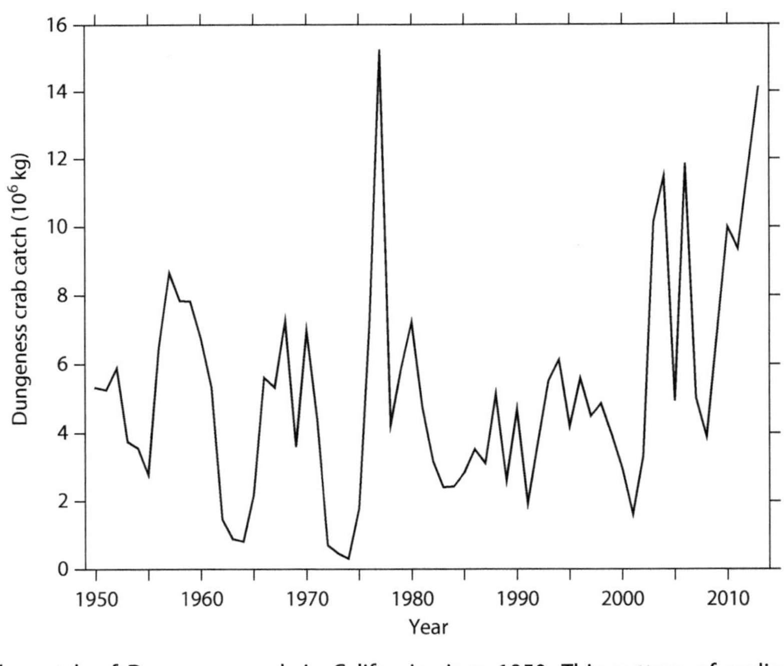

Fig. 1.1 The catch of Dungeness crab in California since 1950. This pattern of cyclic variability was analyzed in the late 1970s to determine its cause. Catch is a reasonably close approximation to male abundance in this species because it is a male-only fishery with high levels of harvest of males greater than a minimum size limit. Data from the NOAA Fisheries Information System https://www.fisheries.noaa.gov/national/commercial-fishing/fisheries-information-system-program

focused on two hypothetical mechanisms: (a) cannibalism of older crabs on new recruits to the benthic habitat, which had been shown to occur; and (b) density dependence in fecundity, which was proposed to exist but had not been documented. Two research groups concerned with determining the cause of the cycles both showed that in age-structured models with density-dependent recruitment, cycles could occur with period roughly twice the mean age of adult crabs. Based on that result, Botsford and Wickham (1978) noted that cannibalism of older crabs on young crabs could cause cycles, but that their period would be less than the period of the observed cycles (i.e. 10 y). However, McKelvey et al. (1980) interpreted the same result from a similar, age-structured model as evidence to reject cannibalism as the cause, and declared density-dependent fecundity to be the cause, because a model with density-dependent fecundity produced cycles of the observed period.

The differences in the two approaches lay in their conclusions and recommended next steps. Both groups rejected cannibalism as modeled, but Botsford and Wickham (1978) recommended further testing of the sub-hypotheses (essentially the assumptions of their model that cannibalism rate would be proportional to individual metabolic demand), before rejecting cannibalism as a possible mechanism. McKelvey et al. (1980), on the other hand, rejected cannibalism as a possible cause, and accepted fecundity as the cause (thus confirming the consequent). They recommended a program of study to show that density dependence of fecundity existed. Botsford and Wickham (1978) focused on the potential for a strong result, i.e. possibly being able to reject the cannibalism hypothesis by further study of the actual age dependence of cannibalism. McKelvey et al. (1980), on the other hand, rejected cannibalism and focused on further study of a different hypothesis based on the deductively invalid argument of confirming the consequent. Their recommended first step would be to show that fecundity was density dependent. The question of the cause of the cycles went beyond scientific curiosity since the practical implication of the different hypotheses for fishery management differed. If the females were responsible for the cycles, fishing would not be affecting the cyclic behavior. However, if males were involved in the density dependence, as in cannibalism, truncation of the age structure of males by fishing would have a destabilizing effect on the population (more on this in Chapter 7).

As we shall see in Section 1.2.2, the logic followed by McKelvey, et al. (1980), while invalid in deductive logic (confirming the consequent), is not unlike that followed in an acceptable inductive argument. In essence their conclusions were based on choosing the model that currently provided the best fit, a common approach in applied ecology.

1.2.2 *Inductive reasoning and practical applications of modeling*

Inductive arguments consist of conclusions drawn from a number of observations. The basic form of an inductive argument is that out of n trials a certain proposition has been observed to be true Z percent of the time, therefore we can conclude that it is true Z percent of the time. This is referred to as an argument by enumeration (Salmon, 1973). Inductive arguments are not described as being "valid" or "invalid," rather they are referred to as being relatively "stronger" or "weaker." An argument by enumeration is stronger for greater n, and conversely weaker for lower n.

In representing population dynamics, we make use of a specific form of an argument by enumeration, referred to as an argument by analogy. The form of that argument is: if an object, A, is the same as another object, B, in characteristic 1, in characteristic 2, in characteristic 3, and so on, up to characteristic n, then one can conclude that A and B will be the same in terms of a new characteristic not yet compared. This argument becomes stronger as the number of comparisons increases. In the case of population dynamics, the argument could be that a population model is similar to the population in terms of the pattern of survival versus age, fecundity versus age, individual growth versus age, and past abundance versus time; therefore one can conclude that the model will be similar to model behavior predicted for the future. Again, this argument by analogy also becomes stronger as the number of comparisons increases. Doing this formally with statistical approaches and information criteria requires stochastic models (but often less formal approaches are useful and appropriate).

Stock assessments made for the purpose of managing fisheries are a good example of turning to arguments by induction for the purposes of management. Examples of these can be seen at websites for the regional federal fishery management councils in the USA (e.g. the Pacific Fishery Management Council (PFMC) for the west coast of the contiguous USA), and other similar sites elsewhere around the world. These are typically based on a model fit to several types of data at once: data on age structure of the catch, a fishery independent survey of abundance, and existing information on growth versus age fit previously to size-at-age data.

1.2.3 *Consequences of deductive and inductive logic for population dynamics*

The question of why we need two different kinds of logic (inductive and deductive), with two different sets of seemingly conflicting rules, may have occurred to the reader, just as we imagine it did to the early Greek students of logic several millennia ago. Deductive logic is very conservative, leading us only to a number of potential causes of phenomena that have been tested and found not to be actual causes; it withholds judgement on phenomena that have not been rejected, and never confirms that a particular cause is correct (this would be confirming the consequent). This does not provide the information needed for practical applications to management. To manage populations in the future, we need to know what the actual causes of the dynamic behavior are, and that requires the approach of inductive logic.

The separate uses of inductive and deductive logic have a long history in ecology in general (e.g. Dayton, 1973). Also, Caswell et al. (1972) long ago separated models into those developed for the purposes of better understanding and those developed for prediction. He was addressing the "problem of validation," which has to do with what one can conclude from positive outcomes of model predictions. As stated previously, these outcomes mean something in inductive logic, but much less in deductive logic.

Much more could be said about inductive and deductive logic, but here we describe them only briefly, solely to assess their requirements of models (see Salmon, 1973 and Chapter 7 of Ford, 2000 for more information). We simply conclude for our purposes here that whether you are employing deductive logic in a scientific use of models or inductive logic in a practical application of models, you will be more effective if the models explicitly contain the actual observable mechanisms in the populations. This would be satisfied by the models having a high degree of what is called "realism" in the trichotomy of Levins (1966). In the use of deductive logic, one ultimately needs specific hypothetical mechanisms that can be tested. In the use of inductive logic, the arguments improve as the number of comparisons increases, and realism is by definition comparability with reality.

1.3 The state of a system

A formal approach to deciding how to construct population models can be found in the early developments of system science (Caswell et al., 1972; Zadeh, 1973; Caswell 2001). In the 1960s, scientists were beginning to grapple with complex systems through mathematical models and computer solutions for those models in an effort called system science. A system was "a collection of objects, each behaving in such a way as to maintain behavioral consistency with its environment" (Caswell et al., 1972). System science required a formal, consistent method for constructing models, which was achieved by carefully defining the *state* of a system. The basic idea was that the behavior of a system would depend not just on the current stimulus from the environment, but also on its history, i.e. how it had responded to past environments. System scientists represented the effect of the past completely by expressing the current state of the system. In order to describe how a system would respond to an external stimulus from the environment, the state and the stimulus–response–state relationship had to satisfy several conditions. The most illuminating condition for our purposes was that the combination of the state variable, the current stimulus from the environment, and the stimulus–response relationship had to *uniquely* determine the response of the system (i.e. the state of the system at the next time instant; see Caswell (2001) for a more extensive discussion). The idea that the state of the system at one time point should completely determine the state of the system at the next time point is analogous to the *Markovian* property of certain stochastic models (Box 1.3).

Box 1.3 MARKOVIAN PROCESSES

One important property of a properly defined state variable is that adding information about the past state of the system provides no more information about the future than the knowledge of the current state. This is the same property shared by an important class of stochastic models called Markov processes. One common example of a Markov process is a Markov chain, which is a random sequence of events (such as the position of a particle in a fluid exhibiting Brownian motion) in which the position at any point in time depends only on the position at the previous time. That is, predictions of the future state of the system based on the current state are not changed by adding information about past states. Markov chains are commonly encountered in statistics, and in Chapter 8 we discuss their use in stochastic population models. Most of the models we discuss in this book are not stochastic, but the state variables nonetheless happen to share this same property of being Markovian, and of containing all of the information at time t needed to predict the state of the system at time $t + 1$.

This simple concept of state may sound obvious and trivial, but it is very powerful. In a sense it determines the organization of the rest of this book. Glancing at the table of contents, you will see that we begin in Chapter 2 with simple models that represent a population in terms of the total number of individuals (i.e. abundance, N), then we move quickly to adding age structure, and representing the population in terms of the number at each age. The reason for doing so is that we show that representing a population by its current abundance N is not sufficient for us to predict, uniquely, how many there will be next year. That depends (at least) on the age structure of the population, i.e. how many of the N were younger than age of maturity, and how many were older and hence reproductive.

While we develop a fairly comprehensive view of population dynamics using age-structured models, we ultimately decided that we had to move on to models that add representation of population state in terms of size structure, i.e. the number of individuals at each age and size. This is because for many taxa, reproductive maturity depends on size, not on age. If all individuals grew along the same plot of size versus age, we would not need a model with both age and size, but could use either an age-structured or a size-structured model. However, many taxa exhibit plastic growth, whereby the growth at any time could depend on population density or the environment (e.g. temperature or food). Also, for many species, individuals at a certain age do not all have the same size.

1.3.1 *Models of i-states and p-states*

The concept of state can apply to either an individual or a population (Metz and Diekmann, 1986). From an individual perspective, the *i*-state variables are the things you need to know about an individual to be able to uniquely predict what its state will be in the next time instant and its response in terms of reproduction and mortality. The so-called *i*-states could include the individual's age, size, reproductive status, level of energy reserves, and so forth.

From a population perspective, one of the ways of formulating a *p*-state, i.e. a description of the state of a population, is to describe the *p*-state as the number of individuals at each *i*-state in the population. The stimulus–response function is then a law of mass action that describes the "flow" of individuals through the space defined by the *i*-state variables. The number at each *i*-state at time t is determined by the number there previously that died, the number reproduced to that state, and the numbers that "grew" or "traveled" to and from that state. One of the conditions necessary for this *i*-state distribution to be a sufficient description of the *p*-state is that individuals with the same *i*-state respond identically to the environment and each other. A second necessary condition is that the population output, the production of new individuals and their *i*-states, can be calculated from the distribution over the *i*-state. Together, these two conditions are sufficient to satisfy the definition of state (Metz and Diekmann, 1986). One might notice that they are also an example of the Markovian property we introduced earlier (Box 1.3).

These definitions were envisioned in a deterministic context. The unique prediction of the future would obviously not be possible if outcomes were stochastic. When populations become locally small, they become inherently stochastic because of demographic stochasticity (see next paragraph). This means that the approach of using the distribution over *i*-states as the state variable will only work if the population abundance remains high enough at all points in the *i*-state. High enough here means such that the outcomes of mortality and/or behavioral interactions do not have to be expressed in terms of discrete numbers of individuals.

This condition can be understood by explaining demographic stochasticity as an example, a phenomenon we will refer to in Chapter 8 on random variability in populations, but which we introduce here to understand the consequences of populations being *locally small*. Suppose we have a population in which, on average, ninety percent of the individuals survive each year. There are two different ways that we could represent that in a model. One would be simply to multiply the number at the beginning of the year by 0.9 to obtain the number at the end of the year. To see how that would work, assume we start with fifteen individuals, as an example. Multiplying that by 0.9 gives us 13.5, which is problematic, since the real population will not have fractions of individuals. We can get around this departure from reality by following a different procedure to obtain the number surviving: for each individual in the population, each year we conceptually flip a coin that comes up heads ninety percent of the time and tails ten percent of the time. Heads means that the individual survives, tails means they die. The result of this will be an integer each year, thus getting around the problem of ending up with fractions of individuals. However, now of course the outcome will not be the same for every sequence of fifteen coin flips. Thus we have introduced a kind of randomness that is actually present in a population. Fortunately this kind of randomness is well studied, and the distribution of outcomes is a binomial distribution (Box 1.4).

Box 1.4 THE BINOMIAL DISTRIBUTION

The binomial distribution is useful in understanding demographic stochasticity and in deciding how many simulation runs to make when trying to compute a probability of an outcome such as extinction.

The binomial distribution describes the outcome of a number of Bernoulli trials. A Bernoulli trial is the process of conducting a random experiment that has one of two possible outcomes. The most common examples are flipping a coin, or drawing a marble out of a bag that contains red and white marbles. In the latter case, the probability of drawing a red marble could range between 0 and 1, depending on the ratio of red to white marbles in the bag. We will use coin flipping as our example, but imagine a coin does not necessarily have a probability of 0.5 of being heads and 0.5 of being tails, but rather has a probability p of being heads (like the bag of marbles, p could be any value between 0 and 1) and probability $1 - p$ of being tails.

Imagine you flip the coin several times in a row. The probability of obtaining k heads in n flips of the coin is $p(k) = \binom{n}{k} p^k (1 - p)^{(n-k)}$, where $\binom{n}{k}$ is the binomial coefficient, which is the number of ways that a sequence of n flips with k heads can occur: $\binom{n}{k} = n!/\,[k!\,(n - k)!]$. The important characteristics for our purposes are the mean and variance of this distribution. The mean, or what is called *expected value* in statistics, of this distribution is np, which makes intuitive sense (the number of flips times the probability of getting heads on each flip), and the variance is $np(1 - p)$. In both of the examples for which we will use this, we will be interested in the relative amount of variability, the standard deviation divided by the mean, which is called the coefficient of variation (CV).

This means that the amount of demographic stochasticity (i.e. variability in the outcome of the Bernoulli trials) declines in proportion to the square root of the number of individuals in the population. This simple result is also useful in other ways; for example, when estimating a probability of extinction by repeated random simulations, the relative standard error of that estimate also declines in proportion to the square root of the number of simulations (Harris et al., 1987).

Using the formulas described in Box 1.4, we can see that the relative amount of variability due to this binomial process (survival) declines with the square root of n. Thus, as long as we start with a high enough number, the error incurred by simply multiplying the number of individuals by 0.9 is small. For example, with fifteen individuals the coefficient of variation would be 0.0775, while if we had 150 individuals it would be 0.0245, and with 1500 individuals it would be merely 0.0077.

Demographic stochasticity is an important source of variability in population dynamics. The example we just used is demographic stochasticity in survival, but it is not difficult to imagine that this kind of randomness can arise in processes other than survival, such as the sex ratio of a mother's offspring, or specific kinds of behavioral interactions between individuals. Demographic stochasticity is the key reason why the *p*-state description in terms of the distribution of abundance at each *i*-state breaks down when population numbers at any *i*-state are small. It is important to emphasize, for example, that in a population with very large total numbers, but small numbers of reproducing individuals, stochasticity will be important even though total numbers are large.

1.3.2 *Individual based models (IBM)*

In the late 1980s and early 1990s, mathematical ecologists began to develop an alternative to forming a representation of a *p*-state as a distribution over *i*-states. They began to realize that distribution-based *p*-states were not consistent with two of the basic tenets of population biology: (1) that individuals are inherently heterogeneous, and are inherently different in more ways than can be described with a few *i*-state variables, and (2) that in some of the most important population situations, one of the causes of problems is that populations become locally small. One example of the latter is low spatial density, but others were less obvious, e.g. behavioral interactions among individuals. Characteristics (1) and (2) are of course not independent, being two ways of stating the same thing: adding more dimensions to a multi-dimensional space makes individual densities per unit space lower, by definition. In other words, adding more *i*-states in order to improve realism will necessarily reduce the number of individuals in each *i*-state category, and thus reduce the precision of *p*-state representations of *i*-state distributions.

Some mathematical ecologists decided to avoid formulating a distribution-based *p*-state model, and instead to write computer programs that just kept track of the *i*-states of all individuals within a population directly (Huston et al., 1988; DeAngelis and Gross, 1992; Judson, 1994; Grimm and Railsback, 2005; Railsback and Grimm, 2011). This meant that instead of analyzing how a distribution over *i*-states changed with time, instead they numerically computed how the *i*-state of each individual changed, then added them up to present results. These models were termed ***individual-based models*** **(IBM)**, which is a bit of a misnomer since the *p*-state distribution models were also ultimately individual based. These are now also called *agent-based models* (Railsback and Grimm, 2011).We do not discuss IBMs further in this book because they tend to depend on brute force computations rather than requiring mathematical descriptions of population dynamics. The fact that computers continue to become faster and IBMs require little background in matrix algebra or partial differential equations has made them popular, and other books address them directly (Grimm and Railsback, 2005; Railsback and Grimm, 2011).

Population viability analyses (PVA) of the red-cockaded woodpecker (*Picoides borealis*) provides a good example of using IBMs in order to accurately represent population dynamics at low abundance. In the 1990s, initial PVA analyses for this species were accomplished with deterministic and stochastic stage-structured models (Heppell et al., 1994; Maguire et al., 1995). However, these models could not adequately represent the breeding behavior of red cockaded woodpeckers. Some red cockaded woodpeckers disperse to acquire breeding positions, but others remain on their natal territory as non-breeding helpers (Walters et al., 1988). These helpers constitute a pool of replacement breeders who can replace breeders that die. This has a buffering effect on breeder mortality that reduces the variability in reproduction in a way that required an IBM to represent. Subsequent analyses showed that while probabilities of extinction increased with a reduction in breeding colonies, greater clumping of territories ameliorated that by maintaining larger potential replacement pools of helpers (Letcher et al., 1998; Walters et al., 2002).

The fact that IBMs or agent-based models are primarily simulation based limits the easy application of analytical solutions, and the potential easy or elegant interpretation of those solutions. It also limits opportunities to make use of various mathematical methods such as stability analysis and optimization. However, simulation approaches can be a useful check on the formulation of analytic models—if the behavior of the analytic model and the simulation approach are similar over a range of parameters, then one has much greater confidence in using the simpler analytic model. Additionally, Metz and Diekmann (1986) emphasized that their approach based on *i*-states and *p*-states was essentially an analytic description of an IBM approach. However, the *i*-state, *p*-state approach is not a panacea as models based on this approach can quickly become complex. Yet, the approach has been shown to be powerful for understanding structured populations (de Roos and Persson, 2013; and the examples we give in the rest of this book).

1.4 Uncertainty and population models

In this book we do not include the topic of estimating population parameters or population states from data, hence we do not directly address aspects of randomness associated with estimation. Nonetheless we will need to refer to the different kinds of variability or uncertainty in population modeling. The three categories of uncertainty in population models are: (1) process error, (2) measurement (or observation) error, and (3) structural error. The first, process error, is variability that we decide not to account for with explicit causal mechanisms in a population model; rather, we simply treat it as noise in the population dynamics. The dramatic variability in recruitment to most fish populations is a good example. We could conceivably describe the multiple interacting physical and biological processes that lead to that variability, e.g. Caselle et al. (2010), but the increase in explanatory power would likely not be worth the huge effort, so we simply treat it as "noise."

Measurement error (or alternatively "observation error") results from the inherent imprecision in the methods we use to estimate population rates or cumulative states such as abundance or biomass. Most observations of populations include a combination of measurement and process error (Fig. 1.2), and a major challenge is to separate them (de Valpine and Hastings, 2002). We try to separate them because process error will affect dynamics, whereas measurement error does not. For example, population variability (e.g. year-to-year changes in the mortality rate) influences the level of risk of extinction, but observation error (e.g. slightly overestimating the number of animals one year and underestimating the next) does not (directly). Therefore, one would not want to use the total variability in an abundance time series to calculate risk, because that would overestimate the process error.

Structural error refers to the effects of using the wrong model formulation. The sequence of chapters in this book from Chapter 3 to Chapter 9 shows efforts to determine the *i*-states that lead to a valid definition of the state of a population, hence they are efforts to reduce structural errors. Another familiar effort to reduce structural errors would be

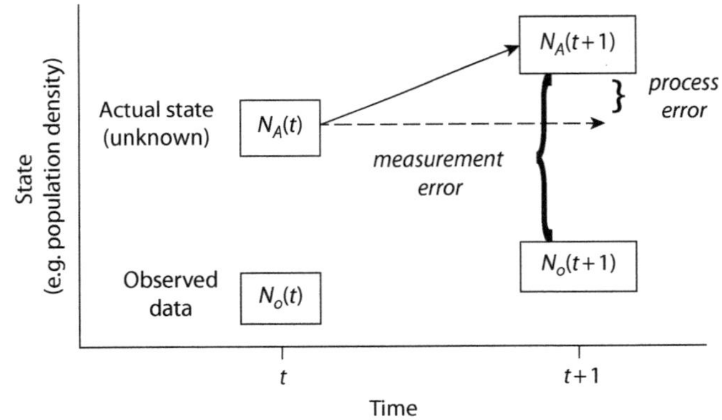

Fig. 1.2 A schematic illustration of process error and measurement error. At time t, the observed data, $N_O(t)$, differs from the actual state, $N_A(t)$, because of measurement error. The actual state is expected to move along the dashed line but does not, rather it moves to a higher value because it has been perturbed by process error (e.g. environmental noise). Again, at $t+1$, the actual state differs from the observed state by the measurement error (which has a different value at $t+1$ than at t).

the call to model whole ecosystems instead of individual populations (ecosystem-based management; Pikitch et al., 2004; Fulton, 2010).

Of the three types of uncertainty, this book will deal with how process error and structural error affect population dynamics. With regard to process error, we have mentioned previously how the presence of demographic stochasticity influences the conceptual foundation for our choice of *i*-state variables. We will also account for a second source of process error: the large amount of environmental variability in population vital rates, especially in younger stages. For example, marine fish populations exhibit order-of-magnitude variability in recruitment. Near the end of the book we will see that age structure of populations causes them to be more sensitive to certain frequencies of variability in environmental drivers (such as temperature or rainfall) than others. Since the frequency content of environmental variability has changed in the past (e.g. Cobb et al., 2003), and will likely change with a changing climate in the future (e.g. Timmermann et al., 1999), we will be interested in how populations respond to such changes.

While we will not deal with population estimation in this book, observation error will influence our efforts. For example, we will see in Chapters 4, 10, and 11 that one of the critical parameters for population persistence is one that is highly uncertain and extremely difficult to estimate. As a consequence, we are forced to pursue modeling avenues that seek to minimize dependence on knowing that parameter.

As previously noted, structural error plays a fundamental role in this book, at least conceptually. It guides our choices among the different age, size, and spatial models throughout the book, together with consideration of measurement uncertainty. The essential tension between these two types of uncertainty as model complexity varies is illustrated in Fig. 1.3. Simple models (Chapter 2), for example, do not possess the age structure necessary to represent the behavior of most populations faithfully, so would be on the left side of that figure. In this book, we generally increase model complexity, thus reducing structural error. However, as this figure indicates, in so doing, we accumulate a greater number of parameters whose values need to be determined, thus nominally increasing measurement uncertainty. This effect becomes increasingly important as one goes from describing population behavior to attempting to describe the behavior of

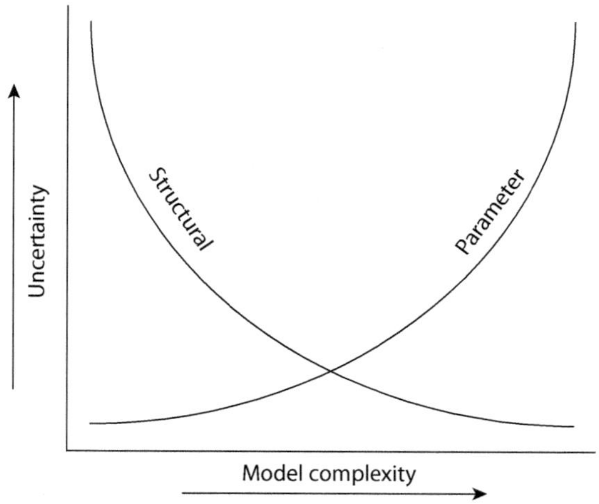

Fig. 1.3 A schematic view of the relative amounts of structural and parameter (measurement) uncertainty as model complexity increases from left to right.

whole communities or ecosystems (Plagányi et al., 2014; Collie et al., 2014). Note that community modelers have evaluated the effects of lumping components in communities (referred to as aggregation error, Gardner et al., 1982), but we know of no similar attempts in population dynamics.

1.5 Levels of integration in ecology

Most readers will have been introduced to the different levels of ecological integration, perhaps even as early as in their secondary education (Fig. 1.4). In this book we will obviously be focused on the population level, and in this chapter we have spent considerable time describing some relationships between the population and the individual levels. Those relationships are a consequence of the more general characteristic of the levels of ecological integration: the explanation of behavior at one level will be found in the rates at the next lower level. That is, population dynamics are determined by the combination of individual reproductive, mortality, growth, and movement rates. By the same token, ecosystem dynamics will be determined by the growth rates of populations and their interactions.

A second important characteristic of the levels of integration in Fig. 1.4 is that both temporal and spatial scales of variability generally increase with increasing levels of integration. This means, for example, that we will generally be better able to study individuals over sufficient time and space to understand their processes, than we are able to study populations or ecosystems. Recent developments in ocean acidification illustrate this idea. Since the 1990s, when it began to be obvious that increasing CO_2 levels in the atmosphere would change the bicarbonate chemistry in the ocean, thereby decreasing pH, there have been many studies of changes in individual survival and growth rates brought about by lower pH (see Kroeker et al. (2010) for a review), but very few studies of the effects of pH on marine populations or communities. The studies at the individual level provide little direction for management, so there is a growing appreciation of the need for modeling studies to "scale up" effects at the individual level to their consequences at the population and ecosystem levels, e.g. Le Quesne and Pinnegar (2012).

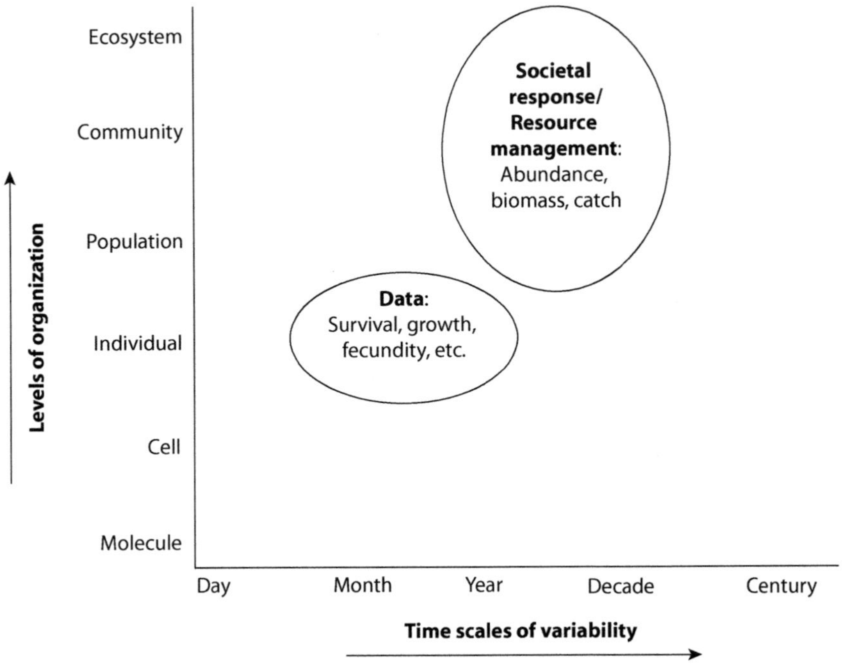

Fig. 1.4 The different scales of ecological integration, showing the differences in time scales of variability. A similar plot could be made for spatial scales. Because individual-level processes occur on shorter temporal and spatial time scales, we can observe them more easily, but there is a need for models that can scale them up to the population, community, or ecosystem level, where decisions are made.

1.6 State of the field

In this chapter, using a historical account of how ecologists have approached population (and other) models, we have distilled a number of desirable features of population models. From Levins' trichotomy and a review of how population models can be used in deductive and inductive models, we concluded that realism was a desirable characteristic of models, i.e. that we wanted models whose components could be compared to the real world in as many ways as possible. From May's and Holling's assessments we realized that there could be two useful categories of model: early, more general, strategic models and later, more specific, tactical models, but that simple, general models often achieved generality by saying less about a broader range of situations. Therefore, we decided that to achieve realism in our models, we should seek models with appropriate definitions of the state of a system. From a description of the three different types of uncertainty, we realized that our use of appropriate definitions of state to achieve realism was an effort to minimize structural uncertainty. We also concluded that even though we are not concerned with estimation in this book, we needed to identify and characterize both process error and observation error. From a review of the concept of ecological levels of integration, we noted that our focus on population dynamics here was, in a sense, translating information at the individual level to its consequences at the population level, and that these two levels operated at different temporal and spatial scales.

We now ask the question, what are others currently thinking about these issues? We review this ongoing conversation as seen in recent publications as of 2017.

Evans and associates (Evans et al., 2013b) asked the question as to whether simple models lead to generality in ecology. They first noted that simplicity can be defined in three different ways: number of model components, brevity of equations, and level of difficulty in analysis. They then noted, as we have previously, that the term general may not mean simply "applies to all or many," but may mean "applies less well to more." They recommended a pathway to generality that involves individually testing components of complex simulation models.

Lonergan (2014) pointed out that Evans et al. (2013b) had neglected the effects on models of the amount of data available. Specifically, he noted that increasing complexity with limited data could lead to overfitting. The response by Evans et al. (2014) was that they had addressed that concern in another publication (Evans et al., 2013a). Evans et al. (2013a) proposed an approach to "predictive systems ecology" to "understand and predict the properties and behavior of ecological systems." They proposed that the best available tools are process-based models, i.e. models that capture the important underlying biological mechanisms driving the behavior of systems. They then described key considerations for systems approaches: uncertainty, complexity, and constraining models with data. In that context, they described several examples in which systems ecology is already being practiced. One example is models that produce robust predictions of community structure in forest ecosystems over time from a combination of ecophysiology, individual-based modeling, and data from long-term forest inventory surveys. A second example is the models of ocean ecosystems with physical, chemical, and biological components, including ECOPATH models based on physiological energetics, and models with universal size structures of many species. These models are also discussed in publications in which some of us participated (Fogarty et al., 2013; Collie et al., 2014). The former is a survey of models recently used to scale up behavior at the individual level to their consequences at the ecosystem level. The latter is a synthesis of what is required of models formulated for the purpose of ecosystem based modeling in fisheries. The third example, given by Evans et al. (2013a) is models enabling the dynamics of ecological systems to emerge at global scales, based on fundamental birth and death, interactions and dispersal of modeled individuals (Purves et al., 2013). The fourth category is models that include both humans and ecosystems. Evans et al. (2013a) listed two major challenges ahead for predictive systems ecology: modeling at the appropriate scale and accounting for evolution.

Cuddington et al. (2013) addressed the use of models to connect data to management systems. They evaluated four classes of possible models: expert opinion, statistical extrapolation, process-based models, and detailed simulation models, and selected process-based models as the most promising. The reasons given for that choice were the transparency of explicit assumptions regarding causality, and the ability to account for uncertainty. For the chosen type of model, they evaluate issues such as whether tactical or strategic models should be used, choice of appropriate scale, whether the system is in flux due to global change, the impact of legacy effects, the potential for threshold dynamics, and including socio-economic impacts in process models.

Marquet et al. (2014) addressed a related, but nonetheless different, question: how ecological theory can best drive scientific progress and address environmental challenges. They describe theories, that: (1) are grounded in first principles, (2) are usually expressed mathematically, (3) are efficient (i.e. generate a large number of predictions per free parameter), (4) are approximate, and (5) provide well-understood standards for comparison with empirical data. Thus efficient theories have some of the characteristics of models favored by us and others, but seem to favor simpler models by including the characteristic of providing a large number of predictions per free parameter, and allowing the theory to be approximate. And indeed most theories in ecology (listed in Marquet, et al., 2014) are not the basis on which one would make management decisions.

In the spirit of providing a broad range of views: related recent comments by scientists concerned more with ecosystem policy than population dynamics (Schindler and Hilborn, 2015) suggest that we accept the fact that ecosystems may never be as understandable as we had hoped, and that we account for the consequent extant uncertainty by formulating robust policies that will work across a broad range of the actual unknown states of nature. They diminish the role of models (particularly complex ecosystem models) to "heuristic tools for communication," with little hope of reliable forecasts. This stems from their belief that "the best forecast models are typically mechanism-free, relying on emergent statistical properties of data to make short-term predictions," and "verification and validation of ecosystem models (i.e. fitting models to data) likely produce overly optimistic impressions of the reliability of forecasts."

From these publications we can at least conclude that the issues described in this chapter are still very much part of the conversation in ecology, and that there continues to be a concern for getting the methodology right. Also, there seems to be a gradual maturation process that leaves us currently less likely to depend on simple models for general answers. Throughout this book, we will describe the historical evolution of different approaches to population modeling while also addressing ongoing challenges. As we will see, some of these current challenges include the role of stochasticity, the role of non-autonomous systems in which there are underlying parameter changes through time, the role of analytical versus computational or simulation approaches, and how complex models should be.

CHAPTER 2

Simple population models

The term "simple population models" here refers to those population models in which the description of population state is a single variable such as abundance or biomass. From the conclusions drawn in Chapter 1, simple models may seem to be of limited utility in gaining a better understanding of real populations, and in managing them. However, because they are widely used in population biology, we include them for the sake of completeness. For example, logistic-type population models are some of the most widely used models in fisheries and resource management, and that trend is continuing with the development of ecosystem models (which often use logistic models to represent each of many species), to meet the call for ecosystem based management. Nonetheless, we must keep in mind the fact that when we assume that the state of a population can be represented simply by total abundance, we are implicitly assuming that all individuals, or all units of biomass, are identical (at least in terms of population dynamics). The limitations of this assumption were even noted by some of the early creators of fishery models (Beverton and Holt, 1957).

Simple models also provide a means of introducing mathematical concepts that will be valuable later in the book. Because they have only a single state variable, the mathematical analyses of stability, sustainable harvesting, and responses to random environments are much more straightforward. We present these three analyses here using simple models, as an introduction to their roles in later chapters. Furthermore, simple models illustrate some of the basic forms of population behavior that appear throughout the book. For example, the initial versions of age-, size-, and stage-structured models we encounter will be linear models, hence they will behave in a fashion similar to the linear simple models of exponential or geometric growth presented here.

In this chapter we will first explore what is known about the behavior of simple models of populations. There are two major classes of these models: (1) models of exponential (or geometric) growth (Section 2.2), and (2) models of logistic, density-dependent growth (Section 2.3). We will cover both discrete and continuous time versions of each of these. After that we use simple models to illustrate several different population dynamic issues (Section 2.4), including the concept of dynamic stability (Section 2.4.1), the calculation of probabilities of extinction (Section 2.4.2), and management of sustainable fisheries (Section 2.4.3). Each of these issues will be discussed further in later chapters.

We begin with the first population model. The development of the first population model, of course, required the development of mathematics. The study of mathematics in medieval Europe was stimulated by Latin translations from Greek and Arabic. These translations took place where there was trade between Christian and Islamic countries, such as in Sicily, southern Italy, Pisa, and Spain. Trade brought arithmetic,

Population Dynamics for Conservation. Louis W. Botsford, J. Wilson White, and Alan Hastings, Oxford University Press (2019).
DOI: 10.1093/oso/9780198758365.001.0001

algebra, and the concept of zero to Europe from Islamic countries, where they had been preserved.

2.1 The first population model—the rabbit problem

Leonardo of Pisa was one of the first mathematicians of the Middle Ages. We know him today as Fibonacci ("son of Bonacci"), which has much greater name recognition than Leonardo of Pisa. This is perhaps the reason that the statue identified only as Leonardo of Pisa draws little attention in among many other medieval relics in the Campo Santo Memoriale in the grounds of the cathedral in Pisa, famous for the leaning tower (Fig. 2.1). You can visit it yourself, without fear of running into crowds of tourists (or most likely, even anyone who knows who Leonardo of Pisa was).

In Leonardo's 1202 book, *Liber Abaci* (*Book of Calculation*; a recent translation is by Sigler, 2002), the first chapter begins with the phrase, "The nine Indian figures are: 9, 8, 7, 6, 5, 4, 3, 2, and 1. With these nine figures, and with the sign 0, which in Arabic is called zephir, any number whatsoever can be written, as is demonstrated below." This introduced thirteenth-century merchants and others to the inclusion of zero, a vital step in European

Fig. 2.1 Author Botsford with the statue of Leonardo of Pisa in the Campo Santo Memoriale, on the grounds of the cathedral in Pisa, with the famous leaning tower. The statue is not identified other than with the name "Leonardo" across the front.

mathematics (also made independently in Mayan culture on the American continent; Schele and Freidel, 1990). The rabbit problem appears near the end of Leonardo's 1202 book, and it appears to be the first population dynamic "model", although it would not be considered to be the beginning of the science of population dynamics, for a couple of reasons. Firstly, a competitor would be the demographic, actuarial calculations made by the Romans in the third century CE (though they focused on mortality, not reproduction; see Hutchinson (1978) for further description). Secondly, it was simply a single problem in a book of solved algebraic word problems, and was not really intended to explain population dynamics. Publications in mathematics at that time consisted of books full of worked problems using the Arabic decimal number system (instead of Roman numerals) for ease of calculations, and the rabbit problem was one of those problems (Devlin, 2011). The development of calculus and the other kinds of mathematics used in the population dynamics of this book did not begin for another four centuries.

The rabbit problem appears near the end of Leonardo's book in a section called *paria coniculorum* ("pairs of rabbits"). It is stated as

How many pairs of rabbits can be bred from one pair in one year? A man has one pair of rabbits at a certain place entirely surrounded by a wall. We wish to know how many pairs can be bred from it in one year, if the nature of these rabbits is such that they breed every month one other pair and begin to breed in the second month after their birth. Let the first pair breed a pair in the first month, then duplicate it and there will be 2 pairs in a month. From these pairs one, namely the first, breeds a pair in the second month, and thus there are 3 pairs in the second month ... (emphasis ours)

The calculations carry on as shown in Table 2.1. The time series of the total number of adult pairs is the well-known Fibonacci series, which appears in many areas in biology and elsewhere (Edelstein-Keshet, 2005). In addition to describing this as the first population model, we could get carried away with identifying the origin of ideas and wonder whether the stipulation that the population be surrounded by a wall shouldn't also be considered the first example of a boundary condition.

Table 2.1 Calculations for Leonardo of Pisa's rabbit problem. Each month, *m*, all rabbits born in $m - 1$ become juveniles, and all juveniles in $m - 1$ become adults, and then all adults spawn new pairs. The series of pairs born is known as the Fibonacci series.

Month	Adult pairs	Juvenile pairs	Pairs born
1	1	0	1
2	1	1	1
3	2	1	2
4	3	2	3
5	5	3	5
6	8	5	8
7	13	8	13
8	21	13	21
9	34	21	34
10	55	34	55
11	89	55	89
12	144	89	144

Box 2.1 CONTINUOUS VERSUS DISCRETE TIME

Mathematical models of dynamic behavior all describe the changes in a system's state as time progresses, but they describe time in one of two different ways: continuous time and discrete time. Continuous-time models keep track of the state of a system at every instant of time (Figure 2.2a). Solutions are referred to as continuous functions, and the fundamental mathematical tools used are from calculus. We write these models as differential equations expressing rates of change of the state variables. For a single-state variable $x(t)$ this would be the statement that the rate of change of variable x with respect to time is a function of the value of x at time t,

$$\frac{dx}{dt} = f(x). \tag{B2.1-1}$$

The function could also contain a dependence on values of x in the past, e.g. $x(t-\tau)$.

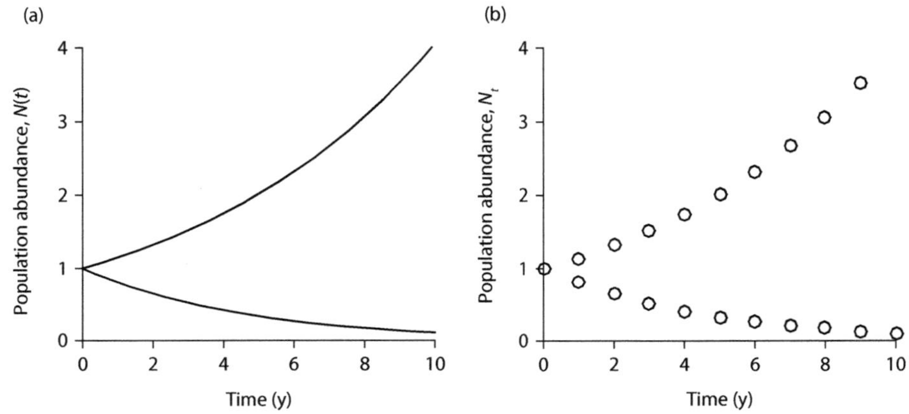

Fig. 2.2 The distinction between (a) continous-time models (Eq. (2.1)) and (b) discrete-time models (Eq. (2.3)). In this example, the time interval $\Delta t = 1$ year. In (a), $r = 0.14\ \mathrm{y}^{-1}$ (increasing) or $-0.22\ \mathrm{y}^{-1}$ (decreasing); in (b) $\lambda = 1.15$ and 0.8, respectively. Note that in both the increasing and the decreasing cases, $\lambda = e^r$, so the continuous and discrete models are equivalent.

Discrete-time models, by contrast, keep track of the state of a system at a series of times, separated by a certain time interval Δt (Figure 2.2b), and the fundamental mathematical tools are those of difference equations. We write difference equations in terms of how the state variable at time t depends on its value at previous times. For example, for a single state variable at time t, X_t:

$$X_t = F\left[X_{t-1}, X_{t-\tau}\right], \tag{B2.1-2}$$

where the integer τ is a constant time interval. For discrete-time models, time is frequently, but not always, written as a subscript rather than in parentheses.

For population biology, models in discrete time would be most applicable to situations in which reproduction occurs once per year over a small time period. This would be more appropriate for populations in the seasonal environments of high latitudes. Models in continuous time are more appropriate for populations that do not experience strong seasonality, or for which the time scale of reproduction is fast relative to environmental variability (e.g. microbial populations).

Box 2.2 DEFINITION OF LINEAR MODELS

Throughout this book, population behavior will be described by the solutions for various models, and we will be interested in any general statements we can make about the behavior to be expected from models with certain characteristics. By "solution" we mean an expression for what the state variable at any arbitrary time t is, given some starting conditions at $t = 0$. Generally a population model is constructed by writing out the various rules for how individuals grow, reproduce, die, etc., such as the expression for the rate of change dx/dt in Eq. (B2.1-1). The solution, then, is an expression for what $x(t)$ would be, given that rate of change and the initial conditions $x(0)$. It might seem that finding that solution would be fairly obvious once one has all of the rates of change written out, but it turns out that it often takes some creative mathematical gymnastics (some might call them "tricks") to find solutions.

For a group of dynamic mathematical models called "linear models" mathematicians have been able to develop an extensive framework of solutions and useful ways of presenting results. The solutions to linear models satisfy the following conditions:

If $x(t)$ is a solution then $ax(t)$ is a solution, where a is a constant.

If $x(t)$ is a solution and $y(t)$ is a solution, then $x(t) + y(t)$ is a solution.

In population biology, the difference between linear models and nonlinear models is typically that linear models are density independent, while nonlinear models are density dependent (see Box 2.3). All populations have some degree of density dependence, so nonlinear models frequently have greater realism. However, there are some times when population dynamics can be adequately approximated by linear models, and linear models have several useful mathematical properties. In fact, the approaches used to analyze linear models are an important part of the analysis of nonlinear models.

We will see later that the special, useful nature of linear models is that their solutions can be written in terms of sums of simple functions: constants to the power t (time) in the case of discrete time, and the exponential function (e^{ct}, where c is a constant and t is time) or sines and cosines in the case of continuous time. This in turn allows linear models to have other useful representations, such as the response of the state variable to temporal variation at particular frequencies (Chapter 4).

The differences between linear and nonlinear models, and the utility of the former, will become clearer as we discuss examples later in the chapter. We have introduced the definition here so that we can show a couple of examples; we will not be repeating all the steps to check if models are linear in later chapters.

Box 2.3 DENSITY DEPENDENCE

In ecology, density-independent models are those for which per capita rates of reproduction and survival do not depend on abundance. In density-dependent models these rates change with population density. The typical density-dependent response to an increase in density is a reduction in per capita reproduction and survival due to competition for limited resources. If the reduction in reproduction or survival is proportional to the increase in density, so that the population tends to return to its original level, the response is termed *compensatory*. If the density-dependent response is stronger, strong enough that an increase in density tends

(*Continued*)

Box 2.3 CONTINUED

to lead to the population being reduced to a point below its original density, it is termed *overcompensatory*. Similarly, a weak density-dependent response that tends to return the population to a point greater than its original density is *undercompensatory*. Finally, density-dependent effects that increase reproduction or survival in response to an increase in density are termed *depensatory* (or sometimes, *inversely density dependent*). Examples of the latter are Allee effects at low population densities, in which reductions in density lead to lower birth rates and lower survival, perhaps because it is harder to find mates or participate in collective defenses against predators.

2.2 Simple linear models (exponential or geometric growth)

The simple model for exponential growth was the first population model developed to explore future population behavior. However, populations were described in terms of a related concept, doubling times, as far back as the seventeenth century (Hutchinson, 1978). The first major formal use was by Thomas Malthus in 1798 (reprinted as Malthus, 1960), who was concerned that the population of the British Isles was growing exponentially, while the resources it required were growing only linearly. Since the rate of growth is continually increasing in exponential growth, he feared that at some point the population would outstrip its required food resources (see Malthus (1960) and Hutchinson (1978) for further historical details).

The basic form of the exponential model in continuous time (see Box 2.1) gives the rate of change of the number of individuals in the population, $N(t)$, in the form of a differential equation,

$$\frac{dN(t)}{dt} = rN(t), \tag{2.1}$$

where r is a constant called the per capita rate of increase (or the intrinsic rate of increase). The units of r are simply time^{-1}, which allows the units of dN/dt to be individuals/time, as one would expect. Note that this differential equation is a linear system (Box 2.2). At the individual level it can be derived by expressing dN/dt as the difference between total birth rate (bN) and total mortality (death) rate (dN), resulting in $r = b - d$. This equation can be solved to obtain an expression for $N(t)$ by the method of separation of variables (i.e. get all of the N's on one side of the equals sign and all of the t's on the other side, then integrate each side; see Box 2.4), to obtain

$$N(t) = N_0 e^{rt}. \tag{2.2}$$

Box 2.4 SOLVING A DIFFERENTIAL EQUATION BY SEPARATION OF VARIABLES

We will solve Eq. (2.1) for the more general case in which r is allowed to vary with time. To solve an equation of the form

$$\frac{dX}{ds} = r(s)X, \tag{B2.4-1}$$

(notice this has the same form as Eq. (2.1), just with different variables as temporary place-holders) we first separate the variables X and s to opposite sides of the equation. We then integrate the equation up to abundance N at time t, beginning from the initial abundance N_0 at t_0:

$$\int_{N_0}^{N} \frac{1}{X} dX = \int_{t_0}^{t} r(s)ds, \qquad \text{(B2.4-2)}$$

which integrates to

$$\ln\left(\frac{N(t)}{N(t_0)}\right) = \int_{t_0}^{t} r(s)ds. \qquad \text{(B2.4-3)}$$

Raising each side to the power of e and then multiplying both sides by N_0 gives

$$N(t) = N_0 e^{\int_{t_0}^{t} r(s)ds}. \qquad \text{(B2.4-4)}$$

If r does not vary with time, and $t_0 = 0$, then Eq. (B2.4-4) reduces to the familiar result

$$N(t) = N_0 e^{rt}. \qquad \text{(B2.4-5)}$$

In this case, the solution exhibits the classic behavior of the simplest linear system (Fig. 2.2a). If $r > 0$ (per capita births exceed deaths), the population grows without bound towards infinity; if $r < 0$ (per capita deaths exceed births), the population declines towards zero (though mathematically $N(t)$ can never reach zero). This type of behavior, in which the rate of increase is proportional to abundance itself (i.e. r times $N(t)$) is termed *exponential* growth or exponential decay (decline).

In discrete time (Box 2.1), both the model formulation and solution are analogous to the continuous case. The basic form of this model is a difference equation instead of a differential equation:

$$N_{t+1} = L_t N_t, \qquad (2.3)$$

where the values of L_t at each time are the fractional growth rate each year (L_t is a unitless proportion), and N_t is the number of individuals in the population at time instant t (often this equation is written with the symbols λ or R instead of L, but we reserve λ for use later in this chapter to refer to a different quantity). Note that this is also a linear model (Box 2.2). The solution is (the capital pi in the following means "the product of")

$$N_t = N_0 L_0 L_1 L_2 L_3 \cdots = N_0 \prod_{i=0}^{t-1} L_i, \qquad (2.4)$$

which, if the L_t's do not depend on time, but rather are a constant L, becomes

$$N_t = N_0 L^t. \qquad (2.5)$$

By substituting the discrete time interval Δt into the continuous time solution, Eq. (2.2), we can see that there is analagous behavior in the discrete and the continuous versions. The solution to the discrete case is a version of the continuous case for $L = e^{r\Delta t}$ (Fig. 2.2b), subsampled at discrete time steps. The behavior of the discrete version is called *geometric* growth, rather than *exponential* growth. Just as the behavior of the continuous version depends on whether $r > 0$, the discrete version exhibits geometric growth if $L > 1$ (note that $e^0 = 1$) and geometric decline if $L < 1$.

The astute reader might have noticed that we have ignored the cases where $r = 0$ or $L = 1$. These values provide the dividing line between the two kinds of behavior of increase and decline. We will return later to a further understanding of the meaning of this particular growth rate, though here we note that explaining steady state behavior of a population with a linear model would require the growth rate to be precisely zero, which essentially can never happen.

Before moving on to nonlinear, logistic-type models, we can evaluate the usefulness of the exponential/geometric models. In terms of realism, at the individual level, all individuals are assumed to be identical, hence both reproductive and mortality rates are proportional to total numbers. In most biological populations, the mere presence or absence of an individual is not a good *i*-state description; we need to know age, at least, to be able to specify current reproduction and mortality. Because of this, simple exponential and geometric growth models do not allow a very realistic portrayal of the contributions of individuals to population behavior. Single-celled organisms such as microbes or phytoplankton might be the exception in which all individuals are essentially identical from the perspective of reproduction and mortality rates.

At the population level, the resulting behavior is exponential or geometric increase, which is only observed in specific situations. In the exponential and geometric growth models, the overall population growth rate is a constant proportion of the current population size; there is no density dependence. In actual populations, density dependence typically arises as populations grow large, intensifying competition for resources. Thus, we might expect to see geometric or exponential growth in situations where abundance is low, such as in recent introductions or in populations declining to extinction. An example of the former is the initial growth of the ring-necked pheasant (*Phasianus colchicus*) introduced to Protection Island in the Strait of Juan de Fuca (Fig. 2.3). An example of

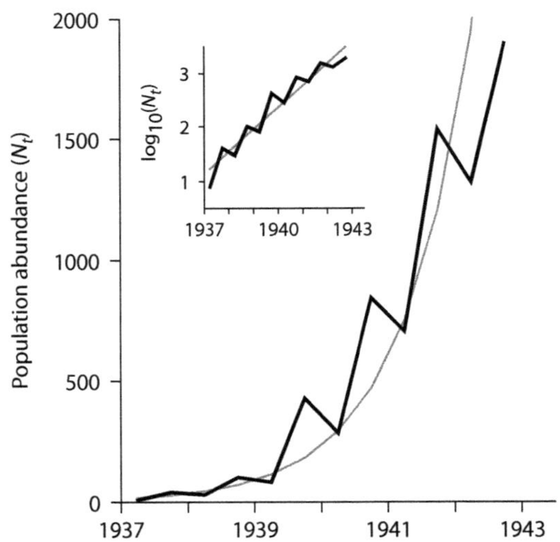

Fig. 2.3 Initial growth of the ring-necked pheasant (*Phasianus colchicus*) population introduced on to Protection Island in the Strait of Juan de Fuca, Washington, USA in 1937. Main plot shows the semiannual inventory of total birds (cocks, hens, and unclassified). Inset plot shows the same data on semilogarithmic axes, illustrating how the slope of increase can be used to estimate λ, the geometric growth rate; the gray curve is a linear regression fitted to the semilog data. In the main plot, the gray curve is a discrete-time exponential model (Eq. 2.3) using the value of $\lambda = 2.58$ obtained from the semilog plot. Data from Einarsen (1945).

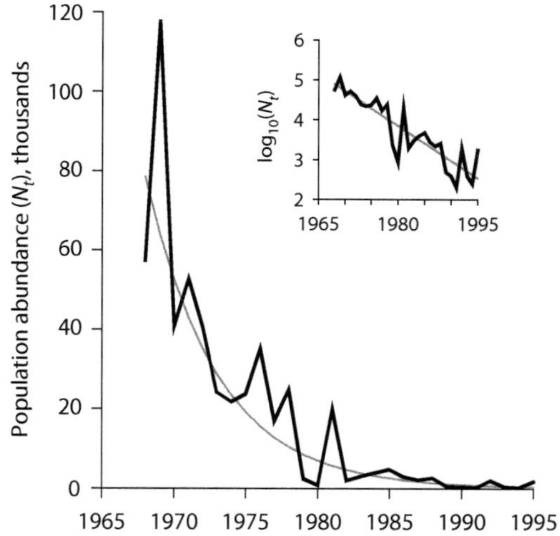

Fig. 2.4 Decline of the winter run of chinook salmon (*Oncorhynchus tshawytscha*) in the Sacramento River, California. Main plot shows the count of salmon traversing the Red Bluff Diversion Dam. Inset plot shows the same data on semilogarithmic axes, illustrating how the slope of decrease can be used to estimate $\log_{10}(\lambda)$, the geometric growth rate; the gray curve is a linear regression fitted to the semilog data. In the main plot, the gray curve is a discrete-time exponential model (Eq. (2.3)) using the value of $\lambda = 0.82$ obtained from the semilog plot. Redrawn from Botsford and Brittnacher (1998).

the latter is the decline of the winter run of chinook salmon (*Oncorhynchus tshawytscha*) in the Sacramento River in California (Fig. 2.4). The insets in each of these examples show the quick and easy way of estimating the per capita growth rate from an abundance time series: when plotted on a semilogarithmic plot, the exponential (or geometric) growth trajectory becomes a straight line, and the slope is r (or $\log[L]$).

In terms of pedagogical value, we have learned that constant per capita rates yield a population whose chief characteristic is that the larger the population, the faster total numbers will grow. Geometric (or exponential) growth is quite relevant to one ecologically important animal population, the population of humans on Earth. The fact that the human population of the world is not only growing at an exponential rate, but that the growth rate $r(t)$ is actually increasing with time is a critical problem (Cohen, 1995). The growing human population—although not the focus of this book—probably influences most of the problems we discuss in the population dynamics of resource management.

Exponential and geometric models also illustrate another point of practical value: that linear models will be of little use in determining optimal harvest (i.e. how much to harvest and when, in order to maximize the total harvest). If the population is declining ($r < 0$ for continuous time or $L < 1$ in discrete time) the population should be completely harvested right away, before it disappears. If the population is growing ($r > 0$ for continuous time or $L > 1$ in discrete time) for a completely unconstrained optimal harvest problem, the solution is to wait until the population is infinitely large, then harvest all of it. This clearly does not make sense.

A more meaningful formulation would be to maximize harvest over a limited time span. However, the answer in this case is similar: for a growing population, the optimal harvest is to wait until just before the end of the time span, then harvest everything. One could possibly find a cost function that leads to a more meaningful solution, but because the standard ways of formulating the optimal harvest problem lead to the problems just stated, and because the model is not considered to be a realistic portrayal of population dynamics, the optimal harvest problem is not approached using this model. See Mendelssohn (1976) for a discussion of this point in the context of age-structured models. We will revisit optimal harvesting with more appropriate models at the end of this chapter, and then in more depth in Chapter 11.

2.3 Simple nonlinear models (logistic-type models)

The simple logistic model (and similar logistic-type models) has been the workhorse of both theoretical and applied population biology. It has been the most commonly used model of population dynamics in population and community theories, and forms the basis of management of many populations, especially in fisheries (Chapter 11; Graham, 1935; Schaefer, 1954; Gulland, 1983). It was originally developed in the early nineteenth century in response to the observation that models with geometric or exponential increase did not hold for some populations of laboratory animals and some human populations. In these, the population growth rate declined at high abundance, thus exhibiting density dependence (Box 2.3). The form we now use most commonly was developed by Verhulst (1838) as the simplest way of modifying the exponential model to produce an upper limit to abundance (see Hutchinson (1978) for further historical information).

Note that the rationale in the modeling efforts associated with the logistic model is basically inductive curve fitting of population-level data. That is, the functional form and parameter values are phenomenological, chosen to mimic trends in population abundance as closely as possible, with (almost) no attempt to represent the individual level mechanism(s) that actually cause changes in population growth rate.

2.3.1 *Continuous-time logistic models*

Verhulst (1838) proposed that the model for exponential growth be modified to approach a constant level by multiplying the model for exponential growth (Eq. (2.1)) by a function that declined from one to zero, as abundance $N(t)$ increased from zero to a new constant, K, the carrying capacity. The basic form of the logistic equation in continuous time is

$$\frac{dN}{dt} = rN(t)\,[1 - N(t)/K]\,, \tag{2.6}$$

where r is a constant, the same intrinsic rate of increase. We can obtain a solution for this equation (i.e. the value of $N(t)$ at any given time) fairly easily by separation of variables (Box 2.4),

$$\frac{K}{N\,(K-N)}dN = r\,dt \tag{2.7a}$$

followed by a partial fraction expansion,

$$\left[\frac{1}{N} + \frac{1}{K-N}\right]dN = r\,dt \tag{2.7b}$$

whose solution, integrated from 0 to t, is

$$\ln\frac{N}{K-N} - \ln\frac{N_0}{K-N_0} = r\,(t-t_0)\,, \tag{2.7c}$$

which can be written

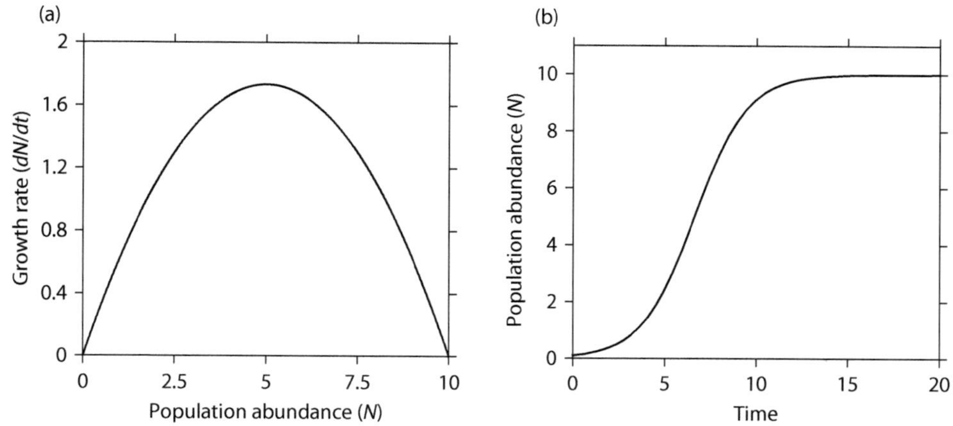

Fig. 2.5 Example of the logistic model (Eq. (2.6)), with intrinsic growth rate $r = 0.69$ and carrying capacity $K = 10$. Panel (a) shows the population growth rate dN/dt as a function of population abundance; note that the growth rate is a maximum at $N = K/2 = 5$. Panel (b) shows the abundance of the population increasing from a low value ($N = 0.1$) to the carrying capacity (Eq. (2.7d)).

$$N(t) = \frac{K}{1 + Ce^{-r(t-t_0)}}, \tag{2.7d}$$

where $C = (K - N_0)/N_0$. Letting t go to infinity, we can see that the solution approaches K asymptotically (Fig. 2.5b).

We can also write Eq. (2.6) in dimensionless form as

$$\frac{d\tilde{N}}{d\tilde{t}} = \tilde{N}\left(1 - \tilde{N}\right). \tag{2.8}$$

Here the tildes (e.g. $\tilde{N}$) indicate that we have rescaled the model by creating new variables or parameters that are combinations of the originals (notice that for simplicity we sometimes write these equations omitting the (t) after N, essentially assuming it is understood that N is the state variable that depends on time). In this case, $\tilde{t} = rt$ and $\tilde{N} = N/K$. In both cases the units cancel out in the new variable or parameter, so the model is now unitless (i.e. has been nondimensionalized). The step of making a model nondimensional often provides a clearer view of how it works, since the dynamical behavior of a model can only depend on nondimensional parameters. That is, the behavior of a model (whether it increases exponentially or reaches a stable equilibrium, for example) never depends on the units in which model parameters or variables are measured, but rather on their magnitude relative to one another. Here in Eq. (2.8) no parameters remain, so the value of this form for the logistic model is that it allows us to see that changes in K simply scale amplitude (i.e. the magnitude of N). Essentially K could be thought of simply as the units associated with population abundance, e.g. thousands of individuals, or population density (per unit area or volume), while changes in r simply scale time (i.e. r merely determines the time it takes the population to reach a particular fraction of K).

In somewhat of a mathematical digression to aid understanding of why some models are linear and some are nonlinear, we can compare how the definition of linear systems in Box 2.2 applies to the linear model of exponential growth (Eq. (2.1)) and the nonlinear model of logistic growth (Eq. (2.6)). For the first condition in Box 2.2, we ask whether $aN(t)$ is a solution of Eq. (2.1), given that $N(t)$ is a solution. That is, whether

$$\frac{d}{dt}(aN(t)) = r(aN(t)). \tag{2.9a}$$

This is obviously true since the a's cancel giving us Eq. (2.1). Applying the first condition for linearity to the logistic model, we substitute $aN(t)$ into Eq. (2.7), presuming that $N(t)$ is a solution. The result is

$$\frac{d}{dt}aN(t) = raN(t)\left[1 - \frac{aN(t)}{K}\right], \tag{2.9b}$$

which is not the same as Eq. (2.6) because the a's do not cancel. To check whether the exponential growth model meets the second condition in Box 2.2, we ask whether $N_1(t) + N_2(t)$ is a solution of Eq. (2.1), given that both $N_1(t)$ and $N_2(t)$ are solutions. The answer again is yes,

$$\frac{d(N_1 + N_2)}{dt} = r(N_1 + N_2) \tag{2.10a}$$

is a solution because $dN_1/dt = rN_1$ and $dN_2/dt = rN_2$ are both solutions. Applying the second condition for linearity to the logistic model we ask whether $N_1(t) + N_2(t)$ is a solution,

$$\frac{d(N_1 + N_2)}{dt} = r(N_1 + N_2)\left[1 - \frac{(N_1 + N_2)}{K}\right] = rN_1\left(1 - \frac{N_1}{K}\right) + rN_2\left(1 - \frac{N_2}{K}\right) - 2\frac{rN_1N_2}{K}. \tag{2.10b}$$

The extra term on the right, $-2rN_1N_2/K$, confirms again that the logistic model is nonlinear.

While we have explored the question of linearity formally here to illustrate it, for most applications in population dynamics, that is not necessary. A good rule of thumb is that if a model includes squared or cubic terms of a state variable it will be nonlinear. Most often this arises when including density dependence. Many other population models have been developed that have essentially the same behavior as the logistic: rapid increase from low abundance to a constant high equilibrium value. We refer to these here as *logistic-type* models. The critical differences between these various models is the manner in which the population grows to reach this high abundance. Choosing which model to use is a challenge since essentially none of these models is developed from an underlying mechanistic approach. Comparisons are typically phrased in terms of the value of N, expressed as a fraction of K, at which the population growth rate is a maximum (cf. Fowler, 1981). In the original logistic model, Eq. (2.6), this maximum growth rate occurs when $N = K/2$ (Fig. 2.2a; later in Section 2.4.3 we see how that result becomes useful in models of harvested populations). However, in the 1960s and 1970s researchers and practitioners became concerned that plots of dN/dt versus N were not always symmetric with a maximum at $K/2$ (as in Fig. 2.2a), but rather were skewed toward higher or lower abundances. Some of the early attempts to obtain a better fit to various data were in a fisheries context, with the models of Pella and Tomlinson (1969),

$$\frac{dN}{dt} = \begin{matrix} HN^m - KN & m > 1 \\ HN^m + KN & m < 1 \end{matrix}, \tag{2.11}$$

with constants H, K, and m, and Fox (1970),

$$\frac{dN}{dt} = N(A - \ln N), \tag{2.12}$$

with constant A. In theoretical ecology, a prominent example is the θ-logistic model of Gilpin and Ayala (1973). In order to obtain a better fit to data from *Drosophila* populations,

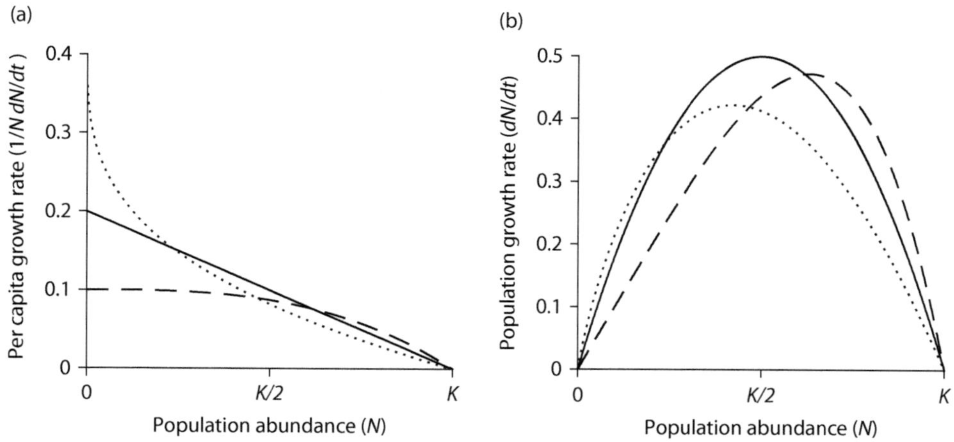

Fig. 2.6 Different life histories that can be represented by the θ-logistic model, as revealed by plots of (a) the per capita growth rate and (b) the overall population growth rate. Shown are a population high reproductive rate (dotted curve, $\theta = 0.33$), presumed to be associated with a short lifespan, and a population with a low reproductive rate (dashed curve, $\theta = 3$), presumed to be associated with a long lifespan, and an intermediate population ($\theta = 1$, equivalent to the simple logistic model). (n.b. these logistic models do not include age structure, hence no direct information on lifespan). Values of r for each population were adjusted to make the maximum growth rate dN/dt similar for each model; $K = 10$ in all cases.

they sought the simplest extension of the logistic that would accurately fit the data. The result was the θ-logistic model,

$$\frac{dN}{dt} = rN\left[1 - \left(\frac{N}{K}\right)^{\theta}\right], \tag{2.13}$$

in which θ is a new constant in addition to r and K that determines whether density dependence tends to be stronger for $N < K/2$ ($\theta < 1$) or for $N > K/2$ ($\theta > 1$; Fig. 2.6).

The relative value (i.e. N/K) at which a population reaches the maximum rate of increase can be interpreted in terms of the relative abundance at which density dependence sets in; this can then be interpreted in terms of life history theory. For example, on the basis of several different data sets, Fowler (1981) argued that species with high reproductive rates and short lifespans and populations held below the limits of environmental resources exhibit most density-dependent change at lower levels (equivalent to $\theta < 1$ in the θ-logistic model) relative to species and populations with the opposite characteristics. This can be seen in a plot of $dN/dt \times 1/N$ (the per capita growth rate, as opposed to the overall population growth rate dN/dt) versus N (Fig. 2.6). For the former type of species with "fast" life histories, per capita growth declines with N very steeply at low N, then flattens; for the latter type of species with "slow" life histories, per capita growth is unchanged with N until N approaches K (Fig. 2.6a). Fowler's review found that mammals tend to show density dependence at higher relative population levels, while insects and some fish tend to show density dependence at lower relative levels.

Another approach that has been followed is to attempt to derive something like the logistic model on the more realistic basis of mechanistic first principles. By making various assumptions about intraspecific competition and energy allocation, Schoener (1973) derived two different types of logistic-type equations, one being identical to the logistic, and the other with a downward concave growth rate with a maximum at zero abundance. He also considered combinations in which populations switched from one type of model to the other.

2.3.2 *Discrete-time logistic models*

There have also been several different discrete-time versions of logistic-type models. May and Oster (1976) listed several different nonlinear functions $F(N)$ (Table 2.2) for the following equation:

$$N_{t+1} = F(N_t). \tag{2.14}$$

As you can see, there are many ways of representing the density-dependent logistic dynamics in discrete time. Two common forms used in population ecology appear similar to the continuous-time logistic equation,

$$N_{t+1} = F(N_t) = N_t e^{\left[r\left(1-\frac{N_t}{K}\right)\right]} \tag{2.15a}$$

$$N_{t+1} = F(N_t) = N_t\left[1 + r\left(1 - \frac{N_t}{K}\right)\right], \tag{2.15b}$$

where r and K are constants (these two models produce similar results because $e^x \approx 1 + x$, for small values of x). These are not equivalent to the continuous-time logistic models because they are written in terms of discrete time, which includes an inherent lag (the time between time t and time $t + 1$). Because of that lag, the behavior of these models may differ from the behavior of continuous-time models that are similar in appearance (Fig. 2.7; May (1974)). Essentially, the lag can produce overcompensatory density dependence; it is then possible for a growing population to "overshoot" K in a single time step, which then leads to a negative growth rate in the next time step, undershooting K, and so forth. This effect, and the divergence between the different versions of the model, are amplified as r increases (we will say more about this type of cyclic behavior in Section 2.4 when

Table 2.2 Nonlinear discrete-time models with logistic dynamics of from $N_{t+1} = F(N_t)$. Adapted from May and Oster (1976).

$F(N)$	Source
$Ne^{[r(1-N/K)]}$	(Moran, 1950; Ricker, 1954; Macfayden, 1963; Cook, 1965)
$N\left[1 + r(1 - N/K)\right]$	(Chaundry and Phillips, 1936; Maynard Smith, 1968; Krebs, 1972; May, 1972; Maynard Smith, 1974; Li and Yorke, 1975)
$\lambda N/\left[1 + e^{-A(1-N/B)}\right]$	(Pennycuick et al., 1968; Usher, 1972; Beddington, 1974)
$\lambda N/(1 + aN)^b$	(Hassell, 1975; Hassell et al., 1976)
$\lambda N/\left[1 + (N/B)^b\right]$	(Maynard Smith, 1974); b = 1: (Skellam, 1951; Leslie, 1957; Utida, 1967)
λN if $N < C$; λN^{1-b} if $N > C$	(Varley et al., 1973)
$\lambda_+ N$ if $N < K$; $\lambda_- N^{1-b}$ if $N > K$	(Williamson, 1972); $\lambda_+ > 1; \lambda_- < 1$
$N\left[1/(a + bN) - \sigma\right]$	(Utida, 1957)
$N\varphi(N)$	$\varphi(N)$ obtained experimentally by Nicholson and others (see Oster et al. (1976))

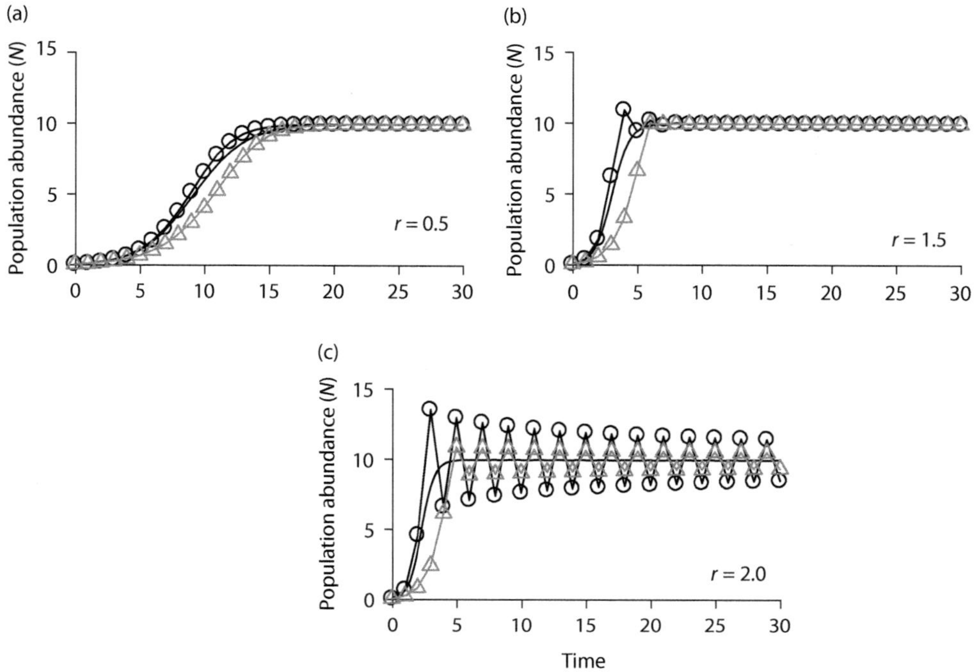

Fig. 2.7 Comparison of the continuous time (solid black curve) and discrete time versions (circles and black curve, Eq. (2.14a); triangles and gray curve, Eq. (2.14b)) of the logistic model. Examples are shown for $K = 10$ and $r = 0.5$, 1.5, or 2 in panels (a, b, c), respectively.

we discuss the stability of these models). The continuous-time version does not have lags and always approaches K smoothly. This difference is in contrast to the (linear) exponential/geometric growth models, which had the same behavior in both continuous and discrete time.

Two common discrete-time analogs of the logistic model used in fisheries are the Ricker (1954) model,

$$N_{t+1} = F(N_t) = \alpha N_t e^{-\beta N_t} \tag{2.16a}$$

and the Beverton and Holt (1957) model,

$$N_{t+1} = F(N_t) = \frac{N_t}{\frac{1}{\alpha'} + N_t/\beta'}, \tag{2.16b}$$

where α, β , α', and β', are constants; these models are shown in Fig. 2.8. These functions are presented here as models of a population, but they were originally used in fisheries to represent the dependence of recruitment (the addition of new individuals to the population, also known as *stock*) on the size of the parental stock (in fisheries these are known as stock-recruitment relationships). In both of these models, α is the slope of the function at the origin, an important parameter determining population persistence as we shall see later in this chapter. In the Beverton–Holt model, the parameter β represents the value that N_t eventually approaches. In the Ricker model, the parameter β determines the equilibrium abundance of the population given the value of α, i.e. $N^* = \beta \ln \alpha$. We use * to indicate the equilibrium abundance, a term we will define later in the chapter, but

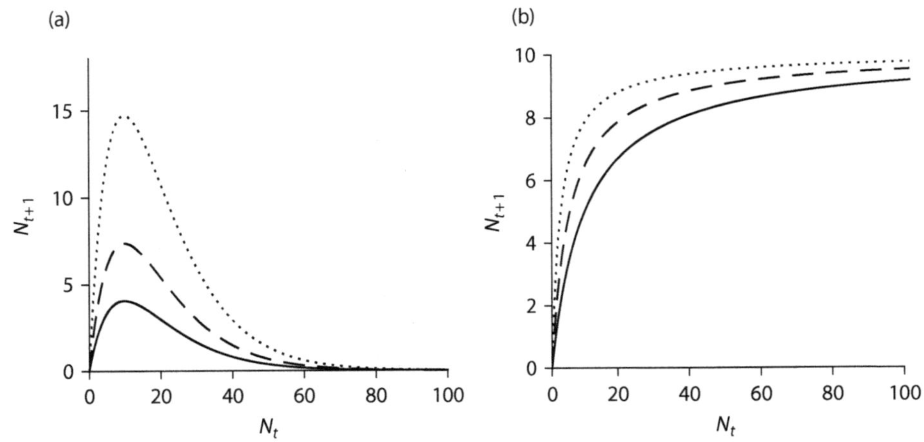

Fig. 2.8 The (a) Ricker (Eq. (2.15a)) and (b) Beverton–Holt (Eq. (2.15b)) models, showing the value of N_{t+1} as a function of N_t for several values of the slope at the origin (α or α'): 1.1 (solid curve), 2 (dashed curve), and 4 (dotted curve) in both panels. In (a), the Ricker β parameter $= 0.1$. In (b) the Beverton–Holt β' parameter $= 10$, so all three curves have the same asymptotic maximum value of 10.

for now the reader can consider it to be the value the population will tend toward as time becomes large.

The dynamic behavior of discrete-time logistic models can be understood most easily by a graphical approach called "cobwebbing." This method was developed and used in the 1940s and 1950s for stock-recruitment equations in fisheries. We will illustrate that procedure here by first following the straightforward procedure of generating a time series of N_t by looking up the next value from the plot of N_{t+1} versus N_t, at each time step (Fig. 2.9). After generating a couple of time steps, we will then point out the graphical shortcut. The first two look-ups in this procedure are shown in Fig. 2.9a. We begin by measuring the distance N_0 along the N_t axis of Fig. 2.9a, then moving up to the function $F(N_t)$ to read the value of abundance at the next time step, i.e. N_1, on the N_{t+1} axis. We plot that value of N_1 on the time series to the right (Fig. 2.9b). To find the next value of abundance, N_2, we measure the distance N_1 along the N_t axis, then move up to the function to read off N_2 so that we can plot it next. At this point we can make the observation that in going from the function to the vertical axis to read off N_1, then marking off N_1 on the horizontal axis, and going vertically toward the function again, we have created a square with all sides equal to N_1, whose upper right-hand corner is on the replacement line. Thus there is no need to measure the distance N_1 along the N_t axis. Instead we can merely draw a line of slope 1 (the dashed line in Fig. 2.9a, called the "replacement line"), then take a short cut by merely going horizontally only to the replacement line in the first look-up, then going vertically to the function, to achieve the second look-up. These are the first steps illustrating the rules of cobwebbing:

1. Start from the initial value on the horizontal line.
2. Go vertically to the function $F(N_t)$, plot that point on the time series.
3. Go horizontally to the replacement line, then repeat step 2.

These steps are illustrated in Figure 2.9c–d. In the case illustrated here, the cobwebs converge on a single point after a few time steps, indicating that the population will approach an equilibrium. This is not always the case, as we will see later.

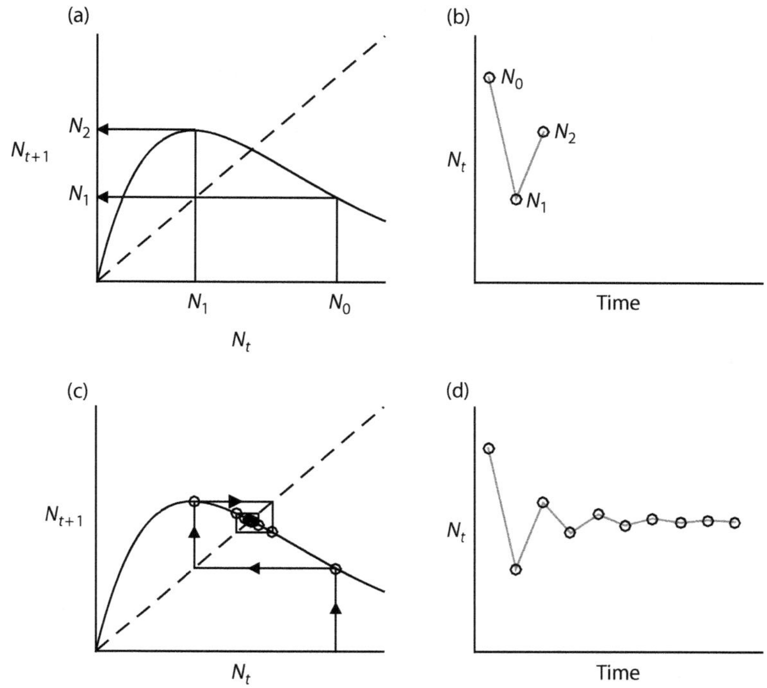

Fig. 2.9 Demonstration of the cobwebbing procedure. (a) The function $F(N)$ (solid curve) and the process of simply looking up the first two values of the next abundance. Also shown is the line of slope = 1 (dashed line), the "replacement line," used to identify the upper right-hand corner of the squares used to avoid looking up values on the horizontal axis. The resulting time series is shown in (b). The shortcut, cobwebbing process can be seen by following the arrows in (c). (d) The process is followed through to equilibrium, using the shortcut of only drawing horizontal lines to the replacement line, then vertical lines to the function.

2.4 Illustrating population concepts with simple models

While simple population models may lack the realism to reflect some important aspects of population dynamics, we can take advantage of their simplicity to introduce fundamental aspects of population behavior that we will be studying later. We first use the simple nonlinear model in discrete time to illustrate the concept of population stability, including stability about zero abundance, as a reflection of population persistence. The second application uses a simple linear model in discrete time (Eq. (2.3)) with random values of the L_t's to illustrate the effects of randomness on population persistence and probability of extinction. The third example uses the continuous time logistic model to give a primitive view of how fishing and density dependence interact when trying to achieve sustainable fishing.

2.4.1 *Illustrating dynamic stability with simple, linear, discrete-time models*

Unlike the simple linear models, for more complex models with density dependence we often cannot write down an explicit formula giving the solution (N_t) as a function of time. Instead, we use stability analysis as a way of determining how a dynamic system will be expected to behave, without simulating each possible situation. Stability analysis can determine what combinations of parameter values will tend to cause the population to increase, decrease, cycle with diminishing amplitude, cycle with increasing amplitude, persist at a constant equilibrium, or go extinct. We will begin by describing the *local stability* of deterministic populations that are responding to being perturbed away from a specific *equilibrium* abundance (See definition of local stability in Box 2.5).

Box 2.5 A GRAPHICAL DEFINITION OF EQUILIBRIA AND LOCAL STABILITY

We can get a clear sense of the mathematical definition of local stability by presuming that model solutions (i.e. the possible values of $N(t)$) define a topographical surface that may have various hills and valleys in it (e.g. Fig. 2.10). The shape of the landscape represents the dynamics of the model. Possible solutions to the model equations can be represented by the location of a ball on the surface. Figure 2.10 is an example of part of such a surface that includes a peak, a valley, and a flat region.

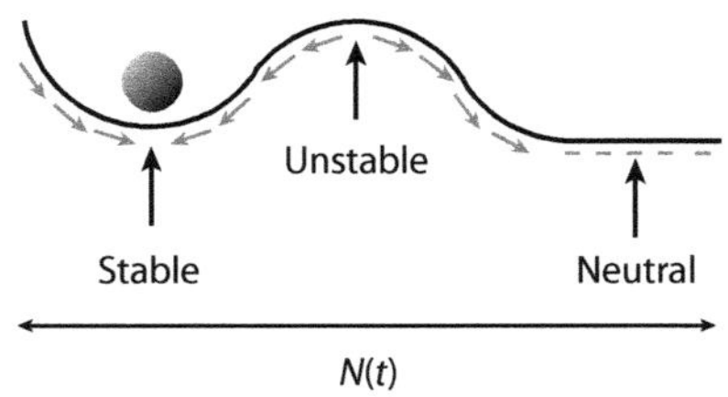

Fig. 2.10 A graphical illustration of dynamic stability. The black curve represents the range of possible model solutions (values of $N(t)$). The black arrows indicate three possible equilibria, i.e. places where $N(t)$ will remain constant unless perturbed. Gray arrows indicate what will happen if $N(t)$ is perturbed away from any of the equilibria: it will tend to be pushed back towards the stable equilibrium and pushed away from the unstable equilibrium. $N(t)$ will not tend to move in either direction if moved a short distance away from the neutral equilibrium point, it will just remain at the new location (thus the entire flat portion of the surface is comprised of neutral equilibria).

First we define an *equilibrium*, which is also referred to as a *fixed point*. It is a point on the surface at which the ball will stay if not perturbed. These are indicated by arrows in Fig. 2.10. The population can be either be locally stable or locally unstable about each equilibrium. It is defined to be locally stable if, when perturbed a little from the equilibrium (e.g. the ball is pushed gently), it tends to return to the equilibrium. If it tends to move away from the equilibrium when perturbed, it is locally unstable. Continuing the topographical metaphor, we refer to the "valley" surrounding an equilibrium as a *basin* of attraction. It is also possible (although probably rare in nature) to have neutral stability, which is represented by the flat surface in Fig. 2.10 (if the ball is pushed, it stays at the new location without rolling back or accelerating away).

A mathematical analysis of the dynamic stability of a system is equivalent to determining whether each equilibrium is on a hill or in a valley, and how steep the surrounding slopes are.

Before beginning our description of population stability, we must place our mathematical definitions in the context of the many uses of the term "stability" throughout ecology (see Table 2.3 for a summary of these definitions). When we describe issues associated with

Table 2.3 Definitions of stability and related concepts. Adapted from Loreau (2010).

Concept	Definition
Local dynamic stability	Property of a system that returns to an equilibrium (or a nonequilibrium trajectory) after a perturbation
Global stability	Property of a system that returns to a single ("globally stable") equilibrium from any set of initial conditions
Resilience	The velocity at which a system returns to equilibrium after perturbation (see Webster et al., 1974)
Resistance	A measure of the ability of a system to remain at equilibrium despite disturbance (see Harrison, 1979)
Robustness	The magnitude of perturbation required to switch a system from the basin of attraction of one equilibrium to another basin of attraction. Similar to Holling's (1973) definition of "resilience".
Amplification envelope	Describes how an initial perturbation from equilibrium is amplified within a system (see Neubert and Caswell, 1997)
Variability	The magnitude of temporal changes in a system property (e.g. abundance)
Persistence	A measure of the ability of a system to maintain itself through time (e.g. at a nonzero equilibrium)

population stability we will use the concept of local dynamic stability (Box 2.5, Table 2.3). This is the meaning used most commonly by physicists and engineers, such as in characterizing electronic filters, or describing the lack of stability arising from holding a microphone too close to the speaker in a public address sound system (the single tone that rapidly increases in volume is a clear sign of instability). We talk of the local stability of a particular equilibrium point, but a population may have more than one equilibrium point, and we may want to evaluate local stability at each of them. Another definition of stability refers to how variable a population is, with high variability being referred to as instability.

Other definitions of stability in ecology are related to the long-standing conversation in ecology about the relationship between diversity and stability, i.e. the idea that diversity (usually species diversity) promotes stability (usually at the community or ecosystem level; Tilman and Lehman, 2002; Loreau, 2010). We describe an example of this in Section 10.8. Attempts to test diversity/stability relationships have been hampered by the many different definitions of the terms stability and diversity (Pimm, 1984; Ives and Carpenter, 2007). Grimm and Wissel's (1997) inventory of the many ways in which the word "stability" is used in ecology yielded 163 definitions of 70 different stability concepts, motivating the authors to conclude that the term is so ambiguous that it is virtually useless. They divided the definitions into three categories: stability implying *constancy* (and conversely instability implying high variability), stability implying return to a resting state (the category used here, which they termed the *resilience* definition), and *persistence* through time. Here we will focus primarily on the second definition, which we will call local stability; we will reserve the term resilience for something else (Table 2.3).

We begin by examining local stability of a simple, linear model in discrete time, as described earlier in Eq. (2.3). The basic approach underlying stability analysis is to make use of our knowledge of the basic forms of solutions to these simple models, i.e. constants to the power t for discrete time (Eq. 2.5), and exponential functions of a constant times t for continuous-time systems (Eq. 2.2d). The first step is to identify the equilibria for which local stability will be analyzed. In a discrete-time model this is done by setting $N_{t+1} = N_t = N^*$, the equilibrium, and solving for N^* (in continuous time one solves the expression $dN/dt = 0$ (i.e. abundance is not changing) for the values of equilibria, N^*). For Eq. (2.3), the only value of N^* that satisfies

$$N^* = LN^* \tag{2.17a}$$

is $N^* = 0$. The fact that the equilibrium is at zero is not surprising, since our solutions to this equation have shown behavior either moving away from zero or moving toward zero, geometrically (Fig. 2.2). To evaluate the stability of this equilibrium we consider what happens when there is a small perturbation ΔN_t away from the equilibrium such that $N_t = N^* + \Delta N_t$. The stability question is, will the perturbation grow with time (instability), or will it shrink, taking the population back towards N^* (stability)? In other words, as time increases, does the solution to

$$\Delta N_{t+1} = L\Delta N_t \tag{2.17b}$$

tend toward $\Delta N_t = 0$?

To find a solution to the equation for ΔN_t, we use our knowledge of linear difference equations (Eq. (2.5)), and guess that the solution is likely to be

$$\Delta N_t = \Delta N_0 \lambda^t, \tag{2.17c}$$

where λ is some constant. First we test whether that solution works, then ask how the value of λ affects population behavior. We test the proposed solution by substituting it back into equation (2.16b),

$$\Delta N_0 \lambda^{t+1} = L\Delta N_0 \lambda^t \tag{2.17d}$$

which we can simplify to obtain what we will call the *characteristic equation* of this system,

$$\lambda - L = 0. \tag{2.17e}$$

We can solve this equation quite easily to obtain the solution, $\lambda = L$, which is called the root of this (the world's simplest) algebraic equation. In other words, our proposed solution to the stability analysis (Eq. 2.17c) only holds if $\lambda = L$. In general, a characteristic equation is a polynomial algebraic expression, with variable λ, set equal to zero. When the characteristic equation is solved, the roots (the values of λ) also provide solutions to the original linear difference equation (in this case, Eq. 2.17c) or differential equation, and the values of those roots tell us about the stability of the system. In this case the characteristic equation is a polynomial of order 1 (λ^1), so there is only one root. In such a simple case this may all seem quite trivial, but keep in mind this is only an introduction to illustrate the idea of stability analysis.

From the form of that solution (Eq. (2.17c)), the stability analysis is straightforward to understand: if $L = \lambda > 1$, the population abundance will increase geometrically away from $N^* = 0$ if perturbed (an unstable equilibrium); if $L = \lambda < 1$ it will decline back towards $N^* = 0$ if perturbed (a stable equilibrium); and if $L = \lambda = 1$, it will be neutrally stable and remain at the level to which it was perturbed. This is the essence of stability analysis: finding a general solution to the equation describing the system, then asking how different values of a variable in that equation, here λ, will lead to different kinds of behavior. Thinking about this in terms of Box 2.5, the value of $N^* = 0$ is an equilibrium because once a population goes to zero, it will remain there. That equilibrium is either in a valley (stable) or on a ridge (unstable), depending on the value of λ, and one can also think of λ as describing the steepness of the slope leading to or away from N^*. For this example, we are only examining positive values of λ because we know that abundance cannot be less than zero. We will see the consequences of λ being negative in the examples in later chapters, for nonlinear systems.

To move incrementally toward more complex examples, we can analyze stability of another linear example, the rabbit problem from Section 2.1. Recall that in the Fibonacci series, which emerges from the rabbit problem, each entry is the sum of the previous two entries,

$$N_{t+1} = N_t + N_{t-1}. \tag{2.18a}$$

Once again there is one equilibrium, $N^* = 0$ (this is generally true for linear models). To analyze small deviations ΔN_t from N^*, we again substitute the expected solution to a linear equation, $\Delta N_t = \Delta N_0 \lambda^t$, to obtain

$$\lambda^{t+1} = \lambda^t + \lambda^{t-1}, \tag{2.18b}$$

which, dividing by λ^t and rearranging, becomes the characteristic equation

$$\lambda^2 - \lambda - 1 = 0, \tag{2.18c}$$

which can be solved using the quadratic formula. The characteristic equation for this linear model is of order 2 (i.e. the highest power of λ is 2), so there are two roots,

$$\lambda_1 = \frac{1+\sqrt{5}}{2}; \lambda_2 = \frac{1-\sqrt{5}}{2}. \tag{2.19}$$

Because there are two roots, the solution will be the sum of two terms, each of which is one of the roots of the characteristic equation to the t power, multiplied by a constant. Thus,

$$\Delta N_t = c_1\left(\frac{1+\sqrt{5}}{2}\right)^t + c_2\left(\frac{1-\sqrt{5}}{2}\right)^t, \tag{2.20}$$

where c_1 and c_2 are constants whose values are determined from the initial conditions (i.e. the starting values) of ΔN_t and ΔN_{t-1}. As time (t) increases the first term will be 1.618^t and the second will be $(-0.618)^t$, with the terms alternating in sign for the latter (because λ_2 is negative, the second term will be positive for even values of t but negative for odd t). Also, as t increases the value of the first term will eventually be much larger than the value of the second term, regardless of what the two constants are, so the value of λ_1 is more important to stability than the value of λ_2. In this case $\lambda_1 > 1$, so the system is unstable at $N^* = 0$ and will increase away from zero without bound. This illustrates a general rule of analysis of stability of a linear system, that stability will be determined by the largest root of the characteristic equation.

These simple examples display the essence of the stability analyses we will employ throughout the book, with more complex models (basically actions at many more lags). For each model we will derive a characteristic equation and find its roots. We will then claim that we can write the solution to the model's equation as we did in Eq. (2.20), that is as the sum of n terms (where n is the order of the characteristic equation) that are each a constant times a root to the power t (for discrete-time models). We will then conclude that, in the long run, the behavior of the population will be determined by the largest root (sometimes the largest two or three roots), as we concluded from Eq. (2.20). Thus to know how a population will behave in future, we will not need to run many simulations, we need only to find the largest root(s) of the characteristic equation.

Stability analysis does not involve using Eq. (2.20) to calculate a solution. That is, we simply look at the values of the largest eigenvalues to determine if the system is stable, but we do not need to find out what ΔN_t will be for a particular value of t. However, some readers may be wondering if Eq. (2.20) is actually a solution. For this particular problem, we know how to generate a Fibonacci series (start with 1 and 1, then add each two values to get the next). How can the weighted sum of two constants to the power t do that? For this case, we will demonstrate that Eq. (2.20) actually produces the model behavior. We first assume that $\Delta N_0 = 1$, and the value of ΔN prior to that is 0. We can use Eqs. (2.17) and (2.20) to solve for the values of the constants c_1 and c_2:

$$c_1 = \frac{1}{\lambda_1}\left[\Delta N_0 - \lambda_2 \frac{\Delta N_0 (2 - \lambda_1)}{\lambda_2 (\lambda_2 - \lambda_1)}\right] = 0.7236 \tag{2.21a}$$

$$c_2 = \frac{\Delta N_0 (2 - \lambda_1)}{\lambda_2 (\lambda_2 - \lambda_1)} = 0.2764. \tag{2.21b}$$

Placing these values into Eq. (2.20) leads to values of ΔN_t of 1.0, 1.0, 2.0, 3.0, 5.0, 8.0, 13.0, 21.0, 34.0, 55.0, ... which is recognizable as the original Fibonacci series in the right-hand column of Table 2.1(with some small numerical rounding errors). Thus the stability analysis correctly predicts the behavior of the system when it is moved away from zero. Again, this is a simple example, but it is comforting to know that writing solutions in terms of weighted sums of eigenvalues to the power t performs as expected in a case for which we know what the correct result should be.

2.4.2 *Dynamic stability of simple nonlinear models*

For nonlinear systems, the solutions could have multiple nonzero equilibria, as seen in Box 2.5. As in the linear systems, we can find the value of the equilibria by solving the model equation for the case when all values of the state variable are the same, regardless of time. For example, for the system in Eq. (2.14), the expression for the equilibrium N^* would be $N^* = F(N^*)$. To determine the local stability about that equilibrium, we are interested in the behavior of small deviations away from it. To do this, we can use the same approach we just developed for linear systems (i.e. taking advantage of what we know about the forms of linear model solutions) *if* we "linearize" the model equation about the equilibrium. The verb linearize means to approximate the system by a straight line through the equilibrium. We can do that by making use of the expression for a Taylor series from calculus (Box 2.6),

$$f(x) = f(x_0) + f'(x_0)(x - x_0) + \frac{1}{2} f''(x_0)(x - x_0)^2 + \ldots \tag{2.22}$$

This expression approximates $f(x)$ near $x = x_0$ as the sum of terms consisting of successive derivatives of f with respect to x, times successive powers of $(x - x_0)$. For our purposes here, we use only the first two terms of the Taylor series to approximate the function $F(N_t)$ as a linear function near the equilibrium N^*, i.e.

$$F(N_t) = F(N^*) + F'(N^*)(N_t - N^*). \tag{2.23}$$

Defining $\Delta N_t = (N_t - N^*)$ (and because it is linear, $F(N_t) - F(N^*) = F(N_t - N^*)$), the expression for the stability of the model in Eq (2.13) becomes

$$\Delta N_{t+1} = \Delta N_t F'(N^*). \tag{2.24}$$

This has the same form as Eq. (2.3) with L constant, i.e. it is a simple linear model, but in ΔN_t, rather than N_t, and with the coefficient being the slope of the function at the equilibrium $F'(N^*)$. Based on our earlier analysis of this model (Eq. 2.17), we already know that the system will be stable about the equilibrium when that slope is between 0 and 1, and unstable about that equilibrium when the slope is greater than 1. When the slope is negative, the magnitude (i.e. the absolute value) of the solution grows in the same way, but with opposite sign. That is, locally stable when the slope is between −1 and 0 and locally unstable when the slope is < -1.

Box 2.6 TAYLOR SERIES

A Taylor series, or Taylor expansion (named after mathematician Brook Taylor, who introduced it in a 1715 publication), is a way of approximating a function as the sum of a series of terms involving increasingly higher-order derivatives of the function. They can be particularly useful in estimating values of complicated nonlinear functions for which we lack a general solution, over a small range.

If we know the value of a function $f(x)$ at a particular value of the variable x_0, we can approximate the function at any other value x as the infinite sum,

$$f(x) = f(x_0) + \frac{1}{1!}f'(x_0)(x - x_o) + \frac{1}{2!}f''(x_0)(x - x_0)^2 + \frac{1}{3!}f'''(x_0)(x - x_0)^3 \ldots, \quad \text{(B2.6-1)}$$

where increasingly higher-order derivatives of f are multiplied by $(x - x_0)$ to the corresponding power and divided by the corresponding factorial. This approximation is increasingly accurate as more and more terms are added to the series, and the smaller the difference $x - x_0$. Also, the smaller that difference, the less important it becomes to add in additional higher-order terms. For this reason, we can use Taylor series in dynamic stability analysis to create linear approximations (order (1) Taylor series, taking just the first two terms in B2.6-1) of the behavior of small deviations to nonlinear systems. In Fig. 2.11 we show Taylor series approximations of order 0, 1, 2, and 3 to the function $f(x) = e^x$.

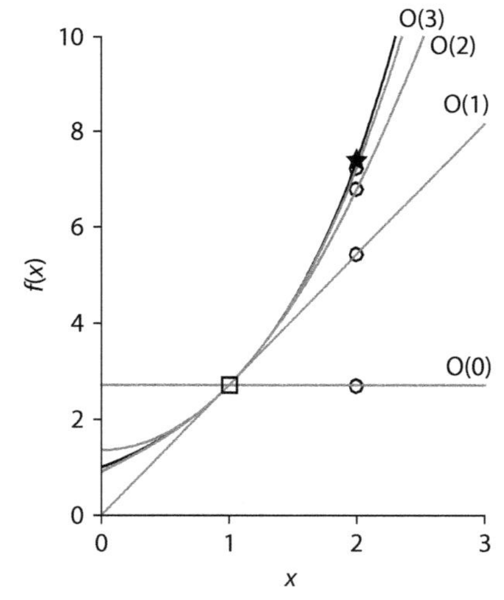

Fig. 2.11 Examples of Taylor series approximation to a nonlinear function, $f(x) = e^x$ (black curve). Here we attempt to estimate $f(x_1 = 2)$ given the value at $f(x_0 = 1)$ using Taylor series of order 0, 1, 2, and 3 (gray curves). The square symbol indicates $f(x_0)$; the star indicates the actual value of $f(x_1)$, and the circles indicate the Taylor series approximations.

Because of this result, that the stability of simple discrete models depends on the slope of $F(N_t)$ at equilibrium, the graphical method of cobwebbing can also be used to examine stability. As an example we will use the Ricker model again (Fig. 2.12). Firstly we observe

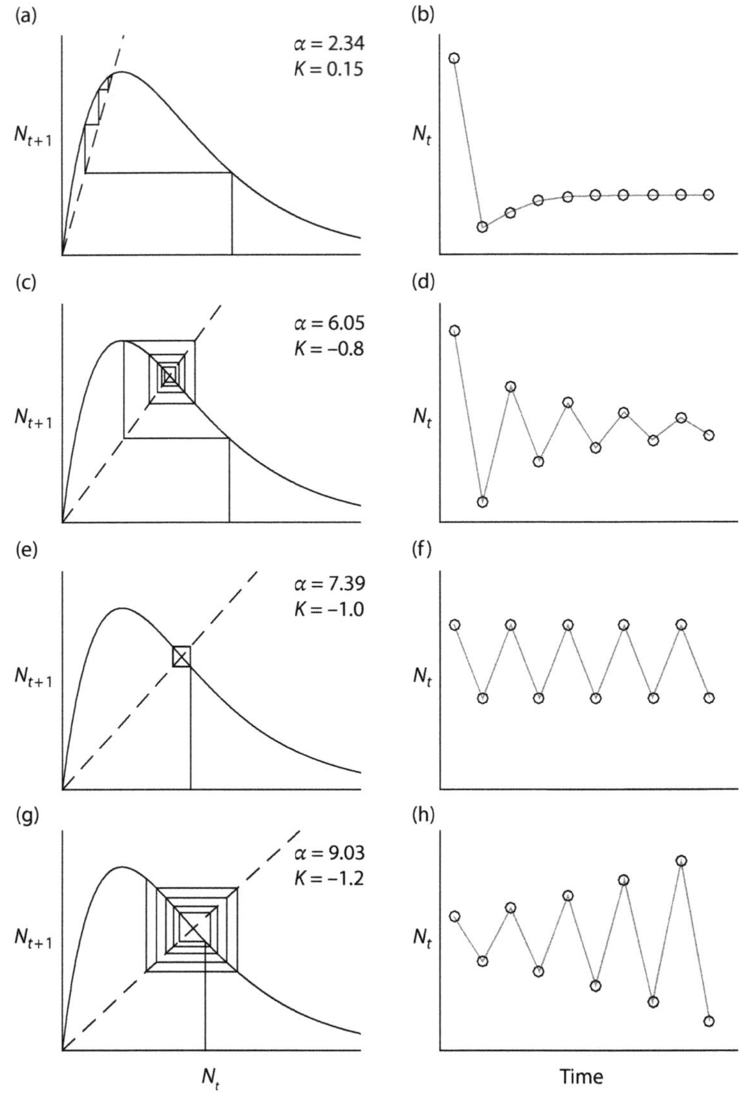

Fig. 2.12 Using cobwebbing to characterize stability of the Ricker model. As in Fig. 2.9, the black curve is $F(N)$, and the dashed line is the replacement line. The cobweb (a, c, e, g) and corresponding time series (b, d, f, h) are shown for a range of parameter values that lead to different dynamic stabilities. Parameter $\beta = 1$ in all panels, but parameter α (which determines the slope, K) varies as indicated. Note that the scale of the vertical axis in the left-hand panel varies in order to show the cobwebbing (the replacement line has slope 1 in all panels).

that the point at which the function $F(N)$ intersects the replacement line is an equilibrium (i.e. $F[N^*] = N^*$; in the Ricker model, $N^* = \ln[\alpha]/\beta$). To examine local stability about that point we define the slope of $F(N)$ at the equilibrium to be K. (This notation, using K as a normalized slope parameter, arises in the analysis of age-structured models. Be careful not to confuse it with carrying capacity. We will use this notation again in Chapter 7.) In the Ricker model, $F'(N) = K = 1 - \ln(\alpha)$. We can then graphically determine solutions for several values of K; these correspond to the solutions we have just presented previously.

For $K > 1$, ΔN_t will increase geometrically without bound. For $0 < K < 1$, ΔN_t will decline geometrically to $\Delta N_t = 0$ (i.e. to the equilibrium; Fig. 2.12a,b), thus exhibiting compensatory density dependence (Box 2.3). Negative values of K lead to cyclic behavior with period 2, because there is overcompensatory density dependence (Box 2.3), causing the population to alternate between overshooting and undershooting the equilibrium. For $-1 < K < 0$, the population is stable, but it approaches the equilibrium cyclically with geometrically declining cycles of period 2 (Fig. 2.12c,d). For $K = -1$, the population is neutrally stable. The cycles of period 2 are of constant amplitude (Fig. 2.12e,f). For $K < -1$, the population is locally unstable, with cycles of period 2 and geometrically increasing amplitude (Fig. 2.12g,h).

A second point also satisfies the equilibrium condition, the value $N_t = 0$. The conditions for stability of this equilibrium are important because they tell us whether the population will persist, i.e. whether it is unstable about zero. Again the stability about this point will depend on the slope of $F(N_t)$ at the equilibrium. We will call that slope α, consistent with the two forms of discrete-time logistic models in Eq. (2.16), the Ricker model and the Beverton–Holt model. If $\alpha > 1$, the population will be unstable about the point $N_t = 0$ (i.e. the population will persist), and conversely if $\alpha < 1$, the population will be stable about 0 (i.e. it will not persist). Thus persistence of a population represented by this model depends on whether the value of α is greater than 1. This concept will recur in different forms throughout the book.

Notice that from a conservation perspective there is a difference in the "desirable" outcome of stability analyses for different equilibria. In general we are interested in determining whether a population is *unstable* at the point $N^* = 0$ (i.e. will a population reduced to low numbers tend to increase?), but *stable* at other, nonzero values of N^* (i.e. will a population tend to return to its equilibrium abundance, or if perturbed will it decline towards zero, or possibly begin to undergo cycles?).

2.4.3 *Quasi-extinction in random environments with a discrete-time linear simple model*

Another fundamental concept we can illustrate with simple models is the effect of random environmental variability on populations. We are usually interested in the effect of randomness on populations at low abundance, where there is a risk of extinction. For example, in Eq. (2.3), what would our answer be if the L_t's varied randomly through time? What would we require of the distribution of the random L_t 's to avoid extinction (e.g. that the mean value be greater than 1)? Lewontin and Cohen (1969) pointed out several somewhat unexpected aspects of this question. To illustrate their results, we will rewrite the solution to the deterministic model given previously (Eq. 2.4) for the case where the L_t's are not identical,

$$N_t = N_0 \prod_{i=1}^{t} L_i. \qquad (2.25)$$

We then use that model to calculate what ecologists call the probability of quasi-extinction. We generate a large number of possible future time series of abundance by randomly choosing from the distribution of L_t values, and use the fraction of those that fall below the threshold N_{QE} as the probability of quasi-extinction (Fig. 2.13). We calculate the probability of abundance going below a threshold, instead of calculating the probability of abundance going all the way to zero, for several reasons. One is that many mathematical models do not decay all the way to zero, rather they approach it asymptotically. A more practical reason is that we recognize that the model may not

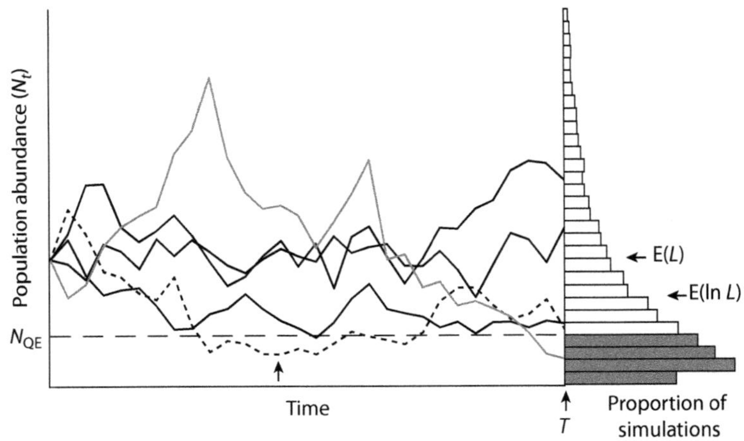

Fig. 2.13 The calculation of quasi-extinction through simulations of a stochastic, linear, discrete-time model. Curves show five example simulated time series; the histogram shows the lognormal distribution of population abundance at time T. Gray bars indicate the proportion of simulations that fall below the quasi-extinction threshold N_{QE} at T. Note that this proportion does not account for simulations like the one dashed curve that falls below N_{QE} (indicated by arrow) but then recovers prior to T. In this example the arithmetic mean of L is 1 (so the expectation at N_T is N_0), but the mean of ln L is −0.02, so most populations will decline. The values of E[L] (the arithmetic mean abundance) and $e^{E[\ln L]}$ (the geometric mean abundance) are indicated by arrows to the right.

accurately describe behavior at very low population levels. There may be low, nonzero levels below which the population may begin to decay rapidly (Allee effects), or are at risk of being extinguished by a random disturbance (environmental stochasticity). A final reason is precaution: we may not know much about how a population behaves at low abundance, so we seek to avoid those levels where extinction may be likely.

For this simple example, following Lewontin and Cohen (1969), we will assume the fraction of abundances at time T in Fig. 2.13 represent the probability of quasi-extinction. This ignores populations that have fallen below the quasi-extinction level and returned to above that level before T. We address the calculation of probability of quasi-extinction again in Chapters 8 and 10, where we include a full account of this possibility.

To describe the distribution of abundances at time T, we initially assume that the L_i's are statistically independent and identically distributed. Then the expected value (i.e. the arithmetic mean value) of the population numbers is

$$\mathrm{E}\left[N_t\right] = N_0\,\overline{L}^{\,t}, \tag{2.26}$$

where $\overline{L} = \mathrm{E}[L_i]$ is the arithmetic mean of the L_i's. In other words, if there were a large number of populations growing according to Eq. (2.4), the average abundance after t years would be $N_0\overline{L}^{\,t}$, such that the number of individuals grows geometrically with growth rate $\overline{L}$. However, as we shall see, the arithmetic mean is not a good representation of likely abundance when we are interested in extinction. This is because the mean is biased upward by the possibility of extremely high abundances, even when there is a low probability of that occurring. There is a much higher probability that the population will be at low abundance. Therefore we are more interested in the distribution of possible population abundances, not just the arithmetic mean, to understand the proportion of populations that will drop below the quasi-extinction threshold.

To determine the actual distribution of N_T, we take the natural logarithm of both sides of Eq. (2.4) and rewrite the solution as

$$\ln\left(\frac{N_T}{N_0}\right) = \sum_{i=1}^{T} \ln L_i. \tag{2.27}$$

Taking the logarithm turns the solution into a sum instead of a product. From the Central Limit Theorem in statistics (Box 2.7), the sum of T independent, identically distributed random variables (e.g. here the $\ln(L_i)$'s) will be normally distributed (i.e. Gaussian) with mean T times their mean, i.e. $T\mu_{\ln L}$, and variance T times their variance, i.e. $T\sigma_{\ln L}{}^2$. Thus, if we want to determine the probability that a population that starts at N_0 is below a certain threshold level N_{QE} at time t, we need only determine the mean and variance of $\ln L_i$, then calculate the proportion of the normal distribution with that mean and variance that is $< N_{QE}$ (similar to calculating a one-tailed p-value). The fact that $\ln(N_T/N_0)$ has a normal distribution means that N_T/N_0 is said to have a *lognormal distribution*.

Box 2.7 THE CENTRAL LIMIT THEOREM

The Central Limit Theorem (CLT) is a result from statistics that describes the results of taking large numbers of random samples from a probability distribution. If the random samples X_i are independently and identically distributed (i.i.d.), and drawn from a distribution with mean μ and variance σ^2, then as the number of samples, n, grows large, the following will be true about the distribution of sample means, $\overline{X} = \frac{1}{n}\sum_{i=1}^{n} X_i$:

1. The distribution will be normal
2. The distribution will have mean μ
3. The distribution will have variance σ^2/n

Likewise, the distributions of sums of the X_i's will also be normal, with mean $n\mu$ and variance $n\sigma^2$.

The theorem is very useful because it means we can use the well-known properties of the normal distribution any time we are dealing with sample means or sums. This is true even if the original distribution of the data is highly non-normal, provided n is large enough. In Fig. 2.14 we show an example in which the underlying data follow a very asymmetrical gamma distribution (Fig. 2.14a); when $n = 5$ the distribution of means is still somewhat skewed (Fig. 2.14b), but by $n = 20$ it is very close to a normal distribution with the expected mean and variance (Fig. 2.14c).

We present the definition of the CLT in the same terms as it would likely have been encountered by an ecology student taking statistics, i.e. in terms of the distribution when estimating sample means. However, it is important to realize that the CLT is more generally about the distributions of sums of random variables, not just about means. From that point of view, when dealing with a random variable that is the sum of n independently chosen values of a random variable from any distribution with mean μ and variance σ^2, as n increases, the distribution of this new variable (the sums) will approach a normal distribution with mean $n\mu$ and variance $n\sigma^2$. This view will be valuable when we discuss the portfolio effect in population diversity in Chapter 9.

Lewontin and Cohen (1969) used a simple numerical example to illustrate these calculations. In their example, the environment experienced by a population was good

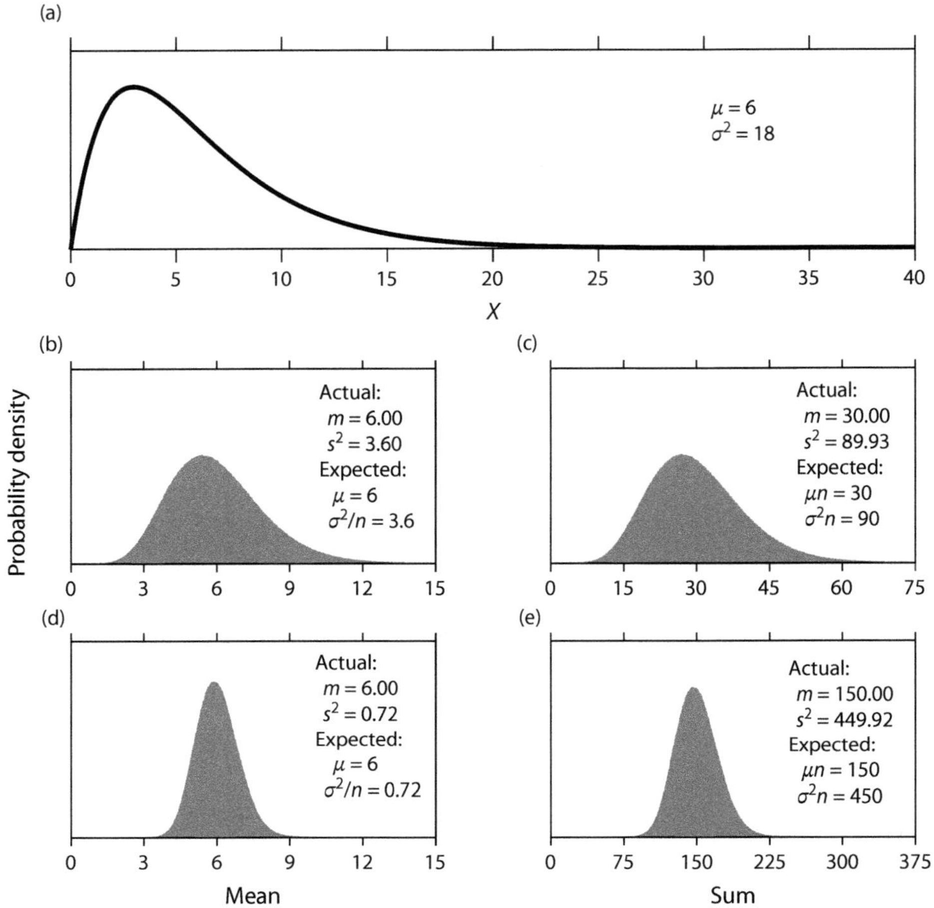

Fig. 2.14 Illustration of the Central Limit Theorem (CLT). 10^7 random samples were drawn from the non-normal distribution in panel (a) (gamma distribution, $a = 2, \beta = 3$). The distribution of the means and sums of the random samples is shown for sample size (b, c) $n = 5$ and (d, e) $n = 25$. The mean and variance of the original distribution, μ and σ^2, are indicated in (a), and the means and variances of the distribution of means and sums, m and s^2, are indicated in (b–d), along with the theoretical expectation for those values from the CLT.

half of the time with a growth rate of 1.7 and bad the rest of the time with a growth rate of 0.5, i.e. that L_t is either 0.5 or 1.7, with equal probability 0.5. The arithmetic mean of L is 1.1, therefore the arithmetic mean of abundance increases geometrically, and, for example, after $T = 100$ time units (e.g. years), the average value of N_T (if we sampled many such populations) would be $E[N_T] = 1.1^{100}N_0$. However, examining the logarithm of the L_t's instead, $\mu_{\ln L} = -0.081$ and $\sigma_{\ln L}{}^2 = 0.374$. The probability that the population has increased at all ($p[N_{100} > N_0]$) is equivalent to asking what proportion of the normal distribution with mean $T\mu_{\ln L}$ and variance $T\sigma_{\ln L}{}^2$ is greater than $\ln(N_0/N_0) = \ln(1) = 0$. In this case, the probability that the population has increased is small, $p[N^{100} > N^0] = 0.092$, despite the arithmetic mean being a more than 10,000 fold increase ($1.1^{100} = 1.3 \times 10^4$).

To understand this more generally, consider the expression for the mean log ratio of abundances, $T\mu_{\ln L}$. If $\mu_{\ln L}$ is negative, the mean will become more negative as time increases, and the probability that $N_T > N_0$ (or that $N_t >$ any threshold N_{T}) will tends towards zero. Thus, $\mu_{\ln L}$ is a key indicator of the probability of a randomly driven population falling below a certain level. Note that $\mu_{\ln L}$ is the logarithm of the geometric mean of L. To see this we simply write the expression for the mean,

$$\mu_{\ln L} = \frac{\ln L_1 + \ln L_2 + \cdots \ln L_n}{n} = \frac{\ln L_1 L_2 \ldots L_n}{n} = \ln (L_1 L_2 \ldots L_n)^{1/n}. \tag{2.28}$$

Since $\mu_{\ln L}$ is the logarithm of the geometric mean, we have shown that if the geometric mean is less than 1, the population will become extinct with certainty after a long time. Thus, in terms of extinction, the *geometric mean* of the growth rates is more important than the *arithmetic mean*, in spite of the fact that the average size of the population is the arithmetic mean (Eq. 2.26). Why is this important? For one thing, the geometric mean is much more sensitive to low values. For example, if five values of L_t are 1.2, 1.4, 0.9, 1.5, and 0.8, and the next value is 0.1, that would have far less of an impact on the arithmetic mean (1.16 → 0.98) than the geometric mean (1.12 → 0.75).

The fact that the arithmetic mean abundance can be large, while the probability of the population being large is small, can be further understood by examining a plot of the lognormal distribution of abundances (e.g. the distribution on the right-hand side of Fig. 2.13, which has a long skinny tail extending to high values). At time T, there is a very low probability of extremely high abundances (these would be populations for which L_t had been high every year). The reason the mean is so high is that the few extremely high abundances draw the mean value well above the median (or modal) value. This emphasizes that the mean abundance is not the most useful quantity when considering extinction risk.

It is important to keep in mind that calculating the probability of extinction using the mean and variance of $\ln L$ (i.e. using the normal distribution approximation instead of simulating population trajectories) does not actually calculate the complete probability of quasi-extinction. It will only calculate the probability in the shaded triangular area to the right in Fig. 2.13, i.e. the probability of being below the quasi-extinction level at the end of the time period T. It would not account for population trajectories that went below the quasi-extinction level, then returned to above the quasi-extinction level (see the single example indicated by the arrow in Fig. 2.13). This requires solving what is called a first crossing problem, which we will discuss in Chapter 8.

A more general way of viewing this problem is as an example of the mathematical principle known as Jensen's inequality, which states that mean of a nonlinear function of a set of values X is not the same as the function of the mean of X. This principle arises frequently in ecology, so we describe it in more detail in Box 2.8. In the context of extinction probabilities, Jensen's inequality is important because it is related to the difference between arithmetic and geometric means. To investigate further the difference between E[ln L], the logarithm of the geometric mean and ln E[L], the logarithm of the arithmetic mean, we can write a Taylor's series expansion about the logarithm of the mean of L, $\mathrm{E}[L] = \mu$ (recall that the derivative of $\ln x$ is $1/x$):

$$\ln L = \ln \mu + \frac{1}{\mu}(L - \mu) + \frac{1}{-2\mu^2}(L - \mu)^2. \tag{2.29}$$

Taking the expected value of these three terms (i.e. the mean over all L), and using the fact that the variance of L^2 equals the mean of the squares minus the square of the means, we see that

$$\mathrm{E}[\ln L] = \ln \mu + 0 - \frac{\sigma^2}{2\mu^2}. \tag{2.30}$$

Thus, the difference between E[ln L] and ln E[L] is half the square of the coefficient of variation (CV, i.e. σ/μ) of L, which means that the ratio of the two is $e^{0.5\mathrm{CV}^2}$. We will make use of this result later, in Chapter 8.

Box 2.8 JENSEN'S INEQUALITY

An important principle in stochastic linear models is that the expectation of the logarithm of λ, E[ln λ], is not equal to the logarithm of the expectation of λ, ln E[λ]. This an example of a general mathematical principle known as Jensen's inequality. The principle is named for the Danish mathematician Johan Jensen, who described it in a 1906 paper (Jensen, 1906). The principle is that for some set of values, X, and some nonlinear function, f, the mean of the function of X, E[$f(X)$], will not be equal to the function of the mean of X, f(E[X]). A simple graphical example illustrates how the inequality arises: consider just two values, X_1 and X_2, and the convex function f in Fig. 2.15. The mean of $f(X)$ is noticeably less than the function of the mean of X. The size of the difference depends on the variance of X (the difference increases with the variance) and the shape of f (a more nonlinear function, i.e. with a greater second derivative, will have a greater difference). It is possible to obtain an expression for the size of the difference using a Taylor expansion (Box 2.6). If the function is concave rather than convex (i.e. curving upwards) then the inequality is reversed, and ln E[λ] > E[ln λ].

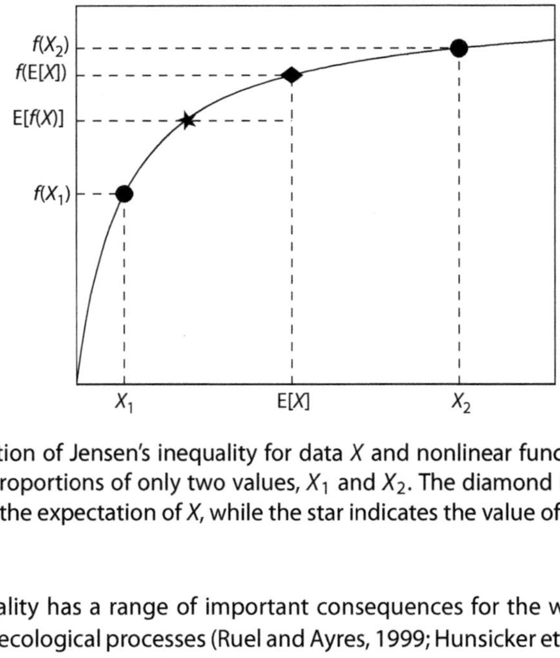

Fig. 2.15 Illustration of Jensen's inequality for data X and nonlinear function $f(X)$. The data consist of equal proportions of only two values, X_1 and X_2. The diamond indicates the value of the function of the expectation of X, while the star indicates the value of the expectation of the function of X.

Jensen's inequality has a range of important consequences for the way environmental variability affects ecological processes (Ruel and Ayres, 1999; Hunsicker et al., 2011; Chesson, 2012). In the context of the stochastic growth rate of a linear population model, we are interested in the logarithm of λ, which is a convex nonlinear function much like the one depicted in Fig. 2.15. The difference between E[ln λ] and ln E[λ] is $\sigma^2/2\mathrm{E}[\lambda]^2$, which is obtained from a Taylor expansion of ln λ, and reflects the influence of both the variance in λ and the curvature of f (the second derivative of ln x is x^{-2}, which is why $\mathrm{E}[\lambda]^2$ appears in the denominator).

2.4.4 *What does the simple logistic model tell us about managing for sustainable fisheries?*

In the early twentieth century, concern arose among wildlife and fisheries biologists because harvested populations were declining. They wondered how much a population should be expected to decline in abundance with removals by harvesting, and whether there was an optimal level at which such extraction should be operated. Fortunately, mathematical ecology was developing in parallel, and researchers in these fields were well aware of each other. The idea of the logistic model was introduced into fisheries and management of other animal resources at that time as a way to represent the growing belief that the greatest yield could be obtained from populations not at their greatest possible abundance, but rather at some lower level. Hjort et al. (1933) described this concept in the context of whaling in the Barents Sea, but they also mentioned bear hunting in Norway. Graham (1935) was concerned with the increase then decline of fish catch in the eastern Atlantic, and he used data from the period of no fishing during the First World War as the basis for his logistic-type models. Both were aware of the parallel development of the logistic model in general ecology (see Chapter 6 in Smith (1994) for more of this history).

The version of the logistic model developed by Verhulst (Eq. 2.6) provided a framework for investigating the question of optimal harvest. One approach would be to assume that harvest exactly balanced the population growth rate, so that $dN/dt = 0$. That would represent the conditions under which yield could be removed from a population while maintaining it at a constant abundance. That consideration led to logistic models in fisheries being referred to as *surplus yield* models. There was naturally interest in determining the maximum yield that would satisfy that relationship. This could be obtained by taking the partial derivative of the expression for dN/dt in Eq. (2.6) with respect to N, and setting it equal to zero to find the maximum growth rate:

$$\frac{\partial}{\partial N}\left(rN - \frac{rN^2}{K}\right) = 0. \tag{2.31a}$$

Evaluating this derivative leads to

$$r - \frac{2rN}{K} = 0, \tag{2.31b}$$

which simplifies to

$$N = K/2. \tag{2.31c}$$

This implies that the growth rate (and thus the possible harvest rate) is at a maximum at $K/2$ (Fig. 2.5a). This characteristic of this simple model is the origin of the oft-quoted result that the best way to harvest a population is to harvest at a rate that reduces it to half of its carrying capacity. This is the point at which it is increasing the fastest (dN/dt is greatest), hence it can endure the greatest harvest without declining. Even in the early 1900s, while some viewed this approach as a universal law, few believed that the straight line representing the decline in dN/dt with N held in general. That motivated the search for other forms of the logistic mentioned earlier as logistic-type models (Section 2.3.1). The valuable lesson observed was that it was not necessarily optimal to keep the population at high abundance, rather that the optimal yield would be at more of a "middling" level. Schaefer (1954) applied the logistic model to management of several fisheries, and in Chapter 11 we describe how he added the assumption that abundance was proportional to catch per unit fishing effort (CPUE), so that catch and effort data could be used to fit the model.

Maximizing harvest by maintaining the population near half of the carrying capacity and removing the annual growth each year became known as managing for *maximum sustainable yield*. It was a very influential idea that dominated fishery management globally through the 1980s, and is still currently used (see Botsford (2013) for a bibliography, and brief history). It is also used in the management of species other than fish. Examples include surplus yield curves for northwest Atlantic harp seals (*Pagophilus groenlandicus*) (Lett et al., 1981), a plot of the production of an unharvested species, grizzly bears (*Ursus arctos horribilis*) versus abundance (McCullough, 1981), a surplus production plot of life history information on gamebirds (Robertson and Rosenberg, 1988), and a demonstration that the maximum sustainable ivory yield for African elephants (*Loxodonta africana*) could be met by gathering tusks from natural mortality (Basson and Beddington, 1991). The logistic model continues to be used as a description of harvest of many kinds of animals (e.g. in a recent book on wildlife and climate change, the discussion of harvesting is based on the logistic model; Boyce et al., 2012).

2.5 What have we learned in Chapter 2?

Before moving on to Chapter 3, we can ask what we have learned in Chapter 2. First we learned some basic definitions regarding how we will represent time in population models (discrete and continuous) and the kind of equations used for each (Section 2.2). We also learned what linear versus nonlinear models are (Section 2.3), an important mathematical distinction with a parallel biological interpretation (nonlinear models contain density-dependent effects, while linear models do not). We learned some simple solutions to differential equation models (e.g. separation of variables), and a graphical solution to some nonlinear discrete-time models (cobwebbing).

The three illustrations with simple models of: (1) the determination of population stability (Sections 2.4.1 and 2.4.2) and (2) the effects of random variability in populations (Section 2.4.3) gave us a simple introduction to concepts we will be dealing with throughout the book. In the first illustration, introduction to stability analysis, we presented a useful way of describing general characteristics of behavior of models and populations. With simple linear, discrete time models (Section 2.4.1) we learned that we could propose solutions of the form $N_t = c\lambda^t$, and come up with an algebraic equation whose solution told us something about how the population would behave. We tried that approach with arguably the first population model, the rabbit problem, and saw that a weighted sum of two constants, each raised to t power, could reproduce the known behavior of the Fibonacci series. From Chapter 3 on, we will refer to these values of λ as eigenvalues, and this approach will underpin much of what we learn about population behavior. For simple, nonlinear, discrete-time models (Section 2.4.2) we saw that populations can have a nonzero equilibrium, and that we can form linear models of how the population behaves near those equilibria. We also learned some of the characteristics of that equilibrium that we will be interested in, such as local stability and cyclic behavior. The stability of a zero equilibrium was also of interest because it told us how populations can continue to persist (essentially the opposite of the topic in the second example, extinction).

In the second illustration (Section 2.4.3), we added the effects of the random environment to our consideration of the zero equilibrium. We learned that when we are interested in population extinction, we will most often be calculating quasi-extinction, rather than complete extinction, when trying to represent extinction risk. From the Lewontin and Cohen (1969) model, we learned that it is the geometric mean of the environments, not the arithmetic mean, that is important for quasi-extinction. The expected value of population abundance grows exponentially (or geometrically) at a rate governed by the arithmetic mean of the randomly varying growth rates, but even populations with a positive mean of the distribution of possible growth rates can have a nearly 100 percent probability of going extinct.

Finally, we saw an initial example of how models are used to determine how a population can be exploited without driving it to extinction (Section 2.4.4). The main message, important in the early twentieth century was that populations should be expected to decline when harvested, and moreover that the best yields could be obtained by fishing populations to an abundance well below the unfished level (Smith, 1994; Botsford, 2013). This example differs from the other two in that this simple model has been widely applied to fisheries management on a global scale, and is still in use today (Smith, 1994; Finley, 2011; Botsford, 2013). It will be the starting point of our description of the management of marine fisheries in Chapter 11.

While dynamically limited, these simple models give us a glimpse of general trends in population behavior upon which we will build with more realistic models in later chapters. Recalling the limited realism of these simple models, one could say that we have learned something about how "populations" could work, but that we have not yet learned about how actual populations work.

CHAPTER 3

Linear, age-structured models and their long-term dynamics

In this chapter we will take a first step toward making population models more realistic portrayals of population dynamics, by adding age structure. A limitation of the simple models we described in Chapter 2 was that they assumed all individuals in the population were identical. Age-structured models relax that assumption, allowing us to describe populations in which demographic rates differ with age, and allowing us to explicitly represent the natural aging of cohorts of individuals. These details produce important differences in population dynamics that simple models cannot portray.

We begin in the least complicated way, with linear age-structured models. That is, the models will not be density dependent. Just like the linear models in Chapter 2, the models in this chapter have the same limitation of eventually either increasing to infinity or decreasing to zero. This means their use is only appropriate for situations in which density dependence is not an important factor, such as recently introduced populations or those that have declined to low abundance and are close to extinction. We will point out practical situations to which linear age-structured models apply, and give examples of their use.

Here population state will depend on its age structure, which will change through time, so we will begin to use partial differential equations (describing changes in abundance through both continuous age and time; Box 3.1), as well as matrices (describing discrete age and time). We will also learn to write solutions to linear matrix equations in a general way, in terms of eigenvalues and eigenvectors, and these solutions will form the basis of our analyses of dynamic stability and other aspects of population behavior throughout the book.

Historically, three types of linear, age-structured models have developed. Their dynamic behavior is essentially the same and they differ mainly in the way they describe time (see Box 2.1). Both Lotka's (1907) renewal equation and the M'Kendrick/von Foerster equation (M'Kendrick, 1926; von Foerster, 1959) are continuous time, continuous age models. The Leslie matrix is their discrete time, discrete age counterpart (Bernardelli, 1941; Lewis, 1942; Leslie, 1945). We will first derive the partial differential equation model called the M'Kendrick/von Foerster equation from basic principles of mass balance (i.e. flow of individuals through age with time), then transfom that into an integral version, which will be Lotka's renewal equation. Finally, we will discretize Lotka's model to obtain the Leslie matrix. Keep in mind that these three models all do essentially the same thing. Practically speaking, the Leslie matrix approach is the primary model used in modern ecology, but we show continuous age models here, and continuous size models in Chapter 5, to illustrate the continuous time dynamics of these two variables, both of which change continuously

Population Dynamics for Conservation. Louis W. Botsford, J. Wilson White, and Alan Hastings, Oxford University Press (2019).
DOI: 10.1093/oso/9780198758365.001.0001

in real populations. These derivations also provide a clear view of how individual-level rates lead to population-level dynamics, a concept that we later develop further for size-structured models (Chapter 5) and which is critical for understanding the problems posed by stage-structured models (Chapter 6). In particular, the M'Kendrick/von Foerster model will help to illustrate what new aspects of population dynamics are introduced when we move from age structure to size structure (Chapter 5).

3.1 The continuity equation and the M'Kendrick/von Foerster model

The continuity equation is a partial differential equation that describes the "flow" of a substance through "space". In physics it describes the flow of an incompressible fluid (such as water) through three- dimensional space. It is useful because the total mass of a parcel of fluid cannot change as it is moving (it all has to go somewhere), and the continuity equation keeps track of how much of the fluid moves in each direction. In population biology, the continuity equation is used to describe the "flow" of individuals in a population through a "space" with dimensions of age, size, both age and size, latitude and longitude, or any other combination of variables describing the state or location of individuals in space and time. Just as in physics, the continuity equation is merely a bookkeeping device that keeps track of the fate of each individual as they move through the *i*-state "space" according to their individual growth or movement rates. In each time instant, individuals must either die, remain at the same size (for example), or grow to another size, and the continuity equation keeps track of these developments.

The earliest known formulation of the continuity equation as a population model is by M'Kendrick (1926). Lieutenant-Colonel A. G. M'Kendrick, from the Laboratory of the Royal College of Physicians in Edinburgh, was concerned with the spread of infectious diseases. His epidemiological models were descriptions of the number of individuals or households that had experienced a certain number of infections, recoveries, or other related events. He represented this process in two dimensions of discrete variables and developed relationships describing the change with time of the associated distribution. As an example, tucked into the back of this publication, he let the two axes represent continuous time and age to obtain the continuity equation description of the age structure of a human population.

The next appearance of this model was in the work of Heinz von Foerster (1959), an electrical engineer studying cell growth at the University of Illinois. His view of the problem of understanding population dynamics is of some interest. He proposed that there were two approaches; one could infer population behavior by: (1) experimentally measuring reproduction and mortality in cohorts as they aged, and (2) observing the distribution of abundance of each age at sequential times. He believed the latter to be the most appropriate for the study of cell growth, and structured his research as an *inverse* problem, i.e. one in which the model is determined from the output patterns of the data (as opposed to a forward problem of starting with input data on individuals and calculating the output data for the population). He and other early developers of age-structured models of cell populations (e.g. Trucco, 1965a; 1965b) developed age-structured models to allow inference regarding individual characteristics from samples of age distributions. von Foerster noted that there were two kinds of one-dimensional representations of the basically two-dimensional (i.e. age and time) age distribution. The *congenerate* sample is a plot of numbers in a cohort versus time, and is non-increasing (the number of individuals in a cohort is always highest at birth and decreases over time as individuals die). The *contemporary* sample is a plot of the age distribution of all the cohorts present at a specific time, and is not necessarily non-increasing (there may be more old individuals than newborns at any given time). These are also referred to as *longitudinal* versus *cross-sectional* observations, and one may see these concepts referred to as *horizontal* and *vertical* life tables, respectively, in demography (see Box 3.1 and Fig. 3.1 for graphical examples).

Box 3.1 MULTI-DIMENSIONAL CALCULUS

Because we wish to keep track of the distribution of the abundance of individuals over age, as abundance changes with time, our independent variable will be a function of two variables (age and time), meaning we will have to make use of concepts from calculus with more than one variable. For example, Fig. 3.1 depicts a system in which n depends on a and t, i.e. $n = f(a, t)$. Understanding how n changes with respect to a and t requires a new definition: a partial derivative. Recall that the definition of an ordinary derivative in single variable calculus, say for abundance $N(t)$ as in Chapter 2, is

$$\frac{dN}{dt} = \lim_{\Delta t \to 0} \frac{N(t + \Delta t) - N(t)}{\Delta t}. \tag{B3.1-1}$$

However, if n is a function of both a and t, then the rate of change dn/dt also depends on a (throughout the book we will use N to indicate total abundance, and n to indicate the density distribution of abundance over another variable, such as size or age). In multivariable calculus we deal with this problem in part by giving the rate of change of n with respect to t, for a particular value of a. In other words, it is like an ordinary (single-variable) derivative in the plane perpendicular to an axis of one of the independent variables at a specified value of that variable (Fig. 3.1a).

If $n = n(a,t)$, the partial derivative of n with respect to a is

$$\frac{\partial n}{\partial a} = \frac{\partial n(a,t)}{\partial a} = \lim_{\Delta a \to 0} \frac{n(a + \Delta a, t) - n(a,t)}{\Delta a} \tag{B3.1-2}$$

and it will depend on a specified value of t (note the stylized ∂ [called "doe"] used in place of d to denote a partial derivative) and the partial derivative of n with respect to t is

$$\frac{\partial n}{\partial t} = \frac{\partial n(a,t)}{\partial t} = \lim_{\Delta t \to 0} \frac{n(a, t + \Delta t) - n(a,t)}{\Delta t} \tag{B3.1-3}$$

and it will depend on a specified value of a.

These are both illustrated in Fig. 3.1a,b. In panel (a), the vertical plane represents a constant value of $a = 7.5$. The rate of change of n with respect to t, the partial derivative, $\partial n/\partial t$, for that value of a, is the slope of the black line traced out by the intersection of the vertical plane with the n surface. Panel (b) illustrates the same concept but for the partial derivative, $\partial n/\partial a$, holding t constant $t = 16$ (the slope of the black line which shows the age distribution of the population at time t).

These two partial derivatives allow us to write derivatives in other directions in terms of these partial derivatives. For example, below we will write the rate of change in n versus time along the lines in Fig. 3.1c. Along those lines, the rate of change of n versus time will be

$$\frac{dn}{dt} = \frac{\partial n}{\partial a}\frac{da}{dt} + \frac{\partial n}{\partial t}\frac{dt}{dt}, \tag{B3.1-4}$$

which says that the rate of change in abundance density with time is the rate at which age changes with time (*da*/*dt*) times the rate at which abundance *n* changes in the *a* direction, plus the rate at which time is changing with time (which equals 1) times the rate at which abundance *n* changes in the *t* direction. This is an example of the chain rule in calculus. We will say more about these lines later. For now, to link this figure to historical sources, we note that Figs. 3.1c and 3.1b represent the contemporary versus congenerate samples of Trucco (1965a, b), respectively, as well as the cross-sectional versus longitudinal observations, and the vertical versus horizontal life tables of demography.

A useful way to understand how a continuity equation in age (and later size in Chapter 5) represents age-structured (and size-structured) populations is to derive them from a "mass balance", or in other words a complete accounting of changes in population "mass" with time. Here the population "mass" is abundance of individuals. We first define the age density $n(a,t)$. Age density is a continuous, rather than discrete concept, so it is not the usual way humans think of age. It is defined as the number (abundance) of individuals with ages in the interval between a and $a + \Delta a$ at time t is the age density $n(a,t)$ times Δa. Thus it is a density per unit age, rather than per unit area, as is the usual meaning of the word density in ecology. Figure 3.1 shows an example of $n(a,t)$ plotted versus both a and t. In Fig. 3.1c, several curves are plotted across the surface n. These curves show the trajectory that the abundance of a cohort would follow through the population over both age and time. If viewed from directly overhead, the curves would appear to be straight lines with slope $da/dt = 1$; in other words, every day an organism gets one day older (assuming age and time are measured in the same units). These curves will be useful later.

To derive the continuity equation, we write the changes in age density over a small age interval Δa in a small time increment Δt, as shown in Fig. 3.2. To describe the change in state of the population, we write the number of individuals between ages a and $a + \Delta a$ at time $t + \Delta t$ in terms of the population state at time t. Recall from Chapter 1 that this will depend on the *i*-state variable, which here is age a. This is easy to do because n is an age density. That means that abundance in any interval will be the value of the density in that age interval times the length of the age interval. In Fig. 3.2 we can see the abundance in the age interval at time $t + \Delta t$ (labeled α in the figure) is equal to the abundance in that age interval at time t (labeled β), minus the abundance that grows out of the interval growing according to the individual rate of growth of age with time (da/dt at age $a + \Delta a$), which we labeled δ, plus the number growing into the interval at the growth rate da/dt at age a, labeled γ. Writing the abundances in each interval as densities times the age interval, we can write

$$\alpha = \beta - \delta + \gamma, \tag{3.1a}$$

where the definition of each of these is in terms of a density times an interval, and is given in Table 3.1. Writing this in terms of the abundances from that table, we obtain

$$n(a, t + \Delta t)\,\Delta a$$
$$= n(a,t)\,\Delta a + \left.\frac{da}{dt}\right|_{a} n(a,t)\,\Delta t - \left.\frac{da}{dt}\right|_{a+\Delta a} n(a + \Delta a, t)\,\Delta t - D(a,t)\,n(a,t)\,\Delta a \Delta t, \tag{3.1b}$$

where we have also subtracted the number dying as the product of the per capita mortality rate, $D(a,t)\Delta t$ times the abundance $n(a,t)\Delta a$. Moving β to the left-hand side of the equation,

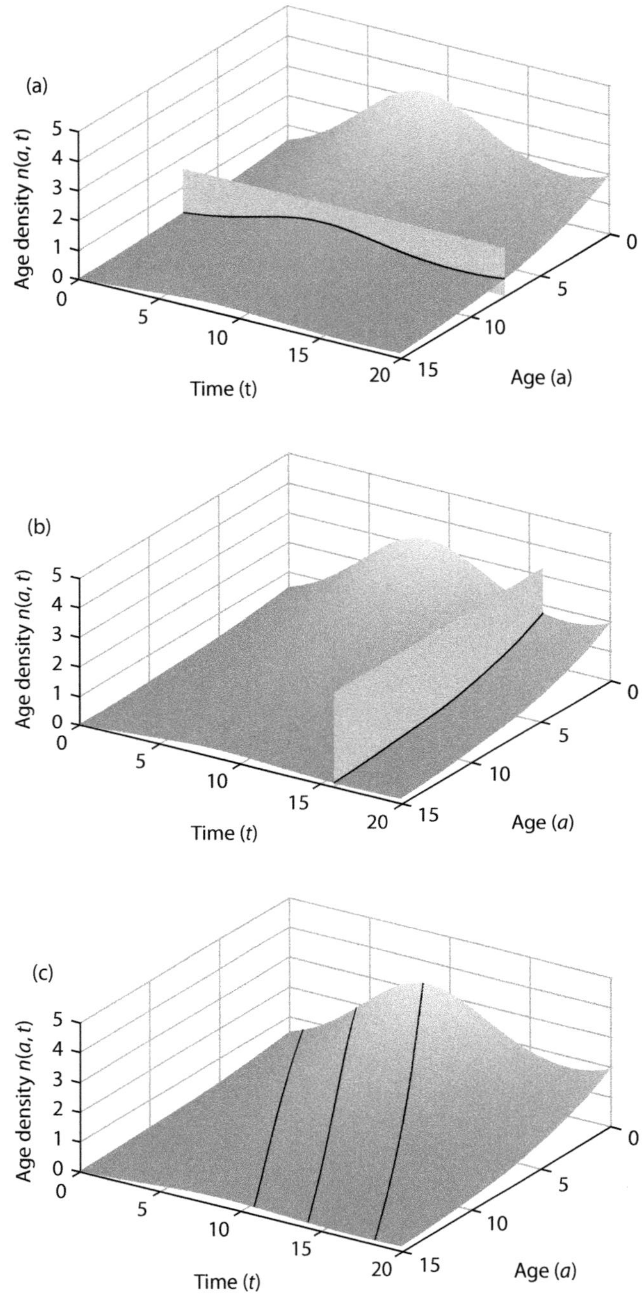

Fig. 3.1 Example of an independent variable, n, that depends on two variables, a and t. The gray surface in all three panels is $n(a,t)$. In (a), the opaque vertical rectangle is the plane within which the partial derivative $\partial n(a, t)/\partial t$ would lie for $a = 7.5$ y. The partial derivative would be the slope of the black line at the intersection of the vertical plane and the function. In (b), the opaque vertical rectangle is the plane within which $\partial n\,(a, t)\,/\partial a$ would lie for $t = 16$ y. Again, the partial derivative is the slope of the black line. In (c), the black lines are the lines along which the abundances of individual cohorts would decline as age and time increased.

Table 3.1 The components of the continuity equation, as shown in Fig. 3.2. The vertical bar in the lower two rows indicates that the derivative is evaluated at the value indicated in the subscript.

Term	Density	Age interval	Function
α	$n(a, t+\Delta t)$	Δa	Final abundance
β	$n(a, t)$	Δa	Initial abundance
γ	$n(a, t)$	$\Delta t \left.\frac{da}{dt}\right\|_a$	Abundance growing in
δ	$n(a+\Delta a, t)$	$\Delta t \left.\frac{da}{dt}\right\|_{a+\Delta a}$	Abundance growing out

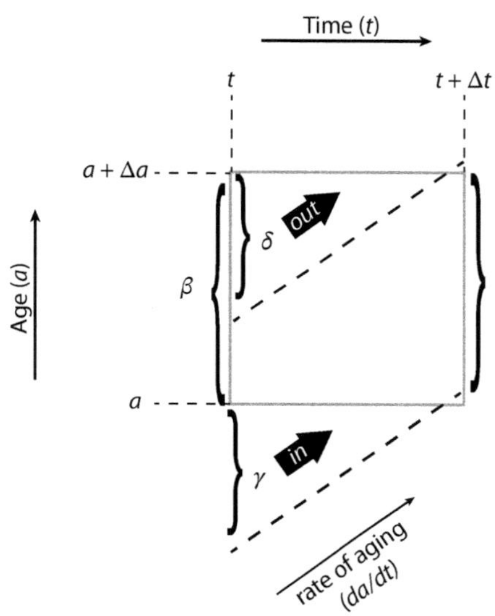

Fig. 3.2 Schematic view of a small part of the surface *n* as a function of age *a* and time *t*, for the purposes of deriving the M'Kendrick/von Foerster equation. This square is a small portion of the surface shown in Fig. 3.1c, viewed from above.

combining it with α, then dividing by both Δt and Δa, and taking the limit as Δt goes to zero, we obtain the expression from Box 3.1 for the partial derivative of $n(a, t)$ with respect to time, Eq. (B3.1-3). Similarly, dividing $-\delta + \gamma$ on the right-hand side of Eq. (3.1a) by Δt and Δa and taking the limit as Δa goes to zero, gives us the partial derivative of $n(a,t)(da/dt)$ with respect to a:

$$\lim_{\Delta a \to 0} \frac{n(a+\Delta a, t) \left.\frac{da}{dt}\right|_{a+\Delta a} - n\left(a, t\right) \left.\frac{da}{dt}\right|_a}{\Delta a},$$

which is

$$-\frac{\partial}{\partial a}\left[n(a, t)\frac{da}{dt}\right].$$

Combining these yields

$$\frac{\partial n(a, t)}{\partial t} = -\frac{\partial}{\partial a}\left[n(a, t)\frac{da}{dt}\right] - D(a, t)\, n(a, t). \tag{3.2}$$

All that remains is to determine the rate of change of age with respect to time for an individual, da/dt. Fortunately, as we noted previously (and show in Fig. 3.1c and 3.2), that rate is simply equal to 1 for all values of a, so we can drop it from Eq. (3.2). Note that the growth rate of other *i*-state variables in future chapters (such as size) will, of course, not equal 1, and they will also not be the same for all values of the *i*-state variable.

At this point most readers generally accept the last term in Eq. (3.2) as a reasonable representation of mortality, but many question the meaning of the first term on the right-hand side, specifically, why is it negative? That term has a simple interpretation that will be valuable later when we get to more complex models. First, from the product rule of differentiation in calculus, this term will be $-n\,(a,t)\,\dfrac{\partial}{da}\left[\dfrac{da}{dt}\right] - \dfrac{da}{dt}\dfrac{\partial}{\partial a}\,[n\,(a,t)]$. Because we know that da/dt is a constant ($= 1$), the first term in this sum is zero, which leaves us with $-\partial/\partial a\,[n\,(a,t)]$. Assuming no mortality for the moment ($D(a,t) = 0$), the first term on the right-hand side of Eq. (3.2) now says that rate of change of age density with time, $\partial/\partial t$ $[n(a,t)]$, is equal to the negative rate of change of age density with age, $\partial/\partial a$ $[n(a,t)]$. This makes sense if viewed as the rate at which a point at a fixed value of a on the age density function moves up or down on a plot of age density versus age, as the age distribution moves from left to right with passing time (Fig. 3.3). The rate of increase of $n(a,t)$ with time at that point will be equal to the negative slope of the age density at that point. A shallow negative slope will cause a slow increase (Fig. 3.3a,b), while a steeper slope will lead to a rapid increase (Fig. 3.3c,d). This happens because a steep negative slope in the

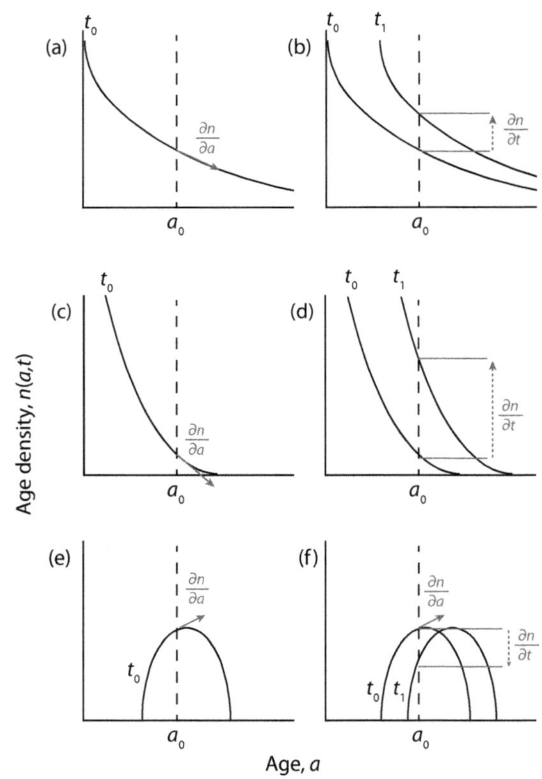

Fig. 3.3 Explanation of why the first term on the right-hand side ($\partial n/\partial a$) in the M'Kendrick/von Foerster equation for $\partial n/\partial t$ has a negative sign (assuming no mortality for clarity). On the left (panels a, c, and e) are several examples of age densities at time t_0 (decaying with age, decaying more rapidly, and a "baby boom," respectively). On the right are responses at age a_0 as time increases from t_0 to t_1.

age density means that there are many more younger individuals about to reach age a. So (again, assuming no mortality), the density of age a individuals is increasing as those young individuals grow older (Fig. 3.3a). Alternatively, if the slope is positive, there are fewer young individuals in the population at time t, and so the density of age a individuals will be decreasing as the population ages (Fig. 3.3e,f, in which there is first an increase, then a decline in $n(a_0, t)$).

3.1.1 *Solving the M'Kendrick/von Foerster model*

To see what the implications of the M'Kendrick/von Foerster equation are for the population dynamics of age-structured models, we must solve Eq. (3.2) to obtain the surface $n(a, t)$ in Fig. 3.1. A straightforward way to do this is to use the calculus technique known as the method of characteristics. Basically this means that we find curves along the two-dimensional surface $n(a,t)$ for which the partial differential equations can be treated as ordinary differential equations (these curves are called *characteristic lines* or *lexis lines*). To use this approach we refer back to the chain rule of the full derivative in Box 3.1 (Eq. (B3.1-4)) to express the full derivative of $n(a,t)$ with respect to t. In that expression we set $da/dt = 1$ to obtain

$$\frac{dn}{dt} = \frac{\partial n}{\partial a} + \frac{\partial n}{\partial t} = \left[\frac{\partial}{\partial a} + \frac{\partial}{\partial t}\right] n. \tag{3.3}$$

To use this result, we need to find an expression that looks like the right-hand side of Eq. (3.3). If we rewrite Eq. (3.2) (with $da/dt = 1$) as

$$\frac{\partial n(a,t)}{\partial t} + \frac{\partial n(a,t)}{\partial a} = -D(a,t)\, n(a,t), \tag{3.4a}$$

or

$$\left[\frac{\partial}{\partial t} + \frac{\partial}{\partial a}\right] n(a,t) = -D(a,t)\, n(a,t), \tag{3.4b}$$

then Eq. (3.3) implies that solving Eq. (3.4b) will be the same as solving the following ordinary differential equation:

$$\frac{dn}{dt} = -D(a,t)\, n(a,t), \tag{3.4c}$$

as long as $da/dt = 1$. This means that as long as we maintain the relationship $da/dt = 1$, that is as long as we stay on the paths shown in Fig. 3.1c, the left-hand side of Eq. (3.4b) will be the full derivative of n and we can integrate the resulting ordinary differential equation (ODE) directly, just like the simple linear model in Eq. (2.1). The lines defined by $da/dt = 1$ (Fig. 3.1c) are called characteristics of the partial differential equation (PDE), because they are the lines along which the PDE can be solved like an ODE (hence the "method of characteristics"). In our age-structured partial differential equation model, they also happen to be the lines along which individuals would grow. This is a valuable result, especially since it will also be true in the size-structured models of Chapter 5, where the characteristic lines are the lines along which individuals would grow in size.

This derivation of the M'Kendrick/von Foerster model to obtain Eq. (3.4c) is not the only way possible. For a different way to conceive of the continuity problem and arrive at the same result, see Box. 3.2 (particularly if the way we have just explained does not make sense to you).

Box 3.2 AN ALTERNATIVE DERIVATION OF THE M'KENDRICK/VON FOERSTER MODEL

Another approach to deriving the M'Kendrick/von Foerster model does not require interpretation of the different inputs and outputs of the box in Fig. 3.2.

In the following we will be somewhat loose in our derivation—we understand that $n(a,t)$ is always a density on age a at time t for the population, but we also use the value of this function at a point when, strictly speaking, this is not completely justified.

The first step in an alternative derivation is to realize that the only way an individual can be age $a + \Delta a$ at time $t + \Delta t$ is to have been age a at time t and survived. Survival here simply means not dying, so if the per capita death rate is $D(a, t)\Delta t$, survival is 1 minus that quantity, and we can write this verbal description simply as

$$n\,(a + \Delta a, t + \Delta t) = n\,(a, t)\,(1 - D\,(a, t)\;\Delta t) \tag{B3.2-1}$$

To get a differential equation, Δt and Δa will both have to go to zero. The next steps in the derivation will allow that, by getting the formulas for the two partial derivatives to appear.

We rewrite Eq. (B3.2-1) to break out the mortality term:

$$n\,(a + \Delta a, t + \Delta t) = n\,(a, t) - D\,(a, t)\;\Delta t\; n\,(a, t)\,, \tag{B3.2-2}$$

which we then rearrange as

$$n\,(a + \Delta a, t + \Delta t) - n\,(a, t) = -D\,(a, t)\;\Delta t\; n\,(a, t)\,. \tag{B3.2-3}$$

At this point if we were to simply divide by Δt we would not know how to deal with the terms on the left-hand side (but the way these appear does provide a hint on how to solve the eventual equation, since this equation implies that the population declines exponentially as we move along any line where the change in a equals the change in t).

So, next we return to Eq. (B3.2-2), and add and subtract the same term to get the differences that appear in the definitions of the partial derivatives to appear:

$$n\,(a + \Delta a, t + \Delta t) - n\,(a + \Delta a, t) + n\,(a + \Delta a, t) = n\,(a, t) - D\,(a, t)\;\Delta t\; n\,(a, t)\,. \tag{B3.2-4}$$

Finally, we divide all terms by Δt, and in the second term introduce $\Delta a/\Delta a$:

$$\frac{n\,(a + \Delta a, t + \Delta t) - n\,(a + \Delta a, t)}{\Delta t} + \frac{n\,(a + \Delta a, t) - n\,(a, t)}{\Delta a}\frac{\Delta a}{\Delta t} = -D\,(a, t)\, n\,(a, t)\,. \tag{B3.2-5}$$

And, noting that Δt and Δa are equal, we take the limit as Δt goes to zero to get the partial differential equation,

$$\frac{\partial n}{\partial t} + \frac{\partial n}{\partial a} = -D\,(a, t)\, n\,(a, t)\,. \tag{B3.2-6}$$

Then, from the results in Box 3.1, we know that the left-hand side of Eq. (B3.2-6) is equal to dn/dt if $da/dt = 1$, so we obtain

$$\frac{dn}{dt} = -D\,(a, t)\, n\,(a, t), \tag{B3.2-7}$$

just as in Eq. (3.4c).

To solve the ODE (Eq. (3.4c)) to obtain an expression for $n(a,t)$, we first note that the characteristics define a line with equation $t = t_0 + a$ (think of t_0 as the birthdate of an organism, so t is the birthdate plus the organism's age). We then define the age density along the characteristic, $n(a,t) = n(a, t_0 + a)$, and for convenience assume a non-time-varying mortality rate, $D(a,t) = D(a)$. Then we actually start by solving the ODE for dn/da (which can be derived in the same way as Eq. (3.4), swapping a for t), holding t constant at t_0. This is essentially solving for the change in abundance with age for any particular characteristic line (value of t_0). We obtain

$$\frac{dn}{da} = -D(a)n\left(a, t_0\right), \tag{3.5}$$

whose solution is

$$n\left(a, t_0\right) = n\left(0, t_0\right) e^{-\int_0^a D(x)dx}. \tag{3.6}$$

This is the same ODE solution as in the simple linear model in Eq. (2.2) and Box 2.4 (except in that model the integral of a time-varying r from 0 to t has been simplified to rt). For convenience we abbreviate the exponential term as $\sigma(a)$, which then represents the fraction surviving from 0 to age a:

$$\sigma(a) = e^{-\int_0^a D(x)dx}. \tag{3.7}$$

Writing Eq. (3.6) in terms of the original density function for any time t,

$$n\left(a, t\right) = n\left(0, t - a\right) \sigma(a). \tag{3.8}$$

Note that with much messier notation, we could write a solution for the case in which mortality rate depended on time in addition to age. The solution we have obtained (Eq. (3.8)) makes biological sense viewed as the development over time and age of a cohort: the number of individuals of age a at time t is the number of individuals we started with at age 0, a time units ago, times the fraction that would survive from age 0 to age a.

The method of characteristics also has a useful biological interpretation. Characteristic lines are the lines along which individuals grow, and the solution along characteristics is the rate of decay of population density (i.e. $n(a,t)$, abundance per unit age) as they grow along the characteristic. This interpretation will be quite valuable when we extend it to size-structured models in Chapter 5.

You may notice that to use Eq. (3.8) we need to know what happens at $t = 0$ and $a = 0$. These are the *boundary conditions* for the continuity equation (i.e. what happens at the edge of the surface in Fig. 3.1). The equation will have one boundary condition indicating the age density at time $t = 0$: $n(a,0) = n_0(a)$ (the initial population age density). It may also have a boundary condition describing individuals entering the population at age $a = 0$. This can either be a specified function of time, $n(0,t) = R(t)$ (some number of newborns appear at each time t), or it can depend on reproduction by individuals in the population. For the latter case, we can write a mass balance similar to Eq. (3.1), but at $a = 0$,

$$\begin{aligned} &n\left(0, t + \Delta t\right) \Delta a \\ &= n\left(0, t\right) \Delta a - \left.\frac{da}{dt}\right|_{0+\Delta a} n\left(0 + \Delta a, t\right) \Delta t - D\left(0, t\right) n\left(0, t\right) \Delta a \Delta t + \int_0^\infty b\left(a, t\right) n\left(a, t\right) da\, \Delta t, \end{aligned} \tag{3.9}$$

where $b(a,t)$ is the number of offspring produced per unit time per adult individual of age a (notice that unlike Eq. (3.1) there is no term for individuals of age $a < 0$ growing into the interval). Letting Δa go to zero, cancelling the Δt's, and noting that $da/dt = 1$, we can rearrange Eq. (3.9) to obtain

$$n(0,t) = \int_0^\infty b(a,t)\,n(a,t)\,da. \tag{3.10}$$

Thus, the age density at age zero is the integral (essentially the sum) over the number of individuals at each age, times their rate of reproduction. In most age-structured models, we are able to use this relatively simple dependence of reproduction on adult numbers because we include only females in the model, and assume there is no variation in their reproduction due to variation in the male population (see Rankin and Kokko, 2007; White et al., 2017) for examples of models that include males also). The combination of Eq. (3.8) and these two boundary conditions constitutes the full solution to the M'Kendrick/von Foerster model. This model is rarely used in modern conservation applications, but as we have seen it is useful in helping us conceptualize several aspects of how age-structured populations should behave (e.g. Fig. 3.3), and we will use these results to arrive at the renewal equation next, which will lead us to the commonly used Leslie matrix.

3.2 The renewal equation—Lotka's model

There are two different ways to completely keep track of the behavior of an age-structured model of a population with non-time-varying survival to age (Eq. 3.8). One is by describing how the distribution of abundance over age varies with time, as in the M'Kendrick/von Foerster equation, and the other is to keep track of recruitment (or births) versus time, as in Lotka's renewal equation. The renewal equation as a model of population growth was developed by Alfred J. Lotka in the early twentieth century. The basic conclusion that the growth rate of a linear, age-structured population would be exponential is in his 1907 paper; a solution to the renewal equation is contained in Sharpe and Lotka (1911); and a description of the stable age distribution can be found in Lotka (1922). As we shall see in the description of the discrete-time linear age-structured models (Section 3.3), the ideas were not really new (e.g. geometric increase and some other relationships went back to the eighteenth century), but their concise formulation in a single, continuous-time model was a significant advance. Lotka (1939) contains an extensive bibliography of developments in the early twentieth century, and the literature cited in Feller (1940) extends that by a few years. Here, we will begin by following solutions by Sharpe and Lotka (1911). The more mathematical reader may want to see the more rigorous development in Feller (1940), as well as more modern papers.

Lotka began with two assumptions: the reproductive boundary condition for non-time-varying fecundity, which we have derived previously (Eq. 3.10) and the solution to the M'Kendrick/von Foerster equation for non-time-varying mortality, which we have also derived (Eq. 3.8). The renewal equation is therefore a special case of the M'Kendrick/von Foerster equation (i.e. for non-time-varying vital rates), and the solution we are about to develop is a continuation of our solution of the M'Kendrick/von Foerster equation for that case.

Substituting Eq. (3.8), the solution to the M'Kendrik/von Foerster equation giving the age distribution at time t as a function of the starting abundance of that cohort at time $t - a$, into Eq. (3.10) (and removing the time dependence in the b term), the number of newborns at time t, we obtain as

$$n(0,t) = \int_0^\infty b(a)n(0,t-a)\,\sigma(a)da. \tag{3.11}$$

Defining the total reproductive rate (i.e. density of newborns produced) to be $B(t) = n(0,t)$, we obtain the renewal equation

$$B(t) = \int_0^{\infty} B(t - a)\, b(a)\sigma(a)da. \tag{3.12}$$

The renewal equation says that to find the number of newborns at time t, you take the integral (sum) of all the individuals born at each successive time in the past, $B(t - a)$, multiplied by their fecundity at their current age, $b(a)$, times the proportion that survived to their current age, $\sigma(a)$. We can combine these reproductive and survival functions to form

$$\varphi_b(a) = b(a)\sigma(a), \tag{3.13}$$

which is the number reproduced at each age per individual that enters the population at age 0 (Fig. 3.4a). Functions such as $\varphi_b(a)$ in Eq. (3.13) have a useful interpretation which we will build upon and make further use of throughout this book. Current reproduction can be thought of as the weighted sum over reproduction in previous years, where the weighting function is the product of age-specific mortality and age-specific reproduction (Fig. 3.4b). Population ecologists have also studied the effects on population dynamics of various other age-dependent influences, representing them with functions similar to $\varphi_b(a)$. Examples of age-dependent influences include cannibalistic behavior, consumption rate, space occupied, and, of course egg production. We will refer to functions such as φ_b in Eq. (3.13) as *influence functions*.

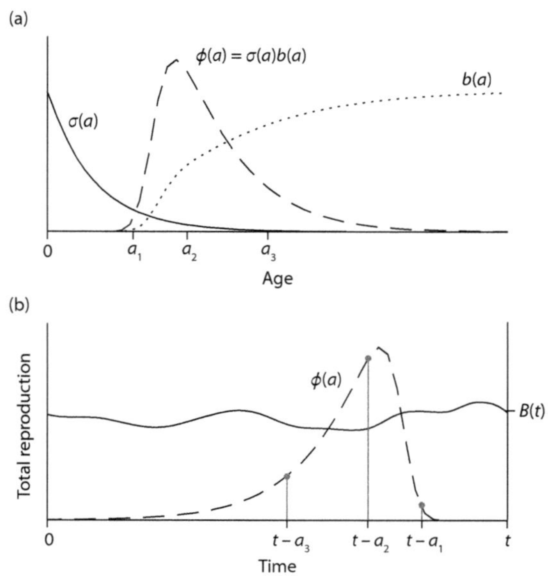

Fig. 3.4 The concept of lifetime reproduction. In panel (a), the product of survival to age versus age, (*a*), (solid curve) and fecundity versus age, $b(a)$, (dotted curve) determines the distribution of total reproduction over age, (*a*), a function of great importance in population dynamics (all three functions are shown on the same axis here, but in reality would have different units). Panel (b) illustrates how the function *(a)* (flipped in time) determine the contribution of each age to the total number of offspring at time t, $B(t)$. The reproduction of the organisms aged a_1 (young), a_2 (mature), and a_3 (old) are determined by the value of $(t - a)$ for their age, times the number of offspring that were born $t - a$ years ago (i.e., the original size of that cohort, times the cohort's reproductive output at age a).

For completeness, we could also write the expression in Eq. (3.12) as

$$B(t) = \int_0^\infty B(t-a)\, b(a)\sigma(a) da + G(t), \tag{3.14}$$

where

$$G(t) = \int_0^\infty n_0(a) \frac{\sigma(t-a)}{\sigma(a)} b(a) da, \tag{3.15}$$

with $n_0(a) = n(a,0)$. This collapses everything that occurred before time $t = 0$ into the expression $G(t)$. However, because $\sigma(a)$ goes to zero beyond the maximum age of an individual (Fig. 3.4a), $G(t)$ also eventually goes to zero. This makes sense since the function $G(t)$ reflects the contribution to reproduction at t by those already alive at time $t = 0$. Therefore, this expression would be important in the short term ($t <$ the maximum age), but we can safely ignore it and use Eq. (3.12) in the long term.

3.3 The Leslie matrix

Although the discrete-time version of the linear, age-structured model is currently called the Leslie matrix, important contributions predate Leslie's (1945) publication. The use of actuarial tables or life tables to express mortality for the purposes of civic planning dates back to the Roman Empire (Smith and Keyfitz, 1977; Hutchinson, 1978). By the eighteenth century, Euler (1760) had recognized that discrete-time, linear, age-structured populations (i.e. mortality *and* reproduction) would, after a certain amount of time, grow (or decay) geometrically, although he did not derive or prove this fact. In his 1760 paper, which was translated and reprinted in 1970 (Euler, 1970), he derived several relationships between total births, deaths, and the rate of increase, which were of substantial value for demographic calculations given the poor availability of statistics of that time.

Another pioneer in this field, Harro Bernardelli (1941), was primarily interested in the periodic waves that propagate through the age structure of populations. The waves he was referring to are now commonly called the *echo effect* in demography (Lee, 1974). We will discuss these in Chapter 4. He used the matrix formulation and noted that whether a population is increasing or decreasing depends on R_0 ("lifetime reproduction rate"). This important observation was the foundation of the concept of replacement, which is an essential mechanism in the population dynamics of sustainability. E.G. Lewis (1942) was also interested in the echo effect. He presented the Leslie matrix and derived basic conditions for increasing and decreasing populations.

We begin by defining the Leslie matrix (throughout the book, we will indicate matrices and vectors with boldface),

$$\boldsymbol{L} = \begin{bmatrix} b_1 & b_2 & b_2 & b_4 \\ p_1 & 0 & 0 & 0 \\ 0 & p_2 & 0 & 0 \\ 0 & 0 & p_3 & 0 \end{bmatrix}, \tag{3.16}$$

where b_a is the number of offspring produced each year per individual at age a, and p_a is the fraction of individuals surviving from age a to age $a+1$. This Leslie matrix is 4×4, which is for a species in which individuals live only to age four. Below we represent the maximum age as A, for which the Leslie matrix would be $A \times A$. We define the age vector $\boldsymbol{n}_t$ to be the number of individuals at each age at time t, which we call $n_{a,t}$:

$$\boldsymbol{n}_t = \begin{bmatrix} n_{1,t} \\ n_{2,t} \\ n_{3,t} \\ n_{4,t} \end{bmatrix}. \tag{3.17}$$

We can then write the basic vector difference equation for the Leslie matrix. The age vector at the beginning of one year is the product of the Leslie matrix and the age vector at the beginning of the previous year:

$$\boldsymbol{n}_{t+1} = \boldsymbol{L}\boldsymbol{n}_t. \tag{3.18}$$

An example of the calculation depicted by Eq. (3.18) illustrates the biological reasonableness of the Leslie matrix:

$$\begin{bmatrix} 0 & 1 & 2 & 2 \\ 0.2 & 0 & 0 & 0 \\ 0 & 0.5 & 0 & 0 \\ 0 & 0 & 0.1 & 0 \end{bmatrix} \begin{bmatrix} 100 \\ 60 \\ 40 \\ 10 \end{bmatrix} = \begin{bmatrix} 0^*100 + 1^*60 + 2^*40 + 2^*10 \\ 0.2^*100 \\ 0.5^*60 \\ 0.1^*40 \end{bmatrix} = \begin{bmatrix} 160 \\ 20 \\ 30 \\ 4 \end{bmatrix}. \tag{3.19}$$

Following the rules of matrix multiplication (Box 3.3), the youngest age class is

---

Box 3.3 MATRIX MULTIPLICATION

Matrices and vectors provide an efficient way of representing the multiplication and addition of many separate quantities. This makes them very useful in age-structured models when one must, for example, multiply the fecundities of each age class by the abundance of that age class, then sum them to obtain the total number of offspring.

The basic rule of matrix (or vector) multiplication is this: the product of an $m \times n$ matrix $\boldsymbol{A}$ and an $n \times p$ matrix $\boldsymbol{B}$ will be an $m \times p$ matrix $\boldsymbol{C}$. The value on row r, column c of $\boldsymbol{C}$, $C_{r,c}$, is calculated by taking row r of $\boldsymbol{A}$ and column c of $\boldsymbol{B}$, multiplying each element together, and summing them. Note that this means that $\boldsymbol{A}$ must have the same number of columns as $\boldsymbol{B}$ has rows.

It is usually easier to understand matrix multiplication using an example written out. First, a row vector and column vector:

$$\boldsymbol{AB} = [a_1 \;\; a_2] \begin{bmatrix} b_1 \\ b_2 \end{bmatrix} = a_1 b_1 + a_2 b_2$$

(note the result in this case is a 1×1 matrix, i.e. a scalar). Now a square matrix multiplied by a column vector:

$$\boldsymbol{AB} = \begin{bmatrix} a_{11} & a_{12} \\ a_{21} & a_{22} \end{bmatrix} \begin{bmatrix} b_1 \\ b_2 \end{bmatrix}.$$

To visualize the procedure, imagine rotating $\boldsymbol{B}$ on top of $\boldsymbol{A}$, multiplying the elements that align, and summing them to obtain the corresponding row of the resulting vector $\boldsymbol{C}$:

$$\begin{bmatrix} c_1 \\ c_2 \end{bmatrix} = \begin{matrix} b_1 & & b_2 \\ \times & & \times \\ a_{11} & + & a_{12} \\ a_{21} & & a_{22} \end{matrix} = \begin{bmatrix} a_{11} b_1 + a_{12} b_2 \\ c_2 \end{bmatrix}$$

(Continued)

Box 3.3 CONTINUED

and

$$\begin{bmatrix} c_1 \\ c_2 \end{bmatrix} = \begin{matrix} b_1 & b_2 \\ \times & \times \\ a_{11} & a_{12} \\ a_{21} + & a_{22} \end{matrix} = \begin{bmatrix} c_1 \\ a_{21}b_1 + a_{22}b_2 \end{bmatrix}.$$

This is the procedure used in discrete-time age-structured models (e.g. Eqs. (3.18) and (3.19)) to advance the age vector.

Multiplying matrices $\boldsymbol{A}$ and $\boldsymbol{B}$ follows the same pattern:

$$\boldsymbol{AB} = \begin{bmatrix} a_{11} & a_{12} \\ a_{21} & a_{22} \end{bmatrix} \begin{bmatrix} b_{11} & b_{12} \\ b_{21} & b_{22} \end{bmatrix} = \begin{bmatrix} a_{11}b_{11} + a_{12}b_{21} & a_{11}b_{12} + a_{12}b_{22} \\ a_{21}b_{11} + a_{22}b_{21} & a_{21}b_{12} + a_{22}b_{22} \end{bmatrix}.$$

Note that unlike scalar (i.e. non-matrix) multiplication, matrix multiplication is not commutative: $\boldsymbol{AB} \neq \boldsymbol{BA}$. Reversing the order of the equation shown above will reveal why this is the case. However, multiplying a matrix by a scalar simply means that every element of the matrix is multiplied by the scalar, and this operation is commutative:

$$c\boldsymbol{A} = c\begin{bmatrix} a_{11} & a_{12} \\ a_{21} & a_{22} \end{bmatrix} = \begin{bmatrix} ca_{11} & ca_{12} \\ ca_{21} & ca_{22} \end{bmatrix}.$$

Because matrix multiplication is non-commutative, an additional bit of terminology is useful. For the multiplication $\boldsymbol{AB}$, we say that $\boldsymbol{A}$ is *right-multiplied* or *post-multiplied* by $\boldsymbol{B}$; whereas for $\boldsymbol{BA}$ we say $\boldsymbol{A}$ is *left-multiplied* or *pre-multiplied* by $\boldsymbol{B}$.

---

calculated as the sum of reproduction over all ages (i.e. number at each age times fecundity at that age, summed over all ages). For the remaining ages, the number at the next youngest age, at the previous time, is simply multiplied by the annual survival and moved to that age.

We can also see from this example that the calculations of the Leslie matrix are essentially the same as those of the renewal equation (Section 3.2). Note that the number surviving to age a is multiplied by a survival p_i at each age, resulting in $n_{a,t} = n_{1,t\text{–}a}p_1p_2p_3 \ldots p_{a-1}$. Defining that product of annual survivals to be s_a,

$$s_a = \prod_{i=1}^{a} p_{i-1}. \tag{3.20}$$

We can write the discrete time analog of Eq. (3.8) as

$$n_{a,t} = n_{1,t\text{–}a}s_a. \tag{3.21}$$

Now we see that the first term in the new age vector can be written as $n_{1,t+1} = b_1s_1n_{1,t} + b_2s_2n_{1,t\text{–}1} + b_3s_3n_{1,t\text{–}2} + b_4s_4n_{1,t\text{–}3}$ (assume $s_1 = 1$), which is the discrete time version of the renewal equation (Eq. (3.12)), if one considers $B(t)$, recruitment at the youngest age (i.e. 0) in the renewal equation to correspond to recruitment to the youngest age in the Leslie matrix equation (i.e. 1), $n_{1,t}$, and $\sigma(a)$ from the renewal equation to correspond to s_a. As in the renewal equation, the meaning of this expression is that the number of offspring at time t is the sum of the number of offspring 1, 2, 3, etc. years ago, times the probability of surviving 1, 2, 3, etc. years, times the fecundity for a 1, 2, 3, etc. year old.

The equivalence between this approach and the renewal equation means that all three of these models, the M'Kendrick/von Foerster, Lotka's renewal equation, and the Leslie matrix are essentially, dynamically the same. Therefore, one can make use of results from any of the three versions. In this book we primarily use age-structured models with discrete time and age, because this allows us to make use of helpful shortcuts from matrix algebra (e.g. Section 3.4) and avoid having to solve complicated partial differential equations.

3.3.1 *Solving the Leslie model*

Based on our experience solving linear difference equations in Chapter 2, the obvious solution to the vector difference equation with the Leslie matrix (Eq. (3.18)) is

$$\boldsymbol{n}_t = \boldsymbol{L}^t \, \boldsymbol{n}_0, \tag{3.22}$$

which is a matrix equation analogous to the solution of the linear, discrete-time model with a single state variable, N (Eq. (2.5)). For any Leslie matrix $\boldsymbol{L}$ and initial age vector $\boldsymbol{n}_0$ we could apply this formula to obtain $\boldsymbol{n}_t$. However, we would like to be able to know what those solutions will look like based on the elements of the Leslie matrix, so that we can understand why populations behave the way they do (at least for age-structured populations at a low enough density that density-dependent effects can be ignored). To do this we will find the roots, λ, of the characteristic equation for the model (just as in the case of a linear difference equation, Eq. 2.17e). We will not show the full derivation of the characteristic equation this time, but it is the same process that we used in Eq. (2.17) and Eq. (2.18): assume that there is a solution of the form

$$\boldsymbol{n}_{t+1} = \lambda \boldsymbol{n}_t, \tag{3.23}$$

then substitute that solution into Eq. (3.22) and simplify. The value of the roots of this equation will tell us about the behavior of the model.

First we will give the result found in a typical ecology textbook. The general form of the characteristic equation for the Leslie matrix model is

$$\sum_{a=1}^{A} \lambda^{-a} s_a b_a = 1, \tag{3.24a}$$

using the notation we have used thus far. More common notation in ecology uses l_a instead of s_a to represent survival from age 1 to age a, and m_a instead of b_a to represent the number of daughters reproduced by a female at age a, at time t. Using that notation we get

$$\sum_{a=1}^{A} \lambda^{-a} l_a m_a = 1. \tag{3.24b}$$

This is the fundamental equation of demography, otherwise known as the Euler–Lotka equation.

Just as in the simple linear model in Chapter 2, the value of λ is the rate at which the population will (eventually) grow geometrically, increasing if $\lambda > 1$ and declining if $\lambda < 1$ (we explain in Section 3.4 why the qualifier "eventually" is needed). This is clear from Eq. (3.23), which resembles the simple linear model in Eq. (2.5). Equation 3.24 then provides a way to calculate what λ is, given the values l_a and m_a in the Leslie matrix. For many years, the standard means of solving this characteristic equation was trial and error, trying values of λ in the left-hand side until the equation worked out to equal 1. Fortunately, there is now a better way, which we explain in Section 3.4. However, even before attempting to find the solution, we know from Descartes' sign rule that this polynomial has only one positive root, and thus there is only one biologically meaningful value of λ (Box 3.4).

Box 3.4 DESCARTES' RULE OF SIGNS

In this chapter (and throughout the rest of the book) we are often interested in finding the roots of polynomial expressions. That is, if we take a polynomial of variable x and set it equal to zero, such as

$$x^3 - 2x^2 + 3x - 1 = 0, \quad \text{(B3.4.1)}$$

what values of x satisfy this equation? Those values of x are the roots.

A simple way of finding the roots is to plot a graph of the polynomial function versus x, and see where the function crosses the horizontal (zero) axis. There are also a variety of algorithms for determining the roots numerically using a computer. The problem with the graphical method is that there may be complex roots that cannot be plotted in the real number plane (complex numbers are those that contain the imaginary number $\sqrt{-1}$; we will spend more time discussing them in Chapter 4 and Box 4.1), and of course computer-based solutions were not available in the early days of population modeling. However, before we start determining what the roots actually are, it is often very useful to determine simply *how many* roots there are, and whether they might be real, complex, and positive or negative.

The philosopher and mathematician Rene Descartes discovered a rule for determining the maximum possible number of positive or negative real roots of a polynomial. To use the *rule of signs*, first write out the polynomial in descending order of exponents, as in Eq. (B3.4.1). Then count the number of times the sign of the coefficient of the terms changes, moving left to right. In this example, the sequence of signs is $+$, $-$, $+$, $-$, so there are three sign changes. This means that there is a maximum of three positive real roots. However, this is only the maximum possible. Some of the roots might be complex. As we will see in Chapter 4, complex numbers always come in pairs, so if there are any complex positive roots in this example, there will be two of them. Thus there could be either three positive real roots, or one positive real root, or none at all.

To determine the maximum possible number of negative roots, substitute $-x$ into the polynomial, and count the sign changes again:

$$(-x)^3 - 2(-x)^2 + 3(-x) - 1 = -x^3 - 2x^2 - 3x - 1 = 0. \quad \text{(B3.4.2)}$$

Now there are no sign changes, so there are no negative roots.

In this chapter we apply the rule of signs to the characteristic equation, Eq. (3.24b):

$$\sum_{a=1}^{A} \lambda^{-a} l_a m_a = 1.$$

Note that if you wrote out the summation to be a polynomial like Eq. (B3.4.1), all of the coefficients would be positive, because l_a and m_a are always positive numbers. However, the last term in the polynomial will be -1 (because you have to subtract 1 from both sides to get zero on the right-hand side). There is only one sign change, and thus a maximum of one positive real root.

Additionally, If one merely wants to know whether the population is increasing or decreasing, and does not need to know the precise rate (λ), one can compute what is typically called the *net reproductive rate*, R_0, from

$$R_0 = \sum_{a=1}^{A} l_a m_a. \quad (3.25)$$

The population will ultimately be increasing ($\lambda > 1$) if $R_0 > 1$, decreasing ($\lambda < 1$) if $R_0 < 1$ and constant ($\lambda = 1$) if $R_0 = 1$ (Bernardelli, 1941). R_0 has the compelling interpretation of being the amount of reproduction in the lifetime of an individual, which means that a population will be growing if the individuals in it reproduce enough to replace themselves. Again, as we shall see in later chapters, this replacement concept is a key concept in population dynamics, concerned with sustainable populations.

3.3.2 *The stable age distribution*

Before we calculate λ, it is useful to think about what the population age structure looks like when the population is growing geometrically. If the population is growing at rate λ, then $\boldsymbol{n}_{t+1} = \lambda \boldsymbol{n}_t$ (Eq. (3.23)) and the age structure does not change from year to year, but instead each age class increases by factor λ (Box 3.3). The shape of the age structure during geometric growth is called the stable age distribution (SAD). We can derive what it looks like simply by noting that the population will be growing at rate λ. We are interested in the number at each age, relative to the number at age 1, at a time t,

$$w_a = \frac{n_{a,t}}{n_{1,t}}. \tag{3.26}$$

We can simplify this using Eq. (3.21), to write

$$w_a = \frac{n_{a,t}}{n_{1,t}} = \frac{n_{1,t-a} l_a}{n_{1,t}}, \tag{3.27}$$

where we have used the symbol l_a found more commonly in ecological textbooks instead of s_a. Since the population will eventually grow geometrically, we can say that $n_{1,t} = n_{1,0}\lambda^t$ ($n_{1,0}$ is the abundance of age-1 individuals at the beginning of the period of geometric growth). Cancelling in the numerator and the denominator then gives us

$$\frac{n_{1,t-a} l_a}{n_{1,t}} = \frac{n_{1,0}\lambda^{t-a} l_a}{n_{1,0}\lambda^{t-a}} = \lambda^{-a} l_a. \tag{3.28}$$

The interpretation of this expression is that the age structure of a geometrically growing population is determined by survival from age 1 to age a, adjusted to account for the fact that the number of births in the past was greater than (if the population is decreasing, $\lambda < 1$) or less than (if the population is increasing, $\lambda > 1$) it is currently. This leads to the common statement in ecology texts that when a population is increasing, its age structure is skewed to younger ages, and vice versa.

The basic behavior of the model and the development of the SAD is illustrated by multiplying an age vector by a Leslie matrix (Table 3.2) successively many times (Fig. 3.5). The initial age vector is shown at $t = 0$ in Fig. 3.5a, and it is multiplied by the Leslie matrix in Table 3.2. The age distribution seems to change haphazardly for the first few years (Fig. 3.5a,b), but approaches a fixed distribution of relative abundance-at-age (as determined by Eq. (3.28)) by about $t = 30$ y, when the population begins to increase geometrically (in this example $\lambda = 1.01$; Fig. 3.5b). The initial, apparently haphazard portion is known as the *transient phase*; once the population has reached the SAD and is growing geometrically we say it is exhibiting *asymptotic* dynamics (the reason for this will become clear later; clearly the population itself is growing geometrically, not approaching an asymptote). Upon closer inspection the initial transient pattern is not totally haphazard; for example, note that the initial age structure was zero from age 6 on, and the lack of reproduction from the missing older age classes causes the abundance of age 1 to drop dramatically in the

Table 3.2 The Leslie matrix used in Fig. 3.5.

0.0	0.0	0.0	0.0	0.8	1.0	1.0	1.0	0.1
0.7	0.0	0.0	0.0	0.0	0.0	0.0	0.0	0.0
0.0	0.7	0.0	0.0	0.0	0.0	0.0	0.0	0.0
0.0	0.0	0.9	0.0	0.0	0.0	0.0	0.0	0.0
0.0	0.0	0.0	0.9	0.0	0.0	0.0	0.0	0.0
0.0	0.0	0.0	0.0	0.9	0.0	0.0	0.0	0.0
0.0	0.0	0.0	0.0	0.0	0.9	0.0	0.0	0.0
0.0	0.0	0.0	0.0	0.0	0.0	0.3	0.0	0.0
0.0	0.0	0.0	0.0	0.0	0.0	0.0	0.3	0.0

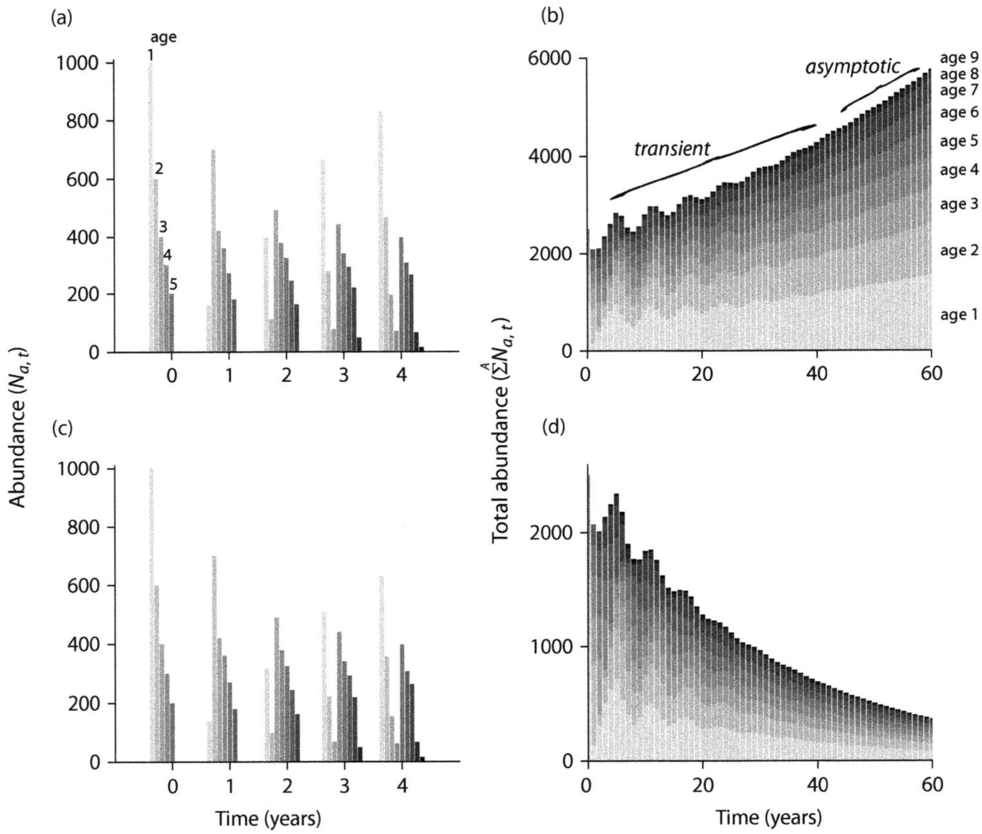

Fig. 3.5 The effects of repeatedly multiplying an age structure by a Leslie matrix model through time. Panels (a) and (b) show the growth of an increasing population ($\lambda = 1.01$), the Leslie matrix shown in Table 3.2, and panels (c) and (d) show a declining population (the matrix altered so that $\lambda = 0.97$). Panels (a) and (c) show the changing age structure in the early transient phase, beginning with the initial age structure at time 0. Panels (b) and (d) show the ages stacked to illustrate how the age structure eventually becomes constant fractions at each age in the asymptotic phase.

second year; that small cohort then propagates through the age structure (Fig. 3.5a) Note that as a result, the total abundance seems to vary cyclically as it approaches geometric growth (Fig. 3.5b). We will examine the transient further in Chapter 4, but for now focus on the asymptotic.

To illustrate the phenomena in Fig. 3.5a,b for a declining population, we can reduce the fecundity for ages 5 through 8 to 0.7 in Table 3.2 (Fig. 3.5c,d). Note the same cyclic approach to geometric growth, but this time the population is declining ($\lambda = 0.97$), and the stable age distribution, $\mathbf{w}$, differs as expected from Eq. (3.28).

3.4 Mathematical theory underlying the Leslie matrix

We will describe this theory here, not for the sake of learning mathematical theory, but rather because some aspects of the theory lead to valuable information regarding the principles of population dynamics. A more comprehensive understanding of the behavior of the solutions to Leslie matrix equations (Eqs. 3.24 and 3.28) will prepare us to do many useful things later in the book, including: (a) evaluating how much each component of the Leslie matrix (i.e. survivals and fecundities at each age) will affect the value of λ, (b) using that information to assess probabilities of extinction, (c) understanding how age-structured populations respond to variability at various frequencies in the environment, and (d) projecting how populations will initially respond to changes in demographic rates, such as reductions in mortality due to harvest. To establish these capabilities we will develop a fuller mathematical analysis of the Leslie matrix.

Box 3.5 EIGENVALUES AND EIGENVECTORS

It happens that for a square matrix $\mathbf{X}$, there are certain vectors $\mathbf{w}_i$ and scalars λ_i for which it is possible to write the equation

$$\mathbf{X}\mathbf{w}_i = \lambda_i \mathbf{w}_i. \tag{B3.5-1}$$

Notice that this equation implies that multiplying by λ_i, a scalar constant, is equivalent to multiplying by the entire matrix $\mathbf{X}$ (although this relationship only holds for the vector $\mathbf{w}_i$). λ_i is termed an *eigenvalue* of $\mathbf{X}$, and $\mathbf{w}_i$ is the corresponding *eigenvector*. *Eigen* is a German word meaning "own", "inherent", or "characteristic." (For the non-mathematical reader, there is no reason that you would suspect that such constants and their associated vectors would exist, so do not be surprised that you did not anticipate this fact.) The eigenvectors $\mathbf{w}_i$ in Eq. (B3.5-1) are the *right eigenvectors* (because the eigenvector is multiplied to the right of $\mathbf{X}$); because matrix multiplication is not commutative (Box 3.3), there are also *left eigenvectors*:

$$\mathbf{v}_i\mathbf{X} = \mathbf{v}_i \lambda_i. \tag{B3.5-2}$$

For an $n \times n$ matrix, there are n pairs of right eigenvectors and eigenvalues and n pairs of left eigenvectors and eigenvalues (although the ith eigenvalue is the same for the ith right and left eigenvectors, if $\mathbf{X}$ only contains real numbers). The set of right (or left) eigenvectors and eigenvalues completely encapsulates the information in $\mathbf{X}$. This means that one can write out the product of $\mathbf{X}$ and any other vector $\mathbf{y}$ in terms of those eigenvectors and eigenvalues:

$$\mathbf{X}\mathbf{y} = \sum_{i=1}^{n} \lambda_i \mathbf{w}_i c_i, \tag{B3.5-3}$$

(Continued)

Box 3.5 CONTINUED

or in other words, the product of each pair of eigenvalue and eigenvector (and the constant c_i), summed over all n eigenvalues. The constant c_i is the ith element of the vector obtained using this equation: $\mathbf{c} = \mathbf{X}^{-1}\mathbf{y}$ (see Box 3.6 for an explanation of the inverse of a matrix). If the matrix were raised to a power, e.g. $\mathbf{X}^t$, then the equation is modified slightly:

$$\mathbf{X}^t\mathbf{y} = \sum_{i=1}^{n} \lambda_i^t \mathbf{w}_i c_i. \tag{B3.5-4}$$

It happens that this way of expressing the *eigendecomposition* of a matrix is very useful for Leslie matrices.

From matrix theory (Box 3.5), we know that for an $A \times A$ Leslie matrix $\boldsymbol{L}$, we can write a general form of the solution to Eq. (3.18) in terms of the eigenvalues λ_i and eigenvectors $\mathbf{w}_i$ of $\boldsymbol{L}$:

$$\boldsymbol{n}_t = \boldsymbol{L}^t\boldsymbol{n}_0 = \sum_{i=1}^{n} c_i\lambda_i^t\mathbf{w}_i. \tag{3.29}$$

Written out more fully, one can see what this equation does: the age vector $\boldsymbol{n}_t$ is the sum of A age vectors (the eigenvectors $\mathbf{w}_i$), each weighted by a constant c_i and the constant λ_i (the corresponding eigenvalue), raised to the power t:

$$\mathbf{n}_t = \begin{bmatrix} n_{1,t} \\ n_{2,t} \\ n_{3,t} \\ \vdots \\ n_{A,t} \end{bmatrix} = c_1\lambda_1^t \begin{bmatrix} w_{1,1} \\ w_{2,1} \\ w_{3,1} \\ \vdots \\ w_{A,1} \end{bmatrix} + c_2\lambda_2^t \begin{bmatrix} w_{1,2} \\ w_{2,2} \\ w_{3,2} \\ \vdots \\ w_{A,2} \end{bmatrix} + \cdots c_A\lambda_A^t \begin{bmatrix} w_{1,A} \\ w_{2,A} \\ w_{3,A} \\ \vdots \\ w_{A,A} \end{bmatrix}, \tag{3.30}$$

where w_{ji} is the jth age component of the ith right eigenvector. The constants, c_i, reflect the initial conditions, and are the elements of vector $\boldsymbol{c}$ obtained from

$$\boldsymbol{c} = \boldsymbol{W}^{-1}\boldsymbol{n}_0, \tag{3.31}$$

where $\boldsymbol{W}$ is a matrix whose columns are the eigenvectors $\mathbf{w}_i$, and $\boldsymbol{W}^{-1}$ is the inverse of $\boldsymbol{W}$ (see Box 3.6). Essentially, the value of c_i corresponds to how similar the initial age vector $\boldsymbol{n}_0$ is to the eigenvector $\mathbf{w}_i$.

Although we have seen eigenvalues and characteristic equations in Chapter 2, and we know that the fundamental equation of demography, the Euler-Lotka equation (Eq. 3.24) is the characteristic equation of the Leslie matrix, seeing Eq. (3.29) for the first time often causes some readers to beg for greater explanation, mathematical or biological, of why such a bold statement is true. A mathematical explanation is beyond the level of this book, and we know of no other way to shore up one's intuition. However, recall that in Chapter 2, we presented the ordinary way of generating a Fibonacci series (i.e. $N_i = N_{i-1} + N_{i-2}$), then also showed that the same series could be generated from the eigenvalues of that model, and an equation in the form of Eq. (3.29).

Box 3.6 MATRIX OPERATIONS: TRANSPOSE, INVERSE, AND DOT PRODUCT

The *transpose* of a matrix is obtained by swapping the rows and columns (first row becomes first column, etc.). It is usually denoted by an apostrophe:

$$\boldsymbol{X} = \begin{bmatrix} A & B \\ C & D \end{bmatrix} \quad \text{(B3.6-1a)}$$

$$\boldsymbol{X}' = \begin{bmatrix} A & C \\ B & D \end{bmatrix}. \quad \text{(B3.6-1b)}$$

Note that the diagonal entries remain unchanged.

The *inverse* of a matrix is defined in a way similar to the multiplicative inverse of a scalar number ($1/x$ is the inverse of the scalar x, so $x \times 1/x = 1$). For a $n \times n$ square matrix $\boldsymbol{X}$, the inverse (also $n \times n$) is labeled $\boldsymbol{X}^{-1}$ and has the property

$$\boldsymbol{X}\boldsymbol{X}^{-1} = \boldsymbol{X}^{-1}\boldsymbol{X} = \boldsymbol{I}, \quad \text{(B.3.6-2)}$$

where $\boldsymbol{I}$ is the *identity matrix*. $\boldsymbol{I}$ is a $n \times n$ matrix with ones along the diagonal and zeros elsewhere, and $\boldsymbol{IX} = \boldsymbol{XI} = \boldsymbol{X}$. Thus multiplying by $\boldsymbol{X}^{-1}$ is the matrix analog of dividing by $\boldsymbol{X}$. There are various numerical algorithms for calculating $\boldsymbol{X}^{-1}$; note that it is not simply one divided by each entry of $\boldsymbol{X}$. It is not possible to invert some matrices (for example, a matrix that is entirely zeros); such matrices are *singular*.

The *dot product*, or *scalarproduct*, of two vectors is simply the sum of the products of each pair of entries. For the $n \times 1$ vectors $\boldsymbol{A}$ and $\boldsymbol{B}$, the dot product is

$$\mathbf{A} \cdot \mathbf{B} = \begin{bmatrix} a_1 \\ a_2 \\ \vdots \\ a_n \end{bmatrix} \cdot \begin{bmatrix} b_1 \\ b_2 \\ \vdots \\ b_n \end{bmatrix} = \sum_{i=1}^{n} a_i b_i. \quad \text{(B3.6-3)}$$

If you recall the rules of matrix multiplication (Box 3.3), you may note that you would obtain the same result if $\boldsymbol{A}$ were a $1 \times n$ row vector multiplied by the $n \times 1$ column vector $\boldsymbol{B}$. So if we transpose $\boldsymbol{A}$,

$$\mathbf{A'B} = \begin{bmatrix} a_1 & a_2 & \cdots & a_n \end{bmatrix} \begin{bmatrix} b_1 \\ b_2 \\ \vdots \\ b_n \end{bmatrix} = \sum_{i=1}^{n} a_i b_i. \quad \text{(B3.6-4)}$$

Writing this operation as the dot product $\boldsymbol{A} \cdot \boldsymbol{B}$ simply conveys that the multiplication is intended to produce a scalar, regardless of whether $\boldsymbol{A}$ and $\boldsymbol{B}$ are row or column vectors. The dot product has useful geometric interpretations that we will revisit later in the book.

Using Eq. (3.30) would give the same result as repeatedly multiplying the age vector $\boldsymbol{n}$ by the Leslie matrix, or simply raising $\boldsymbol{L}$ to the power t and multiplying by the initial age vector $\boldsymbol{n}_0$ (Eq. (3.22)). Either of these things would, of course, be much easier than using Eq. (3.30). The primary usefulness of Eqs. (3.29) and (3.30) is that we can see that as time t gets very large, the solution will be dominated by the term with the largest eigenvalue λ_i, because the eigenvalues are raised to the power t. As we shall see, this is a key useful feature of this equation; it makes age-structured models almost as easy to understand and use as the simple models of Chapter 2. Essentially, the increasing or decreasing solutions to Leslie matrices shown in Fig. 3.5 are the weighted sums of geometrically increasing or decreasing eigenvectors. Because the term with the largest eigenvalue will outgrow the others as t increases, eventually we see only a geometrically growing (or declining) solution with the stable age distribution, which is given by the eigenvector corresponding to the largest eigenvalue. The largest eigenvalue and its corresponding eigenvector are usually called the *dominant*, or leading, eigenvalue and eigenvector. In a minor abuse of notation, we often drop the subscript i for those and simply use λ and $\boldsymbol{w}$ for convenience. Now we also see why we use the term *asymptotic* to describe the long-term behavior of these models: as t gets very large, the growth rate is increasingly determined by λ, but there are always very small but nonzero contributions of the other eigenvectors. Thus the growth rate asymptotically approaches λ.

From Eq. (3.29) we also see that to know whether the population will be increasing or decreasing in the long term, we need only calculate the dominant eigenvalue of $\boldsymbol{L}$. We can find the values of the eigenvalues in several ways. Nowadays, the simplest way is to input the Leslie matrix to a computer program that will solve for the eigenvalues and the eigenvectors of a matrix using an algorithm such as the Cholesky decomposition. This is usually sufficient, although there can be rounding errors that make the result very slightly inexact. A second way would be to find the roots of the characteristic equation (Eq. (3.24)) algebraically, though this is usually impractical.

Both the left and the right dominant eigenvectors (see Box 3.5) of the Leslie matrix have a meaningful biological interpretation. The terms in the right dominant eigenvector are described in Eq. (3.28), as the stable age distribution,

$$\boldsymbol{w} = \begin{bmatrix} 1 \\ p_1\lambda^{-1} \\ p_1 p_2 \lambda^{-2} \\ \vdots \\ p_1 p_2 \dots p_{A-1}\lambda^{-(A-1)} \end{bmatrix} = \begin{bmatrix} 1 \\ l_2\lambda^{-1} \\ l_3\lambda^{-2} \\ \vdots \\ l_A\lambda^{-(A-1)} \end{bmatrix}, \quad (3.32)$$

and the terms of the dominant left eigenvector are the *reproductive value* at each age,

$$\boldsymbol{v} = \begin{bmatrix} 1 \\ \sum_{i=1}^{A} m_i \frac{l_i}{l_2}\lambda^{2-i-1} \\ \sum_{i=3}^{A} m_i \frac{l_i}{l_3}\lambda^{3-i-1} \\ \vdots \\ \sum_{i=A}^{A} m_i \frac{l_i}{l_A}\lambda^{A-i-1} \end{bmatrix}. \quad (3.33)$$

Reproductive value is the total amount of future reproduction that will be produced by an individual at age a. Each element of the vector is the sum over all current and future ages of reproduction at that age (m_i) times survival to that age, given that an individual has already survived to age a (l_i/l_a), times the eigenvalue to a negative exponent, which describes what happens to relative reproductive contribution of older age classes in a population that is growing (it decreases, because younger cohorts will be more abundant and contribute more reproduction) or declining (it increases, for the opposite reason).

Box 3.7 IRREDUCIBLE MATRICES

A square matrix $\boldsymbol{M}$ is *reducible* if there exists a permutation matrix $\boldsymbol{P}$ (a permutation matrix is one in which exactly one element in each row and column is a 1, and all other elements are zeros), such that

$$\boldsymbol{PMP}^{-1} = \begin{bmatrix} \boldsymbol{A} & 0 \\ \boldsymbol{B} & \boldsymbol{C} \end{bmatrix}, \qquad \text{(B3.7-1)}$$

where $\boldsymbol{P}^{-1}$ is the matrix inverse of $\boldsymbol{P}$. This operation, pre-multiplying a matrix $\boldsymbol{M}$ by a permutation matrix and post-multiplying it by the inverse, is the same as interchanging rows and columns of $\boldsymbol{M}$. The result in Eq. (3.7-1) is a rearranged version of $\boldsymbol{M}$ as four submatrices $\boldsymbol{A}$, $\boldsymbol{B}$, $\boldsymbol{C}$, and a matrix of zeros in the upper right corner (a "lower block triangular" matrix). If no matrix $\boldsymbol{P}$ exists that can achieve this, the matrix $\boldsymbol{M}$ is said to be *irreducible*.

3.4.1 *The Perron–Frobenius theorem*

We can now describe the behavior of almost all Leslie matrices in terms of the values the eigenvalues will have. We use an important theorem called the Perron–Frobenius theorem (Perron derived the result for positive matrices in 1907, and Frobenius derived the result for non-negative matrices in 1912; Perron, 1907; Frobenius, 1912; Gantmacher, 1959). The theorem states that "An irreducible non-negative matrix always has a real, positive characteristic value r that is a simple root of the characteristic equation. The moduli of all the other characteristic values do not exceed r. To the 'maximal' characteristic value r there corresponds a characteristic vector with positive coordinates" (Gantmacher, 1959). To unpack some of the terminology in this theorem, *irreducible* is a property of all typical Leslie matrices that we explain in Box 3.7 and discuss later (Section 3.5); *non-negative* means all of the matrix elements are ≥ 0, as a Leslie matrix has (a *positive* matrix has all elements > 0); the term *characteristic value* means eigenvalue; *real* means the number does not contain any imaginary numbers (i.e. it is not *complex*; see Chapter 4); and *modulus* refers to the absolute value. The last sentence means that the eigenvector associated with the dominant eigenvalue will have positive elements. This theorem states that the magnitude $|\lambda|$ of a positive real root of the characteristic equation for a Leslie matrix is greater than or equal to the magnitude of any other root. From Descartes' rule (Box 3.4), we already know that there is only one positive real root. Thus, this theorem assures us that any Leslie matrix will have one dominant eigenvalue that is a real, positive number (so it is a biologically realistic growth rate) and corresponding dominant right and left eigenvectors (the SAD and reproductive contribution) that also have all positive values (again, this means they are biologically interpretable). There is a slight exception to this rule: in some cases there is a second eigenvalue that is complex and equal in magnitude to the dominant eigenvalue. This occurs in particular cases that we will discuss later (Chapter 4), but for now we will deal with cases in which the positive real root is the single dominant root. This may seem somewhat abstract mathematically, but the importance of this theorem is that it means we should focus our attention on the dominant eigenvalue and eigenvector (and ignore the rest) if we want to understand the asymptotic behavior of the model (we will examine the role of the other eigenvalues in Chapter 4).

3.5 Sensitivity and elasticity of eigenvalues: the *Totoaba* example

We noted in Chapter 2 that the number of practical situations to which linear models could be applied would be limited because biological populations typically remain at a roughly constant abundance, rather than declining or increasing geometrically. The exceptions were populations small enough that density dependence was not a factor, such as recently introduced populations and populations at risk. The same comments apply to linear age-structured models, with an additional constraint: linear age-structured models do not exhibit purely geometric growth unless they are at the stable age distribution. As we saw in Fig. 3.5, that does not occur right away, if the population starts from an arbitrary age distribution. Nonetheless these models can be useful in ecology, as we will demonstrate with a few examples.

One of the quantities that applied ecologists are interested in is the *sensitivity* of the population growth rate to the values of the survival and fecundity parameters in the Leslie matrix. That is, how much would the value of the dominant eigenvalue of the Leslie matrix change if there was a specified small change in one of the elements of the matrix? As we have seen earlier in this chapter, this eigenvalue represents the rate of geometric increase or decrease that a population will eventually reach. In Chapter 8 we will see that this sensitivity can be a component of the description of growth of a linear population when the effects of random variability are included.

As an example, we will use an analysis of the sensitivity of a marine fish to the random environment at each of several locations involved in its life history. Totoaba (*Totoaba macdonaldi*) is listed on CITES, the IUCN Red List of Threatened Species, and the US Endangered Species Act. It is the largest fish in the drum family (Sciaenidae), reaching 2 m in length, and it occurs in the Gulf of California where it was heavily fished until 1975 when a ban was imposed. In spite of the fishing ban, the population remains very small, leading to concerns about what actions could best promote recovery. During their lives these fish occur at four different locations in the gulf: reproduction and the youngest juveniles (1–2 years) occur at the north end of the gulf near the outflow of the Colorado River, and pre-adults (3–6 years) and adults migrate south to summer and winter feeding grounds lower in the gulf. These locations (and thus age ranges in the different life stages) are exposed to different stressors, e.g. Colorado River water quality, bycatch of juveniles in shrimp trawls, and poaching of pre-adults and adults (Cisneros-Mata et al., 1995, 1997). Effective conservation required an understanding of which demographic parameters (and thus which life stages) had the greatest influence on the population growth rate, so that management actions could be targeted accordingly.

The life history of totoaba is typical of a long-lived marine fish (Fig. 3.6a). Annual survival is quite low (0.01) during the juvenile stages (ages 1–2), then is higher and constant (0.798) from age 3 on. Fecundity is zero until the age of maturity (age 7), then increases in proportion to individual weight. Growth was assumed to follow a von Bertalanffy relationship (Box 3.8) based on age–length data. Details of the parameterization are given by Cisneros-Mata et al. (1995,1997). The Leslie matrix consists of the survival values, p_a, given previously, and the fecundities, m_a, shown in Fig. 3.6a. As in Cisneros-Mata et al. (1997), the fecundities have been scaled so that $\lambda = 1$, which reflects the apparently stable population size at the time of that analysis. To assess the relative influence of the p_a and m_a values on the population dynamics, we calculate *sensitivity* and *elasticity*.

Box 3.8 THE VON BERTALANFFY INDIVIDUAL GROWTH MODEL

Ludwig von Bertalanffy was an Austrian biologist who contributed substantial conceptual developments in general systems theory and published works on thermodynamics, psychology, sociology, and a range of other fields. For our purposes, his most important work was his model of individual growth (von Bertalanffy, 1934, 1957). The model describes an organism that has continuous, indeterminate growth. That is, it grows continuously throughout its lifetime. Without delving into the mathematical derivation, the basic idea is that the energetic intake of the organism, and its respiration, are proportional to its surface area. That is, they are proportional to its length squared. In a simple case, imagine a sessile bivalve that feeds and respires using its gills; the surface area of the gills will limit both feeding and respiration. As the organism grows, its biomass will increase with the cube of length, while energetic intake only increases with the square. The former will increase more quickly than the latter, and as a consequence the growth rate slows down with age, and the organism's length approaches an asymptotic maximum.

Mathematically, we can write the solution to the growth model for the length of an organism of age a, $L(a)$, as

$$L(a) = L_\infty \left[1 - e^{-k(a-a_0)}\right], \tag{B3.8.1}$$

where L_∞ is the asymptotic maximum length of the organism, k is the rate at which size approaches L_∞, and a_0 is the "age" at which $L(a)$ would equal zero (usually this is a negative number, because organisms begin life with length > 0).

The growth model has been applied mostly widely in fisheries, because it provides an excellent match to empirical age–length relationships for most fishes. It also matches the growth patterns exhibited by many invertebrates and even some mammals (von Bertalanffy 1951, 1957). In the days before computers, it also combined nicely with the exponentials describing survival to lead to simple hand calculations (e.g. Beverton and Holt 1957).

Sensitivity describes how much the geometric growth rate (λ) of a population represented by a Leslie matrix will change per unit change in one of the elements of the matrix. The general expression for that sensitivity to the term in the ith row and the jth colum of a Leslie matrix, l_{ij}, is given by

$$\frac{\partial \lambda}{\partial l_{ij}} = \frac{v_i w_j}{\mathbf{v} \cdot \mathbf{w}}, \tag{3.34}$$

where v_i is the ith term of the dominant left eigenvector $\mathbf{v}$, and w_j is the jth term of the dominant right eigenvector $\mathbf{w}$. The notation in the denominator indicates the "inner product" or the "dot product" of the two vectors (Box 3.6). To interpret this general expression in terms of a Leslie matrix, we recall that the terms of the right eigenvector at each age (w_j) are the values of the stable age distribution at that age (Eq. (3.32)), and the terms of the left eigenvector at each age (v_i) are the reproductive values at that age (Eq. (3.33)). We focus our interpretation on the numerator of Eq. (3.34) because the denominators will be the same for all of the elements of the Leslie matrix.

Since the p_a terms (survival from age a to $a + 1$) are on the subdiagonal of the Leslie matrix, they always appear in column a but row $a + 1$, for any age a. Therefore, the sensitivity of λ to p_a is

$$\frac{\partial \lambda}{\partial p_a} = \frac{v_{a+1} w_a}{\mathbf{v} \cdot \mathbf{w}}. \tag{3.35}$$

This has the interpretation that the sensitivity of λ to the survival from age a to age $a +$ 1 is proportional to the fraction that survive to age a (from the SAD) times the number of individuals an individual at that age would reproduce in the future (the reproductive value). The numerator of the sensitivity of λ to fecundity at age a, b_a is

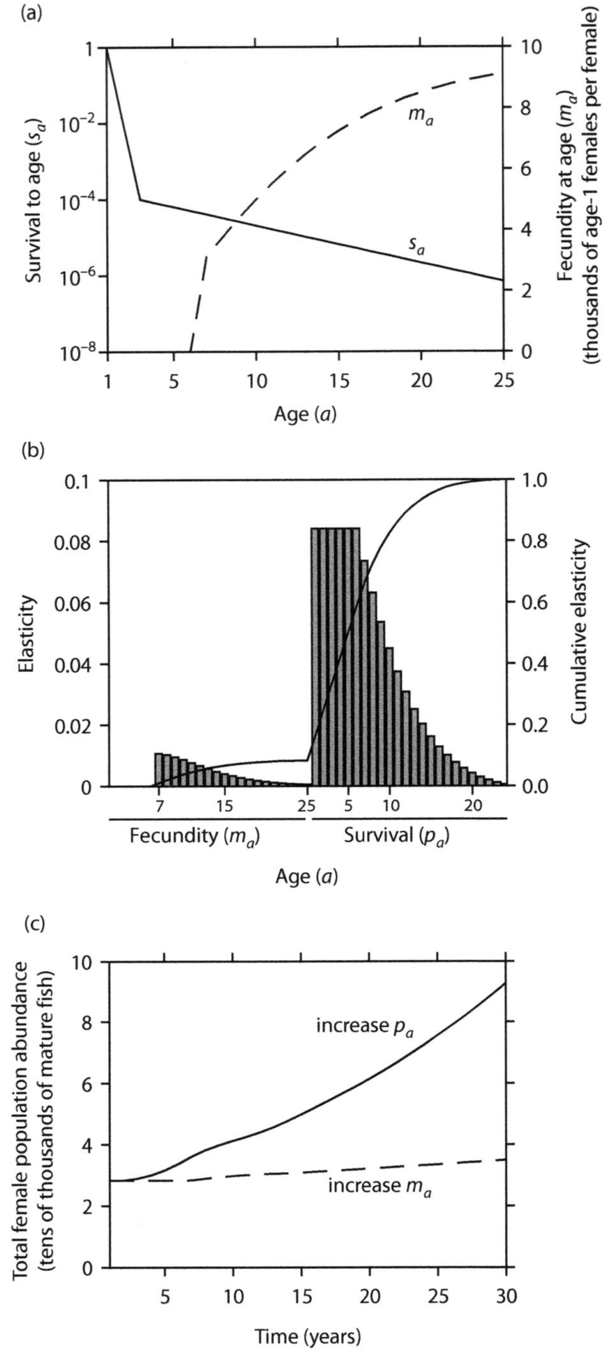

Fig. 3.6 Analysis of an age-structured Leslie matrix model of *Totoaba macdonaldi*. (a) Estimates of survival to age, s_a (solid line, left axis) and fecundity at age, m_a (dashed line, right axis) used in the Leslie matrix. s_a is the cumulative product of survival at each age, p_a, which appears along the sub-diagonal of the Leslie matrix. m_a comprises the top row of the matrix. (b) Elasticity of the asymptotic growth rate, λ, to each model parameter (bars and left axis) and cumulative elasticity (curve and right axis). Model parameters, m_a and p_a, are ordered by age, a, along the horizontal axis. (c) Consequences of different elasticities: projected total (female) population abundance over thirty years if the population begins at the stable age distribution at $t = 0$, with 100 females in the oldest age class, and either all fecundity parameters m_a are increased by 10% (dashed line) or survival of the first five age classes $p_1 \ldots p_5$ is increased by 10% (solid line). We have assumed that the population is currently stable (albeit at low abundance) with $\lambda = 1$, so if no parameters are changed then abundance would also remain unchanged.

$$\frac{\partial \lambda}{\partial m_a} = \frac{v_1 w_a}{\boldsymbol{v} \cdot \boldsymbol{w}}. \tag{3.36}$$

This has the interpretation that the sensitivity of λ to the fecundity at age a is the fraction surviving to age a times the reproductive value of a newly reproduced individual. Thus the interpretations are similar in the sense that they are both the fraction surviving to a certain age times their effect on future reproduction, either through reproduction at that age, or by future reproduction in the remainder of their life.

Sensitivity values are somewhat idiosyncratic because their magnitude depends on the specific values of λ and the life history variable (for example, a change in survival from 0.1 to 0.2 will probably have a greater effect than a change in per capita fecundity from 100.1 to 100.2) To avoid this, ecologists follow economists in using a modified form of sensitivity, termed *elasticity*, which is the fraction of change in λ per unit of fractional change in the life history variable. For survivals p_a this is

$$e_{p_a} = \frac{\partial \lambda / \lambda}{\partial p_a / p_a} = \frac{\partial \lambda}{\partial p_a} \frac{p_a}{\lambda} = \frac{v_{a+1} w_a}{\boldsymbol{v} \cdot \boldsymbol{w}} \frac{p_a}{\lambda}, \tag{3.37a}$$

and for fecundity b_a this is

$$e_{m_a} = \frac{\partial \lambda / \lambda}{\partial b_a / b_a} = \frac{\partial \lambda}{\partial b_a} \frac{b_a}{\lambda} = \frac{v_1 w_a}{\boldsymbol{v} \cdot \boldsymbol{w}} \frac{b_a}{\lambda}. \tag{3.37b}$$

A somewhat surprising result, explained by de Kroon et al. (1986) (although Mesterton-Gibbons (1993) later argued that it was not surprising at all, and followed directly from a little-known theorem of Euler's) that the sum of all of the elasticities of a Leslie matrix is 1.0. Thus elasticity provides the relative contribution of each element of the Leslie matrix to the population growth rate,

$$\sum_{a=1}^{A} \left(e_{pa} + e_m\right) = 1. \tag{3.38}$$

For the totoaba example, Cisneros-Mata et al. (1997) calculated the elasticities for each parameter in the Leslie matrix (Fig. 3.6b). This revealed that the fractional change in λ is much more sensitive to fractional changes in survival than to fractional changes in fecundity, and that approximately 60 percent of the total elasticity lies in the survival rates in the first five years. This suggested that focusing on factors producing variability in totoaba spawning success and larval recruitment was much less important than reducing bycatch and poaching of young fish. The basic reason for this is that survival rates for age-1 fish are very low, so even if one doubled the number of successful larval recruits, most would still die before contributing to reproduction. By contrast, small increases in juvenile survival would lead to substantially more reproductive fish (Fig. 3.6c).

This is a common result in sensitivity analyses of long-lived organisms with high juvenile mortality. For example, sensitivity analysis of a loggerhead sea turtle population model showed that protecting juvenile turtles from trawling bycatch using *turtle excluder devices* would have a much greater effect than efforts to reduce egg mortality on beaches (Crouse et al., 1987; Crowder et al., 1994); although actually these were stage-structured, not age-structured models, which we discuss in Chapter 6. For totoaba, the analysis by Cisneros-Mata et al. (1997) did not lead to any specific management actions. As of this writing the population status of totoaba was unknown (there is no monitoring program) but was still subject to considerable poaching due to lax enforcement (Márquez-Farías and Rosales-Juárez, 2013; Valenzuela-Quiñonez et al., 2015).

3.6 Handling the oldest age classes: age-lumping, terminal age classes, and post-reproductive ages

When constructing an age-structured model, one must decide how many age classes to include. For a species with a well-defined maximum lifespan, such as a semelparous Pacific salmon, this is an easy decision (e.g. White et al., 2014a). For species with an indeterminate maximum age, the choice is more important, because a typical Leslie matrix effectively assumes all adults die after age A. If there is actually some probability of surviving to ages $>A$, then the model will artificially truncate the age structure and thus the reproductive capacity of the model population. The obvious solution is to simply use an arbitrarily large A. Because the probability of survival all the way to age A, s_A, (Eq. (3.21)) will be very small for sufficiently large A, the exact value of A chosen will not matter because so few individuals will occupy the very oldest age classes, and having several mostly empty age classes should not affect the model calculations.

Historically, there have been two practical obstacles to simply using an arbitrarily large A. The first is in model parameterization: it becomes difficult to estimate age-specific survival rates from mark–recapture data (or other methods) for very old ages, because so few organisms can be sampled in those older age classes. In practice one must assume that survival is constant past some age, and that fecundity is either constant or proportional to size (as in the totoaba example; Fig. 3.6a). The second obstacle has been numerical: it is easier to perform eigendecompositions on smaller matrices, particularly by hand. This led to the practice of using a terminal age class, or plus-group: a final age class that includes all individuals age $\geq A$ (e.g. Leslie (1966) used this approach for a model of guillemots, *Uria aalge*).

As an example, Hebblewhite et al. (2003) modeled the population dynamics of black bears, *Ursus americanus*, in Banff National Park, Canada. They constructed the following Leslie matrix using data from radio-collared bears:

$$\begin{bmatrix} 0 & 0 & 0 & 0 & 0.39 \\ 0.64 & 0 & 0 & 0 & 0 \\ 0 & 0.67 & 0 & 0 & 0 \\ 0 & 0 & 0.77 & 0 & 0 \\ 0 & 0 & 0 & 0.77 & 0.84 \end{bmatrix}. \tag{3.39}$$

There are five age classes, corresponding to cubs (age 0), yearlings (age 1), subadults (ages 2 and 3), and adults aged ≥ 4. Adult black bears can reach at least age 20, so this approach compresses many ages into a single class. Notice that to accomplish this, the matrix has a slightly different form from what we used previously: the lower right corner is now nonzero, and that parameter represents the probability of an age-4+ individual remaining alive (and age 4+).

The alternative to using a plus-group in the black bear model would be to choose an arbitrarily large A, say $A = 21$ years (the maximum age observed in the study), and assume that $p_a = 0.84$ and $m_a = 0.39$ for all $a \geq 5$ (essentially extending the matrix in Eq. (3.39) down and to the right, with the diagonal elements all zeros once again).

There are slight, but potentially important differences in the results from these two approaches. First, it is no longer true that $da/dt = 1$: organisms do not move through age at the same rate as they move through time, because in each time step a fraction of individuals remain the same "age" (even though the age is named "A+"). Consequently, the model is no longer identical in form to the M'Kendrick/von Foerster and Lotka models we examined earlier. However, it does still satisfy the assumptions of the Perron–Frobenius

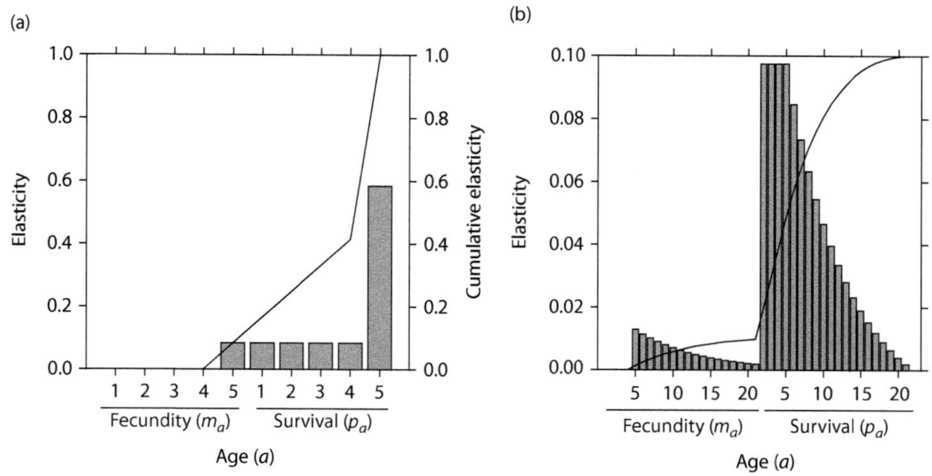

Fig. 3.7 Elasticity of the black bear population model (Hebblewhite et al., 2003) compared to the same model without a plus-group. (a) Elasticity of the asymptotic growth rate, λ, to each age-specific model parameter (bars and left axis) and cumulative elasticity (curve and right axis) for the original model with five age classes, including a ≥ 5 plus group. (b) The same analysis but for a traditional Leslie matrix with maximum age $A = 21$.

theorem, and we can apply those results. Second, speaking more practically, a model with a plus-group will generally have a higher λ than a traditional Leslie matrix, because the plus-group allows some small fraction of individuals to live essentially infinitely long, whereas the absolute maximum lifespan in the traditional approach is A. This longer effective lifespan means more reproduction, and thus greater λ. In the example we used with $A = 21$, the difference is small: $\lambda = 0.95$ versus $\lambda = 0.94$ with a plus-group and $A = 21$ matrix, respectively. The deviation in λ will be greater if A using the traditional method is a smaller number, or if the survival of the plus-group is much greater than the survival of the younger ages (so more organisms accumulate in the plus-group). However, even if λ (the dominant eigenvalue) is quite similar, the remaining eigenvalues may differ substantially, which will alter the transient dynamics (a topic we will address in Chapter 4).

Finally, elasticity analyses will produce different results. In Fig. 3.7 we show elasticities of λ for each model parameter. The conclusion drawn by Hebblewhite et al. (2003) was that the bulk of the elasticity was in the survival of the plus-group (Fig. 3.7a), and conservation efforts should focus on improving conditions for mature adults.

The elasticity analysis for the traditional Leslie matrix without a plus-group ($A = 21$; Fig. 3.7b) reveals a somewhat different pattern; the highest individual elasticities are for the survival rates of the immature ages (just as in the totoaba example, Fig. 3.6b). The sum of elasticities for all ages ≥ 4 still represents the bulk of elasticity (>50 percent), as in the plus-group model, but it is now clear that the younger mature ages have much greater influence on population growth than older mature bears. This nuance could lead to different management strategies. In practice, of course, there is no reason not to just use a realistically large A, given modern computing capabilities. Indeed, one would want to err on the side of choosing A too large rather than too small; only a small proportion of the population will survive to very old age, but if A is arbitrarily small then a noticeable proportion of the population will suddenly drop dead after age A (even if they had relatively high survival from age A-1 to A), and this will unrealistically reduce λ.Using a terminal plus-group is a specific case of the general approach known as *age lumping*, by which multiple age classes are pooled to form stage-structured models. We describe these in Chapter 6.

One final consideration in Leslie matrix development is whether to include non-reproductive age classes. For example, many female mammals experience menopause, and cease reproducing after a certain age. In the Leslie matrix this would correspond to zeros in the upper right-hand corner, indicating that older individuals do not contribute to age-1 in the next time step. In the black bear model, Hebblewhite et al. (2003) considered this possibility, but did not include a separate nonreproductive age class because very few animals would survive to age 20, which was the reported age of senescence. Alternatively, Wielgus et al. (2001) used a Leslie matrix model of a grizzly bear population that did include nonreproductive ages. However, it turns out the debate is moot because of the reducibility of such Leslie matrices.

In Box 3.7 we show how a matrix is said to be *reducible* if it can be permuted to the form

$$\begin{bmatrix} \boldsymbol{A} & 0 \\ \boldsymbol{B} & \boldsymbol{C} \end{bmatrix}, \tag{3.40}$$

where $\boldsymbol{A}$, $\boldsymbol{B}$, and $\boldsymbol{C}$ are block matrices. In general, a matrix is not reducible to this form if it describes a system in which every node (in our case, every age class) can form a connection with every other node over one or more time steps. This is clearly the case with a typical Leslie matrix, because every age class advances to the subsequent age class, and the oldest age class (and likely other age classes too) reproduce, linking back to the youngest age class. Thus, Leslie matrices written in the usual form, in which the oldest individuals still contribute to reproduction, are irreducible (Sykes, 1969). However, when a matrix contains one or more ages past the last age of reproduction, the system then has a dead end: individuals in the oldest age class cannot form a connection back to any other age class. Moreover, the matrix itself has the form of Eq. (3.40) (a zero in the upper right corner), hence it is obviously reducible. This presents a problem in that the Perron–Frobenius theorem no long applies, because it requires an irreducible matrix. However, in this case one can show that for a Leslie matrix, $\boldsymbol{L}$, in the form of Eq. (3.40), $\boldsymbol{L}^t$ can be written as

$$\boldsymbol{L}^t = \begin{bmatrix} \boldsymbol{A}^t & 0 \\ \boldsymbol{X}^t & \boldsymbol{C}^t \end{bmatrix}, \tag{3.41}$$

where $\boldsymbol{X}^t$ is a new matrix, and: (1) $\boldsymbol{C}^t$ goes to zero for sufficiently large t (those too old to reproduce eventually die out), and (2) $\boldsymbol{A}$ contains all of the nonzero eigenvalues of $\boldsymbol{L}$ (i.e. $\boldsymbol{X}^t$ behaves like $\boldsymbol{A}^t$) (Parlett, 1970). So simply writing the model using the matrices $\boldsymbol{A}$ and $\boldsymbol{X}$ is sufficient to describe the long-term behavior, and we need not include age classes older than the oldest reproductive age class in the model. This allows us to use an irreducible matrix and apply the Perron–Frobenius theorem. Of course, if we wished to compare model projections of total population density to data we would have to include those older age classes, but they can be calculated easily by multiplying p_A by the abundance of the oldest reproductive age class.

3.7 What have we learned in Chapter 3?

In this chapter we have taken the first step toward greater realism by adding the dependence on age to our independent variable, abundance. The three different models that represent age-structured populations have essentially the same dynamic behavior, but illustrate different aspects of age-structured dynamics. The derivation of the M'Kendrick/von Foerster equation showed the concept of flow of individuals through age space with time, noting that the velocity of that flow was $da/dt = 1$, and it was the same everywhere. The renewal equation introduced the perspective that current recruitment was the weighted integral (sum) over past recruitment, where the weighting function was the amount of reproduction at each age (Fig. 3.4). The Leslie matrix is the discrete age, discrete time version of these models that we will carry forward for most of our further description of age-structured dynamics. One somewhat trivial, but useful new mode of thinking encountered was that the number of individuals at a certain age a in a population is the number recruited a years ago times the survival to age a.

In exploring the dynamics of age-structured populations, to keep things simple initially we began with linear age-structured models. Keep in mind that while these models lack a key biological component (density dependence), they are still applicable in some useful situations. These are situations where density is low: both recent introductions and populations that have declined to low, risky levels. Both of these are of practical interest.

For the Leslie matrix, we began by reviewing the traditional view in ecology that a population would eventually grow or decline geometrically at a rate λ. Once it began to do that, the age distribution remained constant at the stable age distribution, which each year was merely multiplied by λ. We described how values of λ could be determined, and made note of the fact that if we were only interested in whether it would ultimately be increasing or declining, we could simply calculate R_0. This simple introduction belies the importance of the principle of replacement, which will continue to provide valuable interpretations to other results throughout the book.

We expanded the conventional view to show that this was part of a much richer picture mathematically, hence that we could deepen our understanding of the behavior of linear, age-structured models in useful ways. It turns out that for every $n \times n$ matrix, we can find n eigenvalues and n eigenvectors for which the act of multiplying the Leslie matrix by the eigenvector gives us the same result we would get by simply multiplying the eigenvector by a constant, the eigenvalue (Box 3.3). The value of those numbers and vectors is that they tell us what the behavior of the population will be. We saw that we can express that behavior as the sum of n terms each consisting of a constant times an eigenvector, times an eigenvalue to the power t (Eq. 3.29). So as time passes, there are n constant vectors, not just one, each being multiplied by the eigenvalue raised to the t power. Each term corresponds to one of the eigenvalue/eigenvector pairs. The primary value of this expression stems from the fact that as time goes on, this solution will be dominated by the term with the largest eigenvalue, in the case of the Leslie matrix, λ. The Perron–Frobenius theorem tells us that for a Leslie matrix, that largest eigenvalue will be purely real, not a complex number. It is the eigenvalue that we have been using in the traditional approach to the Leslie matrix. In Chapter 4, we will make further use of some of the other eigenvalues.

Next we learned that eigenvectors come in two flavors, left and right eigenvectors, depending on whether they premultiply or postmultiply the Leslie matrix in the eigenvalue definition (Box 3.5). Each eigenvalue will have both a right and a left eigenvector, and for the largest eigenvalue, both of these have a classical biological interpretation, For the dominant eigenvalue the right eigenvector is the stable age distribution, and the left eigenvector is the reproductive values of individuals at age a.

We then learned that we could determine how sensitive the rate of increase, λ, would be to small changes in each of the elements in the matrix. These turned out to be stated in terms of the components of left and right eigenvectors of the dominant eigenvalue. Our example of a long-lived fish, totoaba, revealed an important general principle: the geometric growth rates of long-lived species are most sensitive to juvenile survival rates. A more useful version of sensitivity was elasticity, the sensitivity of the *fractional* increase in λ to a *fractional* increase in each element of the matrix. Unlike regular sensitivities, elasticities can be used to compare the effects of survival terms and reproductive terms among different species. This is because they are on the same scale; for any species they all sum to 1. We make use of this in Chapter 8 to develop a general approximation of probabilities of extinction.

Last, we learned about a useful way to reduce the dimension of a Leslie matrix of long-lived individuals by the addition of a plus-group, also known as a terminal age class. We were also cautioned against subtle negative effects of nonzero terms on the diagonal in Leslie matrices, something that will pop up again in Chapter 6.

CHAPTER 4

Age-structured models: Short-term transient dynamics

In Chapter 3 we noted that the behavior of the Leslie matrix model would be dominated by a single real eigenvalue, the largest one, λ_1, and we focused our analysis on the behavior of populations when they had reached the stable age distribution (SAD) and were growing or declining at the geometric rate λ_1. However, we could see that when populations started (at $t = 0$) with an age distribution that was not the SAD, they did not initially exhibit uniform geometric growth, but rather underwent an oscillatory transient period before converging on asymptotic dynamics (see Fig. 3.5). Behavior during this transient period obviously differs from the asymptotic behavior at the stable age distribution, and it can last a long time (very long relative to ecological time scales). Because of that, such behavior deserves a chapter of its own.

This transient behavior received little attention prior to the twenty-first century (with notable exceptions, e.g. Taylor 1979), but it is currently drawing much greater interest. Ecologists have begun to appreciate that because both ecological experiments and human management activities operate on relatively short time scales, accounting for short-term transient dynamics can be even more important than long-term asymptotic dynamics (Hastings, 2004). Transient dynamics can lead to different outcomes from those predicted by asymptotic conditions (e.g. extinction versus continued persistence, coexistence versus competitive exclusion; Hastings, 2001). In a management context, there is usually intense interest in whether a management action has "worked," and management actions such as rapid shifts in individual fecundity or mortality rates, as could result from food supplementation or harvest cessation, respectively, will produce immediate transients that must be accounted for to determine the time scale over which one can expect to see evidence of the desired change in the population (e.g. White et al. 2013a). This appreciation has led to the development of a better understanding of transients associated with structured population models. Several reviews have appeared (Ezard et al., 2010; Stott et al., 2011), expanding on the material in Caswell (2001), and there has been some interest in the sensitivity of transient behavior to parameter values (e.g. Caswell 2007). These have been posed in the context of both age- and stage-structured models, so the interested reader may want to read Chapter 6 on stage-structured models before reading this one.

In addition to these transient responses to single shifts in population parameters, continuing frequent environmental disturbances may prevent systems from ever reaching equilibrium, allowing transient dynamics to persist interminably (e.g. Bjørnstad et al. 1999; White et al. 2014a). We address that possibility in the latter half of this chapter. Because we will be interested in the long-term results of these repeated transient

Population Dynamics for Conservation. Louis W. Botsford, J. Wilson White, and Alan Hastings, Oxford University Press (2019).
DOI: 10.1093/oso/9780198758365.001.0001

population responses to constant buffeting by the environment, we will switch from a focus on linear models at low density (where we ignore density dependence) to nonlinear models varying about a nonzero equilibrium level (where we account for fluctuations in the presence of density dependence), as we did in Section 2.4.2.

Because transient behavior involves eigenvalues other than the largest one, and they are often complex eigenvalues, we will begin by explaining the different ways to express and plot complex numbers (Section 4.1), then show how complex eigenvalues can tend to produce cyclic, transient population behavior (Section 4.2). We then address the frequently asked question of how long is this transient behavior going to continue? How long will it take for the population to converge to its asymptotic behavior, geometric population growth at the stable age distribution (Section 4.3)? As noted previously, one possible answer to that question is that if the environment is variable enough, the population may never reach an ultimate asymptotic behavior. For that eventuality, we will describe temporal characteristics of that variability (such as time scales and frequencies), which will require some new tools: Fourier transforms and wavelet analysis. We will use these tools to describe how age-structured populations are more sensitive to certain environmental frequencies than others through a phenomenon termed *cohort resonance* (Bjørnstad et al., 2004; Section 4.4). This will require the switch from linear age-structured models to nonlinear models with compensatory density-dependent recruitment. Cohort resonance is a relatively recent discovery that is improving our fundamental understanding of how age-structured populations with density-dependent recruitment respond to environmental variability.

4.1 The other eigenvalues

In Chapter 3 we described several basic properties of the eigenvalues that were solutions to the characteristic equation of the Leslie matrix (Eq. 3.24). From Descartes' rule of signs (Box 3.4), we saw that there was only one sign change in the characteristic equation, and thus only one positive real root of the equation. From the Perron–Frobenius theorem, we learned that the largest root (the dominant eigenvalue) of a Leslie matrix was positive and real. This implies that the remaining eigenvalues will either be negative, or complex numbers (Box 4.1). In fact, most of the other eigenvalues are usually complex, and occur in pairs termed *complex conjugate pairs* (see Box 4.1). It is these eigenvalues, the complex conjugate pairs, that cause the cyclic abundance that we see during transient periods.

Box 4.1 COMPLEX NUMBERS

In the historical development of algebra, it became clear that some polynomials (cubic polynomials, specifically) had roots that involved the square root of -1. As a simple example, the equation $x^2 + 1 = 0$ has two roots, $x = \sqrt{-1}$ and $x = -\sqrt{-1}$. Since there is no number that when multiplied by itself will equal -1, this presented a problem. In order to handle this problem with polynomials such as this, it was decided to write numbers in terms of two parts: one that involved the square root of -1, and one that did not; then keep track of both parts. Thus, the concept of a *complex number* was introduced. A complex number is represented algebraically as $a + bi$, where a is called the *real part* of the number, and b is called the *imaginary part*. The number i is called the imaginary number, equal to $\sqrt{-1}$.

Often the real and imaginary parts of a complex number z are also denoted Re(z) and Im(z). The development of imaginary numbers began, with some involvement of Leonardo of Pisa (See Chapter 2), in the thirteenth century, but it was not until the sixteenth century that they were more commonly used.

To see how complex numbers could arise, consider the equation $(x+1)^2 = -4$. This equation does not have a real root (you could check this using Descartes' sign rule, or by examining the discriminant of the quadratic equation), but it does have two complex roots: $x = -1 + 4i$ and $x = -1 - 4i$. These two roots form a *complex conjugate pair*, which are numbers with the same real part but imaginary parts with opposite sign.

When working with complex numbers it turns out that it is very useful to plot them graphically in the *complex plane* (Fig. 4.1a), an idea attributed to Leonhard Euler in the eighteenth century. This is an extension of the traditional real number line: the real part of a complex number is plotted along the horizontal (real) axis, and the imaginary part is plotted along the vertical (imaginary) axis. In other words, (a, b) are the Cartesian coordinates of a complex number in the complex plane.

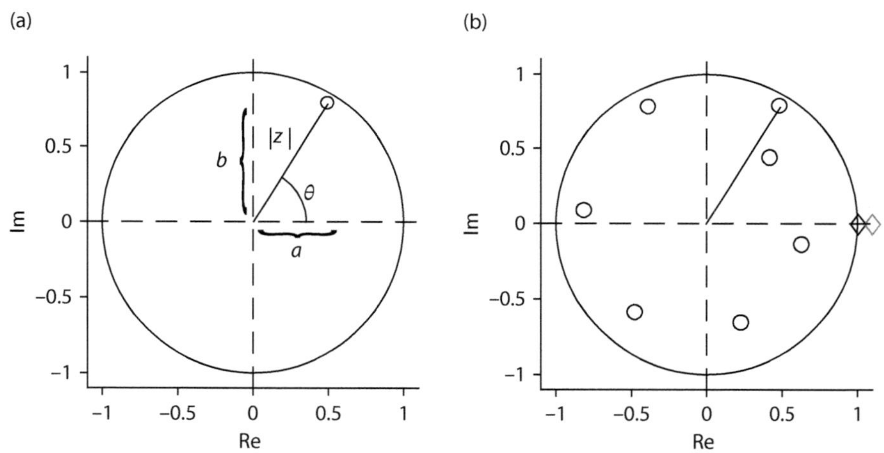

Fig. 4.1 The complex plane. (a) The complex number $z = 0.50 + 0.79i$ in the complex plane. This is the second eigenvalue for the Leslie matrix in Table 4.1. The outer circle is the unit circle (i.e. of magnitude 1). (b) The small circles depict the effects of raising the vector in (a) to the power t, causing the vector to rotate about the origin while declining (i.e. at $t = 0, 1, 2, 3, 4, 5, 6$). The first eigenvalue is a black diamond at $t = 0$, and a gray diamond for $t = 6$.

To understand how to work with numbers in the complex plane, consider the complex number $z = a + bi$. In Chapter 3 we saw that we could write out the behavior of linear age-structured models in terms of eigenvalues, with each eigenvalue raised to the power t (e.g. Eq. 3.29). If most of the eigenvalues are complex numbers like $z = a + bi$, raising them to a power looks like it will be a mess. Fortunately there is an easy way to do this. Using the Pythagorean theorem, we define the *magnitude* of z as $|z| = \sqrt{a^2 + b^2}$; this is the length of z in the complex plane (this notation uses the absolute value symbol; note that the magnitude of a real number is simply its absolute value). Using trigonometry we can also describe the angle that z forms with the complex plane, θ (Fig. 4.1a). For example, $\cos\theta = a/|z|$, and $\sin\theta = b/|z|$. These relationships allow us to rewrite z in terms of trigonometric functions:

(Continued)

Box 4.1 CONTINUED

$$z = a + bi = \mid z \mid (\cos\theta + i\sin\theta)\,. \tag{B4.1-1}$$

Thus, there is a fundamental relationship between complex numbers and sinusoidal trigonometric functions. Additionally, Euler showed that there is a connection to the exponential function, *e*:

$$\mid z \mid e^{i\theta} = \mid z \mid (\cos\theta + i\sin\theta)\,. \tag{B4.1-2}$$

This relationship makes it easier to see that something interesting happens when a complex number is raised to the (integer) power *t*:

$$z^t = (a+bi)^t = \left(|z|e^{i\theta}\right)^t = |z|^t e^{i\theta t} = |z|^t\left(\cos\theta t + i\sin\theta t\right). \tag{B4.1-3}$$

This is referred to as de Moivre's theorem, but Abraham de Moivre noted that it was actually known to Isaac Newton in 1676. It says that a complex number (z) raised to a power (t) will rotate around the complex plane, because the exponent (t in this case) is multiplied by the angle, θ, in the cosine and sine functions. The period of rotation will be $2\pi/\theta$ (the number of additive rotations through the angle θ that it takes to complete the circle, 360°, or 2π radians).

As an historical aside, it is worth noticing what happens in Euler's formula (Eq. (B4.1-2)) if $\theta = \pi$. The cosine of π is -1 and the sine of π is 0, so the formula reduces to

$$e^{i\pi} + 1 = 0. \tag{B4.1-4}$$

This equation is referred to as "Euler's identity" and it is one of the most iconic expressions in mathematics, because it links trigonometry, exponential functions, and imaginary numbers, and contains five fundamental mathematical constants: 0, 1, e, i, and π (Nahin, 2006).

To see how and when the eigenvalues other than the largest one become important to population dynamics, recall how the value of $\boldsymbol{n}_t$ (with maximum age A) can be written out in terms of the A eigenvalues λ_i, right eigenvectors $\mathbf{w}_i$, and constants c_i (Eq. 3.29):

$$\boldsymbol{n}_t = c_1\lambda_1^t\mathbf{w}_1 + c_2\lambda_2^t\mathbf{w}_2 + \cdots c_A\lambda_A^t\mathbf{w}_A. \tag{4.1}$$

Over the long term (i.e. for large values of t), the first (largest) eigenvalue dominates this equation and determines the geometric growth rate. However, for smaller values of t, the smaller eigenvalues will exert a large influence as well. In particular, the second and third eigenvalues can be important. Because they are a complex conjugate pair, they will have equal magnitude. The importance of these other eigenvalues is determined both by the values of the constants c_i (which come from the initial conditions) and the magnitudes of these other eigenvalues relative to the magnitude of the largest eigenvalue.

4.1.1 *An example of cyclic transient dynamics*

As an example of how the smaller eigenvalues affect transient dynamics, consider the Leslie matrix from Table 3.2 and Fig. 3.5. The eigenvalues of that matrix and their magnitudes are in Table 4.1.

The second eigenvalue is shown in polar coordinates in the complex plane in Fig. 4.1a. The behavior of that eigenvalue as it is raised to the powers $t = 2, 3, \ldots, 7$ is shown

Table 4.1 The eigenvalues of the Leslie matrix in Table 3.2.

Eigenvalue	Magnitude	
$\lambda_1 = 1.01$	$\|\lambda_1\| = 1.01$	
$\lambda_2 = 0.50 + 0.79i$	$\|\lambda_2\| = 0.93$	Conjugate pair
$\lambda_3 = 0.50 - 0.79i$	$\|\lambda_3\| = 0.93$	
$\lambda_4 = -0.60 + 0.39i$	$\|\lambda_4\| = 0.72$	Conjugate pair
$\lambda_5 = -0.60 + 0.39i$	$\|\lambda_5\| = 0.72$	
$\lambda_6 = -0.19 + 0.67i$	$\|\lambda_6\| = 0.70$	Conjugate pair
$\lambda_7 = -0.19 - 0.67i$	$\|\lambda_7\| = 0.70$	
$\lambda_8 = -0.38$	$\|\lambda_8\| = 0.38$	
$\lambda_9 = -0.03$	$\|\lambda_9\| = 0.03$	

in Fig. 4.1b. From De Moivre's theorem, the angle of the complex number grows as θt, causing the vector to rotate around the origin. The period of rotation is $2\pi/\theta$; in this case $\theta = \cos^{-1}[\mathrm{Re}(z)/|z|] = 1.00$ radians (or 57.3°), so the period is 6.28; accordingly, after six time steps z^t has nearly returned to its original angle in the complex plane (notice that we were able to calculate the value of θ using the cosine function and the right-angled triangle formed in the complex plane by the real and imaginary parts of z [with z as the hypotenuse] in Fig. 4.1a). In this case $|z| = 0.93$, so the magnitude of z^t also decreases with t. Because the first two eigenvalues are complex conjugate pairs, the third eigenvalue has the same real part, but a negative imaginary part, and as time increases, it will rotate in the opposite direction. By comparison, the first eigenvalue for the same Leslie matrix ($\lambda = 1.01$, the diamonds to the right on Fig. 4.1b), because it is real and has magnitude >1, moves outward along the real axis.

An example of transient behavior of total abundance of this model is shown by plotting all of the eigenvalues in Fig. 4.2a, and the total abundance in Fig. 4.2b, along with plots of the largest two cyclic components. There the population was started from the same initial conditions as in Fig. 3.5: a truncated age distribution with no individuals of age $a > 5$. Note that the abundances due to the complex conjugate pair λ_2 and λ_3, and the complex conjugate pair λ_4 and λ_5 both cycle with a period approximately equal to the generation time of 6.21 years. Although the total population abundance will eventually grow geometrically at rate 1.01, it oscillates at first because of the sub-dominant eigenvalues.

To illustrate how the real time series due to the subdominant eigenvalues arise, we plot the eigenvalues raised to successive powers of t in Fig. 4.3a, b, d, and e. As the members of each pair rotate as mirror images about the imaginary axis, their real parts move from being positive to negative, and then back again, with period $2\pi/\theta_2$ and $2\pi/\theta_4$. Because they are complex conjugates, the imaginary parts always cancel out when they are summed in Eq. (4.1), but the real parts are doubled. Thus, the value of $c_2\lambda_2\mathbf{w}_2 + c_3\lambda_3\mathbf{w}_3$ raised to successive powers of $t = 1, 2, 3, \ldots$ and summed to represent the contribution to the total population abundance, follows a sinusoidal wave that is initially negative before becoming positive (Fig. 4.3c). It also gradually dampens in amplitude because $|\lambda_2| = |\lambda_3| < 1$. This sinusoidal pattern is the major contributor to the early oscillation in the overall population dynamics; we show this as the dashed line in Fig. 4.2b to illustrate its contribution (the initial population abundance at $t = 0$ was added to the eigenvector curves in Fig. 4.2b so that those curves are on the same vertical scale as the

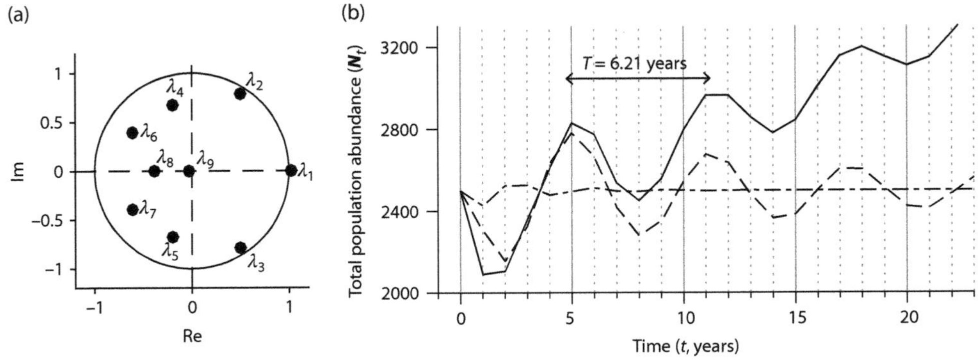

Fig. 4.2 Eigenvalues and transient dynamics of the Leslie matrix in Table 3.1. (a) The eigenvalues for that Leslie matrix, each eigenvalue labeled as λ_i (their values are also listed in Table 4.1). (b) Total population abundance over the first 24 years, beginning at the initial age distribution shown in Fig. 3.5, which has an abundance of juveniles but no adults ages >5. The generation time, T, is shown for comparison to the dominant period of oscillations in eigenvalues. The combined effect of the transient of the 2nd and 3rd eigenvalues (i.e. the sum over all age classes of $c_1\lambda_2{}^t\mathbf{w}_2 + c_3\lambda_3{}^t\mathbf{w}_3$) is shown as a dashed line, the combined effect of the 4th and 5th eigenvalues is shown as a dot-dash line.

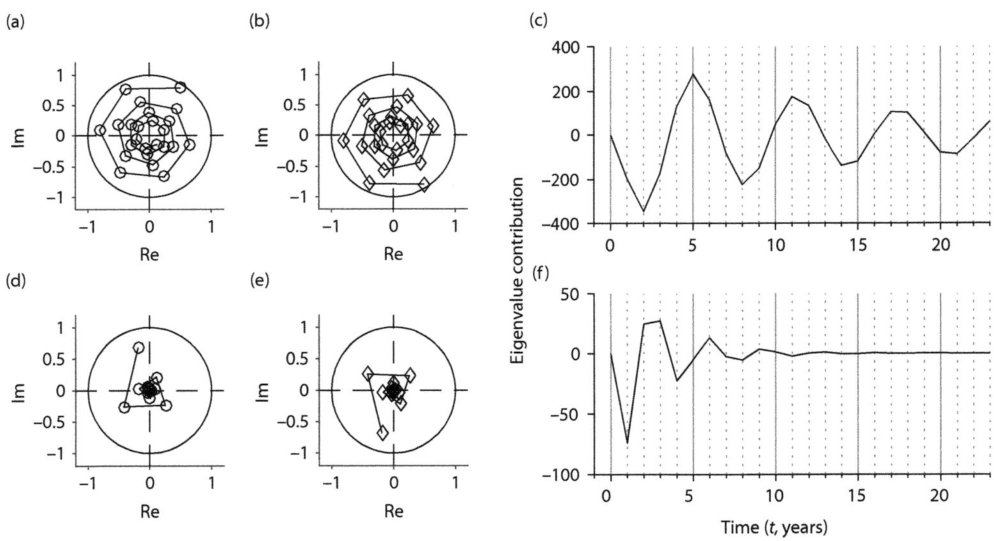

Fig. 4.3 How eigenvalues generate a real cyclic signal: the effects of the 2nd–5th eigenvalues during the transient dynamics depicted in Fig. 4.2b. The top row displays the (a) 2nd and (b) 3rd eigenvalues (complex conjugates) raised to successive powers $t = 1, 2, \ldots, 24$. This corresponds to the Leslie matrix being iterated over $1, 2, \ldots, 24$ years. The combined effect of these two eigenvalues as projected on the real axis is shown in panel (c) (i.e. the sum of $c_1\lambda_2{}^t\mathbf{w}_2 + c_3\lambda_3{}^t\mathbf{w}_3$ for each value of t). The lower row (d–f) shows the same set of results for the 4th and 5th eigenvalues (also a complex conjugate pair).

overall population abundance). The 4th and 5th eigenvalues are the next most important contributors to the overall oscillation (Fig. 4.3d–f), and their contribution to the overall dynamics is shown as a dot-dash line in Fig. 4.2b. The 4th and 5th eigenvalues have a smaller overall effect and their cycles dampen faster, because their magnitudes are smaller.

One can see their effect at $t = 2$, when it slightly reduces the negative effect of the 2nd and 3rd eigenvalues (Fig. 4.2b).

In this example, the smallest two eigenvalues are real but negative (recall there can be only one real, positive eigenvalue, thanks to Perron–Frobenius). The number of negative real roots depends on the number of age classes and the age-specific patterns of survival and fecundity in the matrix. This number can be calculated using Descartes' sign rule (Box 3.4), but for our purposes it is not necessary; for nearly any biologically realistic Leslie matrix, the 2nd and 3rd eigenvalues will be a complex conjugate pair, and they will dominate the transient dynamics. However, if you are wondering what effect the negative real eigenvalues will have through time, consider what happens when you raise a negative number to successive powers 2, 3, 4, etc. (as time increases): it will alternate in sign, thus producing oscillations with period = 2. (In fact these negative numbers can be thought of as behaving like complex numbers with an angle of π and therefore a period of 2!)

4.2 How the dependence of reproduction on age influences these cycles

We can learn something about the nature of these cycles by examining another known characteristic of Leslie matrices, *primitivity*, and reviewing an early numerical exploration of the characteristics of insect age structure that led to cycles (Taylor, 1979).

4.2.1 *Semelparous species and imprimitive Leslie matrices*

A special case of the Leslie matrix that is somewhat informative regarding transient behavior is the case in which there is reproduction only at the oldest age. This extreme version of reproductive age structure is biologically termed an obligate semelparous species, and is quite rare. A *semelparous* species is one in which individuals reproduce only once then die. In *obligate semelparous* species, all individuals spawn at the same age, then die. For example, Pacific salmon, *Oncorhynchus* spp., are famously semelparous, though not always obligate semelparous. Being anadramous, they are hatched in freshwater streams, swim downstream to the ocean in their first year (or second year, in some species), spend some number of years (depending on species and latitude) in the ocean, and then return to their natal stream to spawn and die. For sockeye salmon, *Oncorhynchus nerka*, at the latitude of the Fraser River in southern Canada, most sockeye spawn at age 4. If they were truly obligate semelparous, a linear age-structured model of a Leslie matrix for their population dynamics would be

$$\begin{bmatrix} 0 & 0 & 0 & b \\ p_1 & 0 & 0 & 0 \\ 0 & p_2 & 0 & 0 \\ 0 & 0 & p_3 & 0 \end{bmatrix}. \tag{4.2}$$

A matrix of this form has the property of being *imprimitive*, whereas all of the Leslie matrices we have considered so far are *primitive*. The technical definition of a primitive matrix is one that becomes *positive* (i.e. all entries >0; whereas a Leslie matrix is merely *non-negative*) when raised to a sufficiently high power (you can check that the Leslie matrix in Table 3.1 meets this criterion by raising it to some arbitrarily high powers using a computer). In other words, this technical definition means that if a Leslie matrix is

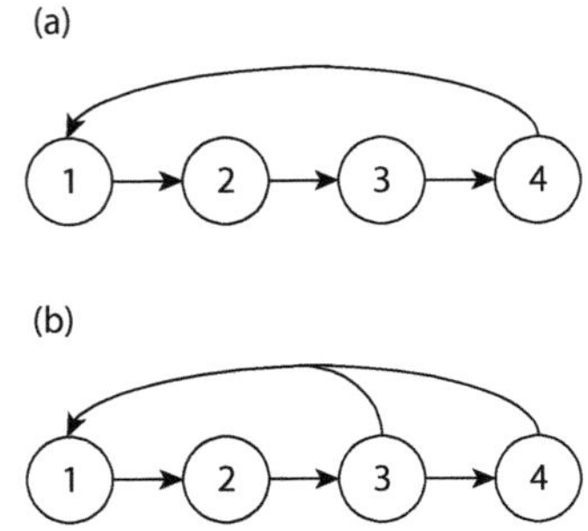

Fig. 4.4 Life cycle graph for two species with four age classes: (a) a semelparous species with an index of imprimitivity = 4, therefore with an imprimitive matrix, and (b) a similar species that is iteroparous with an index of imprimitivity = 1, and therefore with a primitive matrix.

imprimitive, there is a year in the future when an individual in any year class will be the ancestor of an individual in every year class.

We make the heuristic definition of imprimitivity more precise in biological terms by using the life cycle graph of Leslie matrices (Caswell, 2001). The life cycle graph corresponding to the matrix in Eq. (4.2) is shown in Fig. 4.4a. If one draws all possible paths through the life cycle, the greatest common divisor of the length of those paths is termed the *index of imprimitivity*. For Eq. (4.2), there is only a single path: all individuals begin life as age 1, then progress through each age class, and finally new age-1 individuals are only produced by the final age class. The length of the path is 4 (and there is only one path, so the greatest common divisor is also 4) and the index of imprimitivity is 4. If, however, there were reproduction in both age 3 and age 4, then there would be two different possible paths through the life cycle, the greatest common divisor would be 1, and the index of imprimitivity would then be 1. If the index of imprimitivity of a matrix is 1, the matrix is *primitive*. So any Leslie matrix with at least two adjacent reproductive age classes will be primitive. Imprimitive matrices could also be iteroparous, but the reproductive ages would have to be multiples of each other. For example, the matrix

$$\begin{bmatrix} 0 & b_2 & 0 & b_4 \\ p_1 & 0 & 0 & 0 \\ 0 & p_2 & 0 & 0 \\ 0 & 0 & p_3 & 0 \end{bmatrix} \tag{4.3}$$

has an index of imprimitivity = 2, because reproduction occurs in every second age class (we know of no biological examples of this, however).

Thinking back to the heuristic definition of primitivity, if a matrix is primitive there are year classes now which will never be the ancestor of some other year classes—the population is effectively split into different groups. Thus, imprimitive matrices and the populations they describe are interesting because they are the exception we mentioned earlier when discussing the Perron–Frobenius theorem (Section 3.4), in which the dominant eigenvalue does not have a magnitude greater than that of all other eigenvalues. For an imprimitive matrix with index of imprimitivity p, there will be p eigenvalues with the same magnitude. One of these will be the positive real root λ_1, and the others will be either negative or complex conjugate pairs. Because these roots all have the same magnitude, the positive real root will no longer dominate as the matrix is iterated through time and the eigenvalues are raised to successively higher powers; consequently, the cycles associated with the other eigenvalues will not dampen. The period of the cycles associated with these other eigenvalues will be either p or multiples of p, the index of imprimitivity.

Lewis (1942) provided an informative way of thinking about imprimitive Leslie matrices. He invoked the fact that a matrix is the solution to its own characteristic equation (Eqs. 4.5 and 4.6). Recall the characteristic equation whose roots are the eigenvalues of a Leslie matrix (Eq. (3.24)):

$$\sum\nolimits_{a=1}^{A} \lambda^{-a} s_a b_a = 1. \tag{4.4}$$

For a semelparous species, all of the b_a terms would be zero except for the last one, b_A. Expanding the summation given that simplification, then subtracting 1 from both sides and multiplying through by $-\lambda^A$, we obtain

$$\lambda^A - s_A b_A = 0. \tag{4.5}$$

Since the original imprimitive matrix $\boldsymbol{L}$ is a solution to this equation, we can substitute it in for λ:

$$\boldsymbol{L}^A - s_A b_A \boldsymbol{I} = 0, \tag{4.6}$$

where $\boldsymbol{I}$ is the *identity matrix* (a matrix with ones on the diagonal and zeros elsewhere). Notice also that $s_A b_A$ is the net reproductive rate R_0 in this case. The consequence of this, as Lewis (1942) explained, is that multiplying the starting age distribution $\boldsymbol{n}_0$ by $\boldsymbol{L}$ A times (i.e. projecting forward in time A years) is the same as multiplying $\boldsymbol{n}_0$ by R_0. In other words, the original age distribution will just be repeated every A years without alteration. Also, the maximum age A is both the period of the resulting cycles and the index of imprimitivity of $\boldsymbol{L}$.

Imprimitive matrices can be useful for modeling the dynamics of semelparous species in which all individuals reproduce at the same age, such as periodic cicadas (Williams and Simon, 1995) and some plants (Young and Augspurger, 1991). It turns out that most Pacific salmon do not actually meet this criterion, because not all individuals return to freshwater to reproduce at exactly the same age. For example, in sockeye salmon, the Leslie matrix in Eq. (4.3) is not correct: some fish return to spawn at age 3, and others return at age 5 (Groot and Margolis, 1991). The exception is pink salmon, *Oncorhynchus gorbuscha*, which all reproduce at age 2, without exception. The index of imprimitivity of the corresponding Leslie matrix is 2, and many pink salmon populations show clear two-year cycles (Fig. 4.5). Also, because there are only two age classes, the Leslie matrix has only two eigenvalues; they have equal magnitude but one is negative, which as we saw in Section 4.1.1 is associated with period-2 cycles. In fact, pink salmon populations essentially consist of two separate populations ("lines") that spend alternating years in fresh and salt water, like sailors hot-bunking in a submarine. Of course, actual pink salmon populations experience density-dependent feedbacks at high densities and their dynamics are somewhat more complicated than we have described here (Krkosek et al., 2011), but the basic phenomenon of imprimitive matrices is what gives rise to these period-T cycles.

As with many mathematical descriptions, even if the conditions for the model are not exactly met, the results will still be informative. In order to be primitive a matrix must have entries (in our case fecundity rates) which are exactly zero. If all of the fecundities are zero except for the oldest one, we have seen that the population essentially behaves as though it takes an infinitely long time to approach the stable age distribution. If, as would be realistic in many cases, these rates are not exactly zero but are very small, we would expect the time to converge to the stable age distribution would be very long. This intuition is correct, as we will see.

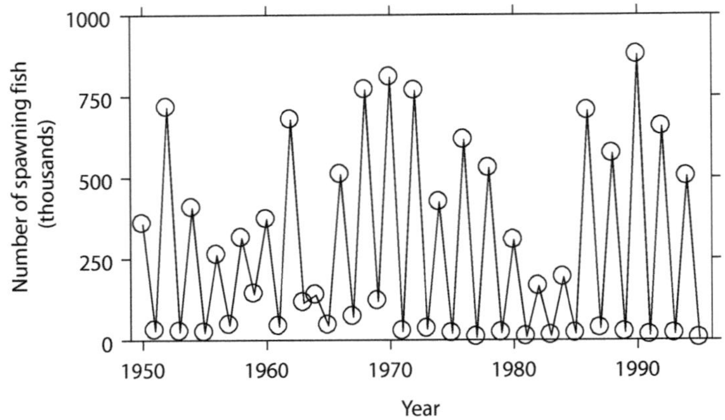

Fig. 4.5 Example of period $T = 2$ cycles in obligate semelparous pink salmon, *Oncorhynchus gorbuscha*, from the Area 2E stock on the north coast of British Columbia, Canada. Data from Dorner et al. (2008).

4.2.2 *Cycle period: the mean age of reproduction and the echo effect*

The period of cycles produced by any eigenvalue λ_i is $2\pi/\theta_i$, where θ_i is the angle formed by λ_i in the complex plane. It turns out that the period corresponding to λ_2 (and thus also λ_3) is approximately equal to an important biological quantity: the *mean age of reproduction*, which is also a common definition of the *generation time*, and is usually abbreviated T (Box 4.2).

Box 4.2 THREE DEFINITIONS OF GENERATION TIME

There are three different quantities that ecologists commonly refer to as "generation time." They may have similar values, under certain conditions, as we explain here. We refer to the three quantities here as T_1, T_2, and T_3. Throughout the rest of the book, if we refer simply to T, we are using the first definition, T_1, and have dropped the subscript for simplicity. If we use a different definition, we will mention it.

The first definition is a weighted average age, with the weights given by the number of offspring produced at each age. Coale (1972) gave the following expression for the mean age of reproduction:

$$T_1 = \frac{1}{R_0}\sum_{a=1}^{A} a l_a m_a. \tag{B4.2-1}$$

Readers familiar with probability will recognize this as the calculation of the mean of the distribution $l_a m_a / R_0$. Those readers who have not studied probability may be more used to thinking of the calculation of a mean as a sum of numbers, divided by the total number: $(1+2+3+1+2+2+2)/6 = 2.17$. However, this can also be expressed as the sum of the numbers multiplied by their frequency, i.e. the number of times each occurs in the list: $(1\times2 + 2\times4 + 3\times1)/6 = 2.17$. The latter is the same form as Eq. (B4.2-1).

The second definition of generation time is the same mean as T_1 in a population that is growing geometrically at the asymptotic rate λ:

$$T_2 = \sum_{a=1}^{A} \frac{a l_a m_a}{\lambda^a}. \tag{B4.2-2}$$

Finally, the third definition of generation time cannot be identified as a mean, rather it is the amount of time it takes for the population to grow by a factor R_0 ($\lambda^T = R_0$). This is usually written as

$$T_3 = \frac{\log R_0}{\log \lambda}. \tag{B4.2-3}$$

The value of T_3 was often close enough to the other values to generate an initial guess in solving the fundamental equation of demography, Eq. (3.24b) for λ by hand. Whether it does depends on the shape of the distribution of $l_a m_a$ over age, and apparently that approximation worked pretty well for human populations. However, to realize that using approximation requires caution, one need only consider that when $R_0 = 1$, hence $\lambda = 1$, T could be any value.

To determine what the period of cycles caused by second and third eigenvalues might be, Coale (1972) considered the special case in which the function $l_a m_a$ (the distribution of reproduction over age) is symmetrical with age: it rises to a peak at the mean age of reproduction, T, then declines (here we are referring to the generation time defined as T_1 from Box 4.2). He solved the resulting characteristic equation (Eq. 3.24b) for the first complex root, rewriting the complex solution in terms of sine and cosine functions (Box 4.1). He then showed that the only solution that would satisfy the characteristic equation (if $l_a m_a$ were symmetrical) was a complex root with period equal to T. For non-symmetrical distributions of reproduction, this solution is not exact, but is still a good approximation (Fig. 4.2).

This relationship between T and the cycle period can easily be demonstrated in our example of the Leslie matrix by adjusting the parameters of that matrix. To do so, we simulated harvesting all individuals age 4 and greater by removing 75 percent of them each year (i.e. multiplying p_4 to p_8 by 0.25). Few individuals will then survive to the older age classes so that nearly all reproduction occurs at age 5, the first reproductively mature age class. In order to keep $\lambda_1 > 1$ for this comparison, we also increased the fecundity, b_i, of all age classes by 950 percent. One result of truncating the age structure and compressing reproduction into fewer ages is to reduce T from 6.19 years to 5.31 years. The effect on the eigenvalues is that the second through fifth eigenvalues all increase in magnitude and shift their angle in the complex plane (Fig. 4.6a). The angle for the second eigenvalue, θ_2, increases from 1.01 to 1.20 rad (58° to 69°), which leads to a shift in period from 6.21 to 5.24. These values closely correspond to the respective values for T, and the dominant cycles of those models are approximately 6 years for the original Leslie matrix (Fig. 4.2b) and 5 years for the modified one (Fig. 4.6b).

The tendency for populations in which reproduction is concentrated into a small range of ages to exhibit period-T cycles is a well-known phenomenon, called the *echo effect*. It is so named because small perturbations in the age distribution tend to echo across subsequent generations. In the case of Fig. 4.6b, the initial age distribution has a large pulse of juveniles and few individuals in the older age classes. This pattern echoes, or recurs cyclically, as those young organisms age, become reproductive, and then spawn a new pulse of offspring just as the original cohort is dying off. This phenomenon can be an important component of cyclic variation in exploited populations, where harvest truncates the age structure and compresses the age distribution of reproduction (Bjørnstad et al., 2004; Worden et al., 2010; Botsford et al., 2014b), as we shall see in Section 4.3. Recall

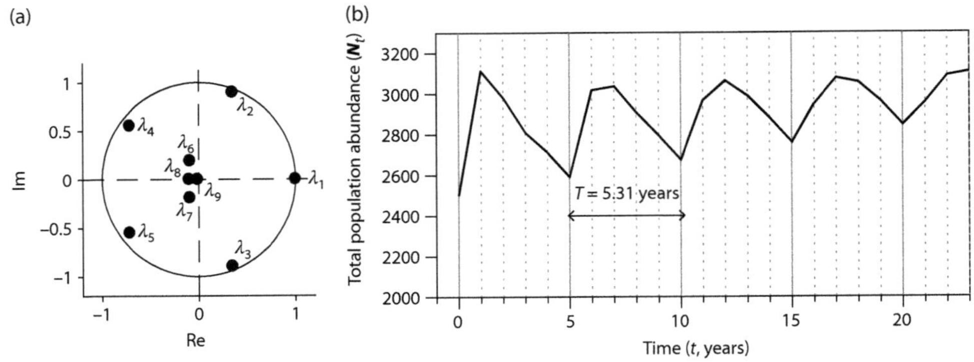

Fig. 4.6 Eigenvalues and transient dynamics of the Leslie matrix in Table 3.1 (and depicted in Figs. 4.2–4.3) with survival for ages ≥4 reduced by 75% to represent harvesting. (a) The eigenvalue spectrum, with each eigenvalue labeled λ_i. (b) Total population abundance over the first 24 years, starting at the initial age distribution shown in Fig. 3.5, which has an abundance of juveniles but no adults ages >5. The generation time, T, is shown for comparison to the dominant period of oscillations.

again that obligate semelparous species can be thought of as extreme examples of the echo effect. Species which are nearly obligate semelparous should therefore be prone to cycles as well.

4.2.3 *How age structure influences the occurrence of cycles*

Having established how life history characteristics determine the dominant time scale of transient variability, a second question is how does life history influence the overall strength of transient behavior? Taylor (1979) was the first to pursue this question by examining how the difference in magnitudes between the dominant eigenvalue and the second largest eigenvalue (a member of a complex conjugate pair) varied as he changed various life history parameters in a linear age-structured model of insect life histories. Taylor (1979) used a continuous-time model (the Lotka renewal equation model, Section 3.2), but a similar analysis using a Leslie matrix yields similar results. In continuous-time models the terms in Eq. (4.1) are exponential functions, $e^{(\text{eigenvalue})t}$, instead of λ raised to the t power. The largest eigenvalue, r, is purely real, and the second and third eigenvalues are $u_2 + iv_2$ and $u_2 - iv_2$. Taylor (1979) represented the strength of the transient effect by the difference $(r - u_2)$. Since

$$e^{(r-u_2)} = \frac{e^r}{e^{u_2}} = \frac{\lambda_1}{|\lambda_2|}, \tag{4.7}$$

where λ_1 and λ_2 are the equivalent first and second eigenvalues of a discrete-time model, Taylor's continuous-time measure of strength is equivalent to the ratio of the magnitudes of the first two discrete-time eigenvalues.

The goal of Taylor's (1979) analysis was to determine whether it was acceptable to represent the growth of insect populations each year in terms of their asymptotic exponential growth. In other words, was the transient behavior strong enough that it had to be accounted for in the annual population development in temperate climates? He tabulated life history parameters for 30 insect species, and noted that all of them begin the growing season with a single pulse of juveniles, i.e. which would be far from the SAD.

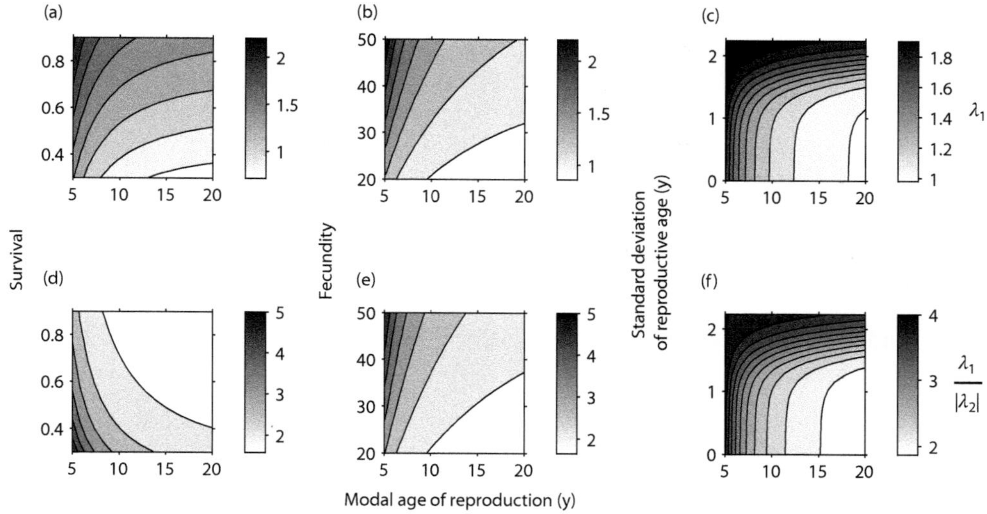

Fig. 4.7 Relative magnitudes of transient effects. Effect of different insect life history parameters in Taylor's (1979) model on (top row) the dominant eigenvalue λ_1 and (bottom row) the rate of convergence to the SAD, $\lambda_1/|\lambda_2|$. Effects of the modal age of reproduction and (a) survival, (b) fecundity, and (c) standard deviation of the distribution of reproductive ages.

Taylor presented plots similar to Fig. 4.7 with isopleths of various values $e^r = \lambda_1$ in the top row, and the difference between the magnitude of the first two eigenvalues in the second row. We recreated a version of this analysis (Fig. 4.7) using a discrete-time Leslie matrix model. Following Taylor's approach, survival was constant with age, and reproduction followed a gamma distribution, with a peak (or mode) at a particular age. Thus reproduction can be described by the modal age (at which most reproduction occurs) and the variance (how much reproduction is spread out over ages). The isopleths (contour lines of equal value) in Fig. 4.7 are plotted across various life history variables on the two axes. Here we focus on the second row because we are interested in the strength of transient effects. The first plot on the second row (Fig. 4.7d), transient strength versus survival and the modal age of reproduction, showed a strong dependence of transient strength on the modal age of reproduction, but little dependence on survival. Earlier reproduction was associated with a larger ratio of λ_1 to λ_2, and thus weaker transient behavior. The second plot of the second row (Fig. 4.7e) is a plot of the strength of transient behavior versus modal age of reproduction (again) on one of the axes, and total fecundity. This one showed the same dependence of transient strength on age of reproduction, but little dependence on fecundity. In the third plot of the second row (Fig. 4.7f), the most interesting one, the two axes are the modal age of reproduction and the standard deviation of the distribution of reproduction versus age. The isopleths indicate that the strength of the transients increases with both increasing age of reproduction and decreasing variance of the distribution, in other words when reproduction is concentrated into a relatively narrower age range. This idea that populations are less stable (i.e. have more cyclic fluctuations due to transient dynamics) with a "skinnier" distribution of reproduction over age pops up again in Chapter 7 on cycles of period $2T$, and also later in this chapter as a condition leading to extreme cohort resonance. Taylor did not make much of this result in 1979, but with hindsight, it was a broadly important result.

With regard to the general field of population dynamics, the work of Taylor (1979) and other investigators of the echo effect are examples of the importance of the function $\varphi(a)$, which we identified in Eq. (3.13) and described as influence functions in Section 3.2 .

Taylor's (1979) overall result was that, assuming a typical length of the growing season in North America, only 18 of the 30 insect species would approach within five percent of the SAD within that time (he accounted for the effects of temperature on insect development by measuring time in degree-days rather than simply days). Thus he concluded that transient dynamics would dominate for most of the growing season in many species. Caswell (2001) later questioned this conclusion, pointing out that five percent is an arbitrary cutoff. This underscores the need for our next topic, to estimate the *distance* (or time) from the SAD, and initial trajectory of the population when starting from non-SAD initial conditions.

4.2.4 *Convergence to the asymptotic dynamics*

Ecologists are typically interested in transient dynamics and the effects of the subdominant eigenvalues in situations where a population that has been perturbed away from the stable age distribution (and usually pushed towards extinction) by a natural disturbance or some human factor such as overexploitation. In a conservation context, managers usually want to take steps to increase λ_1 (cause long-term population increases), but there is a question as to how soon the population will begin growing at that geometric rate. During the transient phase, year-to-year growth, being cyclical, can either be slower or faster than the long-term growth. This question can be viewed in two different ways: (1) how far away from the stable age distribution (SAD) is the population initially, and (2) how quickly will it converge to the SAD?

4.2.4.1 *Rate of convergence to the stable age distribution: the damping ratio*

In order to determine how quickly transient dynamics will converge to asymptotic dynamics, consider the first two terms of the eigenvalue solution to the Leslie matrix model (Eq. (4.1)),

$$\boldsymbol{n}_t \approx c_1\lambda_1^t\mathbf{w}_1 + c_2\lambda_2^t\mathbf{w}_2. \tag{4.8}$$

If we define a new quantity termed the *damping ratio*:

$$\rho = \lambda_1/|\lambda_2|, \tag{4.9}$$

we can divide both sides of Eq. (4.8) by λ_1 to obtain

$$\frac{\boldsymbol{n}_t}{\lambda_1^t} \approx c_1\mathbf{w}_1 + c_2\rho^{-t}\mathbf{w}_2. \tag{4.10}$$

In the limit as t becomes very large, the difference between the left-hand side and right-hand side of Eq. (4.8) approaches zero. Therefore, we can write

$$\frac{\boldsymbol{n}_t}{\lambda_1^t} - c_1\mathbf{w}_1 = c_2\rho^{-t}\mathbf{w}_2. \tag{4.11a}$$

If we take the vector norm of both sides (the norm is simply the length of an n $\times$ 1 vector in n-dimensional space; it is calculated using the Pythagorean theorem in the same way as the magnitude of a complex number [Box. 4.1]), we obtain

$$\left\| \frac{\boldsymbol{n}_t}{\lambda_1^t} - c_1\mathbf{w}_1 \right\| = k\rho^{-t} = ke^{-t\ln\rho}, \tag{4.11b}$$

where $||\boldsymbol{x}||$ is the notation for the norm of the vector $\boldsymbol{x}$ (note that the new constant $k = ||c_2\mathbf{w}_2||$). This equation tells us that the difference between the actual age vector ($\boldsymbol{n}_t$) and the stable age vector ($\mathbf{w}_1$) will decrease over time at an exponential rate given by the log of the damping ratio, ρ. This means that if we want to get some idea of how rapidly the transient oscillations will decline, we can calculate ρ from Eq. (4.9). High values indicate it will decline rapidly. We have already hinted at the role of the damping ratio with the example in Fig. 4.1, and it is essentially the second variable Taylor (1979) was exploring in Section 4.2.3.

4.2.4.2 *The distance to the stable age distribution*

In addition to knowing the rate of convergence to the SAD, it is helpful to know how far from the SAD one has begun. Duncan and Duncan (1955) made the first attempt at this (originally in the context of developing a way to measure the degree of racial segregation in the distribution of white and nonwhite populations in a city); their Δ metric was simply

$$\Delta = \sum_{a=1}^{A} | n_a - \mathbf{w}_a |, \tag{4.12}$$

which is just the summed difference between the current age distribution and the SAD. Keyfitz (1968) later proposed a similar metric that was equal to Eq. (4.12) multiplied by 0.5. The problem with this metric is that the units are arbitrary, they depend on A and the overall magnitude of $\boldsymbol{n}_0$, so Δ is not useful for making comparisons across different populations.

Cohen (1979) took a more comprehensive approach to the problem, and considered the "path" that a population must travel between the initial conditions and the SAD. Looking back at Eq. (4.11b), he noted that the left-hand side approaches zero in the limit as time approaches infinity. The left-hand side of that equation essentially measures the difference between the SAD and $\boldsymbol{n}_0$, much as does Δ, but also accounting for the effect of λ_1 on the size of the population at t. Cohen (1979) proposed simply summing that difference over time as the population approaches the SAD in order to obtain the cumulative distance D_1:

$$s = \lim_{t\to\infty}\sum_{t=0}^{T}\left(\frac{\boldsymbol{n}_t}{\lambda_1^t} - c_1\boldsymbol{n}_0\right) \quad \text{(Eq. 4.13a)}$$

$$D_1 = \sum_{a=1}^{A} | s | . \quad \text{(Eq. 4.13b)}$$

This metric captures the idea that a population that is very far from the SAD will cycle through many oscillations in the approach to asymptotic conditions. D_1 sums up the deviation at each point in that approach, so a transient that has higher-amplitude cycles (larger differences from asymptotic behavior) and longer lasting cycles will have larger values of D_1. However, if you think about how the calculation of D_1 will proceed, notice that D_1 can have either positive or negative values, and positive deviations from SAD in one time step could be cancelled out by negative deviations later on. So some types of oscillations will lead D_1 to be smaller than it should be. To correct this, Cohen (1979) also proposed a second statistic based on the absolute value of the difference:

$$r = \lim_{t\to\infty}\sum_{t=0}^{T}\left|\frac{\boldsymbol{n}_t}{\lambda_1^t} - c_1\boldsymbol{n}_0\right| \tag{4.14a}$$

$$D_2 = \sum_{a=1}^{A} | r | . \tag{4.14b}$$

Unfortunately Cohen (1979) was only able to work out an analytical solution to D_1, not D_2. It involves a quantity $\boldsymbol{B}$, known as the Perron projection matrix:

$$\boldsymbol{B} = \lim_{t\to\infty} \frac{\boldsymbol{A}^t}{\lambda_1^t} = \frac{\boldsymbol{w}_1 \boldsymbol{v}_1'}{\boldsymbol{v}_1 \boldsymbol{w}_1'}. \tag{4.15a}$$

Then the solution is

$$D_1 = (\boldsymbol{Z} - \boldsymbol{B})\, \boldsymbol{n}_0, \tag{4.15b}$$

where

$$\boldsymbol{Z} = (\boldsymbol{I} + \boldsymbol{B} - \boldsymbol{L}/\lambda_1)^{-1}, \tag{4.15c}$$

where $\boldsymbol{I}$ is the identity matrix.

The advantage of Cohen's distances is that they measure distance "as the road turns," capturing the oscillatory trajectory of the population, whereas Δ only captures the one-step "as the crow flies" distance. It is possible that a population could have a relatively small *crow-flies* distance at some point during an oscillatory transient, yet still have considerably more oscillations in its future before reaching asymptotic dynamics. As an example, we calculated Δ, D_1, and D_2 for the transient shown in Fig. 4.2. As we suggested, Δ oscillates greatly (with the same *frequency* and *phase* as the population time series), D_1 is generally declining but also oscillates initially, while D_2 decreases monotonically as the oscillations dampen (Fig. 4.8).

There is one additional crow-flies metric that has a useful application despite a few limitations. White et al. (2013a) suggested decomposing $\boldsymbol{n}_0$ into a component that is parallel to $\boldsymbol{w}_1$ (the SAD) and a component that is orthogonal to it (analogously, imagine if $\boldsymbol{n}_0$ were a two-dimensional vector in a Cartesian plane, one could determine the horizontal and vertical components that comprise that vector). This is done by projecting $\boldsymbol{n}_0$ onto $\boldsymbol{w}_1$ using the formula for the angle between two vectors (see Box 4.3):

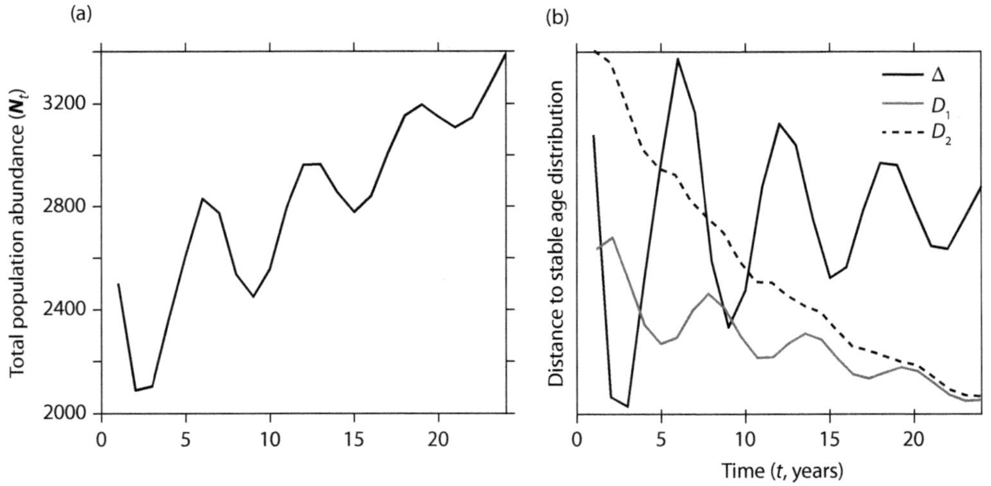

Fig. 4.8 Metrics of convergence to the stable age distribution (SAD). (a) The transient population dynamics depicted in Fig. 4.2b. (b) Duncan and Duncan's (1955) D metric (Eq. (4.13), solid curve) and Cohen's (1979) D_1 (Eq. (4.14), gray curve) and D_2 metrics (Eq. (4.15), dashed curve) calculated for each year in the same time period. The units for the metrics are arbitrary and not shown.

$$\cos\theta = \frac{\boldsymbol{w}_1 \cdot \boldsymbol{n}_0}{\|\boldsymbol{w}_1\| \, \|\boldsymbol{n}_0\|} \tag{4.16a}$$

$$\alpha_0 = \|\boldsymbol{n}_0\| \cos\theta = \frac{\boldsymbol{w}_1 \cdot \boldsymbol{n}_0}{\|\boldsymbol{w}_1\|}, \tag{4.16b}$$

θ is the vector angle (in age-space) between the initial age distribution $\boldsymbol{n}_0$ and the SAD ($\boldsymbol{w}_1$) and α_0 is the scalar projection of $\boldsymbol{n}_0$ onto $\boldsymbol{w}_1$. This means that the component of $\boldsymbol{n}_0$ that is parallel to $\boldsymbol{w}_1$, termed $\boldsymbol{n}_{||}$, is

$$n_{||} = \alpha_0 \boldsymbol{w}_1 \tag{4.17a}$$

and the component orthogonal to $\boldsymbol{n}_0$ is

$$n_{\perp} = \boldsymbol{n}_0 - \boldsymbol{n}_{||}. \tag{4.17b}$$

This solution was also derived by Haridas and Tuljapurkar (2007), although they took a different approach by projecting $\boldsymbol{n}_0$ onto $\boldsymbol{v}_1$ instead. The value of the approach we have shown here is that α_0 (or θ) can be interpreted as the degree of similarity between the initial conditions and the SAD.

Box 4.3 VECTOR PROJECTION

The direction and magnitude of any length-n vector $\boldsymbol{p}$ can be described in terms of its relationship to a second length-n vector $\boldsymbol{q}$ by splitting $\boldsymbol{p}$ into a component that is parallel to $\boldsymbol{q}$ and the component that is orthogonal (i.e. perpendicular) to $\boldsymbol{q}$ (Fig. 4.9). The component that is parallel to $\boldsymbol{q}$ is termed the *projection* of $\boldsymbol{p}$ onto $\boldsymbol{q}$. The length (magnitude) of the projection onto $\boldsymbol{q}$, p_1, can be found using the trigonometric definition of a cosine:

$$p_1 = |\, \boldsymbol{p} \,| \cos\,\theta, \tag{B4.3-1}$$

where θ is the angle between $\boldsymbol{p}$ and $\boldsymbol{q}$. It can be found using the dot product (Box 3.6):

$$\cos\theta = \frac{\boldsymbol{p} \cdot \boldsymbol{q}}{\|\boldsymbol{p}\| \, \|\boldsymbol{q}\|}. \tag{B4.3-2}$$

The magnitude p_1 is also called the *scalar projection* of $\boldsymbol{p}$ onto $\boldsymbol{q}$.

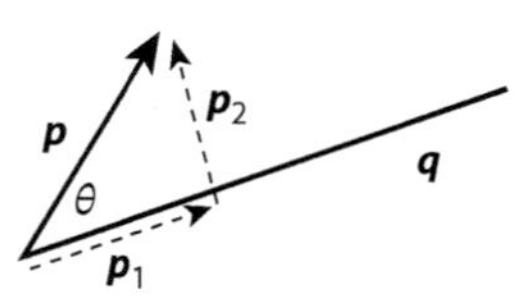

Fig. 4.9 Vector projection of $\boldsymbol{p}$ onto $\boldsymbol{q}$. Dashed arrows indicate the components of $\boldsymbol{p}$ that are parallel to ($\boldsymbol{p}_1$) and orthogonal to ($\boldsymbol{p}_2$) $\boldsymbol{q}$. $\boldsymbol{p}_1$ is the projection of p onto q.

Note that we have seen a vector projection earlier: the real parts of the complex second and third eigenvalues are projections of each two-dimensional vector.

The reason that α_0 is useful as a crow-flies distance is that we can calculate the initial trajectory of the population at $t = 0$. Although the population will asymptotically increase (or decrease) at rate λ_1, the trajectory at any point in time could be much different, and likely even of opposite sign. The initial trajectory at $t = 0$ is calculated by decomposing the growth rate into the asymptotic component (which is proportional to the similarity of $\boldsymbol{n}_0$ to the SAD) and the component due to all of the other eigenvectors:

$$\lambda_{initial} = \alpha_0 \lambda_1 + \sum_{a=1}^{A} \boldsymbol{L}\boldsymbol{n}_{\perp}. \tag{4.18}$$

This is a useful quantity to know, as we see in the example with marine protected areas in Section 4.2.3.

4.2.4.3 *Example: adaptive management of marine protected areas*

When managers close a portion of the ocean to fishing in order to create a marine protected area (MPA), there is an expectation that fish populations in the MPA will increase in abundance. For example, a meta-analysis by Lester et al. (2009) suggested that fish density is, on average, 166 percent higher inside no-take MPAs, and a separate meta-analysis by Halpern and Warner (2002) suggested that MPAs have "rapid and lasting" effects that could materialize in one to three years. These papers created an expectation that fish populations would always show rapid increases immediately after implementation. That expectation was not met in California, where a state-wide network of MPAs was put into place between 2007 and 2013 (Botsford et al. 2014a), and the abundance of many fished species, particularly nearshore rockfishes (*Sebastes* spp.), did not immediately respond (Starr et al., 2015; Nickols et al., 2019). This led White et al. (2013a) to investigate how a sudden cessation of fishing would affect the short-term (transient) dynamics of a fish population.

Presuming that a population has been fished heavily for some time, its abundance would be low enough to avoid density-dependent effects, and a linear age-structured model would be applicable. Most fishing is size, and thus age selective (i.e. either fishing regulations or the fishing method itself limit the harvest to larger, older fish), so the main effects of fishing on a population would be: (a) to truncate the age distribution, so the mean age of reproduction is shifted downward; and (b) to reduce overall abundance and spawning output. When fishing stops, the survival rate for older age classes immediately increases, but it takes time for those age classes to "fill in" and approach the SAD. Because the initial age distribution $\boldsymbol{n}_0$ at the time of MPA implementation will have been truncated by prior fishing, and will thus be quite different from the SAD, we should expect to see noticeable transient dynamics after MPA implementation, including the possibility of an initial decline in overall fish abundance, even if removing fishing leads to $\lambda_1 > 1$ overall.

To illustrate the different way of describing the initial transient phase of growth in age structure, White et al. (2013a) calculated D_1, ρ, θ, and $\lambda_{initial}$ for a representative rockfish species, the kelp rockfish *Sebastes atrovirens*, under a number of different hypothetical pre-MPA fishing rates (Fig. 4.10). They found that higher fishing rates would lead to greater deviations from the SAD (not surprisingly) and correspondingly longer transients with greater oscillations, including decreases in overall population size, before the population reached the SAD after five to ten years. These patterns were exacerbated if the species had a later age of maturity, which compressed reproduction into fewer age classes (consistent with the echo effect and Taylor (1979)). These results helped set expectations for the adaptive management of California MPAs, suggesting that initial flat or even negative trends in abundance could be expected, even if the populations would eventually be increasing. Of course the actual population dynamics are somewhat more complicated than those modeled by White et al. (2013a), so subsequent efforts used more complex models that included year-to-year stochasticity in recruitment to better represent short-term dynamics (White et al., 2016; Nickols et al., 2019).

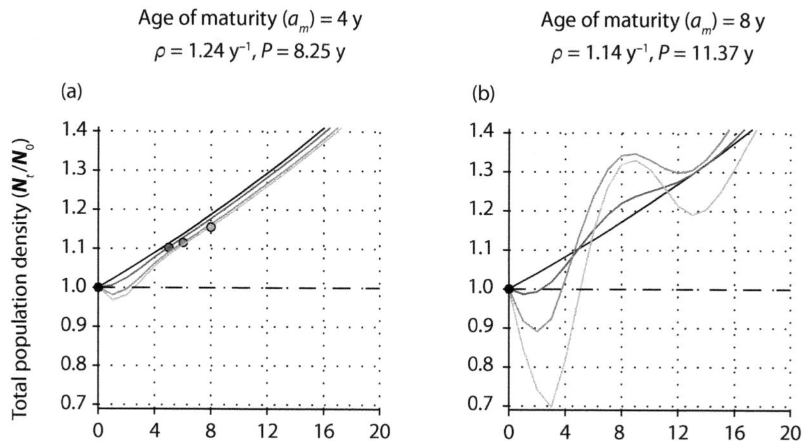

(a)

F	D	θ	$\lambda_{initial}$
0.0	0.00	0.0°	1.02
0.2	0.49	8.7°	1.01
0.8	1.26	26.6°	0.98
1.6	1.58	36.8°	0.97

(b)

F	D	θ	$\lambda_{initial}$
0.0	0.00	0.0°	1.02
0.2	0.53	11.4°	0.99
0.8	1.64	38.0°	0.92
1.6	2.44	55.7°	0.85

Fig. 4.10 Transient dynamic responses to removal of fishing by implementation of an MPA (White et al., 2013a). A kelp rockfish population that was subject to fishing at rate F (y^{-1}) in the past, that ceases at $t = 0$. The table below gives the values of D_2 = the distance from the current age vector to the stable age distribution (Eq. (4.14b)), θ = the angle between the age vector and stable age distribution (Eq. (4.17a), and $\lambda_{initial}$, the initial, transient trajectory of the population after $t = 0$).

4.3 Transient responses to ongoing environmental variability

In a sense, the scheme we have described so far in this chapter, in which populations respond to individual departures from equilibria, then return to a constant quiescent state, does not occur in many populations (Hastings 2001, 2010). Rather, most populations are subject to a constant, nontrivial level of environmental variability, hence we do not clearly see a complete return to equilibrium from each perturbation. However, the consequences of the response(s) to constant buffeting by environmental noise are controlled by the same dynamics as responses to single departures from equilibrium. Thus, these responses will be controlled by the values of the largest eigenvalues in a way that is consistent with our findings thus far in this chapter.

To describe this problem for small enough stochastic influences, we switch to a different mathematical approach: characterizing the responses of systems such as populations to random noise. Mathematically, in this problem, as in the case of stability (Section 2.4) we can get much further if the system is linear, so again for populations not near zero abundance we will linearize the models about equilibrium. This means that we will cover the question of what determines the equilibrium of an age-structured population in this section, for use here and in later chapters (primarily Chapter 7).

4.3.1 *Determining the equilibrium of a nonlinear age-structured population*

In Chapter 3 and the first part of this chapter, we focused on populations that were at sufficiently low abundance that we did not need to account for density-dependent processes, and we could use linear models. Here we consider populations that are not at such low abundances, so we must turn to nonlinear models that include density dependence (as we did for simple models in Section 2.4.1).

In many species, density dependence affects either the production of new offspring (e.g. due to competition for resources among reproductive-age adults) or the survival of those offspring in early life, when they are most abundant, small in size, and most vulnerable. Density-dependent survival could be due to mechanisms such as cannibalism by adults, or competition for refuges from predators. We will discuss more about the implications of the different types of density dependence in Chapter 7. For now it is enough to say that the abundance of new offspring at time $t + 1$ is a nonlinear function of the number of adults present in time t (possibly as a weighted sum over age as in the influence functions, Section 3.2, Eq. (3.13)).

To write out an expression for the equilibrium population size, $\boldsymbol{n}^*$ (the vector of abundances of each age class), we could use the method from Chapter 2 (Section 2.4.2), setting $\boldsymbol{n}_{t+1} = \boldsymbol{n}_t = \boldsymbol{n}^*$ and then solving for the vector $\boldsymbol{n}^*$. However, if density dependence affects only the first age class, there is a somewhat simpler approach. Because the abundance of the older age classes, $n_{a,t}$ will be a linear function of the number of offspring a years in the past, $n_{1,t-a}$. the abundance of that first age class will set the dynamics for the rest of the population, and we can write the model solely in terms of that age class using a renewal equation (as we did for linear models in Section 3.2). New offspring are often termed "recruits" (particularly in fisheries), so we rename the first age class, $n_{1,t}$ to be recruitment, R_t, dropping the subscript related to age. The number of recruits at time t will depend on the number of individuals in each older age class, $n_{a,t}$ times the annual reproductive rate of that age class, b_a. We use a similar per capita constant c_a to reflect the relative effect of an individual of age a on density-dependent recruitment. For example, c_a could reflect the relative propensity of an individual of age a to cannibalize recruitment-aged individuals. Thus we write R_t as the product of annual reproduction, B_t, and a recruitment survival function, f, that also depends on effective population size, C_t

$$R_t = B_t f(C_t) = \left[\sum_{a=1}^{A} b_a n_{a,t}\right] f\left(\sum_{a=1}^{A} c_a n_{a,t}\right), \tag{4.19}$$

where A is the number of age classes, and f is the nonlinear density-dependent survival function, which depends on a weighted sum over age classes, similar to the form of reproduction. We can express the abundance of individuals at age a at time t as

$$n_{a,t} = \sigma_a R_{t-\tau}, \tag{4.20}$$

where we have used σ_a to represent cumulative survival to age a, as we did in Eq. (3.12) in the linear renewal equation. We can then rewrite Eq. (4.19) as

$$R_t = \left[\sum_{a=1}^{A} b_a \sigma_a R_{t-a}\right] f\left(\sum_{a=1}^{A} c_a \sigma_a R_{t-a}\right). \tag{4.21}$$

To obtain the conditions for equilibrium, we set $R_{t+1} = R_t = R^*$,

$$R^* = \left[\sum_{a=1}^{A} b_a \sigma_a R^*\right] f\left(\sum_{a=1}^{A} c_a \sigma_a R^*\right) = \left[\sum_{a=1}^{A} \varphi_{b,a}\, R^*\right] f\left(\sum_{a=1}^{A} \varphi_{c,a}\, R^*\right) = \Phi_b R^* f\left(\Phi_c R^*\right), \tag{4.22a}$$

where we have defined the influence functions for reproduction and effective population size

$$\varphi_{b,a} = b_a \sigma_a \text{ and } \varphi_{c,a} = c_a \sigma_a, \tag{4.22b}$$

as well as their cumulative totals

$$\Phi_b = \sum_{a=1}^{A} \varphi_{b,a} \text{ and } \Phi_c = \sum_{a=1}^{A} \varphi_{c,a}. \tag{4.22c}$$

Φ_b and Φ_c represent the cumulative influence of older ages on reproduction and density dependence, respectively. Notice that Φ_b is actually the same as R_0, the lifetime reproductive rate, which we defined earlier (Eq. 3.25).

From Eq. (4.23a), the equilibrium R^* is found by solving the equation

$$1 = \Phi_b f\left(\Phi_c R^*\right). \tag{4.23}$$

In other words, the equilibrium, R^*, is at the value of R for which the lifetime reproductive rate (>1) is exactly large enough to compensate for density-dependent recruit survival (<1). This relationship is analogous to the Euler–Lotka equation (Eq. (3.24)) for determining the growth rate of a linear population. Equation (4.23) turns out to have a useful graphical interpretation: on a plot of the density-dependent relationship between egg production (on the horizontal axis) and the number of recruits that will survive from those eggs (i.e. a plot of the function $\Phi_b R f(\Phi_c R)$ versus $\Phi_b R$), the equilibrium is found by plotting a line from the origin with slope $1/\Phi_b$ and finding the intersection with f (Fig. 4.11). At that point, each of the R^* recruits will produce Φ_b offspring in its lifetime, and those offspring will survive with probability $f(\Phi_c R^*)$. The curves in Fig. 4.11 are also

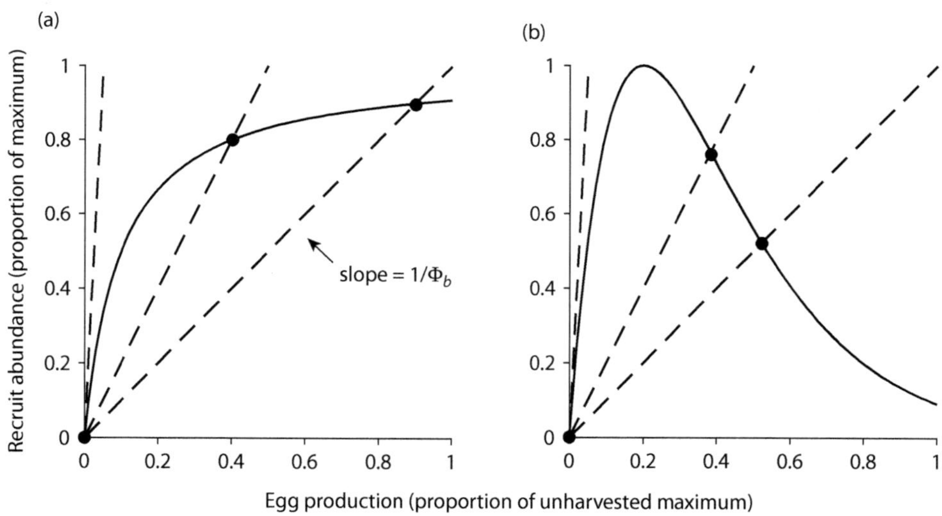

Fig. 4.11 Example of the graphical representation of the equilibrium recruitment for a population with density-dependent recruitment (Sissenwine and Shepherd, 1987). The solid curves are the egg-recruit curves, either (a) Beverton–Holt ($\alpha = 0.5, \beta = 10$) or (b) Ricker ($\alpha = 0.5, \beta = 5$). The dashed lines have slopes of 1/(lifetime egg production), i.e. $1/\Phi_b$. Equilibria are shown by dots, and they move to the left as fishing reduces LEP. When 1/(lifetime egg production) exceeds the slope of the egg-recruit curve at the origin, the equilibrium recruitment $R^* = 0$, meaning the population will go extinct.

called the egg-recruit curve in a fisheries context. In some fisheries, it is easier to estimate the biomass of spawning adults rather than the actual number of eggs produced, and since those two quantities are assumed to be proportional to one another, spawning biomass is taken to be a proxy for egg production and the curve is called the spawner-recruit or stock-recruit curve. We will make further use of egg-recruit curves in Chapter 11.

The graphical interpretation of the equilibrium in Fig. 4.11 was originally described by Sissenwine and Shepherd (1987), and has a very useful interpretation for harvested populations. As the harvest rate increases, the probability of survival to older ages, σ_a, and thus lifetime reproduction, Φ_b, will decrease. This means that the line with slope $1/\Phi_b$ which determines the equilibrium will become steeper, and the equilibrium R^* will move to the left on the density-dependent curve (Fig. 4.11). If the $1/\Phi_b$ line becomes steeper than the slope of the density-dependent curve at the origin, then $R^* = 0$ and the population would eventually become extinct (recall, this a condition for equilibrium, not a statement of the current state).

Note that this is another example of the key concept of replacement referred to for linear models in Chapter 3. Once again replacement is a key concept regarding persistence. The difference is that here R_0 is not required to be greater than 1.0, but rather is required to be greater than a higher value, the inverse of the slope of the egg-recruit function at the origin. We will revisit this idea in the context of harvest in Chapter 11. For now, we will rely on this description of the equilibrium to help to understand how populations will respond to environmental variability.

4.3.2 *The frequency response of a population*

Having calculated the equilibrium (Eq. (4.23)) we can now linearize the model at the equilibrium to determine how it responds to random environmental variability. However, we will be taking a different approach from that used thus far: instead of making statements about actual solutions to model equations in terms of abundance versus time, we will be keeping track of what happens to the amount of variability at each frequency in population abundance and in the environment. We will do this by taking advantage of a mathematical result whose antecedents lie in early Babylonian and Greek mathematics, but which was developed more recently, in the eighteenth century, culminating in 1807 in what is called the Fourier series. This result says that a time series over any specified time period can be exactly duplicated by a weighted infinite sum of sines and cosines, called a Fourier series, where the sum is taken over a set of sines and cosines with integer periods relative to the length of the time series. The weighting factors for each cosine or sine of a certain frequency represent the "frequency content" (i.e. amount of variability at frequency f). Another way to view this is as a linear multiple regression problem where the dependent variable is the observed time series, and the independent variables are the sine and cosine functions of all possible frequencies. We can look at which sine and cosine parameters are most correlated with the data and these tell us the frequency content. We can plot this either as a function of frequency (cycles per unit time) or, equivalently, as a function of its inverse, period ($T = 1/f$, the amount of time taken for one cycle). In ecology there is great interest in the time scales of variability (Levin, 1992), which would correspond to the period content (as opposed to the frequency content) of time series (Box 4.4).

Figure 4.12 is an example of a possible spectral content of a time series. The time series in Fig. 4.12a is the sum of all of the time series in Fig. 4.12c (all with the same means, but plotted in a stacked fashion): three sinusoids (sines or cosines, which one is not important here) of period 30 y, 10 y, and 4 y, and a time series of white noise (uncorrelated in time, with equal variance at all frequencies). The spectral content of the signal in Fig. 4.12a

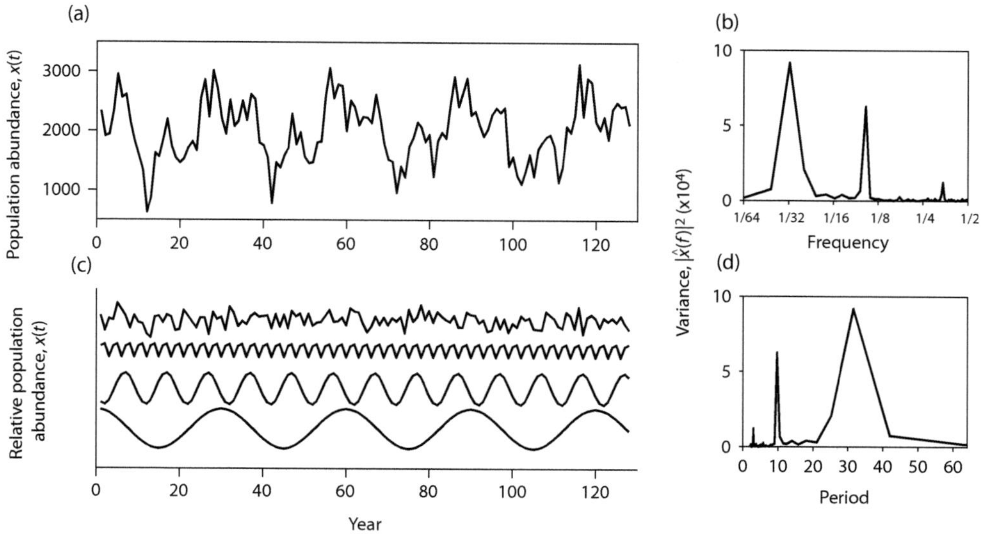

Fig. 4.12 Examples of Fourier spectra. (a) A simulated time series, $x(t)$, consisting of the sum of the three sine waves and one series of normally distributed random variates (white noise) in (c). The signals in (c) are offset for illustration but all had the same mean value. Panels (b) and (d) show the frequency spectrum, $|\hat{x}(f)|^2$, of the time series plotted in terms of both frequency (b) and period (d). Note the peaks corresponding to the three sinusoidal component in (c).

can be plotted either as a function of frequency (Fig. 4.12b) or period (Fig. 4.12d). Note that the sinusoids are present in both, and between them there is low, roughly constant variance from the white noise.

Box 4.4 FOURIER TRANSFORMATION AND THE FREQUENCY DOMAIN

In most of this book we describe population dynamics in terms of change happening over time, e.g. changes to population density n over a specified time period. This is a description in the *time domain*. However, we can also describe n in terms of its frequency content. This is a description in the *frequency domain*. The method for translating between the time and frequency domains is the Fourier transform. For a certain pattern of variability in the time domain, $x(t)$ (this could be population abundance over time, or some environmental variable measured over time), the Fourier transform provides the corresponding pattern in the frequency domain, $\hat{x}(f)$:

$$\hat{x}(f) = \int_{-\infty}^{\infty} x(t)e^{-2\pi i f t}dt, \tag{B4.4-1}$$

where f is the frequency (with units time^{-1}). In Box 4.1 we saw that $e^{if} = \cos f + i \sin f$, so a simple way to understand the Fourier transform is that it essentially decomposes the time-domain pattern $x(t)$ into a weighted sum of sine and cosine waves, with larger weights for waves with frequencies that are strongly represented in the time series $x(t)$. The set of sine or cosine waves that are summed (integrated) to produce $x(t)$ is termed the *Fourier series*.

The frequency domain representation, $\hat{x}(f)$, describes the pattern in terms of the amplitude of the component wave that has frequency f (Fig. 4.9). A plot of $\hat{x}(f)$ versus f is termed the *spectrum* of the signal. In the context of electromagnetic signals, this is usually called

(Continued)

Box 4.4 CONTINUED

the *power spectrum*, because the units on the y-axis are $|\hat{x}(f)|^2$, which is power (energy per unit time). In population dynamics, $|\hat{x}(f)|^2$ is the equivalent of variance. Therefore $|\hat{x}(f)|^2$ can be integrated over a range of frequencies to determine how much of the variance in the original signal $x(t)$ occurs in that frequency band.

We frequently transform time series from the time domain to the frequency domain where we do various frequency-dependent calculations, then we need to transform the results of the calculations back to the time domain (Fig. 4.10). For that we use the inverse Fourier transform,

$$x(t) = \int_{-\infty}^{\infty} \hat{x}(f) e^{-2\pi i f t} df. \tag{B4.4-2}$$

As might be expected, the inverse Fourier transform of the Fourier transform of a function of time will return the function of time itself.

These two expressions are in continuous time and frequency. For discrete time series with values x_t at times $t = 1, 2, \ldots, T$, we use the discrete Fourier transform (DFT), which is

$$\hat{x}_f = \sum_{t=0}^{T-1} x_t e^{-2\pi i f t / T}. \tag{B4.4-3}$$

In modern applications, DFT is implemented using a type of numerical algorithm known as the fast Fourier transform (FFT), a term that readers will typically encounter in the context of frequency analysis.

The primary value of representing a time series as the amount of variability (variance) at each frequency (actually a variance density, variance per unit frequency) is that for linear (or linearized) population models, the effect of random variability on a population can be determined as simply a scalar multiplier at each frequency. In other words, sine waves pass through linear systems undistorted, and for each frequency they are simply multiplied by a different constant, indicating the sensitivity of that population to that frequency. This allows us to view populations as "filters" that have different levels of sensitivity to different frequencies of environmental variability. Different populations are described in terms of their different "frequency responses," also known as "transfer functions," (the ratio of the output variance to the input variance for each frequency; Fig. 4.13). The frequency response describes how sensitive the system (population) is to variance in environmental variability at each frequency. It determines whether and how the population dynamics will dampen variability at some frequencies, but amplify variability at other frequencies. Specific random environmental signals affecting a population can have particular characteristic frequency content (e.g. the El Niño-Southern Oscillation varies with a period of 3–7 y; Wang and Wang, 1996; Torrence and Compo, 1998). A transfer function allows us to determine whether a population will be affected by that environmental signal. Moreover, we can project the effects of changes in the frequency of ENSO under future climate change (see Schmidt et al., 2018 as an example). Such changes are projected by global climate models (Timmermann et al., 1999), and have been seen in the paleo record (Cobb et al., 2003).

Frequency-domain signals are often referred to in terms of color, by way of analogy to the visible light spectrum, that ranges from low-frequency electromagnetic waves (red) to high-frequency electromagnetic waves (blue). Thus a signal in which higher frequencies

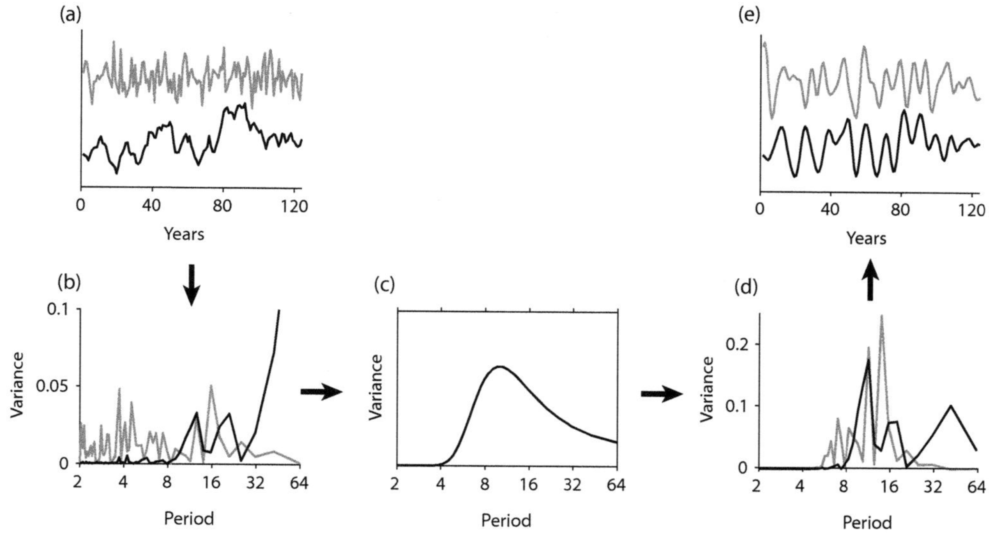

Fig. 4.13 Examples of a filter acting on environmental time series. (a) Two hypothetical time series with random variability, both white noise (gray curve) and red noise (black curve). (b) The corresponding frequency spectra. (c) Both spectra were filtered using a filter with a transfer function with a peak at ten years. (d) The spectra after multiplication in the frequency domain. (e) The filtered result translated back into the time domain.

dominate is termed "blue noise," and those in which lower frequencies dominate is termed "red noise." Correspondingly, a blue noise signal would vary rapidly in the time domain, and a red noise signal would vary slowly. A signal with equal representation of all frequencies is "white noise" (Fig. 4.13). A white noise signal is one with no intra-series correlation in the time domain (i.e. the values at each time are independent). Most environmental variability is red (or "pink"); time series of temperature, wind or water velocity, storm intensity, and so forth are typically described as "$1/f$ noise" because the spectrum $\hat{x}(f)$ is proportional to one divided by the frequency, f, or $1/f$ raised to some exponent (Denny et al., 2004; Vasseur, 2007).

Figure 4.13 illustrates the concepts of frequency domain filtering and use of the Fourier transform and the inverse Fourier transform. The filter there is meant to be a generic example of a filter, not necessarily a population. The two time series to be filtered are in Fig. 4.13a, a white noise signal shown in gray and a red noise signal shown in black. The Fourier transforms of these are plotted versus period (rather than frequency) in Fig. 4.13b. The red noise signal shows substantial variability at long periods, and the white noise signal shows a variable, but basically flat spectrum (the latter is quite noisy because 120 years is a relatively short period over which to expect a randomly varying signal to have exactly equal variance at all frequencies). The *transfer function* of the filter, shown in Fig. 4.13c, does not pass periods lower than 4 y, is most sensitive to period 10 y, and has declining sensitivity to longer periods. Consequently, the spectrum of the output of the filter (Fig. 4.13d) for red noise has no variance below period 4, high variance between periods 8 and 20, and high variance reflecting the large variance in the input for periods greater than 32 y. Theoretically (ideally), the expected shape of the output spectrum is the period by period product of the transfer function and the input spectrum. It does not look exactly like that because this is a single sample of a random process. The output spectrum for the white noise should ideally have the same shape as the transfer function since the input (ideally) has equal variance at all periods. The white noise and red noise output signals look quite similar primarily because of the strong sensitivity to period 10 y in the transfer function: each signal appears to be nearly cyclic with period between 10 y and 16 y. Finally, the inverse transform of Fig. 4.13d clearly shows that the filtering has removed high frequencies from Fig. 4.13a.

When we represent populations as filters, the input can be an environmental signal that drives survival at some age, or causes variability in growth rate, as examples, and the output could be abundance, catch, or recruitment.

4.3.3 *Cohort resonance*

In ecology, there has long been interest in the mechanisms by which environmental variability produces fluctuations in population abundances (e.g. Hjort, 1914; Andrewartha and Birch, 1954). In general, the understanding has been that environmental factors can produce extrinsic variability in the population, such as by producing particularly "good" or "bad" years for a cohort of recruits. However, the effects of internal population dynamics, such as density-dependent interactions intrinsically leading to cyclic population dynamics, have been considered separately (e.g. Nisbet and Gurney, 1982; Turchin, 2003), and ecologists have discussed the relative importance of these two sources of variability (Grenfell et al., 1998; Ims et al., 2008). In the late 1990s attention began to turn to the idea that intrinsic population dynamics might shape the way extrinsic environmental forcing produces variability in the population. For example, Bjørnstad et al. (1999) examined the pattern of variability in abundance of Atlantic cod (*Gadus morhua*) on the southern Skagerrak coast of Norway over the period 1921–94 (Fig. 4.14a). A plot of the spectral density revealed that the variability in the population data was greater at two spectral peaks, one at very low frequency (periods >40 y) and a second at a period of 2–3 years (Fig. 4.14b; Fig. 1d of Bjørnstad et al., 2004). There were no known environmental causes of these peaks. They then calculated the spectral sensitivity (i.e. the transfer function) of this cod population, which, being in relatively warm waters, matured at a relatively young age of 2–3 years. This transfer function also had peaks at both very low periods and near period 2 (Fig. 4.14c; Fig. 3b of Bjørnstad et al., 2004), indicating that the frequency content of the catch time series was entirely due to the population dynamics, presumably being driven by white noise.

Bjørnstad et al. noted that, in general, the spectral sensitivity of age-structured populations have two peaks, one at a frequency of the inverse of the maturation time T, and the other at very low frequencies. They called this effect *cohort resonance*, drawing attention particularly to the high frequency (short period) part. They noted a concern that the greater sensitivity to low frequencies could obscure environmentally induced slow changes in abundance, an effect they called a *cloaking effect*.

Worden, et al. (2010) examined the cohort resonance effect in the context of two species of Pacific salmon, using a matrix model that allowed environmental variability in individual growth rate as well as survival. They showed that in addition to there being two modes of greater sensitivity, reduction of overall survival in a population, such as by fishing or other extra mortality, led to an increase in the frequency selectivity at the two peaks, as well as an increase in overall variability. This was important because it aided in the understanding of observations that variability in fish populations increased with fishing (Hsieh, et al., 2006). They also showed how the frequency response of a population, in addition to depending on life history, depended on the inputs being variation in

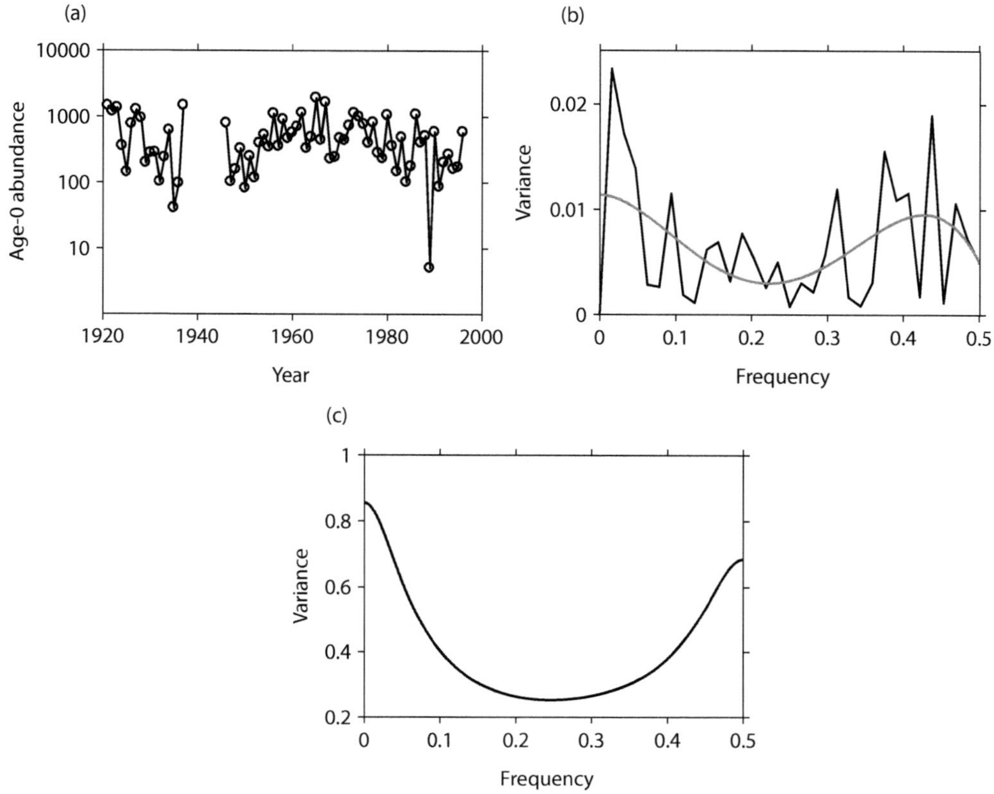

Fig. 4.14 The southern Skagerrak population of Atlantic cod (*Gadus morhua*). (a) Abundance data from annual beach seine surveys in Skagerrak, Norway (note gap during World War II; data from Bjornstad et al., 2004). (b) Frequency spectrum for the post-1946 data (black curve; gray curve is a polynomial spline fit to the spectrum). (c) The transfer function of a population model of this species from Bjornstad et al. (2004).

growth rate or survival at certain ages, and the outputs being abundance, recruitment, egg production, or catch (these were explored in Botsford et al., 2014b).

4.3.3.1 *Analysis of cohort resonance*

To understand the dynamics of this response of a population to random variability via cohort resonance, we consider a population that is at equilibrium (Section 4.3.1, Fig. 4.11), and then perturbed away from that equilibrium by an ongoing series of disturbances. To represent that effect, we take the same approach as in stability analysis, and linearize the model by defining a small deviation from the equilibrium, $\Delta\boldsymbol{n} = \boldsymbol{n}^* - \boldsymbol{n}_t$. We follow the same procedure as in Chapter 2, using a Taylor series expansion to find an expression for the dynamics of $\Delta\boldsymbol{n}$. For the simple nonlinear model without age structure in Chapter 2, $N_{t+1} = F(N_t)$, and the linearized version was $\Delta N_{t+1} = \Delta N_t\, F'(N^*)$, so the stability depended on the slope of the nonlinear function F at the equilibrium. For the age-structured case, we can write out the matrix model in a similar shorthand, $\boldsymbol{n}_{t+1} = F(\boldsymbol{n}_t)$, where the nonlinear function F performs the same dynamics we described previously in the context of recruits (Eqs. 4.17–4.20): the number of new age-1 individuals at time $t + 1$ is a density-dependent function f of number of eggs produced at time t (Fig. 4.8), and the survival of older age classes is density independent. The linearization of this model is similar to the nonstructured case:

$$\Delta \boldsymbol{n}_{t+1} = \boldsymbol{J}\Delta \boldsymbol{n}_t + \boldsymbol{H}\xi_t, \tag{4.24}$$

where $\boldsymbol{J}$ is the Jacobian matrix. A Jacobian matrix fills the role of the single partial derivative in the linearization of a simple model in Chapter 2. It is a matrix containing the partial derivative of each element of a vector function with respect to each of the elements of the vector. In other words, the elements of $\boldsymbol{J}$ are $J_{ij} = \partial \boldsymbol{F}_i / \partial n_j$, with the partial derivative evaluated at the equilibrium $\boldsymbol{n}^*$ (which we determined earlier using the renewal equation for recruits in Eq. (4.23)).

This has a similar effect to the effect of linearization of the nonstructured model: the stability at the equilibrium depends on the slope of $F(\boldsymbol{N}_t)$. Writing out the elements of $\boldsymbol{J}$ (Box 4.5), we see that when we linearize $F(\boldsymbol{n})$ at a particular equilibrium, we are essentially approximating $F(\boldsymbol{n})$ as a linear, Leslie matrix model $\boldsymbol{Ln_t}$, in which the relationship between eggs produced and the number of age-1 individuals produced (i.e. the first row of the Leslie matrix) is given by the slope of F at the equilibrium. This means that the Jacobian, the derivative of $F(\boldsymbol{n})$ with respect to $\boldsymbol{n}$, takes the form of a Leslie matrix. Because the elements of the Jacobian resemble the elements of the Leslie matrix, the eigenvalues of that matrix will have the same effect that we described earlier in this chapter. That is, if a population at equilibrium is perturbed (by a small amount so that the linear approximation is reasonable), the return to equilibrium will be similar to the transient dynamics we described in Section 4.3, including oscillations at a period determined by the second eigenvalue of the linearized Leslie matrix $\boldsymbol{J}$. In other words, the population will tend to have oscillations with a period equal to the generation time, T. Because these oscillations arise, perturbations to one cohort tend to echo T years later when that cohort becomes reproductively mature. That is why this phenomenon is known as cohort resonance. However, the value of the largest eigenvalue of the Jacobian also plays a role in cohort resonance in that its value determines the rate at which perturbations return geometrically to equilibrium, which is related to the greater sensitivity to lower frequencies in cohort resonance.

Box 4.5 THE JACOBIAN MATRIX FOR A LINEARIZED AGE-STRUCTURED MODEL

It is helpful to see an example of the Jacobian matrix for a linearized age-structured model to understand why it is useful. To take a simple example of a nonlinear model with age structure, we use a population with four age classes, with reproduction only in the two oldest ages, density-dependent recruitment to the first age class, and density-independent survival of older age classes:

$$\boldsymbol{n}_t = F(\boldsymbol{n}_{t-1}) = \begin{bmatrix} R(P_t) \\ sn_{1,t-1} \\ sn_{2,t-1} \\ sn_{3,t-1} \end{bmatrix}, \tag{B4.5-1a}$$

where s is adult survival and P_t is the reproductive output at time t. P_t is a function of the abundance of the two reproductive age classes,

$$P_t = \gamma_3 n_{3,t-1} + \gamma_4 n_{4,t-1}, \tag{B4.5-1b}$$

Where γ_3 and γ_4 are the per capita fecundity of age-3 and age-4 individuals, respectively. $R(P_t)$ describes the nonlinear density-dependent mortality of new recruits. We assume it follows a Beverton–Holt function, with initial slope α and asymptotic maximum density α/β:

$$R(P_t) = \frac{\alpha P_t}{1+\beta P_t}. \tag{B4.5-1c}$$

This model has an equilibrium $\boldsymbol{n}^*$ that is simply the stable age distribution multiplied by the equilibrium number of recruits,

$$\boldsymbol{n}^* = \begin{bmatrix} 1 \\ s \\ s^2 \\ s^3 \end{bmatrix} \frac{\alpha - 1/LEP}{\beta}, \tag{B4.5-2}$$

where LEP is the lifetime egg production of a new recruit, which in this case is simply the probability of survival to the two reproductive ages multiplied by the fecundity at each of those ages,

$$LEP = \gamma_3 s^2 + \gamma_4 s^3. \tag{B4.5-3}$$

To obtain the Jacobian, $\boldsymbol{J}$, of this system, each element $\boldsymbol{J}_{i,j}$ is the partial derivative of the ith element of $\boldsymbol{J}$ with respect to the jth value of $\boldsymbol{n}$,

$$\boldsymbol{J} = \begin{bmatrix} 0 & 0 & \gamma_3 \dfrac{1}{\alpha LEP^2} & \gamma_4 \dfrac{1}{\alpha LEP^2} \\ s & 0 & 0 & 0 \\ 0 & s & 0 & 0 \\ 0 & 0 & s & 0 \end{bmatrix}. \tag{B4.5-4}$$

The first row of $\boldsymbol{J}$ is the derivative of $R(P_t)$ with respect to each of the ages, evaluated at $\boldsymbol{n}^*$. In other words, it is the slope of the Beverton–Holt function at the equilibrium point multiplied by the equilibrium reproductive rate. The first two entries on that row are zero because those are nonreproductive ages. The subdiagonal entries are the derivates of sn_i with respect to n_i, so simply s. All other entries are zeros. Thus $\boldsymbol{J}$ has a form very much like a Leslie matrix.

There is one important difference between the Jacobian and the Leslie matrix: the first row of the Jacobian turns out to contain the first row of the Leslie matrix, but each term is multiplied by the derivative of the egg-recruit curve at equilibrium, i.e. $1/(\alpha\ \text{LEP}^2)$. This difference has a great effect on the overall variability about the equilibrium. If lifetime reproduction, Φ_b, is high, then the equilibrium will lie on the asymptotic, nearly horizontal portion of the function (Fig. 4.11). There, the slope of F approaches zero, so the value of the entries on the first row of $\boldsymbol{J}$ (the slope of the egg-recruit relationship) will also be small. Moderate fluctuations in egg production will lead to very small deviations in R_t, and thus small deviations in the other age classes as well. From Eq. (4.25), deviations from equilibrium will be dampened quickly. However, as lifetime reproduction declines, the equilibrium moves down onto the ascending, linear portion of the egg-recruit curve. The elements on the first row of $\boldsymbol{J}$ will grow larger (the slope of F is greater), so the system will not dampen perturbations as quickly. This effect underlies the increase in overall sensitivity caused by a decline in survival, such as through fishing (Worden, et al., 2010, Fig. 2 in Botsford et al., 2014b).

The other part of Eq. (4.24) is the product $\boldsymbol{H}\xi_t$, which is the stochastic variability at time t (for example, this could be a white-noise process, such as uncorrelated, normally distributed random numbers, or colored noise, such as an autocorrelated red-noise process). The matrix $\boldsymbol{H}$, the entries of which describe how ξ_t affects each age class at successive time lags (for example, environmental variability might affect age-1 survival only, so the entries of $\boldsymbol{H}$ corresponding to age class a would describe the lagged effect of a disturbance at time $t - a$). We will not delve here into the different possible forms of $\boldsymbol{H}$ (for examples, see Worden et al., 2010), instead we focus on how the structure of the model influences the Jacobian.

4.3.3.2 *Cohort resonance: effects of life history, fishing, and eigenvalues*

To illustrate how age-structured populations respond to environmental fluctuations via cohort resonance, it is instructive to compare two different types of species. Sockeye salmon (*Oncorhynchus nerka*), as we mentioned earlier, are nearly semelparous, typically reproducing at age 4 (at the latitude of southern Canada), with few individuals spawning at age 3 or age 5. For the purpose of illustration, we will make them obligate semelparous. By contrast, the Skagerrak stock of Atlantic cod can mature as early as age 2, but live for at least eight years and are iteroparous, reproducing annually, with egg production proportional to size at age. We assume that both species have a Beverton–Holt egg-recruit curve, such that the number of age-1 individuals produced is a saturating nonlinear function of the total number of eggs spawned that year (similar to the example model in Box 4.4 but with the appropriate different number of spawning age classes for salmon or cod, respectively). We simulated 100 different instances of population dynamics for both species, subject to random, normally distributed, white noise fluctuations in age-1 survival. Because white noise has equal variance at all frequencies, the resulting frequency responses represent the transfer functions.

As fishing increases, the mean values of population abundance decline (Fig. 4.15a, b), because the equilibrium recruitment would be moving to the left and down to lower values along a curve like Fig. 4.11a. As that occurs, the amount of variability relative to the mean is increasing with fishing (Fig. 4.15a, b), due primarily to the increasing slope of the egg-recruit function (Fig. 4.11a), which appears in the first row of the Jacobian (Worden, et al., 2010). The average frequency spectrum of the simulated time series (obtained by taking the Fourier transform) reveals a strong peak at a period of four years for sockeye but not for cod (Fig. 4.15c and d). The strong peak at period 4 in the spectrum of the salmon is due to a peak in the distribution of reproduction-at-age (Fig. 4.15e). The broad range of spawning for cod (Fig. 4.15f), on the other hand, leads to very little variance at short periods, and quite a bit of variability at longer periods (Fig. 4.15d). One can see in the time series (Figs. 4.15a and b) that the dominant time scale in the salmon remains at four years, while for cod, as fishing increases, there is an increase in variability and a reduction in the period at the short time scale, and some random, decadal-scale meandering. As fishing increases in the cod population (we assumed that cod began to be fished at the age at which they begin to mature), older individuals are removed and the distribution of reproduction at age is truncated to three to four years. Consequently, the frequency spectrum displays a peak near period 4 (Fig. 4.15d) at the highest fishing. The changes in the two species with fishing show up in the plot of the first three eigenvalues (Fig. 4.15h, i). For salmon, both the dominant eigenvalue λ_1 and the complex conjugate pair λ_2 and λ_3 move outward slightly with fishing. For cod, however, with no fishing, the magnitudes of the eigenvalues are smaller than for the salmon, then as fishing increases they move outward, away from the origin, so that their magnitudes are similar to those of the salmon. Essentially, one can

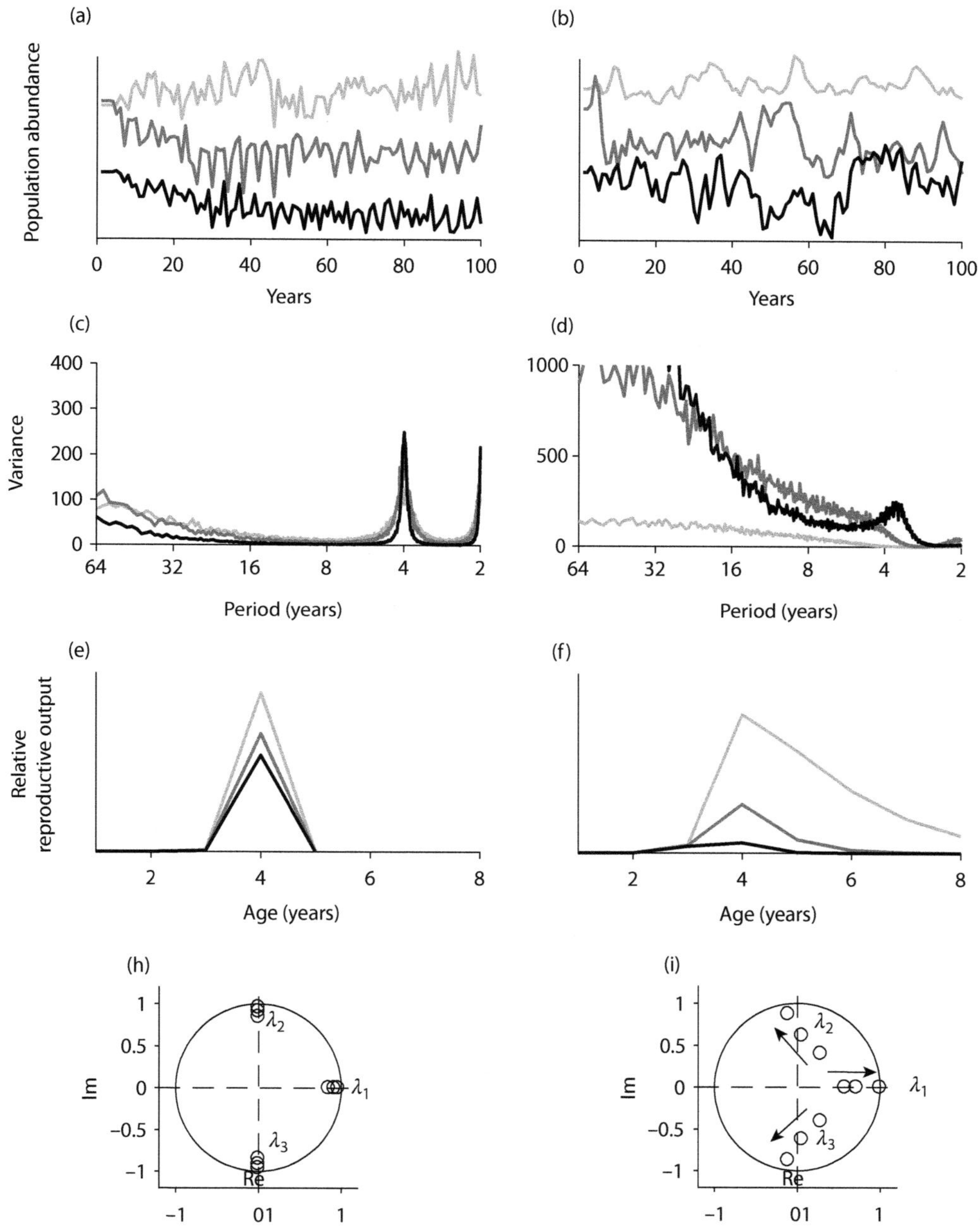

Fig. 4.15 Comparison of cohort resonance when all spawn at the same time (obligate semelparous sockeye salmon), and spawning over a range of ages (iteroparous Atlantic cod). Time series of population abundance were simulated for (a) sockeye salmon and (b) Atlantic cod, with no harvest (light gray curves) or harvest rates that reduce lifetime reproduction to just greater than the persistence threshold (black curves) and halfway between unharvested and collapse (dark gray curves). The corresponding frequency spectra are in (c, d). The distributions of spawning over age are shown in (e, f); note that sockeye are semelparous and harvest essentially only occurs on fish that have undertaken the spawning migration (but before they spawn), so harvest reduces overall reproduction but does not change the distribution very much. For cod, harvest truncates the age distribution and concentrates spawning in a few young age classes. The consequences of this for the first three eigenvalues are shown in the complex plane in (h, i). Arrows show the movement of the eigenvalues with increased harvesting.

view this as fishing making an iteroparous species (cod) more like a semelparous species (salmon). Note in Fig. 4.15g, h that with increasing fishing, the eigenvalues for cod, unlike those for salmon, rotate counter clockwise, decreasing the period of oscillations as the mean age of cod reproduction, but not salmon reproduction, declines (Fig. 4.15e, f).

4.3.4 *Extreme period-T cycles: cyclic dominance in sockeye salmon*

One of the valuable applications of cohort resonance was that it provided another potential explanation for the cycles in sockeye salmon in the Fraser River. The cohort resonance phenomenon we described in Section 4.3.3.2 reflects the effect of small amounts of environmental variability experienced by a population at equilibrium. If the variability is very large, then the linear approximation to the nonlinear model no longer holds; essentially the population is varying so much in a single time step that we have to account for the actual nonlinear shape of the function $F(\boldsymbol{n}_t)$, instead of being able to approximate it as a straight line.

One example of populations that exhibit period-*T* cycles that seem to be explained by cohort resonance effects with high environmental variability is the sockeye salmon populations that spawn in the Fraser River, British Columbia (Fig. 4.16). These populations exhibit an extreme cyclic phenomenon known as *cyclic dominance*. This consists of a four-year cycle (coinciding with the dominant age of reproduction) but with a highly asymmetrical amplitude. Because most fish spawn at age 4, the reproductive population returning to freshwater to spawn each year is essentially a single cohort, so the population can be viewed as consisting of four separate *cycle lines*. In cyclic dominant stocks, one cycle line is dominant, with an abundance several orders of magnitude greater than the

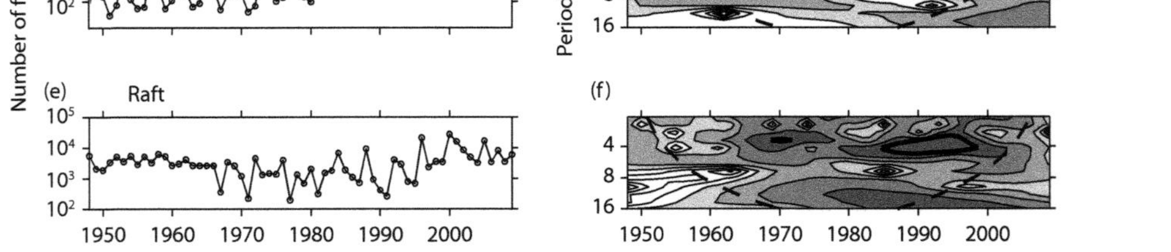

Fig. 4.16 Time series and wavelet spectra for some sockeye salmon stocks. Panels (a, c, e) show the number of spawning females each year in the Late Shuswap, Quesnel, and Raft spawning populations on the Fraser River. Panels (b, d, f) show the corresponding wavelet spectra. Shading indicates the variance at each frequency (period) in each year; thick black contour lines indicate frequencies at which variability is significantly greater than expected for a red-noise spectrum with the same lag-1 autocorrelation as the original time series ($p < 0.05$). The dashed line indicates the cone of influence; only patterns inside the cone are meaningful. Spectra were calculated using the Morlet mother wavelet following procedures in Torrence and Compo (1998). Data courtesy of M. LaPointe; figure redrawn from White et al. (2014a).

others. Typically there is a second, subdominant cycle line that is much smaller than the dominant, but also one to two orders of magnitude greater than the other subdominant lines.

Fraser River sockeye support a large fishery, and cyclic dominance produces large inter-annual variability in that fishery, leading to a lot of interest in understanding the underlying cause. A variety of mechanisms have been proposed over the years, from depensatory harvest in the 1950s (Ricker, 1950) to consumer–resource interactions in the spawning lakes more recently (Guill et al., 2011). Models used to test those hypotheses can produce four-year cycles, and some even produce the extreme amplitudes typical of cyclic dominance. However, because these mechanisms are basically deterministic, none of them could also reproduce an additional key feature of the cycles: they sometimes stop and restart, or change phase (Fig. 4.16). This would be more likely to be a stochastic process, like cohort resonance, rather than deterministic cycles.

The starting and stopping of cycles at a particular frequency is difficult to detect in a typical frequency spectrum of a time series, but it can be seen in a slightly different tool, a wavelet spectrum (Box 4.6), which allows one to see how the frequency content changes over time (Fig. 4.16). Wavelet analysis essentially does this by replacing the sines and cosines in the Fourier transformation (Box 4.4) by periodic functions that decay with time in both directions from the time for which the calculations are being made.

White et al. (2014a) showed that it was possible to generate simulated population time series that matched the observed patterns of cyclic dominance with a density-dependent population model (like that used previously in Box 4.4) and stochastic variation in the survival of the first age class. The conditions under which this would happen were: (a) low lifetime egg production (i.e. at equilibrium the population was on the steep, ascending portion of the egg-recruit curve); (b) narrow spawning age structure; and (c) high stochastic variability in juvenile survival (coefficient of variation near 1). Because the cycles originated due to stochastic perturbations of a population that was otherwise at a stable equilibrium, they would be expected to stop and restart based on the random environmental inputs, just like real populations. The relative simplicity of this explanation is appealing. Because cyclic dominant dynamics emerged in this model when the population was very far down on the linear portion of the egg-recruit curve (i.e. near the point of collapse), a concerning interpretation is that extreme cyclic dominance occurs in populations that are at greater risk of overexploitation and collapse.

Box 4.6 WAVELETS

Wavelet analysis is a method for examining how the frequency response of a signal in the time domain changes over time. Instead of decomposing a signal into stationary cosine and sine waves for the entire time series, the wavelet analysis allows the signal to have different, temporally local frequency content at different times. The analysis begins by defining the "mother" wavelet function, which is essentially a small, localized wave (instead of extending from $-\infty$ to ∞ with equal amplitude, like a sine function, it has a peak at zero and dampens out in either direction). The wavelet is given a particular frequency (e.g. 0.25 y^{-1}) and multiplied with the original signal at each point in the signal; that process is then repeated for each frequency of interest. The result is, if an oscillation of a particular frequency is present at a given time t, that frequency will be amplified by the wavelet of the corresponding frequency; if the signal does not have oscillations at that frequency, the wavelet is dampened. The result is a sort of moving window of the frequency response

(Continued)

Box 4.6 CONTINUED

that produces a two-dimensional view of the frequency domain, with frequency on one axis and time on the other (Fig. 4.16). In choosing the mother wavelet function, one faces a tradeoff between resolution in the time domain (determining when cycles start and stop) and resolution in the frequency domain (determining the exact frequency of cycles). Additionally, one can compare the wavelet spectrum to a random red-noise spectrum to determine which peaks in the spectrum are statistically nonrandom. This test produced the bold outlines of significance in Figs. 4.16b, d, and f. As in any frequency analysis, detecting low frequencies (long periods) requires a time series at least twice as long as that period. Thus the wavelet can only detect low frequency signals in the middle of the signal, away from the edges of the time series. This effect is identified by the dashed line curves at each end of Figs. 4.16b, d, and f. More details can be found in Torrence and Compo (1998).

4.4 What have we learned in Chapter 4?

In this chapter we have focused on how age-structured populations behave before they reach their asymptotic state, be it the stable age distribution for geometrically growing linear populations, or constant recruitment for age-structured models with density-dependent recruitment. This could be because a linear population has been highly disturbed (e.g. by heavy overexploitation) and it is returning to the stable age distribution (e.g. an unharvested population included in an MPA, as in Fig. 4.8), or it could be that a nonlinear population (e.g. age structured with density-dependent recruitment) is near equilibrium but is being constantly perturbed by random environmental variability (as in Fig. 4.15). What binds these two cases together in population dynamics is that, in both cases, we can understand what will happen to the population by examining the second (and third) eigenvalues of the Leslie matrix (actually the Jacobian matrix in the nonlinear case, but it is quite similar to the Leslie matrix). These eigenvalues are a complex conjugate pair, and have an angle in the complex plane that is approximately equal to 360 degrees divided by the average generation time, T, thus they lead to period-T oscillations. These oscillations could either dampen out (in the case of the deterministic return to the stable age distribution) or constantly recur (in the case of the nonlinear population experiencing recurring environmental variability). Note that the cycles in nonlinear populations described in this chapter would not occur without environmental variability. Because of that they are referred to as stochastic cycles, to distinguish them from the deterministic cycles we describe in Chapter 7. In both the linear and the nonlinear case, the link between the eigenvalue and its biological interpretation provides a quantitative prediction (the period and duration of cycles) linked to a general biological phenomenon (a cohort that is larger or smaller than usual will tend to echo or resonate through time as individuals reproduce and generate a new cohort that is also unusually small or large). Moreover, a signal that is close to being periodic with period T will be reinforced by this mechanism.

The findings in this chapter underscore one of the major theses of this book, the value of describing population dynamics as a broad coherent whole. For example, recent increases in our understanding of the responses of age-structured populations to random environments via cohort resonance were aided by the characterization, almost 40 years ago by Taylor (1979), of how the second and third eigenvalues affected insect population dynamics. The level of generality in this information is notable from another point of view, in that Taylor was interested in the annual development of insect populations on sub-annual time scales, while we have been interested in fish, on decadal scales.

It is important to keep in mind that the more recent phenomenon, cohort resonance, involved a dominant eigenvalue (of the Jacobian, in this case), in addition to the second and third eigenvalues. The value of this eigenvalue, which governs the rate of return to equilibrium, seems to be associated with the peak in cohort resonance at low frequencies, whose practical implications were emphasized by Bjørnstad, et al. (2004) by referring to it as a *cloaking effect*. Their basic concern was that in an age when a changing climate may be causing slow, permanent changes in population parameters, those may be difficult to detect when increased fishing has magnified the sensitivity of a population to low frequencies.

In this chapter we focused on scenarios in which the population was either small enough to ignore density dependence, or fell on the ascending or saturating portion of the density-dependent egg-recruit function. In Chapter 7 we will revisit density-dependent age-structured populations, but consider the type of cycles that arise when the population is on the right-hand, descending part of an overcompensatory density-dependent function.

So far in this book we have used age as the *i*-state variable to describe how populations are structured. This is usually successful, because in many cases population dynamics have characteristic time scales, particularly the age of maturity and the average generation time. Using age-structured models is thus essential to representing those time scales (such as in the two-year cycles of pink salmon or four-year cycles of sockeye salmon) and how populations will respond to environmental variability that has particular frequencies (such as the Pacific decadal oscillation; Botsford et al., 2011). However, in some species, size is a better predictor of demographic rates than age. Thus in the next two chapters we examine models that use size (or "stage") as the *i*-state variable. However, as we do so, the reader should keep in mind that many of the important features of population dynamics, such as the presence of cycles, are the result of various time lags, so the differences in dynamics due to size or stage occur because of their effects on the age structure.

CHAPTER 5

Size-structured models

In Chapters 3 and 4 we provided an extensive description of various aspects of age-structured population models. Such models can be straightforward to construct and analyze because individuals advance through age at the same rate that time advances in the model (e.g. $da/dt = 1$ in the continuous-time M'Kendrick/von Foerster model, Eq. 3.2). However, in many populations, vital rates depend on size rather than age. For example, in many fishes and invertebrates, reproductive output scales with female size because large females can brood more eggs. Mortality rates often decline with size; mortality due to predation is expected to decline with size because the number of species large enough to eat an individual declines. The opposite is true in many harvested populations, where size-based regulations commonly restrict harvest to larger individuals only. If all individuals were to grow in the same way in a population, we could simply convert size to age, and use an age-structured model, but this is never the case for several reasons. One reason is that growth rate can be time-varying because of a time-varying environment or food supply. A slightly different reason is that individuals could grow along slightly different growth trajectories. A third scenario is that individuals could experience slightly different environments. All of these mean that it is quite possible to have two individuals of the same size but different ages, and vice versa. If we recall our definition of the *state* of a system from Chapter 1, we required a state variable that uniquely determined the future response of a system. If a population has size-dependent vital rates (e.g. mortality, fecundity), and individuals of the same size can be different ages, then size, not age, uniquely specifies the population dynamics, and would be the appropriate state variable to use.

An added benefit of modeling size-structured dynamics is that it is frequently easier to make comparisons to field observations. For most wild populations it is considerably easier to observe the size of an individual than its age, because estimating age typically involves either long-term tagging (e.g. in birds and fish) or examining growth records in hard parts (e.g. bones, teeth, tree rings, fish otoliths and scales, or whale ear plugs). Most of these methods can only be applied to dead organisms, which makes them less desirable to apply at a large scale to populations one is trying to conserve (not to mention the labor involved in processing and examining them).

While these are good reasons to use a size-structured model, such models are more complex. A key difference in using size- versus age-structured models is that age distributions are relatively easy to interpret, while size distributions are not. In fact, as we shall see, the intuitive, "common sense" interpretations of size distributions can be misleading. A basic difference between age and size models is that because individuals die as cohorts age, an age distribution will always decline with age, but the same is not always true for a size distribution. If individual growth rates slow down with age, then individuals can pile up at certain sizes, and there will be more individuals in larger size classes than in smaller size

Population Dynamics for Conservation. Louis W. Botsford, J. Wilson White, and Alan Hastings, Oxford University Press (2019).
DOI: 10.1093/oso/9780198758365.001.0001

classes, even though all the individuals in the larger size class passed through the smaller size class at some point.

A simple, practical effect of this difference is in estimating survival rates. We saw in Chapter 3 that the stable age distribution of a population was given simply by $\lambda^{-a} l_a$, where λ is the geometric growth rate and l_a is the probability of survival to age a (Eq. 3.29). If the population is not growing or declining (so λ is approximately 1), plotting the logarithm of that expression versus age will produce a line with a slope equal to the log of the survival rate. This makes it straightforward to estimate survival (in fact this idea is the basis of the catch-curve analysis used to estimate mortality rates in fished populations; Robson and Chapman, 1961), assuming the population is at the stable age distribution. However, one cannot follow the same procedure for a size distribution. Even if the survival rate is constant with size, a change in the growth rate can cause distortions in the relationship between age and the logarithm of the size distribution; the end result is that one cannot directly estimate the survival rate (or otherwise interpret the size distribution) without also knowing something about age and growth (Smith et al., 1998). In this chapter we will use size-structured models to see more precisely why this is so, and we will identify the information we *can* accurately extract from size distributions, despite these complications.

In spite of the challenges inherent in analyzing size distributions (or perhaps because those challenges are not always well understood), there is a long history of using them to infer size-dependent mortality rates (and, of necessity, growth rates). This can be done in two ways. The first is to analyze a *cohort distribution*, which is a size distribution of a group of individuals that all start out at the same time. A classic example of this is the development of the self-thinning "law" in plant populations, which states that for a field of identical-aged plants (i.e. all in a single cohort), a plot of the logarithm of plant biomass versus plant density (number of stems) will have a slope of $-3/2$, because some plants are lost to competition even as others grow in size (Yoda et al., 1963; but see Weller, 1987 and Taubert et al., 2015 for more nuanced and up to date views on this law). The second, and more common, approach is to analyze a *stand distribution*, which is the size distribution of the entire population at any given time. A common, but deceptive, way of interpreting stand distributions is to assume that size classes at low abundance in size distributions are low because mortality is high for those sizes, while more abundant size classes in the size distributions are more abundant because mortality is low for those sizes. For example, Pollock (1979) found that the population of mussels (*Aulacomya ater*) on Robben Island, South Africa, was dominated by very large individuals. He explained this by invoking the observation that the primary mussel predator, the lobster *Jasus lalandii*, was only able to consume smaller mussels (except for very large lobsters, which were rare). Thus very few mussels appear to survive the high predation at small sizes, but once they reach larger size they are relatively invulnerable and accumulate in the size distribution. This logic is very appealing, but in this chapter we will see that a more careful analysis is required to understand how size distributions actually emerge.

We begin this chapter by revisiting the M'Kendrick/von Foerster model, but using size instead of age as the state variable. We then use the lessons from that model to describe how individual rates determine the shape of both stand and cohort distributions. In the simplest form of the size-structured M'Kendrick/von Foerster model, individuals share a common growth pattern, so the analyses of stand and cohort distributions will provide different ways of incorporating the abovementioned sources of greater variability in size. Ultimately, the numerical calculations required to model size-structured populations for future projections are more challenging than those needed for age structure, so we will close by discussing some mathematical tools that have been developed to accomplish this. These include the integral projection model, a recent approach that is very useful because, while more complex, it has a lot in common with the age-structured models we examined in Chapters 3 and 4.

5.1 The size-structured M'Kendrick/von Foerster model

Following the same logic we used to derive the age-structured M'Kendrick/von Foerster model we described earlier (Section 3.1), we can write a continuity equation for the change in the number of individuals in the size interval m to $m + \Delta m$ over the time period Δt (we also present an alternative derivation, analogous to the one we developed in Box 3.2, in Box 5.1; readers may also wish to consult Metz and Diekmann (2014) for more detailed explorations of the derivations presented here. We do not repeat the derivation of the continuity equation here because the steps are the same, but substituting m for a. However, the graphic we used earlier to explain the continuity of flow of individuals through age (Fig. 3.2) would be slightly different. For the size model, the dashed lines defining the abundances growing in and out of a particular size interval Δm (i.e. γ and δ, respectively, in Fig. 3.2) would have different slopes because the growth rate dm/dt varies with m, whereas the rate of aging, da/dt, is always 1. This difference leads to the size-structured form of the von Foerster equation:

$$\frac{\partial n(m,t)}{\partial t} = -\frac{\partial}{\partial m}\left[n(m,t)\, g(m,t)\right] - D(m,t)\, n(m,t)\,. \tag{5.1}$$

Comparing this equation to the age-structured version (Eq. (3.2)), the second term on the right-hand side is the same; it is simply describing the loss of individuals to death, with a size-specific mortality rate $D(m, t)$. The first right-hand side term describes the growth of individuals in and out of the size interval, and it is slightly different from the age-structured version. In place of the da/dt term in Eq. (3.2) (which was equal to 1, making things much simpler), we have now written the rate of change of size over time, dm/dt as $g(m, t)$. This is because the change in individual size with age (i.e. their growth rate) can depend on both size and time, and unlike the rate of change in age does not simply equal 1.

Box 5.1 A DERIVATION OF THE SIZE-STRUCTURED CONTINUOUS TIME MODEL

We can also derive the size-structured M'Kendrick/von Foerster model using an approach analogous to the one we presented for the age-structured model in Box 3.2. However, for the size-structured model we need to be a bit more careful and consider intervals.

The first step is to specify the growth rate of an individual of size m at time t as $g(m,t)$ and then to realize that the change in the number of individuals between sizes m and $m + \Delta m$ can occur in three ways: individuals can enter this size interval by growing, individuals can leave this size interval by growing, or individuals can die.

This leads to the following equation, in which the three terms on the right-hand side correspond to growing in, growing out, and dying, respectively:

$$\frac{\partial}{\partial t}\int_{m}^{m+\Delta m} n(s,t)\, ds = g(m,t)\, n(m,t) - g(m+\Delta m,t)\, n(m+\Delta m,t) - \int_{m}^{m+\Delta m} D(s,t)\, n(s,t)\, ds. \tag{B5.1-1}$$

Notice that here we are temporarily using s as the dummy variable for size in the integrand on the left-hand side, because m and $m + \Delta m$ denote the size interval under consideration.

Approximating the integrals by multiplying the integrands by the length of the small interval Δm (and returning to using m as the size variable), this becomes

$$\frac{\partial}{\partial t}\Delta m\, n(m,t) = g(m,t)\, n(m,t) - g(m+\Delta m,t)\, n(m+\Delta m,t) - \Delta m\, D(m,t)\, n(m,t). \tag{B5.1-2}$$

Dividing by Δm, we get

$$\frac{\partial}{\partial t}n(m,t) = \frac{g(m,t)\, n(m,t) - g(m+\Delta m,t)\, n(m+\Delta m,t)}{\Delta m} - D(m,t)\, n(m,t). \tag{B5.1-3}$$

If we then take the limit as Δm goes to zero, we get

$$\frac{\partial}{\partial t}n(m,t) = -\frac{\partial}{\partial m}\left[n(m,t)\, g(m,t)\right] - D(m,t)\, n(m,t). \tag{B5.1-4}$$

This is the size-structured continuous time equation (see Eq. 5.1).

We can begin to see the consequence of changes in growth rate with size by using the product rule from calculus to expand the first term on the right-hand side of Eq. (5.1):

$$-\frac{\partial}{\partial m}\left[n(m,t)\, g(m,t)\right] = -\frac{\partial n}{\partial m}g(m,t) - \frac{\partial g}{\partial m}n(m,t). \tag{5.2}$$

The first term in this expression is the growth rate, g, times the rate of change of density with size. This negative term has the same interpretation as the analogous term in the age-structured model depicted in Fig. 3.4. Essentially, if the slope of the size distribution ($\partial n/\partial m$) at a given size m is negative, there are a lot of individuals smaller than m about to grow into the interval between m and $m + \Delta m$, so the overall effect on $\partial n/\partial t$ is positive (Fig. 5.1a). That is, the density of individuals at size m tends to increase. The difference between this term and the depictions in Fig. 3.4 is that while the shape of $n(a,t)$ does not change with time (ignoring mortality for the moment), the shape of $n(m,t)$ does change with time because the growth rate changes with size (Fig. 5.1a).

The second term in Eq. (5.2) is the size density multiplied by the rate of change of growth with size; it is absent in the age case because $\partial a/\partial t = 0$. If the growth rate slows down with size (as it usually does) then this slope will be negative, again having a positive effect on $\partial n/\partial t$ in the interval between m and $m + \Delta m$. This is because a slowing growth rate will cause individuals to stay in that interval longer (on the right in Fig. 5.1b). Alternatively, in the case where the growth rate speeds up with size (on the left in Fig. 5.1b), individuals will be growing out of the size interval at m more quickly, and the overall effect on $\partial n/\partial t$ in that interval will be negative. This effect of $\partial g/\partial m$ that either causes the size distribution to pile up (as seen in Fig. 5.1a) or stretch out at particular sizes is called *deterministic dispersion*. The consequence is that a population can have broad peaks and valleys in the size distribution due solely to the growth rate slowing down or speeding up with size, without any effect of size-dependent mortality. Of course, usually there is also variability in the growth patterns among individuals, and we will address that effect a little later.

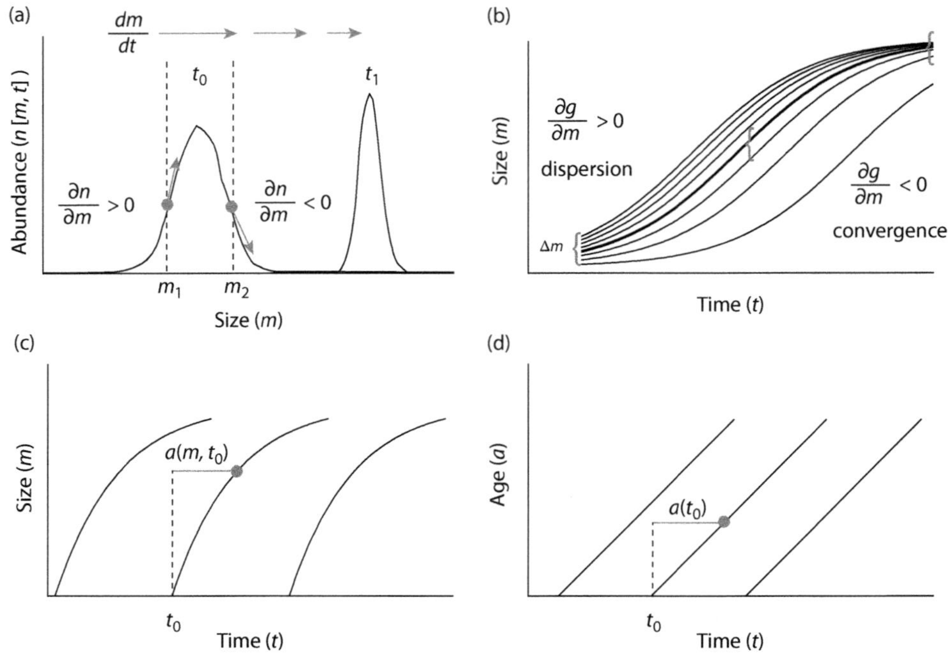

Fig. 5.1 Mechanics of the M'Kendrick/von Foerster model of size-structured growth. (a) Effects of growth in the first term on the right-hand side of Eq. (5.2) (n.b. similar to Fig. 3.3): a pulse in a size distribution at two different times as it moves from left to right with a growth rate that decreases with size (gray arrows along top). Gray arrows on pulse indicate the rate of change in abundance with size ($\partial n/\partial m$, i.e. the slope of the size distribution) at the indicated points. The effect on $\partial n/\partial t$ is negative at m_1 (the pulse is passing) but positive at m_2 (the pulse is approaching). The second pulse at t_2 contains the same number of individuals (assuming no mortality for clarity), but is narrower because of deterministic dispersion, which is illustrated in (b). (b) Deterministic dispersion: the effects of the second term on the right-hand side of Eq. (5.2). Individuals in a small size increment Δm disperse when $\partial g/\partial m > 0$, so that there are fewer within Δm as their size increases. When $\partial g/\partial m < 0$, they converge so that more are within Δm as time goes on. (c, d) A comparison of the solution by the method of characteristics for a size-structured model (c) and an age-structured model (d).

5.1.1 *The solution to the size-structured M'Kendrick/von Foerster model*

To find the solution we again use the method of characteristics, but things are more complicated here. In the age-structured version, we followed the characteristic lines that had the equation $t = t_0 + a$, which trace out the path of an individual through time from t_0 until age a (Fig. 5.1d). With size structure, the characteristics have the equation $t = t_0 + a(m)$, where $a(m)$ is the age of an individual of size m. This is not a straight line, but traces out the path of an individual through size over time (Fig. 5.1c).

To apply the method of characteristics, we start by using the chain rule to write out the expression for dn/dt (compare this to Eq. (B3.1-4)):

$$\frac{dn}{dt} = \frac{\partial n}{\partial m}\frac{dm}{dt} + \frac{\partial n}{\partial t}\frac{dt}{dt}. \tag{5.3}$$

As we did in Chapter 3, we can now substitute the von Foerster equation for the last term of Eq. 5.3, to obtain

$$\frac{dn}{dt} = -n\frac{\partial g}{\partial m} - Dn. \tag{5.4}$$

In the age-structured case this equation included only $-Dn$ on the right-hand side, whereas here we have a term involving how growth rate varies with size. This is our first clue that if we want to learn something about the mortality rate from the size distribution, we will have to know something about the growth rate, too. Specifically, it appears that a high rate of change in growth rate with m will tend to counter a high mortality rate (and vice versa) making it difficult to distinguish the effects of these two processes.

We find a solution in the same way as before, holding t constant at t_0 and using separation of variables (Box 2.4). To keep things simple we again assume that both $g(m)$ and $D(m)$ depend on size (m) but not time (t). We start by separating the variables,

$$\frac{1}{n}dn = -\frac{\partial g}{\partial m}dt - D\,dt. \tag{5.5}$$

Integrating both sides over time along the characteristic line from m_0 to m, as in Chapter 3, we obtain

$$\ln\left(\frac{n(m)}{n\,(m_0)}\right) = -\ln\left(\frac{g(m)}{g\,(m_0)}\right) - \int_{m_0}^{m}\frac{D(x)}{g(x)}dx. \tag{5.6a}$$

In this integration, the term on the left-hand side and the last term on the right-hand side in Eq. (5.5) play the same roles as they did in Eqs. (3.7) through (3.10) for the age-structured M'Kendrick/von Foerster model: abundance at a point on the characteristic is the amount at the beginning of the characteristic times the fraction surviving to each size. The difference is that the last term on the right-hand side in Eq. (5.6a) now has g in the denominator, because faster growth will reduce the mortality experienced between m_0 and m (organisms get to the larger size faster). The first term on the right-hand side of Eq. (5.5) ends up with the growth rate g in the denominator when written in terms of size (because $g = dm/dt$, $1/dm = 1/gdt$), so it integrates to a logarithmic form similar to the term on the left-hand side of Eq. (5.5). Its function differs, however, as it reflects the effect of deterministic dispersion (Fig. 5.1b): density on the characteristic line will be greater at m if $g(m) < g(m_0)$, i.e. if the growth rate has been declining with size.

Finally, exponentiating both sides of Eq. (5.6a), rearranging again, adding back in the m and t notation, and recalling that by the method of characteristics we follow solutions on the line $t = t_0 + a(m)$, we have the solution

$$n\,(m,t) = n\,(m_0, t - a(m))\left(\frac{g\,(m_0)}{g(m)}\right)e^{-\int_{m_0}^{m}\frac{D(x)}{g(x)}dx}. \tag{5.6b}$$

As with Eq. (5.5), this is similar to the age-structured version (Eq. (3.10)) but now there are terms with $g(m)$. We will see how these differences produce counterintuitive effects on size distributions when we analyze both stand distribution in Section 5.2 and cohort distributions in Section 5.3.

5.1.2 *Adding reproduction to obtain a complete population model*

To obtain a complete model we have to define a boundary condition to account for the reproduction of individuals at the smallest size, m_0:

$$n\,(m_0,t)\,g\,(m_0,t) = \int_0^{\infty} b\,(m,t)\,n\,(m,t)\,dm, \tag{5.7}$$

where $b(m, t)$ is the per capita reproduction by an individual of size m. The left-hand side of this equation is multiplied by the growth rate at the smallest size, which essentially makes the left-hand side a rate of change (the equivalent boundary condition for the age-structured version was also multiplied by the growth rate da/dt, but since that is equal to 1 we did not include it when writing out Eq. (3.12)). Just as in the age-structured case, we can substitute the solution in Eq. (5.6c) into Eq. (5.3), to write a renewal equation (see Section 3.2) that expresses the current number of new recruits, $B(t) = n(m_0, t)g(m_0)$, in terms of the number recruited in the past, $B(t - a(m))$:

$$B(t) = \int_0^\infty B\left(t - a(m)\right) b(m) \frac{1}{g(m)} e^{-\int_{m_0}^{m} \frac{D(x)}{g(x)} dx} dm. \tag{5.8}$$

Unfortunately this renewal equation does not allow exactly the same simple interpretation in terms of reproduction now being a weighted sum over reproduction in past years (Fig. 3.4), because the time lag is $a(m)$, the age of a size-m individual, instead of simply a; $a(m)$ depends on g, which is usually nonlinear.

Often we assume that there is a constant birth rate that does not vary with time or depend on the current population size:

$$n\left(m_0, t\right) g\left(m_0, t\right) = R(t) = R_{\text{const}}. \tag{5.9}$$

When this is the case, and growth rate g and mortality rate D do not vary with time, then eventually $n(m, t)$ will be constant with time, t, because there is a constant number of recruits entering the population and thus eventually a constant number of individuals at each size. This is unlike the model in Eq. (5.7), which will eventually either increase or decrease geometrically. Assuming that g and D do not vary with time allows us to drop the t from the equation and just examine dn/dm as a full derivative of a single variable instead of as a partial derivative. This yields an ordinary differential equation, which is much easier to interpret:

$$\frac{dn}{dm} = -\frac{1}{g}\left(g_m + D\right) n, \tag{5.10}$$

where g_m is the partial derivative of growth rate with size ($\partial g/\partial m$), which is merely a function of m here, like the mortality rate D. We will use this version of the model with constant recruitment to understand how we can interpret stand distributions, in Section 5.2.

5.2 Stand distributions

As we mentioned at the beginning of this chapter, ecologists frequently wish to draw inferences regarding the size dependence of mortality rates from the shape of a size distribution of a population. This is usually a stand distribution, i.e. it is composed of individuals of all different sizes at one time. A classic example is the work of Tegner and Dayton (1981), who examined the size structure of sea urchin populations in California kelp forests. Among other findings, they described a bimodal size distribution in the population of red sea urchins, *Strongylocentrotus* (now *Mesocentrotus*) *franciscanus*, with a peak at very small sizes (<40 mm) and a peak at larger size (~110 mm; Fig. 5.2). Tegner and Dayton (1981) explained the pattern as resulting from shifts in mortality with size. Juvenile urchins seek refuge under the spine canopies of adult conspecifics, where they are protected from predation and have higher survival. Slightly larger urchins are

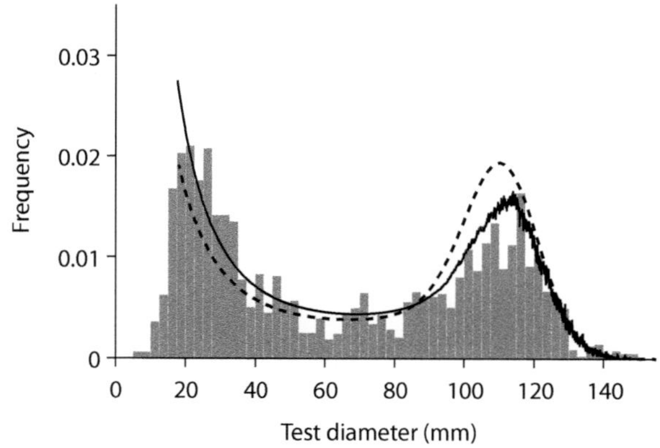

Fig. 5.2 Size distribution of red sea urchin, *Strongylocentrotus franciscanus*, sampled at 18 m depth at Pt. Loma, California (bars). Solid black curve is the solution to the M'Kendrick/von Foerster model with constant recruitment, variability in L_∞, and mortality declining exponentially with size. Dashed black curve is the equilibrium distribution from an IPM with similar parameters. Redrawn from data and model parameters in Botsford et al. (1994), data originally presented in Tegner and Dayton (1981).

vulnerable to predation from spiny lobsters (*Panulirus interruptus*) and a fish, the California sheephead (*Semicossyphus pulcher*), but the largest urchins appear to have a size refuge, based on laboratory observations that the long spines of large urchins deter predators and the observed distribution of dead urchin tests at the field site. Thus the proposed explanation was that the initial peak is a single cohort of juvenile urchins; the trough represents high mortality on the medium-sized urchins, and the second peak represents the accumulation of multiple cohorts of older urchins who have survived the gauntlet of high predation. This explanation makes intuitive sense, but we will test it with a model, following the analysis of Botsford et al. (1994).

We first consider what sort of patterns in $g(m)$ and $D(m)$ would be necessary to produce a bimodal stand distribution. In the sea urchin case, we assume that we are examining data from a single small population that receives a relatively constant larval supply from other kelp forests along the coast, so we use the model with constant external recruitment, R_{const} (Eq. 5.10).

First, consider that in order to have a peak or a valley in the size distribution, there has to be some size m at which $dn/dm = 0$ (recall from calculus that setting the derivative of a function equal to zero is the way to find maxima or minima, i.e. peaks or valleys). Looking at the terms in Eq. (5.10), the sign of dn/dm will depend on the relative values of D and g_m. D is always positive (negative mortality is not possible), so for dn/dm to ever reach zero, g_m must be negative (i.e. growth must slow with age). Otherwise the size distribution would always be decreasing with size (just like an age distribution). When the slope of the size distribution, dn/dm, is positive, $-g_m > D$, and we say the conditions are *growth dominated*: the slowdown in growth causes a buildup in the size distribution, despite the effect of mortality. If the reverse is true, $-g_m < D$, conditions are *mortality dominated* and abundance decreases with size.

Based on this observation about Eq. (5.10), we can see that the shape of the size distribution will depend on how the relative values of both D and g_m vary with m. There are many possible combinations, so to explore how this works we will look at an example relevant to the red sea urchin. Sea urchins have indeterminate growth that slows with age, so we use the von Bertalanffy equation (Box 3.8) to describe $g(m)$:

$$g(m) = k\,(L_\infty - m), \tag{5.11}$$

Where L_∞ is the asymptotic maximum size and k is the rate of convergence to that size. Usually this equation is expressed with length given as l, but we have used m to represent length to be consistent throughout the chapter. For von Bertalanffy growth, $g_m = -k$, so the change in growth rate with m does not vary with m. We assume that mortality has a quadratic shape, with low values for small and large urchins and a peak at intermediate sizes (Fig. 5.3a), as Tegner and Dayton (1981) suggested based on the peaks at high and low sizes. We can see that, given these two patterns of size dependence, the size distribution will actually decline strongly over the midrange of sizes (Fig. 5.3b), rather than remaining at a constant low value over that range as in the actual size distribution (Fig. 5.1). Thus, the magnitude of mortality seems to govern the rate of change of the size distribution, rather than the magnitude of the size distribution (just as it does in age models). Figure 5.3a indicates it will be growth dominated at very small and very large sizes, and mortality dominated at intermediate sizes. The size distribution will have a mode or a valley wherever the dominance is shifting from one to the other (i.e. wherever

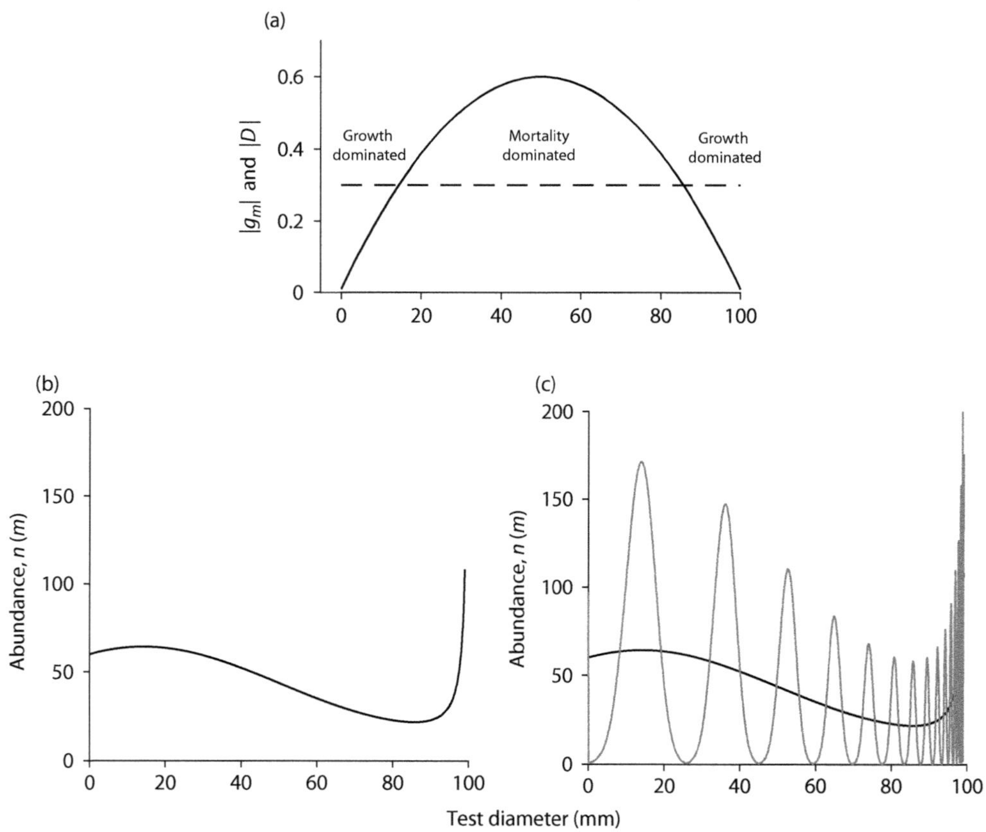

Fig. 5.3 (a) Rate of change in growth with size, g_m, for von Bertalanffy growth (dashed curve) and the pattern of mortality, D, proposed by Tegner and Dayton (1981) for red sea urchin (solid curve). Both quantities are shown as their absolute value, because D is always positive and g_m is always negative for this species. (b) Equilibrium size distribution for a model with the patterns of growth and mortality shown in (a). Other parameters are $L_\infty = 100$ mm *and* $R_0 = 60$ recruits y^{-1}. (c) Size distribution predicted by a model with periodic recruitment (gray curve). Annual recruitment is the same as in (a) but occurs in a peak centered on June 30, with a standard deviation of 0.15 y. The size distribution predicted for Jan 1 is shown.

$-g_m = D$). Where mortality dominates it declines as an age-structured model would, and where growth dominates it behaves as an age-structured model does not.

Our depiction of the sea urchin size distribution can be improved even further. While the size distribution in Fig. 5.3b has modes at low m and high m, the first mode is very broad and the second mode is very steep, with its peak at the maximum size, L_∞, rather than having symmetrical humps at both ends, like the actual red sea urchin size distribution (Fig. 5.2). Thus, high mortality at intermediate size does not seem to be able to produce the observed pattern on its own.

Another factor that could play a role in the bimodal size distribution is seasonal recruitment: if recruitment tends to happen only over a few weeks or months each year, then sampling at a different time of year could lead to observation of a large peak corresponding to the youngest, most recent recruitment cohort, with earlier cohorts lumped together in the second, larger mode. To simulate this, we assume that recruitment, $R(t)$, is no longer constant, but follows a normal distribution within each year, with mean = 0.5 year and standard deviation = 0.15 year (in this convention, Jan 1 = 0, Dec 31 = 1). When Eq. (5.6c) is used to solve the model, substituting $R(t - a(m))$ for $n(m_0, t - a(m))$ (since $R(t) = n(m_0, t)$), the result (Fig. 5.3c) has obvious cohort pulses, with the most recent several cohorts most distinct in the size distribution. As a result, the distribution now has too many modes, and the amplitude of the pulses traces out the same basic pointed shape at high sizes as the distribution without pulses (Fig. 5.3b). Thus, the addition of pulsed recruitment also does not, on its own, produce the observed pattern.

An additional factor that we have not yet accounted for is the fact that organisms within a population may vary in their growth rates. This could be due to genetic variation or simply stochasticity in the precise environmental conditions that each organism encounters. We can represent this by assuming that individual organisms have unique, independent values of the growth parameters k and L_∞. As an example, we assume that k is drawn from a gamma distribution (a gamma distribution has all positive values with a long tail, so it is a good choice for representing biologically reasonable parameter values), and L_∞ is drawn from a normal distribution. The possible growth trajectories (size at each age) that could result from varying L_∞ only, k only, or both parameters simultaneously, are shown in Fig. 5.4.

We simulated several thousand independent solutions to Eq. (5.6b), each with different growth parameters, then summed the results to obtain a population-level distribution. To examine the effect of growth variability alone, we now keep mortality, $D(m)$, constant with m, at a value less than $-g_m$. This means that the size distribution is growth dominated everywhere, and we expect the size distribution to be sloping upwards over all sizes (i.e. $dn/dm > 0$ for all m; solid black curve in Fig. 5.5a). Adding variability in k alone does not change this pattern (dashed gray curve in Fig. 5.5a), but adding variability in either L_∞ or both k and L_∞ causes the peak at large sizes to spread out into a symmetrical mode (solid gray and dashed black curves in Fig. 5.5a). This suggests that perhaps the second mode in the red sea urchin distribution could be explained simply by growth variability in L_∞.

If we then combine pulsed recruitment with growth variability in the same model (Fig. 5.5b), we see one possible explanation for the mode at small sizes: if there is no growth variation or just variation in L_∞ (solid black curve and solid gray curve in Fig. 5.5b), we see the same train of cohort pulses as in Fig. 5.3c. But if there is variation in k or both k and L_∞, then the initial pulses are smoothed out, except for the first one (dashed black and gray curves in Fig. 5.5b). So it appears that, in a purely growth-dominated system with pulsed recruitment, variation in the growth rate could produce an initial mode at small sizes, corresponding to the most recent cohort, while variation in the maximum

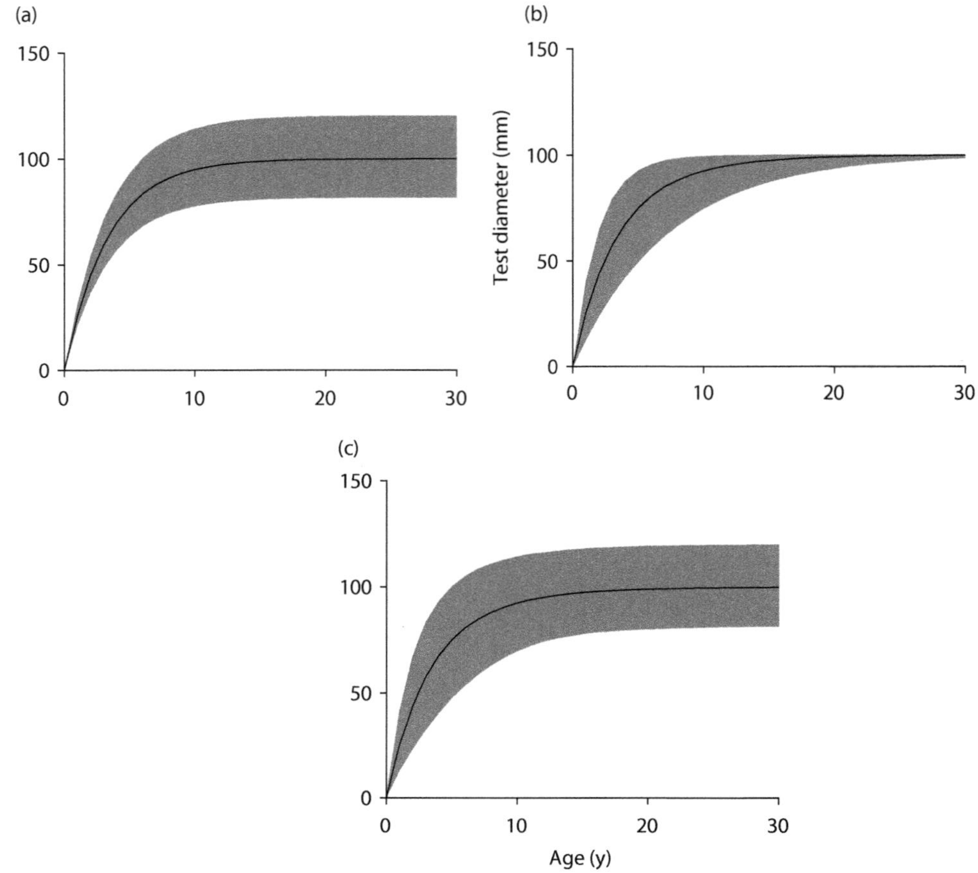

Fig. 5.4 Range of growth trajectories possible when (a) L_∞ varies among individuals, (b), k varies, or (c) both parameters vary. In each panel, the black curve shows the mean trajectory ($k = 0.3\ y^{-1}, L_\infty = 100$ mm), and the gray region indicates the middle 95% of the distribution of 10^4 random trajectories, with parameters drawn from a gamma distribution with variance of 0.01 for k and a normal distribution with a standard deviation of 10 for L_∞.

size would produce a symmetrical mode at large sizes. This provides an important lesson for interpreting size distributions: peaks and valleys can arise solely from the action of variability in growth and pulsed recruitment (as long as the growth rate slows down with size)—no pattern of size-dependent mortality is required!

The result in Fig. 5.5b is not completely satisfactory as an explanation for the red sea urchin pattern, however, because the first peak is not as high as the second peak, and it occurs at about 10 mm, when it should be more like 25 mm. This calls for a re-examination of the possible role of size-dependent predation. First, assuming constant recruitment for the moment, consider the effect of an upper size refuge from predation: $D(m) = 1.0\ \mathrm{y}^{-1}$ for $m < 80$ mm and $0.1\ \mathrm{y}^{-1}$ for $m \geq 80$ mm (k is $0.3\ \mathrm{y}^{-1}$). This means that the size distribution will be mortality dominated ($D > k$) up to 80 mm, and then growth dominated afterward. As a result, the size distribution would have a negative slope all the way to 80 mm, and then slope upward afterwards (with a symmetrical hump due to variability in L_∞; dashed curve in Fig. 5.5c). Adding periodic recruitment to this model produces a mode at small sizes, but as in Fig. 5.5b the mode is too far to the left (gray curve in Fig. 5.5c).

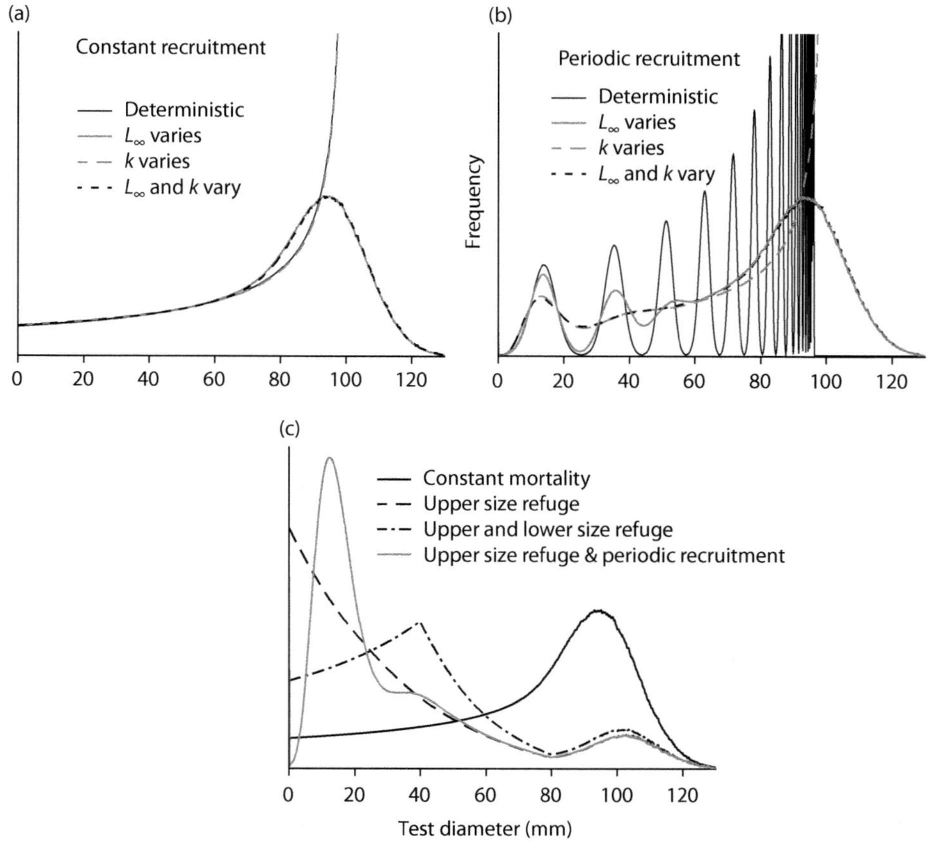

Fig. 5.5 Solutions to the size-based model for red sea urchin under a variety of growth and recruitment scenarios. Each panel shows the mean size distribution of 10^5 solutions, each using randomly selected von Bertalanffy growth parameters (drawn from the distributions shown in Fig. 5.4), or the size distribution from a deterministic solution. (a) Solutions with constant recruitment (as in Fig. 5.3b) and different combinations of variable growth parameters. The solid gray and black curves are overlapping and difficult to distinguish. (b) Solutions with periodic recruitment (as in Fig. 5.3c) and different combinations of variable growth parameters. (c) Trial solutions for red sea urchins, with variable asymptotic maximum size (L_∞) and either constant mortality for all sizes (black curve), low mortality at large sizes only (dashed curve), size refuges for both small and large urchins (dot-dash curve), and the upper size refuge scenario with periodic larval recruitment (gray curve).

If there were a size refuge for urchins <40 mm as well (due to sheltering in the spine canopy), we would have a scenario where $D(m) = 0\ \mathrm{y}^{-1}$ for $m < 40$ mm, but otherwise the same as the last scenario. This means that the distribution would be growth dominated until 40 mm, then mortality dominated until 80 mm, then growth dominated again. As expected, the resulting size distribution has a peak at 40 mm and a valley at 80 mm (dot-dash curve in Fig. 5.5c). The problem with this—and the problem with the spine canopy explanation for the first mode in the size distribution—is that the peak is at 40 mm instead of 25 mm, and the ascending slope of the peak is too gentle. In fact there is no way to make it steeper, because mortality cannot drop below zero!

The mismatch between this last scenario and the actual red urchin distribution underscores an important lesson about the interpretation of size distributions: size-dependent changes in mortality lead to differences in the *slope* of the size distribution, not to differences in its *abundance at size* (or amplitude, as Botsford et al. (1994) refer to it). If mortality is relatively low for a range of sizes, that will lead to an increasing slope in the size distribution, with a peak at the point where mortality starts to increase again. Conversely, if mortality is very high, it will lead to a decreasing slope in the distribution, with a trough at the point where mortality begins to decrease. This is completely different from the usual intuitive way of interpreting size distributions, which is to assume that the peak corresponds to the center of the size range where mortality is low, and the trough corresponds to the center of the size range where mortality is high. Consequently, there is a tendency to bias upwards our estimates of the size range where mortality is high or low; we should be looking to the left of the peak or valley, not on either side of them.

Based on the problems with the location and shape of the first mode in the dot-dash curve in Fig. 5.5c, Botsford et al. (1994) rejected the spine canopy protecting urchins <40 mm as the source of the first mode in the real size distribution. They proposed an alternative explanation: noting that the mode was at 25 mm, they suggested that this was a sampling artifact: it is difficult to sample urchins <25 mm, so those size classes have lower abundance in the sample than they do in the real population, and a combination of variability in L_∞ and mortality declining with size is sufficient to produce the observed pattern (e.g. the dashed curve in Fig. 5.5c). In Fig. 5.2 we show a model solution that includes those two factors and excludes smaller urchins from the sample; it provides a better fit to the bimodal data than the models with a lower size refuge.

5.3 Cohort distributions

Cohort distributions are size distributions of a group of organisms that all began life near the same time. In an age-structured model, a cohort always has the same age, so they travel together through age over time in the age model. The same is true in a size-structured model, unless there is variability in growth rates (as we saw in Section 5.2 by comparing the cohort pulses with and without growth variability in Fig. 5.5b). In that case, we have a situation in which some members of the cohort will reach larger size sooner than other members. One scenario in which this type of variability is important is the timing of the larval stage of species with complex life histories, such as tadpoles, or marine fish and invertebrates that have a planktonic larval stage.

Hjort (1914) was the first to write about the timing of larval development for marine organisms. After noticing that there was a lot of interannual variability in landings of North Atlantic species such as herring and cod, he proposed that this variability arose at the time of *recruitment*, i.e. when larvae (or young juveniles) entered the adult population. He suggested that high interannual variability in recruitment was due to the timing of larval release and larval development, because larvae, which have very limited energy reserves, require abundant planktonic prey in order to survive. Thus larvae have to be in the right place at the right time to encounter a bloom of planktonic productivity at the critical age at which they run out of their yolk sac resources. This basic insight of Hjort (1914) was later generalized to the broader *match/mismatch* hypothesis (Cushing, 1990).

Later authors also recognized that not only do mortality rates tend to be very high during early developmental stages of marine organisms, making this a critical stage to understand, but that larval survival and growth are inextricably intertwined. The first of these is the *bigger-is-better* hypothesis (Miller et al., 1988), which points out that because many aquatic predators are gape limited (they can only eat things smaller than the diameter of their mouth), the mortality rate of a planktonic larva should decrease with their size. Therefore, faster growing larvae should experience less mortality. Even if mortality does not change with size, fast growth can be beneficial. If there is some minimum size at which the organism leaves the larval stage (e.g. metamorphosis, settlement to the benthos), then faster growth will mean less time spent as a larva. The daily mortality rate will then be integrated over fewer days, leading to lower net mortality; this is the *stage-duration* hypothesis (Houde, 1989; Leggett and DeBlois, 1994). We can use a size-structured model to investigate how both of these ideas would affect larval survival in a field population.

Moloney et al. (1994) were interested in the cohort dynamics of Dungeness crab, *Metacarcinus* (formerly *Cancer*) *magister*, along the Pacific coast of the United States. This is an important fishery species in northern California and the Pacific northwest, and it is somewhat unusual in that only males greater than a certain size may be retained (discard mortality is low, so females found in traps are tossed back in). Harvest rates are very high, such that nearly all males of legal size are caught within a fishing season. This means that males only live long enough to grow to legal size (about four years), and thus the harvest in any year is approximately a single male cohort. Therefore, the catch possible in any given year is set by the number of larval recruits settling to the benthos four years previously (Shanks and Roegner, 2007). This has led to great interest in the processes affecting the year-class strength of crab recruitment (we introduced the Dungeness crab fishery in Chapter 1 (Fig. 1.1), and we discuss it further in Chapter 7).

To understand the processes affecting larval survival and recruitment, Moloney et al. (1994) repurposed the size-structured M'Kendrick/von Foerster model. Following their modeling, we are concerned with the transition from size $m = 0$ (newly spawned larvae) to size $m = 1$ (larvae that have reached the size of metamorphosis into megalopae, which is a post-larval stage immediately prior to settlement). If we define the rate of larval release (i.e. spawning) to be $R(t)$ and the rate of larval metamorphosis (or settlement) to be $S(t)$, then we can define boundary conditions:

$$R(t) = n(0, t)\, g(0, t) \tag{5.12a}$$

$$S(t) = n(1, t)\, g(1, t)\,. \tag{5.12b}$$

Then we can rewrite the solution to the model (Eq. 5.6b) as

$$S(t) = R(t - a(1))\, \frac{g(1, t)}{g(0, t - a(t))}\, e^{-\int_0^1 \frac{D(x,t)}{g(x,t)} dx}\,. \tag{5.13}$$

Notice that unlike in Section 5.2, we are now allowing g and D to vary with time. This is because growth and mortality may both depend on environmental conditions (e.g. temperature), which could vary over a spawning season. Thus larvae spawned at slightly different times could experience different conditions.

From Eq. (5.13) we can see how the different factors in the model affect the rate of larval settlement $S(t)$. The first is the rate of larval production, R. This could be constant over a season, or pulsed in some way. Also, settlement at t depends on the reproduction $a(1)$ time units in the past; this is the development time, so if development times speed up over the course of a season (e.g. due to warming temperatures) then larvae from successive days of larval production should pile up into a larger settlement pulse. The number of settlers also depends on mortality (the exponential term), which as before (Eq. 5.6b) has mortality in the numerator and growth in the denominator. Thus faster growth reduces mortality, which is the stage-duration effect. However, that is only assured if mortality is constant. If, for example, growth and mortality both increase with temperature (e.g. if the activity rate of larval predators also increases) then the faster growth will have no effect on the overall survival rate. Alternatively, mortality could also decrease with size, in a bigger-is-better effect.

To see how all of these factors work together (or in opposition), it is helpful to use an example. Consider a scenario where temperature increases over the course of a spawning season (Fig. 5.6a). We have modeled larval growth using the lab-derived relationship with temperature, T, used by Moloney et al. (1994):

$$g\,(m,t) = 0.01\ \ln T - 0.015 \tag{5.14}$$

and larval mortality is either constant over the spawning season or also varies with temperature. Notice that the "spawning season" in our example spans 1300 days, this simply spreads out the temperature transitions, making it easier to illustrate how the model works.

Using this model, we can examine what happens in a few hypothetical scenarios. Given a constant larval release over the spawning season (Fig. 5.6b), we first consider the effects of growth only, with $D = 0$. In the figure, batches of larvae are labeled (A, B, C, D, E) according to time periods in which larvae are spawned under similar temperature conditions.

With *no mortality* (Fig 5.6c), we see a slight peak early, and a slight dip later. The two factors affecting $S(t)$ are $R(t - a(1))$ and the ratio of the growth rates at the beginning and end of development. When temperature is constant throughout development (e.g. period A_1), development time remains constant and the ratio of growth rates is unity, so $S(t) = R(t - a(1))$ (Fig. 5.6c). However, if the larvae are spawned at a low, constant temperature and then temperature increases during development (period A_2), larvae grow faster at the end of development than at the beginning, so from Eq. (5.14), settlement increases over time despite constant larval production. The total number of larvae has not increased; rather the later-spawned larvae speed up development, so the same number of larvae settle in a shorter window (notice the shaded area corresponding to A_2 is slightly narrower; Fig. 5.6c), leading to a greater number of settlers on any one day in that window. Next, if temperature is increasing for all of the larvae in a cohort (period B_1, Fig. 5.6c) then the development of all larvae is continually increasing throughout their larval period, and there is no pile-up of settlers; consequently the peak in $S(t)$ declines. Later, when larvae are spawned at a constant high temperature and then temperature cools during development, the process is reversed and settlement gradually decreases over time (period C_2, Fig. 5.6c). Later still, when temperature is decreasing during spawning but a constant low temperature at settlement, the ratio of growth rates in Eq. (5.14) is again gradually increasing, and so settlement increases (period D_2, Fig. 5.6c).

With *constant mortality rate* (Fig 5.1d), we see that the pulse in temperature causes a substantial peak due to the stage-duration effect (Fig. 5.6d): settlement is much higher during periods of faster development. The only tricky aspect to this is that mortality is integrated over the larval development period. So larvae spawned during A_2 experience partly a period of slow growth and also a period of fast growth, so settlement is elevated for the tail end of that cohort. Similarly, the cohort spawned during the peak in temperatures experience some cooler temperatures at the end of the pulse, and their mortality is slightly elevated. These lags are a consequence of having to consider the time $t - a(m)$ in the model equation.

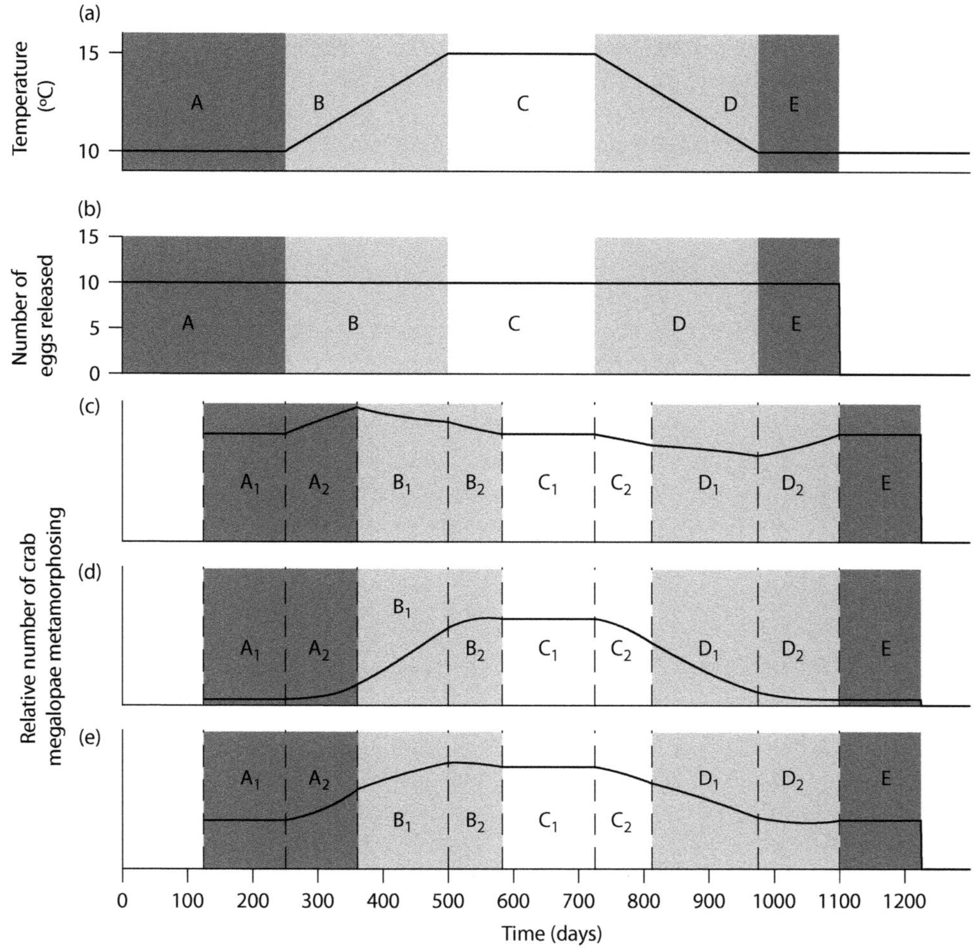

Fig. 5.6 Solutions to the cohort distribution model for Dungeness crab. (a) Water temperature during a hypothetical spawning period; lettered, shaded panels indicate different temperature regimes. (b) Pattern of egg release. (c–e) Pattern of metamorphosis into late stage megalopae, which immediately precedes larval settlement to the benthos. Lettered and shaded panels indicate groups of larvae that experienced similar temperature regimes during larval development. Growth is temperature dependent in all panels; mortality is (c) zero, (d) constant, or (e) temperature dependent.

With *temperature-dependent mortality rate*, the stage-duration effect is reduced, evening out settlement over time (Fig. 5.6e). In fact, if mortality and growth both change with temperature in the same way, the two effects cancel in Eq. (5.14) and settlement does not vary with temperature. In this example, we let growth depend on the natural log of temperature but mortality on the $\log_{10}$ of temperature, so the functions differed slightly. As a result, settlement remains highest during the period where larvae experience a rapid increase in temperature during their larval duration, causing the piling-up effect (Fig. 5.6e), but the difference between high- and low-settlement periods is much smaller.

Moloney et al. (1994) used this modeling framework to explain variation in the timing and magnitude of larval settlement of Dungeness crabs across a latitudinal gradient. In central California, warmer waters that cooled over the duration of the spawning season (due to upwelling) led to relatively high larval survival (due to the stage-duration effect) and a very evenly spread settlement pulse (because increasing development times due to cooling water tends to spread out the settlement, as in Fig. 5.6). Further north in Washington, water was cooler and warmed gradually over the season. This led to much lower settlement rates (because development times were very long) and a narrower peak (because development times became shorter as the spawning season progressed).

5.4 Numerical methods

In Sections 5.2 and 5.3 we have shown how variation in individual growth rates, in addition to mortality and reproductive rates, influence the shape of size distributions. Because of the additional complexity required to account for the role of growth, it is not as straightforward to use models such as the M'Kendrick/von Foerster model to project populations into the future, numerically. The method of characteristics can be used when all individuals develop in the same way, but that is seldom the case. Partial differential equations (PDE) are not trivial to solve. Because of these considerations, a number of methods have developed for projecting size-structured models into the future. We will explain the predominant approaches here.

5.4.1 *Grid-based method*

This is the method used in the red sea urchin example to represent variation in growth parameters. Rather than a single model with constant parameters, one simulates a large number (e.g. 1000) of independent models of individuals, each of which has different values of the growth parameters, drawn from a statistical distribution. The sum (or average) of all of the model solutions then represents the behavior of a population that includes inter-individual differences. The limitation of this method is that it requires considerable brute force: a large number of independent simulations to represent a simple process.

5.4.2 *The escalator–boxcar train*

The escalator–boxcar train (EBT) is a generic method for projecting size-structured populations (Goudriaan, 1986; DeRoos, 1988) that de Roos and colleagues (DeRoos et al., 1992) later extended to handle what they referred to as *physiologically structured* populations—essentially the same thing as the size-structured populations we have described in this chapter using the M'Kendrick/von Foerster equation. De Roos et al. (1992) noted that this model is particularly difficult to solve in the case of a variable environment, when growth and mortality can vary over time (note that in Section 5.2 we dropped the time dependency of g and D to avoid this).

The EBT approach essentially mixes discrete-time and continuous-time methods to keep track of the population in terms of individual cohorts. The idea is that at a given time t, a particular cohort of individuals is recruited, all of whom have the same initial size. In between t and the time when the next cohort is born ($t + 1$), all of the existing cohorts grow and survive according to continuous time processes (just as in the original M'Kendrick/von Foerster formulation, Eq. 5.6b). Each cohort is tracked independently, so within a time interval, all of the individuals in each cohort follow the same size trajectory. This removes the time dependence of the growth and mortality rates and turns the PDE into several ODEs, which are much easier to solve. Thus the entire population is represented by a series of ODEs for each cohort for a given time interval. The ODEs are integrated from time t until time $t + 1$, at which point each cohort increases in age by 1, a new cohort is born, and the process is repeated. In the next time interval, the growth and mortality functions can change due to environmental variability. The name comes from the idea that each cohort is a boxcar traveling along the train tracks in continuous time, then takes an escalator up one floor at each time interval.

The process is somewhat more complicated for a population with continuous, instead of pulsed, births. Essentially the solution is to evaluate reproduction as an ODE over the time interval (assuming that over that interval new individuals are being born and those already born begin to die), and then at the next time step all of those newborns are pooled together into the newest cohort.

5.4.3 *Integral projection models*

A more recent development in structured population modeling is the integral projection model (IPM), which is essentially a continuous-size analog of the discrete-time age-structured matrix models we examined in Chapters 3 and 4. It was first introduced by Easterling et al. (2000), and a recent book by Ellner et al. (2016) summarizes most of the developments since, and provides a handbook for implementing IPMs. Just as in the models we have already considered in this chapter, the state variable is population density at size m and time t, $n(m, t)$. The difference is that time is now discrete, so instead of a PDE we write out an integrodifference equation that describes the transition of individuals at size l in time t to size m in time $t + 1$:

$$n(m, t+1) = \int_{y_0}^{\infty} K(m, l)\, n(l, t)\, dl. \tag{5.15}$$

The quantity $K(m, l)$ is termed the *kernel*, and it describes the probability density of transition from size l to size m. The integral is over all possible sizes, reflecting the idea that n is continuous in size, so to determine the density at size m in time $t + 1$, one must integrate over all possible sizes in time t and take the probability of moving from that size l to m in one time step.

The key to IPMs is the kernel, which represents the continuous size surface of transition probability densities as the density in a large number of infintesimally small size increments (i.e. width dl). Thus it is analogous to the projection matrix in a Leslie model, which can use age increments of width one year in their transitions. The kernel consists of two pieces: (1) $P(m, l)$, which describes the probability density of growing from each size y to each size x in a single time step (e.g. one year), multiplied by the size-dependent probability of surviving that transition; and (2) $Q(m, l)$, which is the density function for the number of offspring of size m produced by an individual of size l. $\boldsymbol{K}$ is the sum of $\boldsymbol{P}$ and $\boldsymbol{Q}$. To see the similarity to a Leslie matrix, Fig. 5.7 is a contour plot of an IPM kernel for a hypothetical species with von Bertlanffy growth and fecundity increasing with size. The $\boldsymbol{P}$ component (growth and mortality) shows up as the quasi-diagonal ridge in the plot, falling just below the diagonal at low sizes where growth is high, then surrounding the diagonal at greater sizes; this represents the probability density of being size l at time t and surviving and growing to size m at $t + 1$—very much like the subdiagonal values in the age-structured Leslie matrix. In this example, the asymptotic maximum size is 112 mm, so small organisms (e.g. <50 mm) grow quickly to near 50 mm in a single time step (one year); organisms near 112 mm grow very little, and organisms >112 mm would tend

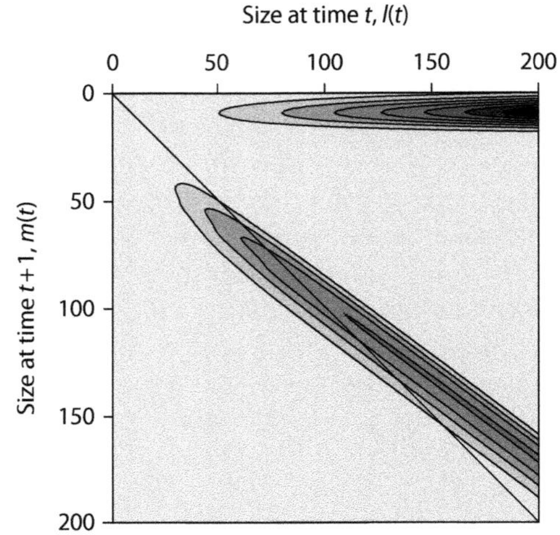

Fig. 5.7 An example of an IPM kernel. Darker shading corresponds to higher probability density.

to shrink back towards the asymptotic maximum somewhat (notice that the kernel grid extends all the way to 200 mm to prevent large individuals from being "evicted" by the model, but in practice the probability of growing much above the asymptotic maximum is very small). The width of the ridge indicates variability in the growth parameters within the population. The $\boldsymbol{Q}$ component of the kernel (reproduction) shows up as the horizontal ridge near the top of Fig. 5.7—analagous to the first row of the Leslie matrix. In this example, new offspring have a mean size of 10 mm (with a standard deviation of 2 mm), so $\boldsymbol{Q}$ is a ridge centered on 10 mm and increasing to the right, as larger individuals have high per capita fecundity.

One of the great advantages of IPMs is that the kernel is comprised of continuous, size-dependent relationships for mortality, growth, and fecundity. This requires many fewer individual model parameters than a discrete stage-structured model (Chapter 6), in which each individual transition from stage to stage (or size to size) must be estimated separately. Also, it is typically easier to find size-dependent relationships with vital rates than it is to find age-dependent relationships. Ellner et al. (2016) provide a useful guide to constructing these models.

In spite of being an approximately continuous model, in practice one must discretize $\boldsymbol{K}$ into a mesh or grid in order to perform calculations on a computer. This leads to decisions about the resolution of the grid (how finely spaced are the size intervals?) and the extent of the grid (what are the minimum and maximum values?) in order to avoid numerical artifacts. If the grid is too coarse to accurately approximate the shape of the continuous functions (for example, if the distribution of new offspring size is very narrow, making the ridge along the top of Fig. 5.7 a knife-edge), then the integration of Eq. (5.15) will be incorrect. If the maximum size of the mesh is too small, then it is possible to "evict" large individuals from the population when the model takes a time step, because they have some probability of moving to a size that is larger than the maximum size in the kernel (Williams et al., 2012). Additionally, one must choose a numerical integration method to compute the size density in each increment. Ellner et al. typically recommend the simplest option, the midpoint rule, but higher-order methods may be more efficient or more precise if the kernel is not very smooth (White et al., 2016). Ellner et al. (2016) describe these and other model-building decisions in depth. The upshot of these numerical considerations is that an IPM is essentially a discrete-size model with a very, very large transition matrix. In some ways this makes it analogous to the stage-structured models we describe in Chapter 6 (with "size" used as the variable-defining stage). However, the extremely high resolution of the IPM kernel, and its basis in continuous size-based functions, allows it to avoid the fundamental problems from which stage-structured models suffer.

One important nuance in the comparison of $\boldsymbol{K}$ to the Leslie matrix is that entries in the latter describe the probability of moving from age a to $a+1$, while any point on the surface of the kernel describes the *probability density*, not the probability, of moving from size l to size m. To obtain the corresponding probability, one must integrate the kernel; so $K(m,l)$ gives the probability density but $K(m,l)dl$ gives the probability. This is a minor detail that should not get in the way of understanding the conceptual similarity between the two, but it means that one must be careful in describing the kernel, and implementing it numerically.

An advantage of the similarity of $\boldsymbol{K}$ to the Leslie projection matrix is that, because it describes a linear operation, we can perform similar calculations on it, including evaluating the dominant eigenvalues and eigenvectors and the sensitivities and elasticities of those eigenvalues. One can also calculate important demographic quantities that we used previously in the context of age-structured models, such as the net reproductive rate, R_0, and the generation time, T. Calculating R_0 in a size-structured case is not as simple as tallying up the number of offspring produced by a cohort over time, because members of a single cohort may vary in initial size and thus their probability of survival and later, fecundity. The IPM framework makes the calculation relatively straightforward, though (Ellner et al., 2016). Starting with an initial cohort distribution n_0, the distribution of offspring produced in the first year of the cohort's life would be

$$r_0 = \int_{y_0}^{\infty} Q(m,l)\, n_0(t_0)\, dl. \tag{5.16a}$$

The distribution of that cohort surviving to the next year would be

$$n_0(t_1) = \int_{y_0}^{\infty} P(m,l)\, n_0(t_0)\, dl, \tag{5.16b}$$

and then the distribution of offspring produced by those survivors would be

$$r_1 = \int_{y_0}^{\infty} Q(m,l)\, n_0(t_1)\, dl = \int_{y_0}^{\infty} Q(m,l)\, P(m,l)\, n_0(t_0)\, dl. \tag{5.16c}$$

Similarly, the distribution of offspring produced by the surviving cohort after two years would be (dropping the m and l indices for clarity)

$$r_3 = \int_{y_0}^{\infty} Qn_0(t_2)\, dl = \int_{y_0}^{\infty} QPPn_0(t_0)\, dl = \int_{y_0}^{\infty} QP^2 n_0(t_0)\, dl. \tag{5.16d}$$

For subsequent years, the solution involves terms of Q times P raised to sequentially higher powers. Summing over all of the years, one arrives at the expression

$$r_0 + r_1 + \cdots r_\infty = \int_{y_0}^{\infty} Q\left[I + P + P^3 + P^4 + \cdots\right] n_0(t_0)\, dl, \tag{5.16e}$$

where I is the identity matrix. Notice that this is conceptually similar to Eq. (3.25), in which age-specific fecundity was multiplied by the sum of the probability of survival to each sequential age in order to calculate R_0 from the Leslie model. The term in brackets in Eq. (5.16e) is a Neumann series, and in the limit as the exponents approach infinity, it converges on $1/(I - P)$. Thus Eq. (5.16e) can be rewritten as

$$r_0 + r_1 + \cdots r_\infty = \int_{y_0}^{\infty} \frac{Q}{I - P} n_0 (t_0)\, dl = \int_{y_0}^{\infty} R n_0 (t_0)\, dl. \tag{5.16f}$$

The kernel $\boldsymbol{R}$ is termed the *next generation kernel* because it projects the initial cohort n_0 into the distribution of offspring in the next generation (keep in mind that this will comprise offspring of many different ages, but all produced by the same parental cohort). The net reproductive rate R_0 is the dominant eigenvalue of $\boldsymbol{R}$ (Ellner and Rees, 2006; Ellner et al., 2016). One can use a similar approach, using the next generation matrix (rather than kernel) to calculate R_0 for stage-structured models as well (Section 6.5).

Once one has a way to calculate R_0 in the context of size (rather than age) structure, as well as the annual population growth rate, λ (i.e. the dominant eigenvalue of the kernel), it is possible to apply one of the formulas for generation time T (Box 4.2),

$$T = \frac{\log R_0}{\log \lambda}. \tag{5.17}$$

The other definitions of T in Box 4.2 involve computing the weighted average of reproduction over all ages. Ellner et al. (2016) point out that attempting to make the equivalent calculation in an IPM context is not satisfactory, because one has to make the calculation for a particular initial size at birth (as opposed to age 0 at birth in the age-structured context).

As an example, we constructed an IPM for the red sea urchin, using the same parameters as the M'Kendrick/von Foerster model depicted in Fig. 5.2. Iterating that IPM to the stable size distribution produces the dashed curve in that figure (we also imposed the same assumption that urchins <17 mm were not observable). The two model size distributions are nearly identical, with small differences arising due to the discrete-time versus continuous-time basis of the two models. The growth–mortality part of the IPM kernel is the same as that depicted in Fig. 5.7; for the simulation in Fig. 5.2 we did not include the reproductive portion of the kernel in Fig. 5.7, but instead introduced external recruitment, as in the M'Kendrick/von Foerster model:

$$n(m, t+1) = \int_{y_0}^{\infty} K(m, l)\, n(l, t)\, dl + r_t \rho(m), \tag{5.18}$$

where r_t is the total abundance of recruits in year t and ρ is the size distribution of new recruits. The IPM version of the model has a slightly higher second mode, but that mode is better centered on the mode in the data.

So far we have considered models in which vital rates depend on age (Chapters 3–4) and now size (this chapter). IPMs provide a way to model a species in which vital rates may depend on both quantities. One approach is simply to create distinct kernels, K_a, for each age class, and sum over them (Ellner et al., 2016):

$$n(m, t+1) = \sum_{a=1}^{A} \int_{y_0}^{\infty} K_a (m, l)\, n(l, t)\, dl. \tag{5.19}$$

For example, Chu and Adler (2014) examined size-dependent demographic rates for a variety of grassland plants, and found that in most cases both size and age were important predictors of growth and mortality. They then used IPMs like Eq. (5.19) to describe the trajectories of each population.

5.5 What have we learned in Chapter 5?

In this chapter, we have seen that we can describe size-structured populations using much of the same framework that has been developed for age-structured populations. The key differences are that, first, because size is a continuous variable, we must rely on continuous-state models, such as the M'Kendrick/von Foerster model or the integral projection model (although the latter is conceptually similar to the discrete-age Leslie matrix). Second, whereas age advances at a constant rate with time, somatic growth can change with size and over time. Thus, unlike age distributions (which are always decreasing with age, assuming roughly constant recruitment), size distributions can increase over size ranges where individual growth rate declines with size. We describe this as deterministic dispersion, which can be either spreading out over size, or piling up. When growth slows with size (the most common scenario) this means that the size distribution will pile up into a peak at higher sizes.

Our examination of the red sea urchin example revealed a few potential pitfalls of inferring patterns from *stand distributions*. Contrary to the intuitive tendency to assume that modes in the size distribution correspond to size ranges with low mortality, we saw that the modes should occur at points where there is a switch between growth-dominated dynamics (the rate at which growth rate changes with size > mortality rate) and mortality-dominated dynamics (the rate at which growth rate changes with size < mortality rate), or vice versa. In other words, modes in the size distribution are driven by the rates of change (with respect to size) of mortality and growth, not their minima or maxima. Additionally, we saw that the solution to the model contains the ratio of mortality rate to growth rate, which implies that we cannot estimate the mortality rate directly without knowing something about the growth rate, and vice versa (see Banks et al. (1991) and Smith et al. (1998) for methods for estimating both rates simultaneously).

In the red sea urchin example, we saw that if vital rates did not vary with time, the equation for dn/dm (the rate of change in abundance with size) could be used directly to produce the stable size distribution. However, if growth rate slows with size, eventually to a growth rate of zero, it is necessary to incorporate variability in the growth parameters in order to obtain a reasonable representation of the broad upper mode that occurs in most population size distributions. We saw that seasonality in reproduction can lead to periodic modes in size distributions, which are broadest at low sizes. The degree of variability in growth will also determine the degree to which young cohorts show up as distinct modes in the lower part of the size distribution; greater variability in growth tends to occlude them.

Turning to *cohort distributions*, we examined several of the proposed phenomena affecting the survival of early stages of marine larvae: stage-duration effects and bigger-is-better effects. Using the example of temperature-dependent Dungeness crab larval development, we saw that a period of fast growth can speed a cohort out of a stage with high mortality. Thus if reproduction is constant over time, periods of increasing developmental growth rates will lead to peaks in juvenile recruitment. If mortality is also size dependent, those peaks will be accentuated. However, if mortality rates covary with growth rates (e.g. both increase with temperature) then the stage-duration effect will be reduced.

We then examined several numerical techniques useful for implementing size-structured models. The escalator–boxcar train allows vital rates to vary with size and over time. As such it has proven especially useful for modeling populations in very dynamic environments, such as aquatic invertebrates that experience large changes in temperature and food availability over the course of a growing season.

Finally, we considered integral projection models, a reasonably recent innovation in population modeling that has exploded in popularity in the past decade. This approach represents transitions between age distributions using quasi-continuous functions discretized on a very fine grid. The result is a transition kernel that is analogous to the familiar Leslie matrix, which suggests that we can make many of the same calculations of demographic statistics and elasticities that we described for age-structured models in Chapter 3. These, of course, would not apply to IPMs that included density dependence or assumed constant recruitment. Given the computing capacity of most modern desktop computers, IPMs offer an adequately fast, flexible, and conceptually straightforward way to analyze the dynamics of size-structured populations.

In Chapter 6 we will discuss models that are structured by stage rather than age or size. As we outline some of the potential problems with that approach, we will revisit IPMs as a potential solution (in some cases).

CHAPTER 6

Stage-structured models

In this chapter we move on from age- and size-structured models to describe population models in which *stage* is the *i*-state variable, and the set of abundances at each stage is the *p*-state variable. Stage is defined to be a period of fixed duration within an individual life history through which survival and "growth" to other stages are specified. This movement and survival is represented by a matrix model in which a stage vector (composed of the abundance at each stage) is multiplied by a projection matrix that describes the transitions between stages over time. The projection matrix is similar to the Leslie matrix for age-structured models (Chapter 3), except it has entries other than just those in the first row and the sub-diagonal.

We introduced age-structured models in Chapter 3 with both a continuous-age, continuous-time model (M'Kendrick/von Foerster) and a discrete-time, discrete-age matrix model (Leslie matrix). In Chapter 5, we introduced size-structured models in a continuous-size, continuous-time form (M'Kendrick/von Foerster again) and a continuous-size, discrete-time form (integral projection models), but we did not present a corresponding discrete-time, discrete-size matrix model with annual calculations. Many of the stage-structured models we describe in this chapter were proposed to fill that role (i.e. to be an approximation of individual growth in size, in terms of a discrete-time, discrete-size "equivalent" of the continuous-size and time model). However, stage models can also represent different life history stages in species with complex life histories (e.g. seed, dormant seed, sprout, reproductive plant).

An important point underlying this chapter is that "stage" in these stage-structured models does not satisfy the requirements for an adequate expression of state as described in Chapter 1. We will demonstrate why this is so. In Chapter 5 we presented several versions of numerical solutions to size-structured models, which, in most cases, would be the preferred alternatives to the stage-structured growth models described here. We include this chapter on stage-structured models because they continue to be widely used in ecology, in spite of the structural errors that are inherent in solutions to a model with an inadequate description of state (e.g. Caswell, 2001; Heppell et al., 2000; Morris and Doak, 2002; van Groenendael et al., 1988).

6.1 Biological processes

Stage-structured models represent one of two different biological mechanisms: (1) natural biological stages and (2) continuous growth or development. The best example of the former is the various instars in arthropod growth. Arthropods grow in discrete size increments, by ecdysis (molting; Fig. 6.1). Molting consists of increasing a length dimension of their exoskeleton by a certain fraction that is fairly constant from molt to molt. Molts

Population Dynamics for Conservation. Louis W. Botsford, J. Wilson White, and Alan Hastings, Oxford University Press (2019).
DOI: 10.1093/oso/9780198758365.001.0001

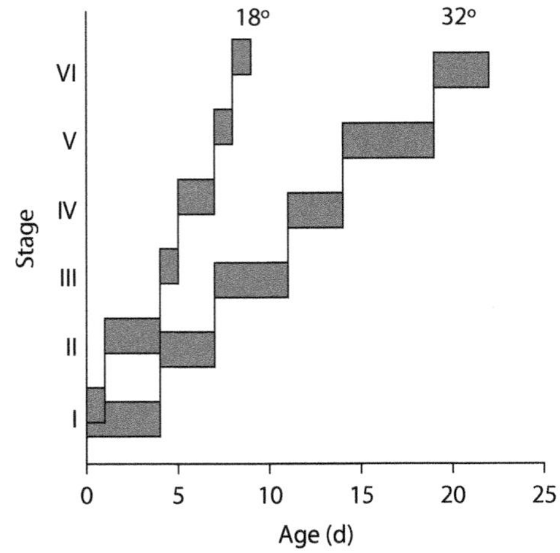

Fig. 6.1 Example of the transitions between instars (stages) of an arthropod that could be represented by a duration-stage model. The gray bar represents the duration of the instar; vertical black lines between bars indicate molting and thus the transition to the next instar. Shown here is the duration of the first seven aquatic larval instars for the mosquito *Anopheles gambiae* (the primary vector of malaria) at 18° C and 32° C, based on data in Bayoh and Lindsey (2003).

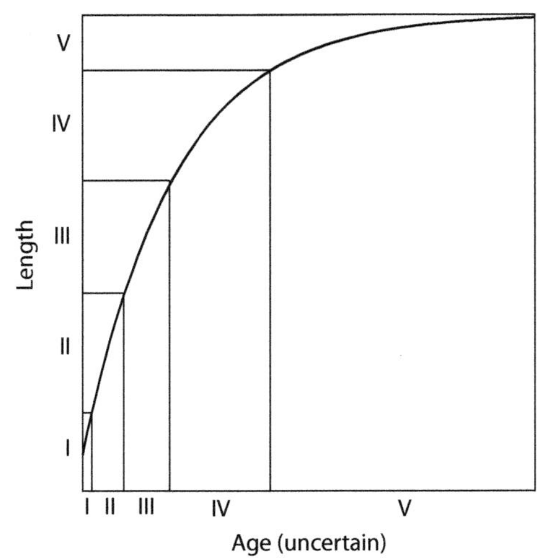

Fig. 6.2 Example of using a growth-stage model, in which continuous growth of an organism is approximated by a number of discrete growth stages. In this case the organism's growth is divided up into five size stages of approximately equal width (indicated by Roman numerals). Those size stages correspond to age intervals of varying duration.

are separated by an inter-molt period, which typically increases with age, but also often depends on temperature or other factors (Fig. 6.1). Development of actual body tissue does not increase in these sharp steps, rather the animal essentially fills the new shell with fluid shortly after each molt, then grows to replace the fluid with new tissue before the next molt. Natural categories in stage-structured models would also include the dramatic changes in morphology, habitat, or trophic level associated with changing life history stages, such as going from an egg to a free-swimming pelagic larval stage to a benthic juvenile stage of a marine crustacean, or the number of leaves in some orchid plants. We will call this type of stage model a *duration-stage* model.

The second kind of stage-structured model approximates continuous growth, as shown, for example, in Fig. 6.2. In this example, an organism grows rapidly while young and small, after which growth slows and a large size is reached asymptotically (as would be represented by the von Bertalanffy growth function; Box 3.8). We call this type of model a *growth-stage* model. Use of this kind of stage-structured model is motivated by the same considerations that motivate size-structured models in Chapter 5 (i.e. in many organisms, size is a better predictor of vital rates such as survival and fecundity than is age), as discussed in Section 1.3 and at the beginning of Chapter 5.

6.2 History of development of stage-structured matrix models

To understand the motivation for the development of stage-structured models, consider the basic goal: given either the discrete developmental stages in Fig. 6.1 or the continuous growth (and size-dependent demographic rates) in Fig. 6.2, how could we fashion a matrix model that multiplied a stage vector once a year to produce the correct stage vector in the following year? If this were possible, it would allow the use of all of the analytical tools for Leslie matrices that we described in Chapters 3 and 4. However, in both cases we encounter problems. For the *growth-stage model* (e.g. Fig. 6.2), the fact that all movement between stages must occur over the same time interval (e.g. one year) means that for the model to be correct, the size in stage 1 would have to be the size at age 1, and the size at stage 2 would have to be the size at age 2, and so on. But this would essentially be an age-structured model.

For the *stage-duration model*, with a constant time interval for the model calculation (e.g. one year, or even one day for the mosquito in Fig. 6.1) and different durations for each stage specified, an accurate matrix model of reasonable dimension is impossible. This is because the model time step would have to be chosen as the shortest stage duration, and then longer stages would have to be split up into multiples of that time interval in order to capture the correct progression of individuals through stages over time. For this reason, the usual solution has been to resort to a delay equation model for duration-stage situations. We describe these in Section 6.5, near the end of this chapter. With these limitations in mind, we now describe the historical development of stage-structured models.

6.2.1 *Early development of stage models*

We have just explained how growth-stage models cannot be constructed properly without simply making them into an age-structured model. This raises the question of how these models developed in ecology in the first place. The answer is that they developed by the somewhat arbitrary adoption of a matrix that had a specified fraction of individuals grow from one stage to the next in each year, while a different fraction remained in the same stage. For example, a model with four stages could have the transition matrix L:

$$L = \begin{bmatrix} p_1 & f_2 & f_3 & f_4 \\ g_1 & p_2 & 0 & 0 \\ 0 & g_2 & p_3 & 0 \\ 0 & 0 & g_3 & p_4 \end{bmatrix}. \tag{6.1}$$

Where p_i is the fraction of individuals in stage i that survive and remain in stage i each year, g_i is the fraction of individuals in stage i that survive and grow to stage $i+1$ each year, and f_i is the number of offspring created per year by each stage i individual. In these models, $p_i + g_i \leq 1$. Notice that this is very similar to a Leslie matrix, except that in a Leslie matrix the diagonals (the p_i terms) are all zeros, because individuals of age i always advance to age $i+1$ the next year and none stay behind at the current age.

Matrices like Eq. (6.1) are often called a Lefkovitch matrix, though the matrix proposed by Lefkovitch (1965) actually allowed nonzero elements at locations other than the diagonal, the sub-diagonal, and the first row of the matrix. We will refer to models like Eq. (6.1), with nonzero values in the first row, in the diagonal, and in the sub-diagonal, as *some-grow/some-stay* (SG/SS) stage models.

In his original paper, Lefkovitch (1965) developed a stage-based model for laboratory populations of the cigarette beetle, *Lasioderma serricorne*. He divided the population into four stages (egg, larva, pupa, adult), and estimated the transition probabilities in the matrix $\boldsymbol{L}$ using two years' worth of observations with a three-week census interval. Interestingly, the initial attempt at estimation produced a number of negative values for the elements of $\boldsymbol{L}$, which Lefkovitch acknowledged was biologically impossible, even though the resulting matrix had an asymptotic growth rate that matched observations. To resolve that problem he made all of the negative entries of $\boldsymbol{L}$ zero (and adjusted several others based on external information), to obtain the matrix

$$L = \begin{bmatrix} 0 & 1.98 & 0 & 0 \\ 1.20 & 0.99 & 27.78 & 15.31 \\ 0.36 & 0.19 & 0 & 0 \\ 0 & 0.84 & 0 & 0 \end{bmatrix}. \tag{6.2}$$

This matrix also had a dominant eigenvalue that approximately matched the rate of geometric growth in the laboratory populations. However, it is unusual in that it suggests that only larvae (stage 2) produce eggs, and that pupae and adults (stages 3 and 4, respectively) produce larvae directly, skipping the egg stage. Lefkovitch explained this as an artifact of some stage transitions happening more quickly than the three-week census interval, but it foreshadows some of the problems with biological realism in stage-based models that we will see in later applications.

In another early example of stage-based modeling, Usher (1966) applied a SG/SS stage-structured model to a problem involving the harvest of a pine forest. His motivation was that renewable resources would be more likely to be measured in size attributes rather than age. His model followed the convention in forestry modeling that trees that were dying would be removed by harvest, hence separate natural mortality and harvest mortality need not be included in the model. As a consequence, the sum of the fraction growing and the fraction staying was set equal to 1 for each stage. The value of the dominant eigenvalue λ for the matrix was determined from the model parameters, and harvest was determined as the removals that would reduce it to 1.

A transformative publication of a growth-stage matrix was Bosch (1971); transformative because it drew rebuttals that described many of the problems encountered when trying to represent continuous development with a SG/SS matrix. Bosch formulated a 3×3 stage matrix model of California coastal redwoods (*Sequoia sempervirens*) with the three stages: 0–200 years of age ("young"), 200–800 years ("mature"), and >800 years ("old"). He determined parameter values from available information on redwoods to obtain a SG/SS matrix, $\boldsymbol{L}$:

$$L = \begin{bmatrix} 12 & 26 & 6 \\ 0.30 & 0.92 & 0 \\ 0 & 0.18 & 0.67 \end{bmatrix}. \tag{6.3}$$

This matrix has a value of $\lambda = 1.973$, which implies very rapid geometric growth, so Bosch's conclusions concerning the ability to sustain very high harvests while satisfying conservation goals were very optimistic. In their rebuttals (or "debunking", as the editors of *Science* described it) of Bosch's model, Brussard et al. (1971), Halbach (1971), and Diem and McGregor (1971) identified a number of ecological problems with Bosch's approach.

Some were simple errors in the parameterization; for example, $p_i + g_i > 1$ for stage 2, implying that new trees were created out of thin air. However, the most salient criticism was Diem and McGregor's observation that the model was "conceptually wrong," because the fractions moving from one stage to the next in a single year would depend on the age structure *within the stage*. That is, the number of 200–800 year-olds moving into the >800-year-old stage depends on whether that stage is mostly 200–400 year-olds or mostly 400–800 year olds. That within-stage structure will, in turn, depend on how many 0–200 year-olds there had been in the previous time step. Consequently, they noted that to predict future populations with such a matrix (and for λ to be a valid estimate of the asymptotic growth rate), the abundance at time $t + 1$ needed to depend only on the abundance at time t, not on abundances at prior times (this is essentially the same as our description of the definition of state being Markovian in Chapter 1).

Bosch's attempt to represent a population of >800-year-old trees with a three-stage model is an extreme example of a problem that Vandermeer (1975) more fully demonstrated conceptually. Lefkovitch (1965) had proposed that the stage matrix could be derived from an age-structured Leslie matrix, so Vandermeer (1975) formulated an age-structured model that was equivalent to a stage-structured model with stage increments small enough that each stage could consist of several ages. His model clearly demonstrated that only the individuals at the ages near the end of each stage would move to the next stage in a given time step, and that the others, the most recent arrivals in the stage, would stay in that stage. This means that (as Diem and McGregor (1971) had observed) the transition probabilities in the matrix, i.e. the proportions of each stage staying and advancing, would have to vary depending on the within-stage age structure. From that he concluded that it was only appropriate to apply the model to a population that had already reached a stable stage distribution, because only then would the within-stage structure be constant over time. Doing otherwise would not be possible, because the stage-structured model does not keep track of the true age structure within each stage, which would cause an error in representing true population behavior during the approach to the stable stage distribution.

Later, Vandermeer (1978) noted that the errors incurred in choosing stage categories were of two types: "errors of estimation" and "error of distribution." The former was the error due simply to having a small sample size for estimating parameter values if the category was narrow with respect to age (this is the sort of error that led Lefkovitch (1965) to obtain negative estimates for some entries in his projection matrix). The latter was the error that Vandermeer had identified in his 1975 paper, in which the stage includes individuals who had just entered the stage, but in reality would not be moving out of that stage in a single time step. This error would increase as the stage category became wider. He proposed a method of choosing the optimal category size given this tradeoff, but it depended on cumbersome simulations; there was no simple rule of thumb.

In spite of these structural problems with the stage-structure approach, identified within a decade after Lefkovitch's original 1965 paper, development of stage-based models continued to increase. In Section 6.2.2 we describe several examples in which stage-structured models appear to match or outperform age-structured alternatives.

6.2.2 *Early successes in stage-structured modeling*

One of the most analyzed stage-structured models is that of the small, herbaceous, semelparous plant, teasel (*Dipsacus sylvestris*). Teasel seeds sprout into leafy rosettes, which live for one or several years until maturing into a flowering stalk, which produces seeds and then dies (Fig. 6.3). Motivated by the fact that Werner (1975) had shown that whether a teasel plant would remain vegetative or flower in a particular year was better predicted by rosette size than rosette age, Werner and Caswell (1977) compared the performance of a (non-SS/SG) stage-structured model (Fig. 6.3a) to that of an age-structured model (Fig. 6.3b) in a situation for which they had both age and stage data. The model was non-SS/SG because any of the rosette stages can immediately mature into the flowering stage. The data were from experimental growth of a number of teasel-free plots into which teasel seeds were sown. These populations were all monitored over time, and seed production was counted, but seeds were not allowed to feed back into the population (in order to follow a single cohort and be able to track age). Because seed production is not the same as actual population growth, values of λ could not be compared to observations. Instead, Werner and Caswell (1977) compared the age- and stage-structured models in terms of predicted times to the first occurrence of flowering. The stage-structured model was more accurate than the age-structured model, which consistently overestimated the time to first flowering.

It is difficult to tell exactly what caused this difference. The age and stage matrices for each plot both had the same age-structured growth for the first two ages/stages (i.e. sub-diagonal terms only, with identical values for the corresponding terms in each model). After those two ages, individuals in the age-structured model had a certain probability of jumping to the flowering stage (this probability increased with age), while in the stage-structured model they had a number of more circuitous routes, because individuals in the smallest rosette stage could move to the largest rosette stage in a single year (and the probability of flowering increased with size). These additional paths allowed the stage-structured model to correctly predict earlier ages at first flowering. The fact that rosette size

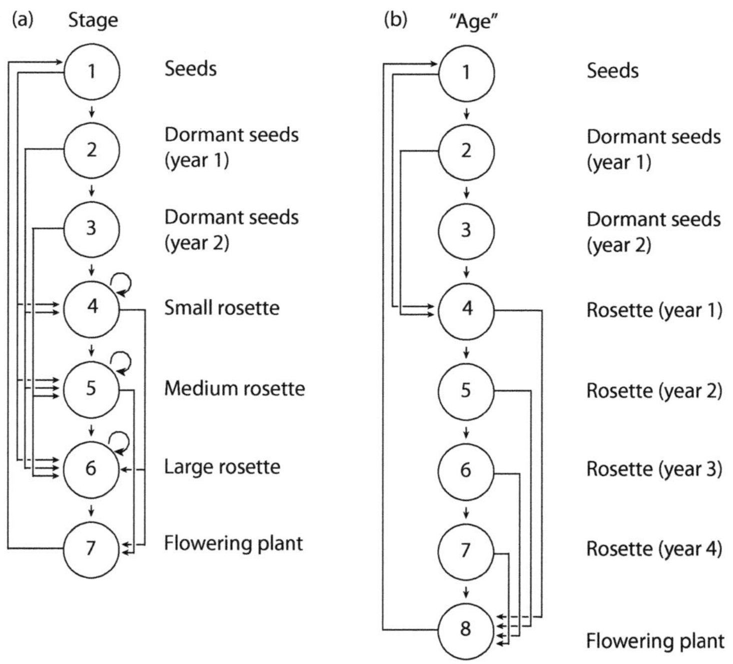

Fig. 6.3 Life cycle diagrams of the stages and possible stage transitions in the (a) stage-based and (b) age-based versions of the teasel model (Werner and Caswell, 1977).

was a better predictor of reproduction than age in this species, and that rosette growth is variable enough to allow small rosettes to grow into the largest class in one year, implies the differences among individuals in size at age would lead to inaccuracies in any age-structured model of this species.

Sauer and Slade (1987) also compared the outcomes of age- and stage-structured models in an effort to see whether stage-based models, which required less field sampling effort to parameterize (because individuals need not be aged), could adequately describe population dynamics (at less cost). In a series of papers, they developed stage-structured models for a number of small mammal populations (Sauer and Slade, 1985; 1986a; 1986b). In one case they had both age-specific and size-specific (body mass) data for Uinta ground squirrels (*Spermophilus armatus*), so they formulated both a size-based stage-structured model and compared it to an earlier age-structured model of the same population (Sauer and Slade, 1987; Slade and Balph, 1974). The models consisted of five size categories (based on body mass) and six ages, respectively. For the age-structured model they used a typical Leslie matrix. The models were parameterized from tagging, which resulted in the stage-based matrix being fully populated, reflecting both positive and negative growth (shrinkage). They constructed their stage-based model using two separate matrices, $\boldsymbol{A}$ and $\boldsymbol{B}$, for reproduction and growth, respectively:

$$\boldsymbol{A} = \begin{bmatrix} p_{11} & p_{12} & p_{13} & p_{14} & p_{15} \\ p_{21} & p_{22} & p_{23} & p_{24} & p_{25} \\ 0 & 0 & 1 & 0 & 0 \\ 0 & 0 & 0 & 1 & 0 \\ 0 & 0 & 0 & 0 & 1 \end{bmatrix} \tag{6.4a}$$

$$\boldsymbol{B} = \begin{bmatrix} 0 & 0 & 0 & 0 & 0 \\ 0 & 0 & 0 & 0 & 0 \\ f_{31} & f_{32} & f_{33} & f_{34} & f_{35} \\ f_{41} & f_{42} & f_{43} & f_{44} & f_{45} \\ f_{51} & f_{52} & f_{53} & f_{54} & f_{55} \end{bmatrix}, \tag{6.4b}$$

where p_{ij} is the per capita production of size i by size j (the first two size classes were small and large offspring), and f_{ij} is the proportion of size j that moves to size i in one year (squirrels were not observed to shrink back to size classes 1 or 2, only to size 3 (yearlings)). The age vector was updated each year using the product of the two matrices,

$$\boldsymbol{N}_{t+1} = \boldsymbol{ABN}_t. \tag{6.5}$$

Notice that this approach makes it straightforward to write out the reproduction and growth components separately (Eq. (6.4)), but introduces an unusual artifact in the model. The matrix $\boldsymbol{AB}$ ends up being very unlike a usual Lefkovitch matrix, because growth probabilities now appear in the first row (you can see this by writing out the multiplication for the first few entries). Conceptually, this is equivalent to making the model have two steps; the first multiplication $\boldsymbol{BN}_t$ accounts for one year's worth of growth, including the growth of young-of-the-year into the third stage (yearlings). Then the second multiplication by $\boldsymbol{A}$ performs reproduction. This means that if (hypothetically) the population consisted only of stage 1 individuals (recently born young-of-the-year) in year t, within the year those individuals would grow into stage 3 (the yearling class, and possibly some into the larger classes as well) and reproduce at the stage 3 rate, producing a new cohort of stage 1 offspring in year $t + 1$. In reality, squirrels would not breed until their second year (when they were stage-3 yearlings), and the new cohort would not be observed until $t + 2$. Thus the model has a new problem with accounting for time lags, in addition to the problem introduced by stage-structure itself. However, Sauer and Slade (1987) focused on asymptotic dynamics, so they did not encounter any issues that their structural error would produce in the transient dynamics.

Sauer and Slade (1987) found that the age- and stage-structured models showed few differences in estimated population growth rates (λ). Given that alignment, they recommended that stage models be used because population dynamics could be described with less sampling effort by determining only the masses and inter-stage movements of size classes, without aging individuals. In one paper (Sauer and Slade, 1987), they even went so far as to conclude that "one need not know the ages of the animals in the population to examine the population dynamics." Of course this would not be strictly true in the sense that rates of individual growth over time must be known to estimate the transition probabilities for the stage matrix in the first place, even if we ignored the problems representing non-asymptotic transient dynamics with this type of model.

6.2.3 *Early applications*

The field of stage-structured models eventually "matured" to the point that researchers no longer focused on fundamental questions regarding their accuracy in portraying real populations, but rather concentrated on applications to practical problems. There are many examples of these applications, but we demonstrate such applications using efforts by Deborah Crouse, Larry Crowder, and colleagues to address the problem of declining numbers of loggerhead sea turtles (*Caretta caretta*) on the southeast Atlantic coast of the USA.

Lacking reliable age-specific demographic information for loggerhead turtles, Crouse et al. (1987) formulated a SS/SG Lefkovitch matrix model. Their goals were to determine how the declining loggerhead populations might best be sustainably managed, and to reveal important sensitivities of loggerhead population dynamics. Their modeling approach was to create stages that included individuals that shared the same fecundities and survivals. Being aware of the fact that abundances of age classes within stages were not identical (from Vandermeer, 1975; 1978), and that stages differed in duration, they formulated an ad hoc method to populate the elements of the Lefkovitch matrix from point estimates of survival and growth, while allowing stage transitions to include only the oldest age class within the stage. To do this, they noted that if the duration of stage i is d_i and the annual survival in stage i is p_i, then the stable age distribution within the stage would be $1, p_i, p_i^2, \ldots p_i^{d_i-1}$ (note that this presumes a stable stage distribution with $\lambda = 1$). Then the proportion remaining (and surviving) in stage i would be

$$P_i = \left(\frac{1 + p_i + p_i^2 + \cdots p_i^{d_i-2}}{1 + p_i + p_i^2 + \cdots p_i^{d_i-1}} \right) p_i, \tag{6.6}$$

which simplifies to

$$P_i = \left(\frac{1 - p_i^{d_i-1}}{1 - p_i^{d_i}} \right) p_i. \tag{6.7}$$

Using similar logic, the proportion surviving and growing into the next stage is

$$G_i = \left(\frac{1 - p_i}{1 - p_i^{d_i}} \right) p_i^{d_i}. \tag{6.8}$$

This meant that the survival (P_i) and growth (G_i) elements of their Lefkovitch matrix incorporated the effects of stage duration, in addition to survival and growth. It also reflected the fact that they actually had enough information to formulate an age-structured model (perhaps with lower quality parameter estimates).

Crouse et al. (1987) analyzed the elasticities (as in Section 3.5 for age-structured models) of their seven-stage model to determine how changes to the survival of each stage would affect the asymptotic growth rate. Their primary conclusion was that management should focus on reducing the mortality of large juveniles (stage III) through the use of turtle excluder devices (TEDs; devices placed in shrimp trawl nets that eject turtles through a flap in the net) rather than another proposed management action, protecting eggs (stage 1) on nesting beaches. This peak in elasticities was likely due to the mortality between the egg and large juvenile stages being very high, so that marginal increases in egg survival had little benefit.

Crowder et al. (1994) refined this model approach to focus on the consequences of the use of TEDs. They simplified the model into a five-stage matrix, consisting of eggs, small juveniles, large juveniles, subadults, and adults, with projection matrix $\boldsymbol{L}$:

$$\boldsymbol{L} = \begin{bmatrix} 0 & 0 & 0 & 4.665 & 61.896 \\ 0.675 & 0.703 & 0 & 0 & 0 \\ 0 & 0.047 & 0.657 & 0 & 0 \\ 0 & 0 & 0.019 & 0.682 & 0 \\ 0 & 0 & 0 & 0.061 & 0.8091 \end{bmatrix}. \tag{6.9}$$

To estimate the transition probabilities for this matrix, they used a slightly different approach from Crouse et al. (1987) that also accounted for the effect the population growth rate, λ, will have on the within-stage age structure. If $\lambda \neq 1$, the stable age distribution within the stage would now be 1, 1, p_i/λ, $(p_i/\lambda)^2, \ldots, (p_i/\lambda)^{d-1}$. Based on that distribution, the proportion of individuals growing into the next stage in a given year, γ_i, is

$$\gamma_i = \frac{(p_i/\lambda)^{d_i-1}}{1 + p_i/\lambda + (p_i/\lambda)^2 + \cdots (p_i/\lambda)^{d-1}}, \tag{6.10a}$$

which simplifies to

$$\gamma_i = (p_i/\lambda)^{d_i} - \frac{(p_i/\lambda)^{d_i-1}}{(p_i/\lambda)^{d_i}} - 1, \tag{6.10b}$$

where p_i and d_i are the annual survival and duration of stage i, as before. Then, $P_i = p_i(1 - \gamma_i)$, i.e. the probability of surviving and not moving to the next stage, and $G_i = p_i\gamma_i$, i.e. the probability of surviving and moving to the next stage (these formulas reduce to the same ones as Eqs. (6.7) and (6.8) if $\lambda = 1$; Caswell (2001)). They then calculated elasticity for each parameter in the matrix $\boldsymbol{L}$. The profile of elasticities to survival, p_i, was similar to that in Crouse et al. (1987), indicating protection of large juveniles would provide much greater proportional change in λ than a similar improvement in survival of eggs (Fig. 6.4a).

To project the short-term transient response to implementation of TEDs, Crowder et al. employed an age-structured Leslie matrix instead of their Lefkovitch matrix. Their motivation for doing this was that the turtles take 20 years to mature, so there would be time lags in the response to TEDs that the stage-based model would not capture. To create this model they simply expanded the stage-based matrix into a 54×54 matrix, with each stage expanded into the number of ages corresponding to its duration, and the

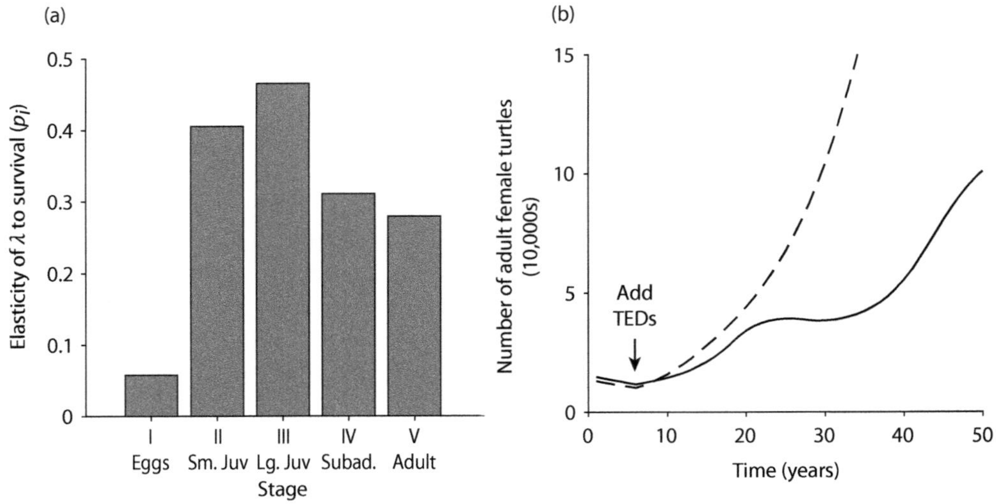

Fig. 6.4 Characteristics of the stage-based loggerhead sea turtle model (Eq. (6.8)). (a) Elasticities of the population growth rate, λ, to the survival of each stage. (b) Transient dynamics of the five-stage model (dashed curve), compared to the age-structured equivalent (solid curve) after turtle excluder devices (TEDs) are introduced.

sub-diagonal elements equal to the corresponding p_i terms. The transient behavior of the model after the mortality rate of large juveniles, subadults, and adults was reduced by 37% by the TEDs (assuming the population was declining at the stable stage distribution prior to adding TEDs) is oscillatory (solid curve in Fig. 6.4b), reflecting the behavior of a species with greatest reproduction near 20 years, and a fairly narrow reproductive age distribution (as we would expect based on the results in Section 4.2.4 on transient dynamics, specifically the interpretation of Fig. 4.7f, which mimics Taylor (1979), indicating that ρ increases as the modal age of reproduction declines and the standard deviation increases). By contrast, the five-stage matrix model has a very short transient and quickly approaches exponential growth (dashed curve in Fig. 6.4b), because it does not contain the same time lags as the age-structured version (additionally, recall that the transition probabilities in the stage-structured model were calculated based on the assumption of a stable age distribution within each stage (Eq. 6.10), so any analysis of transient dynamics would violate that assumption).

6.2.4 *Stochastic stage-structured models*

In 2009, Perry de Valpine identified the key role that stochastic individual development (i.e. variable stage durations) could play in the dynamics of stage-structured populations (de Valpine, 2009). He redefined the state variable in stage-structured models by assuming that there was a range of development schedules (i.e. the duration of each stage) that individuals could follow. That is, there was a specified number of stages, but different individuals randomly following different schedules would progress through the stages at different rates. The state variable he used was $N(\mathbf{s}, a, t)$, which is the density of individuals on development schedule $\mathbf{s}$ (a vector of the time spent in each stage) in age a at time t. The state of the entire population at each time is calculated by integrating over all development schedules, $\mathbf{s}$, then advancing them to the next age at each discrete time. de Valpine (2009) implemented his formulation in an integral projection model, and provided methods for easily calculating values of λ, and sensitivities and elasticities of λ, from simple Monte Carlo cohort simulations.

For example, de Valpine (2009) used a three-stage model of a long-lived desert cactus, *Coryphantha robbinsorum*, which had the following projection matrix (based on data from Fox and Gurevitch 2000):

$$\boldsymbol{A} = \begin{bmatrix} 0.67 & 0 & 0.56 \\ 0.02 & 0.84 & 0 \\ 0 & 0.14 & 0.97 \end{bmatrix}. \tag{6.11}$$

This projection matrix has growth rate $\lambda = 0.9998$ and stable stage distribution equal to (0.58, 0.07, 0.34). He then showed that if the development schedule for stage 2 was represented as a gamma distribution of stage duration times, there was a huge range of different possible distributions of duration times that would all produce the same stable stage distribution. This is shown in Fig. 6.5a as the different possible combinations of means and standard deviations of the stage-2 duration time that produce the same stable stage distribution. Taking two of those combinations as an example (mean durations of approximately 6.5 years and 10 years; points b and c in Fig. 6.5a), we see that the corresponding age distribution within stages differs greatly: Fig. 6.5b and 6.5c show the age distributions within each stage for the mean and standard deviations indicated by points b and c, respectively, in Fig. 6.5a. Longer duration in stage 2 leads to a broader age distribution in that stage, and (somewhat counterintuitively) a stage 3 distribution that is shifted to younger ages (Fig 6.5c). The latter is because most of the stage 3 individuals are those that moved quickly through stage 2, rather than the smaller proportion of individuals that survived staying in stage 2 for >ten years. From our experience with age-structured models we can see that these two populations would have very different transient dynamics, despite the same stable stage distribution.

This example highlights the difficulty in estimating even a probability distribution of stage durations: the data do not constrain the estimate to a single distribution, rather to a range of means and variances. de Valpine et al. (2014) later reviewed more fully the ways that developmental variability can affect stage-structured models.

6.3 Problems with stage-structured models

The examples with the SG/SS models, the most common version of the Lefkovitch matrix, leave one in a quandary. We know that maintaining the correct numbers at each size-based stage is a good idea when fecundity depends more closely on size than age, but stage is not a valid state variable. Lack of knowledge about growth rates or developmental times (or variability in those rates) was the most common reason why some turned to stage-structured models, but we know from models with continuous size and time in Chapter 5 that the nature of the growth rate of individuals has an important impact on population dynamics. In some cases, even if researchers have both size and age information, they choose to use SG/SS models anyway, as though there is something special about them, or perhaps for simplicity because they allow smaller matrices (which was particularly important due to computational limitations prior to the 1990s).

The primary, most general problem with stage-structured models is that stage is not a state variable because it does not satisfy the requirement of containing the information necessary to predict the future of an individual within it. That is, it is not Markovian

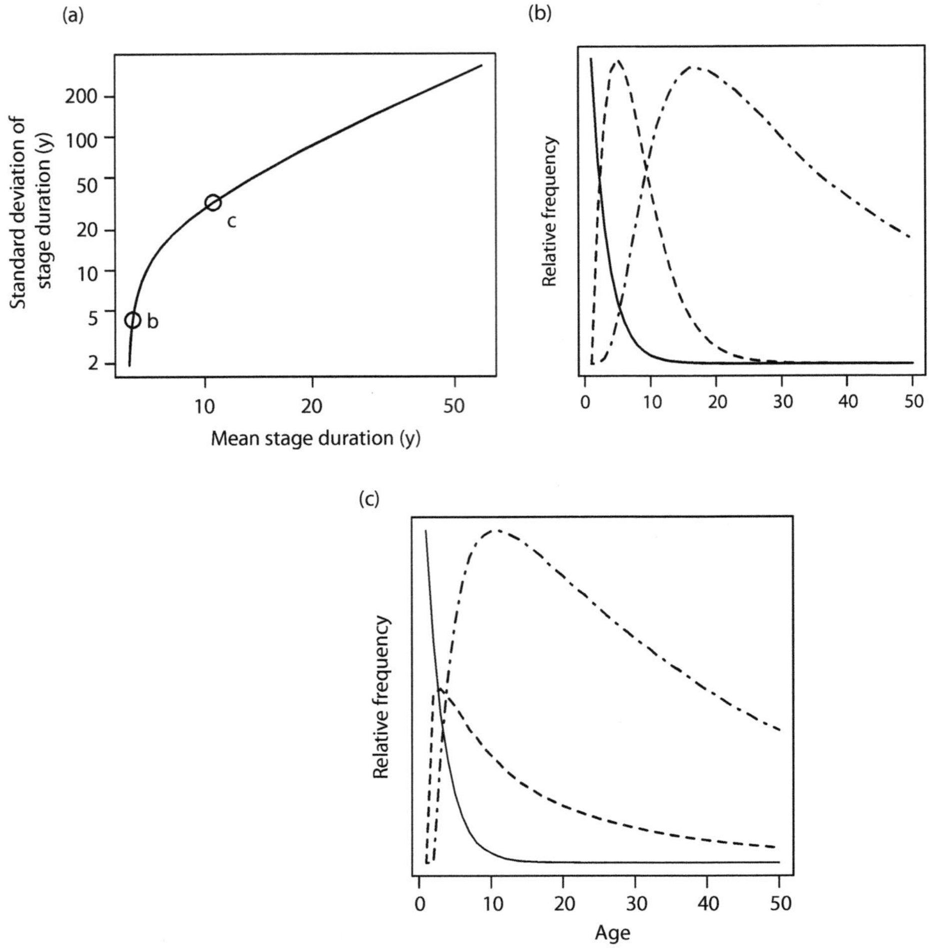

Fig. 6.5 The range of age distributions in a stochastic stage model of a long-lived desert cactus. (a) The curve indicates the combinations of mean and standard deviation of the stage 2 duration distribution that produce identical stable stage distributions. (b, c) The age distributions within each stage (stage I, solid curve; stage II, dashed curve; stage III, dot-dash curve) for models with the stage duration distributions labeled b and c, respectively, in panel (a). Created using code and data provided by de Valpine (2009).

(see Chapter 1). This is essentially a restatement of the prescient comment by Diem and McGregor (1971) regarding Bosch's (1971) model of redwoods. It also follows from the observation of Vandermeer (1975) regarding the presence of individuals of at least two different ages within a stage, one just having entered the stage, and the other about to grow out. This problem has been addressed by describing the age distributions within each stage in a number of different ways (see our description of the loggerhead turtle model in Section 6.2.3, and further discussion in Caswell 2001). However, all such approximations have to assume specific population conditions (e.g. having reached a stable age distribution), hence would not apply to population growth in general. Indeed, most populations are not at equilibrium or asymptotic conditions, so having models that correctly represent transient dynamics is probably more important than correctly estimating λ (Hastings, 2001; 2010).

The fact that stage-structured models lead to errors in depicting somatic growth because stage is not an adequate state variable is somewhat ironic since the definition of state and the fact that size was a better *i*-state variable in some cases were primary motivations for the use of stage-structured models. The errors arise because when stage-structured models are developed, attention is usually focused on the choice of what Vandermeer (1975) called the categorization variable (i.e. the *i*-state variable, size in this case), rather than how the categorization variable itself developed (or grew) in the modeled individuals. The emphasis was on having the most appropriate variable to describe the adult reproduction, rather than on how individuals developed through earlier stages to become adults. This was largely driven by the practical consideration of the expense of determining individual growth rates (primarily the cost of determining age).

The fact that stage is not an adequate state variable leads to model behavior being artificially controlled by the details of model structure, rather than by the values of the parameters that are supposed to represent growth. Specifically, a SS/SG approach leads to a gradual spreading of the abundance of a single cohort over all stages, rather than a developmentally accurate progression of a cohort through successive stages. We can see this in the SG/SS stage-structured model used by Crowder et al. (1994; Eq. 6.9) by comparing development of a cohort of an arbitrary number of individuals starting from stage 1 or age 1 at time $t = 1$ in a stage-structured model with an age-structured model (Fig. 6.6). At time $t = 2$ there will be individuals in stages I and II, then at $t = 3$ there will be individuals at ages I, II, and III, and so on. As time goes on, the maximum stage occupied by members of the cohort will equal the age of the cohort (i.e. the number of matrix multiplications to date). The minimum occupied stage will always be stage II (or the smallest stage that has a "stay" entry on the diagonal of the matrix) (e.g. in the loggerhead model all hatchlings grow into small juveniles in one year, but then turtle abundance in that stage will decay geometrically into the future, indefinitely). Note that this spreading of a cohort over stage is not the same as the spreading over size of cohort size distributions that we saw in Chapter 5. That spreading is determined by the nature of individual growth rates. In summary, stage-structured models have the structural feature that after *a* years the cohort size distribution will always range between the smallest possible size and the size of the *a*th stage. This is an unavoidable (and nonbiological) artifact of the SG/SS model structure: with a nonzero diagonal element for stage 1, some individuals will remain in stage 1 indefinitely, and with nonzero sub-diagonal elements, some individuals can advance through adjacent stages in a single year, regardless of the actual biological development times involved. This lack of realistic behavior is why stage-structured models are usually presented with the caveat that they only apply to populations that are already at the stable stage distribution.

To compare this behavior of a stage-structured model to actual growth through the designated stages, we also plotted the stage distribution of the age-structured version of the loggerhead model. To do this we simply noted the stage the cohort was in for each year of the simulation (Fig. 6.6). The result is simply that the cohort stays in each stage for its duration (noted on the abscissa of Fig. 6.6). The contrast with the stage-structured model is stark: after six years, the cohort is still entirely within the small juvenile stage, which is supposed to last seven years. By contrast the stage-structured model already has some individuals in the mature adult stage. Later, after 52 years, the age-structured model correctly places all individuals in the mature adult stage, while the stage-based model has individuals that are still in all of the younger stages (except hatchlings).

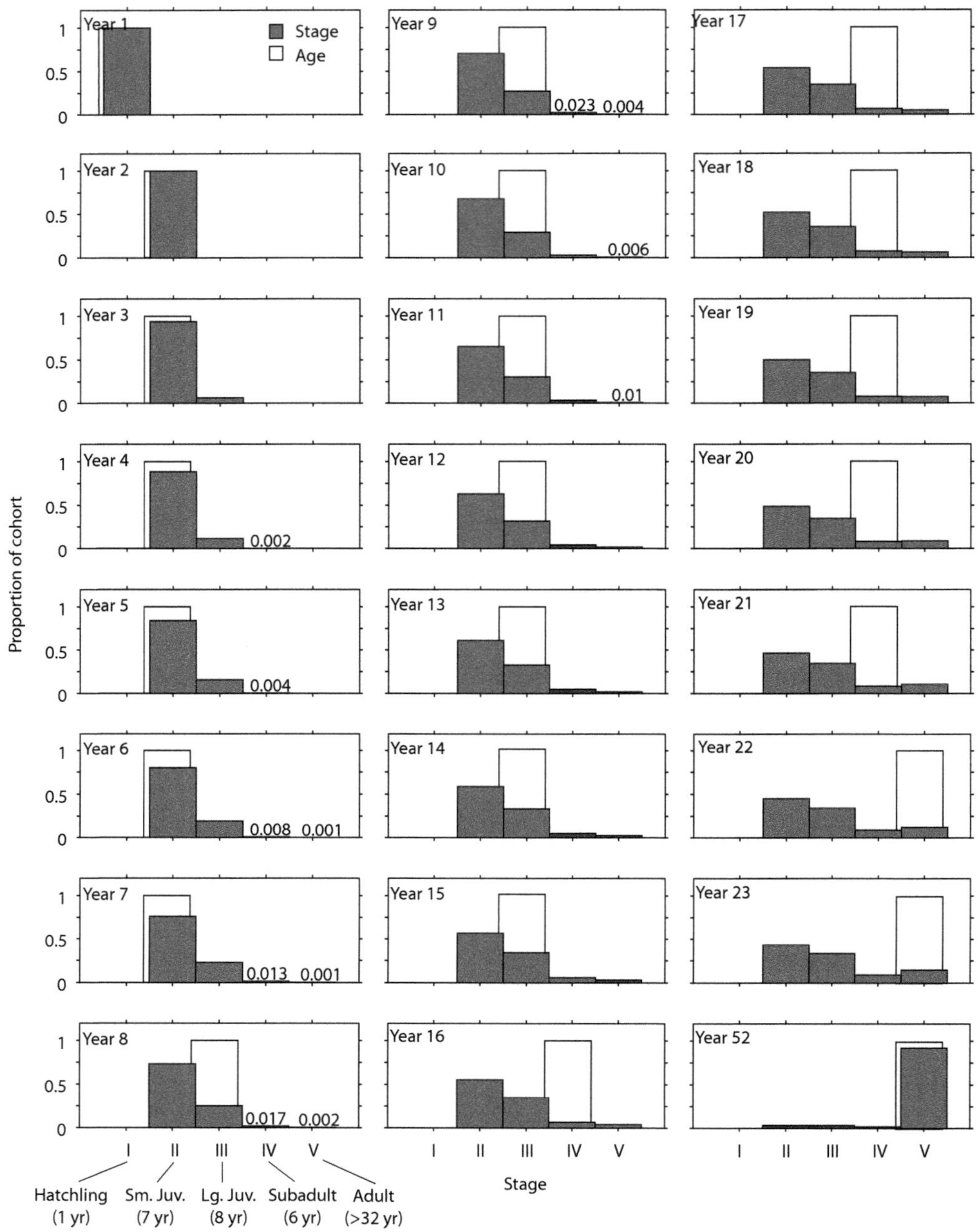

Fig. 6.6 Comparison of the stage distribution produced by the stage model of the loggerhead sea turtle (Eq. (6.9); dark bars) to the stage distribution produced by the actual age-structured model (white bars), beginning with a single cohort of age-1 individuals in stage I and projecting forward through 52 years (the maximum age in the age-structured model). The stage model and the age durations of each age are from Crowder et al. (1994). Numbers indicate the height of small, difficult-to-see bars in the stage-structured model, and the white bars are slightly offset to the left for visibility.

6.4 Possible better alternatives to stage-structured models

Werner and Caswell's (1977) model and others led to a general perspective that plants with highly variable growth rates and size-dependent individual rates are best modeled by stage instead of age (Caswell, 2001). Of course, if size is truly the relevant state variable, then a better solution would be to create a model that uses size as a state variable, but also preserves the age structure so that individuals pass through sizes over time in a realistic way. Law (1983) proposed a size-age-structured approach to this problem. In his model, the probability of surviving one year and moving from size class j at age a to size class k at age $a+1$ is p_{ajk}. He collected the p_{ajk}'s into a transition matrix $\boldsymbol{P}_a$ that gives the probability of an age-a individual of any size moving to any other size the following year:

$$\boldsymbol{P}_a = \begin{bmatrix} p_{a11} & p_{a21} & \cdots & p_{aX1} \\ p_{a12} & p_{a22} & \cdots & p_{aX2} \\ \vdots & \vdots & \ddots & \cdots \\ p_{a1X} & p_{a2X} & \cdots & p_{aXX} \end{bmatrix}, \tag{6.12}$$

where the size classes range from 1 to X. He also constructed matrices $\boldsymbol{B}_a$ that gave the probability of age-a individuals of each size producing size-1 offspring (this matrix had entries only on the first row and zeros elsewhere). Then he defined a transition matrix $\boldsymbol{T}$,

$$\boldsymbol{T} = \begin{bmatrix} \boldsymbol{B}_1 & \boldsymbol{B}_2 & \boldsymbol{B}_3 & \cdots & \boldsymbol{B}_A \\ \boldsymbol{P}_1 & 0 & 0 & \cdots & 0 \\ 0 & P_2 & 0 & \cdots & 0 \\ \vdots & \vdots & \ddots & \ddots & \vdots \\ 0 & 0 & \cdots & P_{A-1} & 0 \end{bmatrix}. \tag{6.13}$$

This matrix is multiplied by the matrix $\boldsymbol{N}_t$ (which consists of column vectors containing the size structure in each age class) to advance the population through size and age each year. Law (1983) called this a Goodman matrix, after Goodman's (1969) original formulation. Notice that the Goodman matrix is essentially a Leslie matrix, so it preserves the correct time lags in the population. This provides a way to model size-dependent demography without the artifacts introduced by using a non-Markovian state variable such as stage. The cost, of course, is that there are a huge number of matrix entries to parameterize, and performing matrix operations would have been computationally challenging (in 1983).

A better way to accomplish Law's (1983) goal is to utilize an integral projection model (Chapter 5), which allows the same type of analysis (discrete time, calculation of eigenvalues and sensitivities) but is much simpler to parameterize. For example, Chu and Adler (2014) examined size-dependent demographic rates for a variety of grassland plants, and found that in most cases both size and age were important predictors of growth and mortality. They then constructed IPMs with size- and age-dependent kernels to represent population trajectories. To do this they built a model in which each age class is represented as a distribution of sizes within that age, so the density of age-a individuals of size x at time t, $n_a(x,t)$ is updated by

$$n_a(x,t) = \int_\Omega K_{a-1}(x,y)\, n_{a-1}(y,t-1)\, dy, \tag{6.14}$$

where $K_{a-1}(x,y)$ is the IPM kernel for the transition from age $a-1$ to age a (see Section 5.4.3) (Childs et al., 2003; 2004; Ellner and Rees, 2006).

6.5 Replacement in stage-structured models

The reader may have noticed that in the examples of stage-structured models thus far, we have referred to the calculation of λ and various elasticities in the examples, as we did in Chapter 3 for Leslie matrices. However, we have not referred to calculation of the net reproductive rate, R_0 (Eq. (3.25)) despite the importance of lifetime reproduction in indicating whether a population will be persistent. The definition of R_0 is not clear-cut for stage-structured models. So instead, here we focus on defining a quantity that we can use to determine if a population is persistent, that has an interpretable biological meaning, and is not simply the growth rate of the population. We will first describe a process that is analogous to the calculation of R_0 for stage-structured models, but not exactly the same. This process centers on the concept of replacement, and was derived from the mathematics of matrices (Hastings and Botsford, 2006b), though we will not show the full derivation here. We will also describe an alternative approach developed by de-Camino-Beck and Lewis (2008) that does allow a direct calculation of R_0.

From Chapter 3, the value of R_0 in the Leslie matrix tells us whether the population will ultimately be increasing geometrically, because it is the replacement rate, or the lifetime reproductive success of each age-1 individual. If that number > 1, then $\lambda > 1$, and the population will grow. This question is more complicated in a Lefkovitch matrix than in a Leslie matrix for several reasons. First, individuals do not advance through the stages at the same rate (e.g. some stay, some go), so it is difficult to define what the "lifetime" is. Second, there may be multiple pathways to progress through the stages. For example, in the teasel model, seeds could advance to become either small, medium, or large rosettes in a single year, rather than having to pass through the stages sequentially. This means that there are multiple paths for replacement; e.g. seed → small rosette → flowering plant versus seed → small rosette → large rosette → flowering plant. By definition, persistence of any replacement path (or "loop" because it loops back to the initial stage) would mean that the population was persistent. Trivially, replacement could occur if any of the terms on the diagonal were greater than 1.

Hastings and Botsford's (2006b) persistence condition, which depended on the matrix being non-negative, was simply that the population would persist if the sum of the persistence of all possible closed loops in the stage model was greater than 1 (even if no single loop was persistent). This is illustrated with a life-cycle diagram of Thomson's (2005) model of an evening primrose population (Fig. 6.7). The full model is in Fig. 6.7a, and the five possible replacement loops are shown in Fig. 6.7b. The population will be persistent if one or more of the five loops is persistent. The contribution of each loop to persistence is determined by dividing the replacement around the loop (the product of all of the transfer terms around the loop) by the shortfall in the self-loops (the product of the absolute values of 1 minus the fraction staying in each stage). If the sum of the contributions to replacement over all of the loops is greater than 1, the population is persistent. Thus, if any single loop is persistent, the whole population would persist.

Mathematically, the calculation of replacement and a persistence condition for a loop is fairly simple to write out for a small number of stages. If there were only two stages (e.g. the lower two loops in Fig. 6.7b), the Lefkovitch matrix would be

$$\mathbf{A} = \begin{bmatrix} a_{11} & a_{12} \\ a_{21} & a_{22} \end{bmatrix}. \tag{6.15}$$

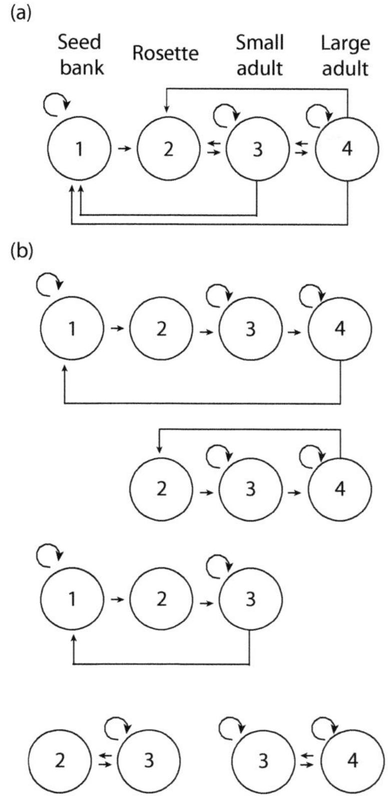

Fig. 6.7 Life-cycle diagrams for Thomson's (2005) model of evening primrose, Panel (a) is the complete model, and (b) is five closed loops within that model, any one of which would cause the population to persist if it satisfied the replacement criterion of Hastings and Botsford (2006b). The condition for persistence for each loop is the product of all straight line arrows, divided by the product of all absolute values of 1 minus the curved line arrows.

Then the persistence criterion would be

$$\frac{a_{21}a_{12}}{|\,(1-a_{11})\,(1-a_{22})\,|} > 1. \tag{6.16}$$

The interpretation of this expression is that the numerator is the rate at which stage-1 individuals make stage-2 individuals, times the rate at which stage-2 individuals make stage-1 individuals (this is the replacement loop). To understand the denominator, it helps to think of $(1 - a_{ii})$ as the probability of an individual leaving stage i in each year (because a_{ii} is the probability of staying). Thus we can think of $(1 - a_{ii})^{-1}$ as the average time spent in stage i, much in the same way we think of a flood that occurs with probability 0.01 as a 100-year flood. Therefore, the persistence criterion (Eq. (6.16)) is the rate of movement through the loop from 1 to 2 and back, multiplied by the average time spent in each stage. That is, similar to the original R_0 definition, Eq. (6.16) is the amount of reproduction per lifetime. Indeed, if $\boldsymbol{A}$ were a Leslie matrix, then $a_{22} = 0$, and Eq. (6.16) could be simplified and rearranged to $a_{11} + a_{12}a_{21} > 1$ (assuming $a_{11} < 1$), which is the same as the formula for R_0 (Eq. (3.26)).

The advantage of this approach in seeing whether a stage-based model is increasing is that it makes the effect of each element of the Lefkovitch matrix on persistence very clear, in a way that simply calculating the eigenvalue λ does not. The disadvantage is that the criteria get increasingly very complicated as the number of stages—and thus the number of possible loops—increases. Hastings and Botsford (2006a) write out the expressions for three and four stages, and provide the general calculation for any number of stages.

One situation in which the Hastings and Botsford (2006b) approach provides a simple view of the replacement persistence condition is with the common SG/SS stage models. For these models the persistence condition is the sum over a number of loops, one for each reproductive stage. Each of those terms consists of a fraction, with the numerator being the increase around that loop, i.e. the product of the fecundity at that age and the product of the growth terms up to that age (the G_i terms from Eq. 6.8), and the denominator being the product of the $(1 - P_i)$ "stay" terms from that loop (from Eq. 6.7). For example, the persistence condition for the loggerhead sea turtle model with the parameter values in the matrix in Eq. (6.9), is the sum of persistence of the two replacement loops,

$$R_0 = \frac{4.665 \times 0.675 \times 0.047 \times 0.019}{(1-0.703)(1-0.657)(1-0.682)} + \frac{61.896 \times 0.675 \times 0.047 \times 0.019 \times 0.061}{(1-0.703)(1-0.657)(1-0.682)(1-0.809)}$$
$$= f_4 0.0186 + f_5 0.00594, \tag{6.17}$$

where f_4 and f_5 are the fecundities at stages IV and V, respectively. As an example of an evaluation of sensitivity, if one asked whether it would be better to change fecundity by a certain factor at stage 4 or stage 5, the answer would clearly be to choose the term with the larger constant, stage 4. Ecologically, this is because relatively few individuals survive to stage 5, despite the higher per capita fecundity in that stage. This calculation yields $R_0 = 0.45$ (not persistent) which is consistent with the persistence condition based on the dominant eigenvalue: $\lambda = 0.95$.

de-Camino-Beck and Lewis (2008) later showed that it was possible to calculate R_0 directly (without writing out all of the loops) an alternative quantity that also can be used to check for population persistence if one decomposed the projection matrix $\boldsymbol{A}$ into a matrix of stage transitions, $\boldsymbol{T}$ (with elements t_{ij} that give the probability of moving from stage j to stage i in one time step), and a matrix of fecundities, $\boldsymbol{F}$ (with elements f_{ij} giving the number of stage i individuals produced by each stage j individual): $\boldsymbol{A} = \boldsymbol{T} + \boldsymbol{F}$. Then, using the same logic we applied to calculate R_0 in an IPM (Section 5.4.3), we can calculate the total number of offspring produced by a single cohort over each year of the cohort's life,

$$\boldsymbol{F n}_0 + \boldsymbol{F T n}_0 + \boldsymbol{F T}^2 \boldsymbol{n}_0 + \cdots = \boldsymbol{F}\left(\boldsymbol{I} + \boldsymbol{T} + \boldsymbol{T}^2 + \cdots\right)\boldsymbol{n}_0 = \boldsymbol{F}(\boldsymbol{I} - \boldsymbol{T})^{-1}\boldsymbol{n}_0, \tag{6.18}$$

where $\boldsymbol{n}_0$ is the initial stage distribution of the cohort. Each term in Eq. (6.18) represents the survival and growth of the cohort for one additional year (multiplication by $\boldsymbol{T}$) multiplied by the per capita fecundity of each stage ($\boldsymbol{F}$). The matrix $\boldsymbol{F}(\boldsymbol{I} - \boldsymbol{T})^{-1}$ is the *next generation matrix*; multiplying it by $\boldsymbol{n}_0$ gives the total number of offspring produced in the lifetime of one cohort. As with the next generation kernel in Section 5.4.3, we will refer to this matrix as $\boldsymbol{R}$. The quantity we seek is the dominant eigenvalue of $\boldsymbol{R}$.

Additionally, de-Camino-Beck and Lewis (2008) showed that this quantity could be written as the sum of the values corresponding to each of the replacement loops in the life-cycle diagram (e.g. Fig. 6.7). For the simple example we used above (Eq. (6.15)), the $\boldsymbol{A}$ matrix would be rewritten as

$$\boldsymbol{A} = \begin{bmatrix} a_{11} & a_{12} \\ a_{21} & a_{22} \end{bmatrix} = \begin{bmatrix} t_{11} & f_{12} \\ t_{21} & t_{22} + f_{22} \end{bmatrix}. \tag{6.19}$$

The initial cohort $\boldsymbol{n}_0$ could consist of individuals in both stage 1 and stage 2, so R_0 is the sum of reproduction happening in the loop from 1 to 2 and the self-loop in stage 2,

$$R_0 = \frac{t_{21} f_{12}}{(1 - t_{11})(1 - t_{22})} + \frac{f_{22}}{(1 - t_{22})}. \tag{6.20}$$

The expression on the right-hand side of Eq. (6.20) is algebraically equivalent to the similar expression with a_{ij} on the left-hand side of Eq. (6.16), and it has the same interpretation: the amount of reproduction produced by each replacement path, divided by the average time spent along the path. As with Eq. (6.16), the population will persist if $R_0 > 1$. So this persistence criterion is equivalent to that developed by Hastings and Botsford (2006b); in fact using the formula in Eq. (6.20) for the loggerhead turtle model yields the same solution that we gave earlier in Eq. (6.17). The advantage of examining persistence using de Camino-Beck and Lewis's R_0 calculation is that the calculation remains relatively simple as the number of replacement loops gets very large.

Finally, this calculation can be used to estimate an approximate generation time (again, just as in the IPM example in Section 5.4.3),

$$T_3 = \frac{\log R_0}{\log \lambda}. \tag{6.21}$$

Recall from Box 4.2 that this is one of the three definitions of generation time; it is not necessarily a mean age of reproduction like T_1 and T_2, but rather is defined as the time needed to increase by a factor of R_0 after asymptotic growth of a stable age distribution has been reached. Actually in a stage-structured model, this equation could only provide an approximation for actual mean age of reproduction, T_1, anyway, because the SS/SG formulation produces a lot of variability in the time individuals spend in each stage and thus the time it takes to mature and reproduce. de Camino-Beck and Lewis (2008) derive the exact formulas for the mean and variance of the generation time; Steiner et al (2014) also provide an alternative derivation of these quantities.

6.6 Delay equations

At the beginning of the chapter we described the problem encountered with duration-stage models (Fig. 6.1), that different stages have different durations, making it difficult to choose a reasonable time step for a discrete-time model. One solution to this problem is to represent the population in continuous time, using a *delay differential equation* (DDE) (Smith, 2011). This is simply a differential equation in which the rate of change at time t depends on the state of the system at some time $t - \tau$ in the past.

For example, Nisbet (1997), using an approach described in Gurney et al. (1983), modeled the dynamics of cladocerans, which are freshwater planktonic crustaceans ("water fleas"). He used a two-stage model, with an immature juvenile stage, n_J, and a reproductive adult stage, n_A. The duration of the juvenile stage was τ, and the adult stage lasted until death. The two stages had mortality rates D_J and D_A, respectively, and adults produced juveniles at a per capita rate β. The differential equation for the adult stage, n_A, is

$$\frac{dn_A}{dt} = Qn_A(t - \tau) - D_A n_A, \tag{6.22a}$$

where

$$Q = \beta e^{-D_J \tau}. \tag{6.22b}$$

The first term on the right-hand side of Eq. (6.22a) is the maturation of juveniles into the adult stage. This depends on the adult abundance τ time units ago, times the per capita reproductive rate (β) times the survival during the juvenile stage ($e^{-D_J \tau}$). The second term on the right-hand side of Eq. (6.22a) is the mortality of the current adult population.

Equation (6.22) can be solved like a normal ODE, as long as one specifies either the prior history of n_A from time $t_0 - \tau$ up to t_0 (Nisbet, 1997) or the starting stage distribution at t_0 (de Roos, 1997); the latter would be more appropriate when modeling a laboratory population that is started by an experimenter and does not have a prior history. The stability analysis of this kind of model can be approached in a brute force way by assuming that the system is linearized near the equilibrium and that solutions grow (or decline) exponentially in time (Smith, 2011).

In principle, the delay equation approach could be used to model a population with any number of stages; there would simply be a separate DDE similar to Eq. (6.22) for each stage, and the DDEs would be solved simultaneously like a system of ODEs. Note that this approach does correctly allow for stages of different duration. In practice, this approach is mostly used for two-stage populations like the cladocerans in Nisbet's model. In that context, the model is particularly useful for examining the effects of environmental conditions that speed up or slow down the development time, τ. For example, Amarasekare and Coutinho (2013) used a modified version of the model in Eq. (6.22) to examine how the population growth rate and stable stage distribution of invertebrate and reptile populations would shift as climate warming shortened development times.

Some authors have represented populations with different developmental stages of certain durations using continuous-time ODEs, but without a delay. For example, Li et al. (2017) used a system of two ODEs to represent the larval and adult stages of mosquitoes to investigate different strategies for control using introductions of sterile males. However, this approach suffers the same problem we described in discrete-time stage-structured models (Section 6.3), that the population dynamics will depend on the age structure within each stage (e.g. how many larvae just hatched out of eggs versus how many are about to metamorphose), and that detail of population state is not captured in a non-delayed ODE model. This is particularly problematic for Li et al., who are interested in the timing of pulsed introductions of sterile males, i.e. in transient dynamics rather than the stable stage structure. Other models in the field of mosquito control use more appropriate DDE models, e.g. White et al. (2010).

Another way of using continuous-time models without delay is better than using simply a single variable for each stage, but does not completely eliminate the problem. This approach, known as the *linear chain trick* (Smith, 2011), replaces a delay that is assumed to be a single time, with a distribution that is created by having one or more "stages" the organism needs to go through. If it is just a single delay, the distribution is exponential and suffers from the problems we have emphasized (Section 6.3). If the organism has to go through multiple stages, the distribution of times the organism spends in the whole set of stages is a gamma distribution with the order of the gamma distribution determined by the number of stages. As the number of stages approaches infinity, the distribution becomes one corresponding to a fixed delay, but even if only a few stages are included, this represents a substantial improvement over a single stage.

6.7 What have we learned in Chapter 6?

In this chapter we have reviewed the way stage-structured matrix models are used to represent stage-structured population dynamics, whether they be actual biological stages or approximations to continuous growth. Such models are widely used, often referred to as Lefkovitch matrices, and many are of the "some go/some stay (SG/SS)" form. However, we have shown that these stage-structured matrix models have serious structural flaws because stage is not an adequate description of state. In general, a stage-structured matrix model will only produce correct dynamics when the population is at the stable stage distribution. They lead to incorrect predictions of the transient (non-asymptotic) phase of population dynamics. Stage-structured models have had some successes, such as with teasel and ground squirrels. In those cases, however, explicitly size-based models, such as integral projection models, should provide better performance by preserving a proper definition of state and thus a realistic depiction of transient dynamics.

We have seen the basic problem with stage-structured models in the comparison of the stage distribution over time of a cohort of loggerhead sea turtles growing according to the SS/SG stage matrix model of Crowder et al. (1994) to the stage distribution produced by the equivalent Leslie matrix with the same survivals and reproductive rates (Fig. 6.6). There the cohort growing according to the actual age-structured model remains in each stage sequentially for its duration, while the cohort growing according to the stage-structured model begins to spread out over the stage structure in a non-biological way. At age 22, all of the cohort growing according to the age model is in the reproductive stage V, while the cohort growing according to the stage model is still predominantly in stages II–III, with small proportions in stages IV–V.

If one knew the age structure within each stage, then one could produce a stage-structured model that would produce a stage-structured population model that correctly produced the dynamics of the stage structure after the population reached the stable stage distribution. But knowing the age structure within each stage would mean that one would already have the information needed simply to formulate an age-structured model.

Stage-structured models were properly motivated by the concerns that for many species, prediction of reproduction by adults is best based on size, rather than age, following the definition of *i*-state in Chapter 1. The problem is that that focus on adults neglected the time lags involved in reaching the adult stages from recruitment.

We have seen how we could get around the fundamental problem of stage-based matrix models: form a matrix model with both age and size (Law, 1983). We have also seen how complicated another approach turned out: letting the growth to adulthood randomly follow many different growth stages (de Valpine, 2009).

Given the fundamental flaws of stage-structured matrix models, we concluded that we would be better off using non-matrix approaches. These were the models that use the numerical methods covered in Section 5.4. However, in spite of the lack of fidelity in stage-structured models, some have spent the effort to derive an expression for replacement in R_0, hence for persistence. We will see the mathematics in Hastings and Botsford (2006) again applied to persistence of spatial models in Chapter 9, where it has been more useful.

Finally, we have seen that stage-structured models can also be constructed in continuous time rather than using matrices. If these models are constructed with lags to represent the time spent in each stage (using DDEs or the linear chain trick), they can correctly describe transient dynamics; un-lagged ODEs will suffer from the same conceptual flaws as stage-structured matrix models.

CHAPTER 7

Age-structured models with density-dependent recruitment

In this chapter we return to the description of age-structured models with density-dependent recruitment that we began in Chapter 4 (Section 4.4). There we discussed the behavior of age-structured models with equilibria on the left-hand side of the peak of the density-dependent egg/recruit (or stock/recruit) function, where the slope becomes shallower with increasing density in a compensatory fashion, but is still greater than zero (Fig. 4.11a). The mechanism governing density-dependent population dynamics in that chapter was cohort resonance; basically the fact that age-structured models are more sensitive to environmental variability on cohort time scales (i.e. the mean age of reproduction) and also on very long time scales. In this chapter we describe population behavior near an equilibrium on the right-hand side of the density-dependent recruitment relationship, where the slope is declining in an over-compensatory way (Fig. 4.11b). We saw in Chapter 2 (Section 2.4.1, Fig. 2.12) that over-compensatory density dependence can give rise to dramatic population cycles in simple, unstructured models. In this chapter we will see how and when similar dynamics can arise in age-structured models. Of course both of these kinds of behavior (cohort resonance and over-compensatory cycles) could be occurring in a population, with large amplitude variability leading to signals transiting both sides of the stock recruitment relationship, but that situation has not been studied.

From Chapter 2, recall that to analyze the behavior of nonlinear models, we can make use of what we know about the behavior of linear models by linearizing our nonlinear model about an equilibrium. The first step in that approach is to determine the conditions for an equilibrium. For age-structured models with density-dependent recruitment, we described that step in the last half of Chapter 4 (Section 4.4.1): graphically, the equilibrium occurs at the intersection of the density-dependent egg/recruitment function, with a line through the origin with a slope equal to the inverse of lifetime reproduction (Fig. 4.11). In this chapter we proceed to analyze deterministic population stability about such an equilibrium when it falls to the right of the peak, as it could in a Ricker-type stock recruitment function (Fig. 4.11b). We then describe how those stability conditions are related to the age dependence of vital rates of the population, and to the shape of the egg/recruitment function. In doing so, we develop a general understanding of what makes such populations display unstable, cyclic behavior. We then demonstrate those principles in several examples in which such behavior has been analyzed for a number of species by others.

The dynamic mechanism causing the population cycles described in this chapter differs from the cyclic mechanism described in the second half of Chapter 4, where we dealt with

Population Dynamics for Conservation. Louis W. Botsford, J. Wilson White, and Alan Hastings, Oxford University Press (2019). © Louis W. Botsford, J. Wilson White, and Alan Hastings.
DOI: 10.1093/oso/9780198758365.001.0001

stochastic responses to environmental variability with period equal to one generation time ($1T$), in models with a (deterministically) locally *stable* equilibrium. Here instead we will be concerned with models where the deterministic equilibrium is unstable, and there are stable cycles of period roughly equal to twice the generation time ($2T$). Another difference of some interest is the historical timing of the intellectual development of each topic. We noted in Chapter 4 that our understanding of cohort resonance has mostly developed during the early years of the twenty-first century. The increase in understanding of the deterministic cycles of period $2T$ described in this chapter took place from the late 1970s to the late 1980s. An exception to that statement is that Ricker (1954), one of the most famous early fishery biologists, reported the basic results for $2T$ cycles, even before modern day electronic computers were available (i.e. probably working with a Frieden electromechanical calculator that could only add, subtract, multiply, and divide (noisily)). For another synthetic discussion of both examples and the causes of cycles in populations, including species interactions, which we do not cover here, see Kendall et al. (1999). In a more recent, similar review, Barraquand et al. (2017) discussed a category of cycles that are "noise-sustained oscillations," which would include the cycles in cohort resonance described in Chapter 4.

As a starting point for the description of population behavior that leads to period $2T$ cycles, recall that the equilibrium condition obtained in Chapter 4 had a convenient, valuable graphical interpretation (Fig. 4.11) that allowed us to understand the effects on the equilibrium of human activities that diminish population survival (such as harvesting). In that figure, the equilibrium is determined by a line with slope $1/\Phi_b$, so as harvest truncates the age structure, it moves the equilibrium from high egg values on the right side of the egg/recruit curve to lower values to the left (Fig. 4.11). This decline in lifetime egg production, Φ_b (a measure of replacement), occurs because fewer individuals will be allowed to survive and spawn into their old age. When $1/\Phi_b$ becomes less than the slope of the egg/recruit relationship at the origin, the equilibrium recruitment will be zero, which would lead to an eventual collapse of the fishery. In this chapter, as in Chapter 4, as the equilibrium changes with declining Φ_b, the consequent changes in the slope of the egg/recruit relationship at the equilibrium (Fig. 4.11) will also influence the behavior of the population near that equilibrium.

7.1 Local stability and $2T$ cycles

Here we continue with the model used to derive the equilibrium conditions in Chapter 4 (Section 4.4), a renewal equation model that is age structured with discrete time and age, with a single population variable, recruitment (R_t), rather than an age vector. Recruits are simply age-1 individuals. The effects of age in the model are captured by a number of *influence functions* (as introduced in Section 3.2) that describe how each cohort contributes either to reproduction (the reproductive influence function, φ_b) or to density-dependent mortality of recruits (the compensatory mortality influence function, φ_c), as it ages.

7.1.1 *Local stability analysis*

The first step in understanding the dynamics of this model is to perform a local stability analysis. The procedure will be similar to that followed in Chapter 2 for an unstructured model (Eq. 2.16). We begin by restating the renewal equation defining the model (cf. Eq. 4.19),

$$R_t = B_t f(C_t) = \sum_{a=1}^{A} [\varphi_b(a) R_{t-a}] f\left(\sum_{a=1}^{A} [\varphi_c(a) R_{t-a}]\right). \tag{7.1a}$$

For some populations, recruitment into the population is not determined by its own reproduction, but rather originates in other nearby populations in a random way (e.g. through a dispersing larval stage). Under those conditions, population models often assume that recruitment depends on a constant input of pre-recruits instead of endogenous reproduction. Such models are termed "open population" models (Roughgarden et al., 1985; Hixon et al., 2002). Accordingly, models that explicitly account for pre-recruits coming from local reproduction are called "closed population" models. The open population version of the model in Eq. (7.1a) is

$$R_t = B_c f(C_t) = B_c\, f\left(\sum_{a=1}^{A} [\varphi_c(a) R_{t-a}]\right), \tag{7.1b}$$

where B_c is the assumed constant source of pre-recruits. Recall that in these equations, $\varphi_b(a)$ equals $b_a \sigma_a$, the birth rate at age a times the probability of survival to a, and $\varphi_c(a)$ equals $c_a \sigma_a$, where c_a is the effect of age a on recruit survival (see definitions in Section 4.3.1). The function f is the nonlinear recruit survival function, which has a value ranging from 0 to 1.

The equilibrium is found by setting all R's to R^* (the equilibrium value of recruitment), then solving this relationship for R^*. For the closed population model this is

$$1 = \Phi_b f\left(\Phi_c R^*\right), \tag{7.2a}$$

where Φ_b and Φ_c represent the sums of $\varphi_b(a)$ and $\varphi_c(a)$, respectively, over all ages. Φ_b is lifetime reproduction, as defined in Chapter 3 as R_0, and Φ_c can be defined similarly for other age-dependent, density-dependent recruitment mechanisms (e.g. Φ_c is the "net area function" in an example described later in Section 7.4, where the density dependence arises from limited space available for occupation by settling barnacles; Roughgarden, et al., 1985). This expression underlies the graphical method for determining equilibrium in Fig. 4.11. The recruitment curve is $R^*\Phi_b f[R^*\Phi_c]$, and the egg production on the abscissa is $R^*\Phi_b$, so the line through the origin with slope $1/\Phi_b$ is simply R^*, thus the intersection is $R^* = R^*\Phi_b f[R^*\Phi_{c,}]$, Eq. (7.2a). The equilibrium expression for the open population model is

$$R^* = B_c\, f\left[R^*\Phi_c\right]. \tag{7.2b}$$

Similar to the closed population case, this can be graphically envisioned as the point on a plot of recruitment, equal to B_c times the recruitment survival function, $B_c\, f\, [R^*\Phi_c]$ $vs. R\Phi_c$, where a line through the origin with slope $1/\Phi c$ intersects. (This is shown later in the barnacle example in Section 7.4.)

To analyze stability of the closed and open population models we define a variable that is the distance of recruitment from the equilibrium, $\Delta R_t = R_t - R^*$. We then use a Taylor series expansion of the renewal equation (Eq. (7.1)) to obtain the linear approximation for the dynamics of ΔR_t.. For the closed population model this leads to

$$\Delta R_t = \sum_{a=1}^{A} \left[\tilde{\varphi}_b(a) + K\tilde{\varphi}_c(a)\right] \Delta R_{t-a}, \tag{7.3a}$$

and the corresponding equation for an open population is

$$\Delta R_t = \sum_{a=1}^{A} \left[K\tilde{\varphi}_c(a)\right] \Delta R_{t-a}. \tag{7.3b}$$

The tilde (∼) over each influence function indicates that it has been standardized to sum to 1 (i.e. $\tilde{\varphi}_b(a) = \varphi_b(a)/\Phi_b$). A simple example of the influence function for birth, $\tilde{\varphi}_b(a)$, would be to assume that there is a constant probability of survival at each successive age (with a higher probability at older ages due to senescence), that length-at-age follows an asymptotic growth trajectory, and that fecundity is proportional to length cubed (Fig. 7.1a). Similarly, a simple example for the influence function for effective population size, $\tilde{\varphi}_c(a)$, would be to assume that compensatory mortality occurs via adults cannibalizing new recruits, and that the cannibalism rate is proportional to the cannibal's metabolic rate, which in turn is proportional to biomass raised to the 0.8 power (Fig. 7.1b). Botsford and Wickham (1978) used this influence function to describe cannibalism in Dungeness crab (*Metacarcinus magister*) populations; the 0.8 exponent approximately matches the 3/4-power scaling law that arises in the metabolic theory of ecology (Brown et al., 2004).

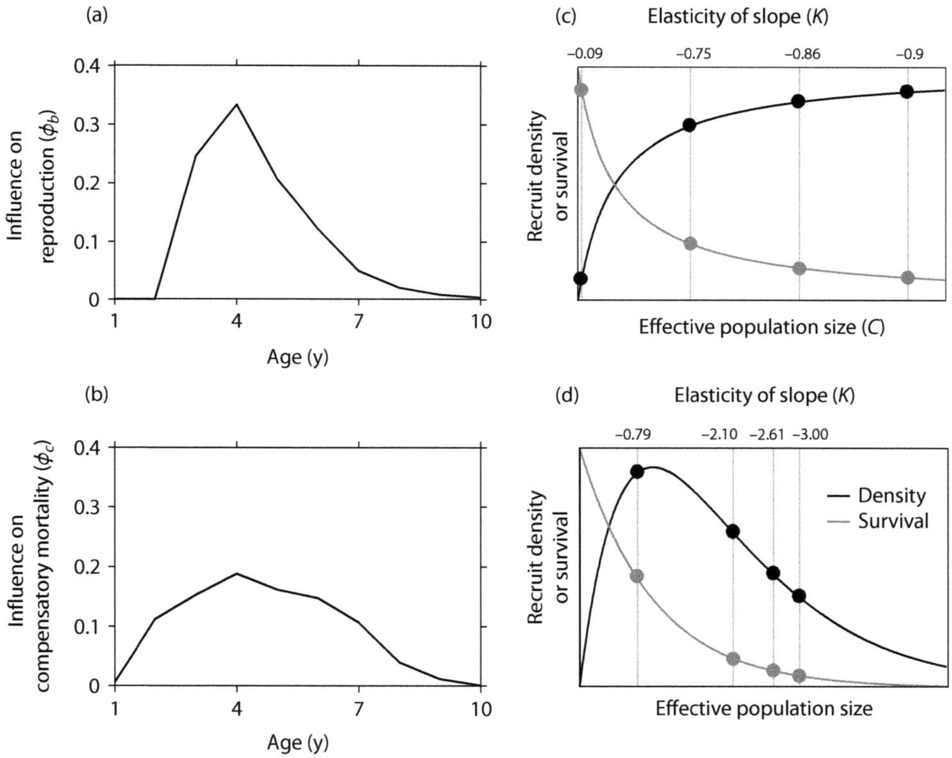

Fig. 7.1 Descriptions of the age-structured influence functions (a,b) and elasticities (c,d) in the linearized model. (a) An example of a possible influence function for reproduction, $\varphi_b(a)$, the product of survival to each age and fecundity-at-age. (b) An example of a possible influence function for compensatory mortality, $\varphi_c(a)$, the product of survival to each age and metabolic demand at each age. (c–d) Two examples of the recruit survival function, *f*: a Beverton–Holt survival function (c) and a Ricker survival function (d). In both panels, the black curves indicate recruitment, which is egg production multiplied by recruitment survival, *f*. The gray curves indicate survival, *f*, alone. In (c, d), dots and vertical lines indicate equilibrium values of recruitment and recruit survival for evenly spaced values of lifetime egg production ranging between 100% and 11% of the maximum. Values above each line are *K*, the normalized slope, or elasticity, of *f* at those equilibria.

The other important term in Eq. (7.3) is K, which is the normalized slope of the recruitment survival function, f at equilibrium,

$$K = f'\left(\Phi_c R^*\right) \frac{\Phi_c R^*}{f\left(\Phi_c R^*\right)}. \tag{7.4}$$

Because this is the fractional change in f per fractional change in $\Phi_c R^*$, this can also be interpreted as the elasticity of the compensatory density dependence (compare the form of this expression to the elasticity of λ in Eq. (3.37)). We will see that the value of K is very influential in determining stability. (Do not confuse this variable, K, with carrying capacity in the logistic model. We apologize for this accident of history.)

To gain an understanding of how the value of K can affect population behavior, we examine its values for the two most common stock-recruit relationships: a Beverton–Holt style of dependence of recruitment on adults (i.e. compensatory density dependence, but not over-compensatory; Eq. 2.16b; black curve in Fig. 7.1c), and a Ricker style dependence (i.e. a dependence that is compensatory at low abundance, then becomes over-compensatory at high abundance; Eq. 2.16a; black curve in Fig. 7.1d). As simple, complete popuation models, both of these models include abundance as N_t, representing the propensity to increase because of reproduction. The black curves in Fig. 7.1c and d include that factor N_t, and are the same as the curves in Fig. 2.18. Because our age-structured model for a closed population (Eq. 7.3) already includes the influence of age structure on reproduction, we remove that factor from each of the stock recruitment functions, f, keeping only the part representing density-dependent survival through recruitment. This gives us a recruitment survival function for the Beverton–Holt case (using the same parameters α' and β' as in Chapter 2), $f\,(C_t) = \alpha'/(1+\beta' C_t)$, which gradually declines with density (gray curve in Fig. 7.1c). Together with the explicit reproduction in the closed population model, this will produce a Beverton–Holt-style density dependence in the overall age-structured model. The elasticity of f in this case is $K = -\beta' C_t/(1+\beta' C_t)$. For values of C_t greater than zero, K decreases from 0 to a minimum value of -1 (Fig. 7.1c). To produce density-dependent recruitment that would look like a Ricker function (also using the same parameters α and β as in Chapter 2), we use $f\,(C_t) = \alpha e^{-\beta C_t}$, which declines with density in an exponential way (gray curve in Fig. 7.1d). In this case, the elasticity is $K = -\beta C_t$, which decreases from 0 towards $-\infty$ as C_t increases (Fig. 7.1d). Thus we see that while we may think of K as a slope of the recruitment survival curve (since it contains that derivative), it is also heavily influenced by the values of the equilibrium survival, and the equilibrium value of effective population size. In Fig. 7.1c and d, we have plotted values of K for a range of possible equilibrium levels of recruitment.

To proceed with the stability analysis, we take the usual step of assuming that because Eqs. (7.3a,b) are linear, each has a solution of the form $\Delta R_t = \lambda^t$. Substituting that into equations (7.3a,b) leads to the characteristic equations

$$\sum_{a=1}^{A} \left(\left[\tilde{\varphi}_b(a) + K\tilde{\varphi}_c(a)\right]\lambda^{-a}\right) = 1 \tag{7.5a}$$

$$\sum_{a=1}^{A} \left(\left[K\tilde{\varphi}_c(a)\right]\lambda^{-a}\right) = 1. \tag{7.5b}$$

There are A solutions ("roots") λ to these equations (A is the number of age classes), and they have the usual interpretation. The root λ with the largest magnitude (i.e. distance from zero in the complex plane) will dominate; if $|\lambda| < 1$ then the equilibrium is locally stable when perturbed, and if $|\lambda| > 1$ then the equilibrium is locally unstable and ΔR_t

increases with time. In many cases, the dominant eigenvalue is a complex conjugate pair, so unstable populations will tend to exhibit locally increasing cycles (see Section 4.11). Locally stable populations could also cycle, with the cycle amplitudes declining with time.

There is no analytical expression for the solution to Eq. (7.5), but we can begin to understand this model by considering particular cases in which it is possible to determine whether $|\lambda|$ is greater than 1. For the closed population model, the two influence functions each sum to 1, so the value of K is quite important in determining the shape of the age dependence. If $K \geq 0$, there is no compensatory density dependence, so the model is linear, Perron–Frobenius conditions apply, and behavior is dominated by a single real eigenvalue with magnitude $|\lambda| > 1$ (i.e. recruitment will increase geometrically away from the equilibrium). Continuing with the closed population, for K between 0 and -2, the quantity in square brackets in Eq. (7.5a) is either between 0 and 1 (if birth influence $\tilde{\varphi}_b(a)$ is greater than compensation influence $\tilde{\varphi}_c(a)$ for age a), or between 0 and -1 (if the opposite is true). In either case the roots λ will be real and $|\lambda| < 1$, so the solution will decay geometrically to equilibrium. We saw in Fig. 7.1c that the compensatory density-dependent functions that lead to a Beverton–Holt style model always have $0 < K \leq -1$, so they do not produce unstable oscillations. When K decreases to below -2, as we can see in the Ricker style model (Fig. 7.1d), there will be a threshold value of K at some point, termed K', below which λ will be complex and $|\lambda|$ will be greater than 1, so the equilibrium will be locally unstable, which leads to persistent oscillations. A similar threshold exists for the open population case (Eq. (7.5b)); because that expression lacks the term $\tilde{\varphi}_b(a)$ inside the brackets in Eq. (7.5a), the critical value K' will be at some value ≤ -1 instead of ≤ -2. In both cases, stability is lost when the largest eigenvalues, a complex conjugate pair, have $|\lambda| \geq 1$. When $|\lambda|$ increases above 1, the system undergoes what is termed a bifurcation (Box 7.1). This simply means that there is a qualitative change in system behavior; in this case a transition from a stable equilibrium to a periodic oscillation about the equilibrium.

Box 7.1 BIFURCATIONS

A "bifurcation" occurs in a model system when there is a qualitative change in its behavior as a parameter is varied (see Iooss and Joseph (1990) for a relatively elementary mathematical discussion). The bifurcation we discuss in this chapter (for the system described by Eq. (7.1)) is termed a Neimark–Sacker bifurcation. This bifurcation occurs when the complex conjugate pair of dominant eigenvalues transition from $|\lambda| < 1$, representing a stable equilibrium, to $|\lambda| \geq 1$, where the complex conjugate eigenvalues produce a periodic oscillation about the equilibrium. The amplitude of the cycles increases from zero (initially very rapidly) as $|\lambda|$ increases away from the stability threshold. The Neimark–Sacker bifurcation occurs in discrete-time models; the continuous-time equivalent is the Hopf bifurcation, which also describes the transition from a stable equilibrium to a periodic limit cycle.

There are two possibilities for the behavior of a system as it crosses a bifurcation: supercritical or subcritical bifurcations. For supercritical bifurcations, the small-amplitude cycles occur for those parameter values where the equilibrium is unstable and the cycles are a stable solution (these are called *stable limit cycles*). This means that if the initial conditions of the system are far from the equilibrium, it will eventually converge on the limit cycle. In contrast, for subcritical bifurcations, the small amplitude cycles occur for those parameter values where the equilibrium point is stable and the cycles are an unstable solution. (These are also sometimes known as backwards bifurcations.) This means that if the system starts far enough away from the equilibrium it will approach a very different solution. In theory

(Continued)

Box 7.1 CONTINUED

it is possible to distinguish between the subcritical and supercritical cases analytically, but in practice knowledge of the different possibilities is used as a way to interpret numerical solutions. The Neimark–Sacker bifurcation in the models discussed in this chapter is supercritical.

It is important to note the difference between the limit cycles produced by the complex conjugate pair of dominant eigenvalues in the models discussed in this chapter and the dampened oscillations we described in the Leslie matrix models in Chapter 4. In those models, the dominant eigenvalue was real and the second and third eigenvalues were complex and produced the cycles; those cycles were driven by ongoing environmental variability that kept the system from staying at the stable equilibrium.

The ecological interpretation of these stability conditions for both open and closed populations is that the occurrence of cycles depends on the relative importance of two opposing forces, density-dependent recruitment and age structure. The former force is the strength of over-compensatory density-dependent recruitment, as represented by the change in survival f with density Φ_C, which is encapsulated in the value of K (Eq. (7.4)). Greater over-compensation causes the population to alternate between overshooting and undershooting the point equilibrium. For most density-dependent survival functions, K becomes more negative with increasing effective population size (e.g. Fig. 7.1d). This means that harvesting a population will tend to have a stabilizing effect on the dynamics, by reducing K. The latter force is the stabilizing effect of the age structure; as reproduction and density dependence are spread over more ages, the tendency to cycle is reduced. Thus harvest could also have a destabilizing effect by truncating the population age distribution (particularly if harvesting targets older individuals). Whether harvesting is ultimately stabilizing or destabilizing will depend on the details of the influence functions and the recruitment survival function.

7.1.2 *An example: 2*T *cycles in Dungeness crab*

Botsford and Wickham (1978) derived several general conclusions about the behavior of this model. First, they showed that the period of oscillations when the population is locally unstable about the equilibrium is approximately twice the mean age of the quantity in brackets in Eq. (7.5). That is, it is twice the mean age of the combined influence of individuals on both reproduction and compensatory mortality (i.e. period $2T$). The biological interpretation of this cyclic mechanism is that because it is due to over-compensation (represented by a highly negative K), an unusually large recruited cohort will, at age T, produce high compensatory mortality in their own offspring. T years later the resulting (unusually small) cohort will exert much lower compensatory mortality on their own offspring, leading in turn to recruitment of a large cohort. Then this process will repeat itself yielding cycles of period $2T$. Note the contrast to the cohort resonance phenomenon, in which there is not over-compensatory density dependence, so a large cohort produces a large cohort of offspring T years later, producing a period T cycle (but not without environmental variability).

Botsford and Wickham (1978) also provided an approximation for the threshold value of K (i.e. K') below which the population is unstable. They expressed the threshold in terms of the "coefficient of variation" of the distribution of the combined influence function term $\left[\tilde{\varphi}_b(a) + K\tilde{\varphi}_c(a)\right]$ in Eq. (7.5). The approximate value of this threshold, K', for the closed population is

$$K' \approx \frac{-1}{1 - \frac{\pi^2 v^2}{2}} - 1, \tag{7.6}$$

where v is the coefficient of variation, which is the standard deviation, s, of the age distribution, divided by its mean, μ:

$$v = \frac{s}{\mu} \tag{7.7a}$$

$$\mu = \sum_{a=1}^{A} a\tilde{\varphi}(a) \tag{7.7b}$$

$$s = \sum_{a=1}^{A} (a - \mu)^2 \tilde{\varphi}(a). \tag{7.7c}$$

The term $\tilde{\varphi}(a)$ is the term $\left[\tilde{\varphi}_b(a) + K\tilde{\varphi}_c(a)\right]$ from Eq. (7.5). As v increases in Eq. (7.6), the critical value K' becomes more negative; that is, instability is less likely, as we suggested earlier. This effect of the relative width of the influence function is similar to that seen in cohort resonance, in that as the reproduction by adults spreads out over more age classes (larger v) there is a lower tendency to cycle. Interestingly, this harkens back to one of the figures in Taylor's (1979) study of transient behavior in insects. Here again, relatively narrow spawning age distributions tend to lead to unstable behavior. This could also be referred to as a lack of diversity of spawning ages (e.g. Barnett et al., 2017).

As an example of period $2T$ cycles and stability for closed population age-structured models, we calculated the solutions to Eq. (7.5a) for the Dungeness crab model analyzed by Botsford and colleagues (Botsford and Wickham, 1978; Botsford, 1986; Botsford and Hobbs, 1995; Botsford, 1997). This model uses the influence functions shown in Fig. 7.1a–b, and a density-dependent survival function f that has constant survival up to some level of effective population size, then declines steeply to zero (this represents so-called "density-vague" regulation, in which density dependence only becomes important at high densities; Strong (1986)). This survival function will cause the relationship between equilibrium and K to differ from those in the model in Fig. 7.1 because f'' is zero at low effective population sizes, and becomes negative only at high values. We varied the annual harvest mortality rate from $F = 0.0$ to $F = 1.2\ \mathrm{y}^{-1}$ to show the effects of fishing truncating the age structure.

In the Dungeness crab fishery, only males are harvested, and we use that case to illustrate the relative roles of density-dependent recruitment, age structure, and fishing in determining stability, because it allows us to see these separate roles most clearly. Greater clarity is possible because harvesting females changes the equilibrium (Eq. 7.2a), thus changing K, whereas the equilibrium remains constant under harvesting when harvesting only males. This can be seen in the equilibrium condition (Eq. 7.2a), where the equilibrium value of recruitment survival f is $1/\Phi_b$. When females are harvested, equilibrium recruitment survival will increase when the density dependence is overcompensatory, and the system will tend to move to more stable, less negative values of K. Conversely, when only males are harvested, harvest does not change the value of K, allowing an unfettered view of the effects of changes in age structure on stability.

The effect of the strength (slope) of density-dependent recruitment on stability is illustrated in Fig. 7.2, with three increasingly steep survival functions (f) (Fig. 7.2a, d, g). With greater (negative) values of the change in survival with density, the value of K

becomes more negative, eventually leading to cyclic behavior (Fig 7.2c, f, i). Note that in this case the value of K is proportional to the value of f', because the equilibrium values of survival f and therefore density $R_e\Phi_c$ are constant at the value of $1/\Phi_b$ (Eq. (7.4)). As the strength of density dependence K increases (i.e. becomes more negative), the summand of the characteristic equation (Eq. (7.5a)) and, by extension, the combined influence function $\tilde{\varphi}(a)$ is basically more negative (Fig. 7.2b, e, h). Once K falls below the stability threshold K', the population exhibits cyclic behavior (Fig. 7.2c, f , i).

The stabilizing/destabilizing effects of age structure on the tolerable stable strength of density dependence (K') can be seen by adding different levels of harvest to the model in Fig. 7.2. In Fig. 7.3a, the three arrows on the left axis correspond to the three values of K for the different density-dependent survival functions in Fig. 7.2. Harvesting increases from left to right, and the contour plot of $|\lambda|$ contains a line indicating K', the boundary between stable cyclic behavior and locally unstable cyclical behavior. Note that the values of K' fall below -2, as we concluded they would from the analyses of Eqn. (7.5). Also, unless K is

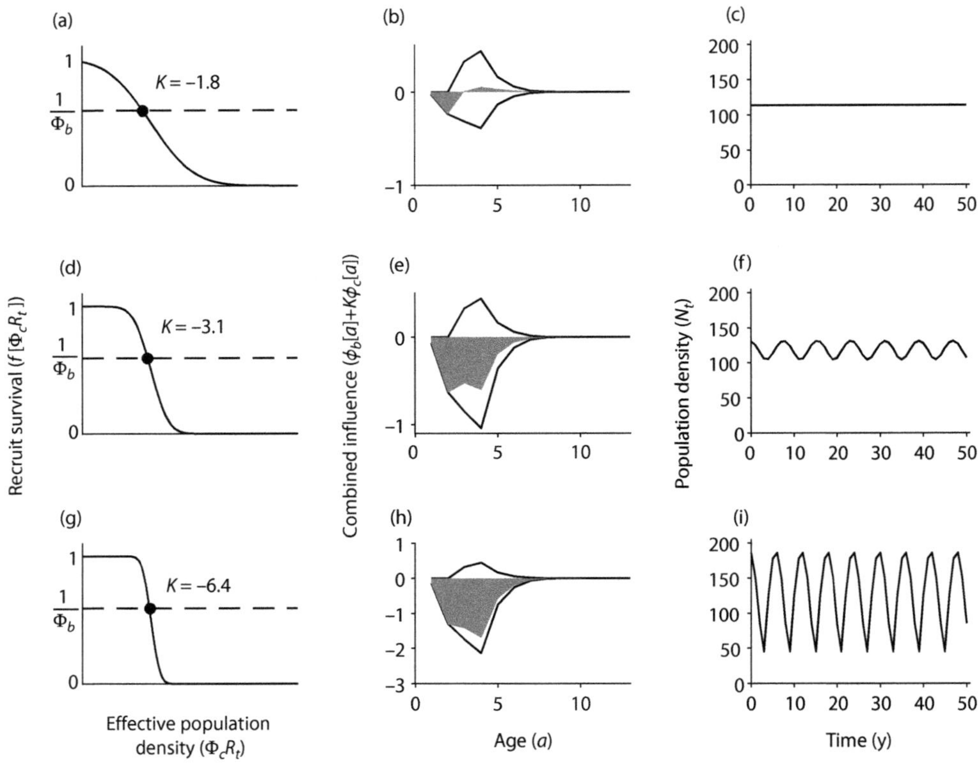

Fig. 7.2 Effect of the strength of the naturally occurring density-dependent recruitment (i.e. different values of the slope of the recruit survival function, f) on the stability of an age-structured model with density-dependent recruitment. The first column (panels a, d, g) shows the recruit survival function, f, with an increasingly steeper slope from top to bottom (indicated by the increasingly negative values of K). The equilibrium, as indicated by the dashed line, at survival $= 1/\Phi_b$. The second column (panels b, e, h) are corresponding plots of the two influence functions (φ_b[a] and $-K\varphi_c$[a]; black curves) shown in Fig. 7.1, and the combined influence function $\varphi_b[a] + K\varphi_c[a]$ (gray shading; note that the scale of the ordinate varies). The right column shows corresponding time series of population abundance for each scenario. The generation time in this example is 3.45 y, leading to cycles of an approximate period of 7 y when the system is unstable.

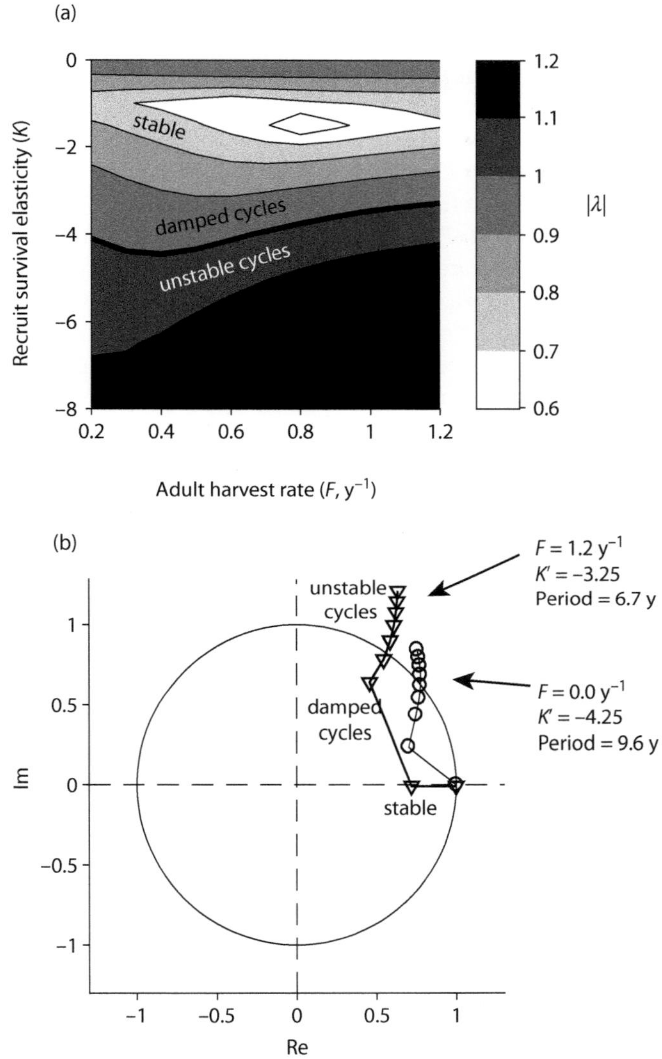

Fig. 7.3 Effects of changes in the age structure due to harvest on stability of the age-structured model with density-dependent recruitment in Fig. 7.2. (a) The effects on the magnitude of the dominant eigenvalue of the combination of the elasticity of density dependence (the value of K) and the shape of the age structure (as influenced by fishing mortality rate, F). The thick black line where $|\lambda| = 1$ indicates the value of K', which becomes less negative with fishing (i.e. the dynamics becomes less stable). (b) The location of the dominant eigenvalue (actually one member of a complex conjugate pair) on the unit circle, as the value of K moves from 0 (values on the real axis) to -8 (in steps of -1), with a fishing rate of 0.0 y^{-1} (circles), corresponding to the left-hand side of (a) and 1.2 y^{-1} (triangles), corresponding to the right-hand side of (a).

greater than −2, the value of $|\lambda|$ increases with increasing harvest, a destabilizing effect of the truncation of the age structure by harvesting.

The nature of the path to instability and, more importantly, the period of the cycles, are illustrated in Fig. 7.3b. As the values of K increase along the two strings of symbols from 0 to −8 (indicating populations with increasingly stronger density dependence), the value of K at which they cross the unit circle, thereby becoming unstable (i.e. K') and the point on the circle at which they cross (determining the period of cycles) both depend on their level of harvest. With no harvest ($F = 0$), as the value of K declines from 0 to −8, the values of (one of) the complex conjugate pairs (the circles) cross to outside the unit circle at a smaller angle with the imaginary axis than the case with greater harvest ($F = 1.2$, triangles). Recalling that the period of the cycles is inversely proportional to the angle of intersection with the unit circle (i.e. the period is $2\pi/\theta$), this indicates that the case without harvest will have cycles with a greater period ($2T$) than the case with greater harvest (i.e. 9.6 y versus 6.9 y). This is consistent with the mean age (T) being greater when age structure is less truncated. Also, the value of K at which the eigenvalue crosses the unit circle is more negative (−4.25 versus −3.25), indicating the unfished age structure is more stable.

The stability analysis for an open population would involve the same steps as we have explained for a closed population, so we do not repeat them here. The results are the same except that the value of K below which the population may become unstable is −1 rather than −2, as it is for the closed population (Botsford and Wickham, 1978). Thus, in this sense, closed populations are inherently more stable than open populations.

We have presented this general model in the framework developed by Botsford and Wickham (1978) and Botsford (1997), but it is not the only way to approach this problem. For example, Levin and Goodyear (1980) used an age matrix and solved for the Jacobian, rather than using a simple renewal equation in recruitment, in their investigation of the Hudson River striped bass population. They also evaluated the dependence of stability on the width of the spawning age distribution. Bergh and Getz (1988) reviewed the basic issues we have covered here, and investigated the consequences of several different potential forms of the stock–recruit relationship (beyond the Beverton–Holt and Ricker that we have shown).

Interestingly, two decades before this type of model was formally analyzed, Ricker (1954) had used hand simulations to work out the basic results, presented in a publication entitled simply "Stock and recruitment." First he noted that for the case of a semelparous species with no age structure, when the slope of the stock–recruitment curve was less than a threshold of −1, cycles of period 2 would be produced (this is consistent with our stability results for the unstructured model in Section 2.4.1). He contrasted that with the iteroparous case, which was a little more stable, with the requirement that the slope be less than a threshold that is itself less than −2 for unstable cycles to be produced, and noted that the period of the cycles was twice the median time from the egg to the production of the next generation. He also noted that the amplitude of these cycles was greater as the maturation delay increased or the number of spawning age classes decreased. These are all consistent with the analytical results worked out much later by Botsford and Wickham (1978). Last, he showed that harvesting reduced the strength of these cycles. This would occur because harvest decreases the lifetime reproduction, thus increasing the slope of the line (i.e. $1/\Phi_b$) and thereby moving the equilibrium towards the more stable part of the stock–recruit curve.

With this general understanding of cyclic behavior of period $2T$ in hand, we turn to a number of examples that draw parallel conclusions for a number of species (we summarize these examples in Table 7.1).

Table 7.1 Summary of the models described in this chapter

Species, model	Density dependence, Elasticity of	Influence function	Time scale (period)	Closed/ open
General (applied to fin whale)	General $f[N]$	Geometric decay in age structure	N/A	Closed
Dungeness crab, predatory worm	Egg production (+) Cannibalism, worm Predation (−)	Metabolic demand versus age	10 years	Closed and open
Barnacle	Limited free space	Size versus age	2 years	Open
Flour beetle	Cannibalism	Abundance at age	10 days	Open

7.2 The simplest general model of age-structured density dependence

The interpretation of the general stability analysis results did not expose a new result (after all, Ricker had already figured out the basic pattern), but it illustrated the value of a general model in drawing general conclusions. Population biologists often ask questions such as "would a lower (or higher) value of survival (or fecundity, etc.) cause a population to be less (or more) persistent (or stable, or variable, etc.)?" The answer always ultimately depends on what you hold constant when you vary the parameter in question. The goal in asking such a question should be to ask it in a way that makes the answer as general as possible. In Section 7.1.2 we saw that there was a relationship between the width of the age distribution (actually, the width of the influence function) and dynamic stability. However the width of the influence function depends on a variety of parameters (survival, fecundity, K, etc.), making it potentially difficult to interpret. In this section we will ask a somewhat simpler and more general question: does decreasing adult survival (as by harvesting, for example) make an age-structured population with density-dependent recruitment more or less stable?

To address this question, we use the delayed recruitment model, which is the simplest model of age structure with density-dependent recruitment. If N_t is the number of mature adults at time t, the dynamics are

$$N_{t+1} = f\left(N_{t-\tau}\right) + pN_t, \tag{7.8}$$

where τ is the age of maturity, $f(N_t)$ is the density-dependent recruitment at time t, and p is the annual survival. In words, the adult population consists of the number of offspring that were produced and experienced density-dependent survival τ years ago and are now maturing, plus the number of adults surviving from the previous year. Per capita fecundity is assumed to be the same for all adults. Clark analyzed the stability of this model in an application to Antarctic fin whale (*Balaenoptera physalus*) populations. We will use his solution to see how changing p (i.e. changing adult survival, thus changing the amount of age structure) affects stability.

Clark (1976) first derived the characteristic equation of the model at the equilibrium N^*,

$$\lambda^{\tau+1} - p\lambda^{\tau} - f'\left(N^*\right) = 0. \tag{7.9}$$

He then asked what values of $f'(N^*)$ (the slope of the recruit survival function) were associated with stability. First, we can see that if $\lambda = 1$, $f'(N^*) = 1-p$. Thus if $f'(N^*) > 1-p$, λ will be real and greater than 1 and the equilibrium will not be stable. If we decrease

$f'(N^*)$, the equilibrium will be stable, but as $f'(N^*)$ continues to decrease, λ may become complex with $|\lambda| > 1$, producing instability. Our goal here is to find the value of $f'(N^*)$ at which that transition occurs. Because the roots of Eq. (7.9) can be complex, we assume there is a solution of the form $\lambda = \cos\theta + i\sin\theta$ that has magnitude $|\lambda|$. Without writing out all the steps, the basic process is to substitute that expression for λ into Eq. (7.9). Realizing that $f'(N^*)$ is real, we assume all the resulting complex terms in Eq. (7.9) must sum to 0, which would allow us to solve for θ. We then substitute that value into the remaining real terms in the expression to find $f'(N^*)$. From this we can obtain the values of $f'(N^*)$ below which the population would be unstable ($|\lambda| > 1$) for each combination of p and τ (Fig. 7.4a). From Fig. 7.4a we can see that in the Clark model, for $\tau = 0$, increasing p is stabilizing; that is, more negative values of $f'(N^*)$ (i.e. steeper, over-compensatory density dependence are required for the equilibrium to be unstable). This is the result we would expect from the analysis of the general model in Section 7.1: increasing adult survival spreads out the influence of older age classes on reproduction and density dependence, stabilizing the equilibrium. By contrast, increasing p makes no difference to stability for $\tau = 1$, and for $\tau > 1$, it is actually destabilizing (you can see how this switch happens in Eq. (7.8) by substituting $\tau = 0$ and solving for λ, and then assuming that τ is large enough for $\tau \approx \tau + 1$, and solving for λ again). This latter result, that increasing adult survival is destabilizing, appears to contradict what we found in Section 7.1.

To reconcile this difference in results—and illustrate the danger of varying just one parameter (in this case, p) without accounting for its effects on other parts of the model—Botsford (1992) reformulated Clark's (1976) model in the form of the general renewal equation in R_t from this chapter,

$$R_t = f\left(\sum_{a=\tau}^{\infty} R_{t-a} p^{t-a}\right). \tag{7.10}$$

The equilibrium of this model is

$$R^* = f\left[\frac{R^*}{1-p}\right]. \tag{7.11}$$

$(1-p)^{-1}$ is the average lifespan of a recruit, so the term in brackets is simply the average lifetime reproduction in this model. Expressing this in terms of the influence function for each age, $\tilde{\varphi}(a)$, we have

$$\begin{aligned} \tilde{\varphi}(a) &= (1-p)\,p^{a-\tau} \quad & a \geq \tau \\ \tilde{\varphi}(a) &= 0 & a < \tau \end{aligned} \tag{7.12}$$

and the elasticity of the recruitment survival function, K, is

$$K = \frac{f'(N^*)}{1-p}. \tag{7.13}$$

Then for each combination of p and τ we can calculate the values of K below which the population will be unstable (Fig. 7.4b). This panel shows clearly that increasing survival, p, is always stabilizing. Decreasing the age at maturity, τ, is also always stabilizing, essentially because long demographic time lags tend to destabilize populations.

In Fig. 7.4, both panel (a) and panel (b) are correct. However, panel (b) provides a more consistent view of the sensitivity of stability to maturity delay and adult survival, in the sense that stability decreases with age of maturation for all values of adult survival, and also increases with adult survival for all values of age of maturity. Changing to this

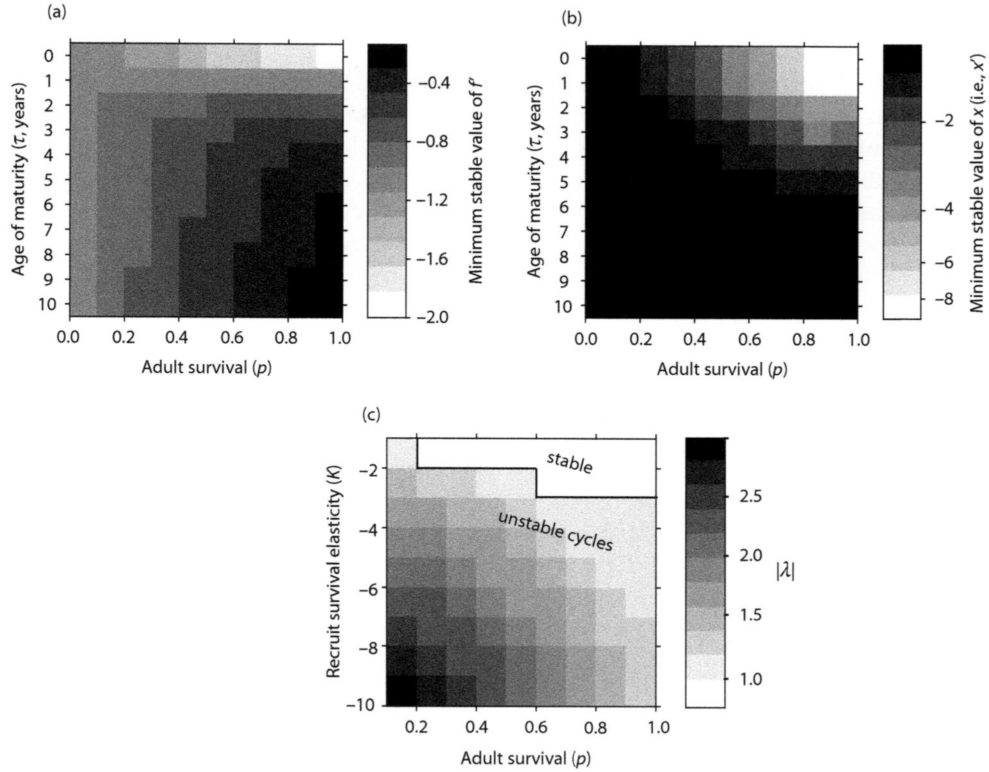

Fig. 7.4 Different ways of viewing the effects of: (1) annual adult survival and (2) age of maturity on stability of age-structured populations. Panel (a) is in terms of a stability threshold in the slope of the recruitment function of Clark (1976), and panel (b) is in terms of a threshold in the elasticity of recruitment survival (Botsford, 1992). Larger values (closer to zero) of both f' and K' indicate lower stability. (c) For comparison with the previous model in Fig. 7.3a, the effects of combinations of the strength of density dependence (K) and the shape of the age structure (as determined by adult survival, p) on stability. The thick black line indicates the stability boundary, $|\lambda| = 1$.

point of view involving elasticity also makes the dependence of stability on adult survival consistent with our conclusions from Section 7.1. Increasing p in the original Clark model adds age structure, but also increases the equilibrium population size. If the derivative of $f(N)$ is approximately the same, then this pushes the equilibrium further to the right on the recruit survival curve, making instability more likely when $\tau > 1$. Expressing density dependence in terms of the elasticity, K, which requires normalizing the influence functions, removes the effect of the equilibrium value of N_t, N^*, allowing us to isolate the effect of increasing age structure from its effect on equilibrium population size. That analysis shows how addition of age structure is generally stabilizing, as we also saw in the context of cohort resonance (Chapter 4), and which was evident in the transient analysis of Taylor (1979).

The behavior of this simple, two-parameter model shows essentially the same dependence on survival elasticity and adult survival as the previous example (compare Fig. 7.4c with Fig. 7.3a). Of course they would not be expected to be identical, but they do reflect the generality of our approach to period $2T$ models in age-structured populations.

7.3 Cycles in Dungeness crab: models and data

We will now describe an application of the general age-structured density-dependent model that illustrates different ways of making use of data in conjunction with the model results to answer questions regarding causes of cyclic behavior. Model-data comparisons are often difficult for questions regarding the dynamic behavior of exploited populations because the time scales of variability (especially for $2T$ cycles) can be a decade or more. There may be long-term fisheries data available on these time scales for some marine resources, but even those would present the problem of determining abundance based on catch data. An insect example of determining mechanisms causing cycles from long-term data is the pine looper moth (*Bupalis piniarius*), a pest of forest trees for which there were more than 30 years' worth of data to assess five-year cycles (Kendall et al., 2018). The Kendall et al. approach was to use nonlinear forecasting to fit various mechanistic models to several time series, and choose the mechanism on the basis of the closest fit. We took a similar approach for Dungeness crab (Higgins et al., 1997), using an age-structured density-dependent deterministic model (similar but not identical to the one we have described in this chapter) that also included stochastic environmental perturbations. The model produced remarkably close fits to 40-year time series of crab fishery landings, for eight different ports, leading to the conclusion that accounting for both intrinsic non-linear mechanisms (i.e. age dependent density dependence) and extrinsic environmental variability were essential to reproduce long-term dynamics correctly. A very dramatic aspect of this model was that in the absence of environmental variability, the underlying deterministic model had a stable equilibrium in seven cases and a small-amplitude limit cycle in one case. Adding environmental variability produced cycles with about a ten-fold difference between minimum and maximum.

Here we demonstrate a different approach to the use of data to try to learn something about causal mechanisms, one that can be useful when lengthy data sets are not available. From our modeling results we know that whether or not a population behaves cyclically depends on characteristics of its life histories (i.e. the shapes of influence functions, and the strength of density dependence), which can be determined on shorter time scales. This makes these models a useful way to "scale up" measurable observations at the individual level of organization to their population-level consequences (see Section 1.5, Fig. 1.3). In this example, we had only four years worth of data regarding a density-dependent mechanism to ask whether it could be responsible for cycles occurring on a decadal time scale. Four years is of course undesirably short, but it illustrates the scaling-up approach. In these times of rapidly changing physical environment, we may have to resort to such methods and model results to project long-term dynamics. These two approaches could be examples of the contrast in inductive versus deductive approaches described in Chapter 1. While use of the long-term population level time series is obviously inductive, comparison of individual components could fit into a deductive scheme.

A problem facing resource managers along the west coast of the United States in the 1970s was the fact that the abundance of Dungeness crab (*Metacarcinus magister*; at the time named *Cancer magister*) could vary by a factor of 15 or so from year to year, and this variability appeared to be cyclic with a period of approximately ten years, particularly from the 1950s to the 1970s (see Fig. 1.3 for an example; Botsford and Wickham, 1978). Managers wanted to know the cause of the cycles so that they might be able to reduce their magnitude and avoid the associated economic disruption of the fishery. Density-dependent recruitment was one potential cause of this cyclic behavior, the other two being environmental variability (Botsford and Wickham, 1975) and a predator–prey interaction (with humans as the predator; Botsford et al., 1983). The category of density-dependent recruitment included three possible mechanisms: (1) density-dependent fecundity, (2) cannibalism (mentioned in Section 1.2.1), and (3) a nemertean worm, *Carcinonemertes errantia*, that preys on crab eggs. Here we will describe the analysis of the worm's effects, as well as the combination of the worm and cannibalism.

The model used to analyze the effect of worm predation on the potential for cyclic behavior was similar to the renewal equation for density-dependent recruitment (Eq. 7.1), but involved an additional equation to represent the dynamics of the worm (Hobbs and Botsford, 1989). Recruitment was still egg production (just as in Eq. 7.1) multiplied by a survival term, but in this case, survival depended on worm abundance, W_t:

$$R_t = \sum_{a=1}^{A} [\varphi_b(a)R_{t-a}]\, g(W_t) = B_t g(W_t). \tag{7.14}$$

We use the variable B_t to represent the number of crab eggs spawned in year t, and the nonlinear function $g(W_t)$ describes the density-dependent effects of worms on crab egg survival (the worm feeding rate depends on worm density). The abundance of worms in a given year was assumed to be proportional to the number of eggs eaten in the previous year (adult worms feed on crab eggs at the time of oviposition, then spawn larvae that disperse and settle on adult crabs; juvenile worms remain attached to the crab carapace and mature into egg-consuming adults in the next reproductive season):

$$W_t = [1 - g(W_{t-1})]\, B_{t-1} = B_{t-1} - R_{t-1}. \tag{7.15}$$

This leads to an influence function, $\varphi_w(a)$, that describes the influence of each crab age class on egg production in year $t-1$ and thus worm abundance in year t:

$$R_t = \sum_{a=1}^{A} [\varphi_b(a)R_{t-a}]\, g\left(\sum_{a=1}^{A} [\varphi_w(a)R_{t-a}]\right). \tag{7.16}$$

This model has essentially the same form as the model in Eq. (7.1), and its stability analysis proceeds in the same way, by examining the elasticity of egg mortality due to the worm near equilibrium, termed K_w. As that elasticity declined from 1 to -3, the dominant behavior of ΔR_t went from a geometric decline to the equilibrium, to a decaying oscillation about the equilibrium, to an unstable oscillation about the equilibrium at $K_w = -2$ (Fig. 7.5a,b). These transitions depended slightly on the value of worm survival at equilibrium, $g(W^*)$. Hobbs and Botsford (1989) had a tiny amount of data relating egg survival to worm abundance (Fig. 7.5c), which led to an estimate of $K_w = -1.1$. This was greater than the value of -2.0 below which the worm mechanism would cause cyclic behavior (Fig. 7.5a,b). Thus the worm alone could not produce the cyclic population dynamics.

The next question was whether the presence of the egg-predator worm influenced stability of the Dungeness crab population, even though it could not, by itself, cause cyclic behavior. This question involves analyzing a model that includes both the nonlinear, density-dependent effect of the worm on egg survival, g, (as in Eq. (7.11)) and of adult cannibalism on eggs, f (as in Eq. (7.1)). The combined model is

$$R_t = \sum_{a=1}^{A} [\varphi_b(a)R_{t-a}]\, g\left(\sum_{a=1}^{A} [\varphi_w(a)R_{t-a}]\right) f\left(\sum_{a=1}^{A} [\varphi_c(a)R_{t-a}]\right). \tag{7.17}$$

The stability analysis of this model involves both the elasticity of worm predation, K_w, and cannibalism, K_c. It turns out that as the values of both elasticities became increasingly negative, their combination could cause cyclic behavior (when $g(W^*)$ is held constant at

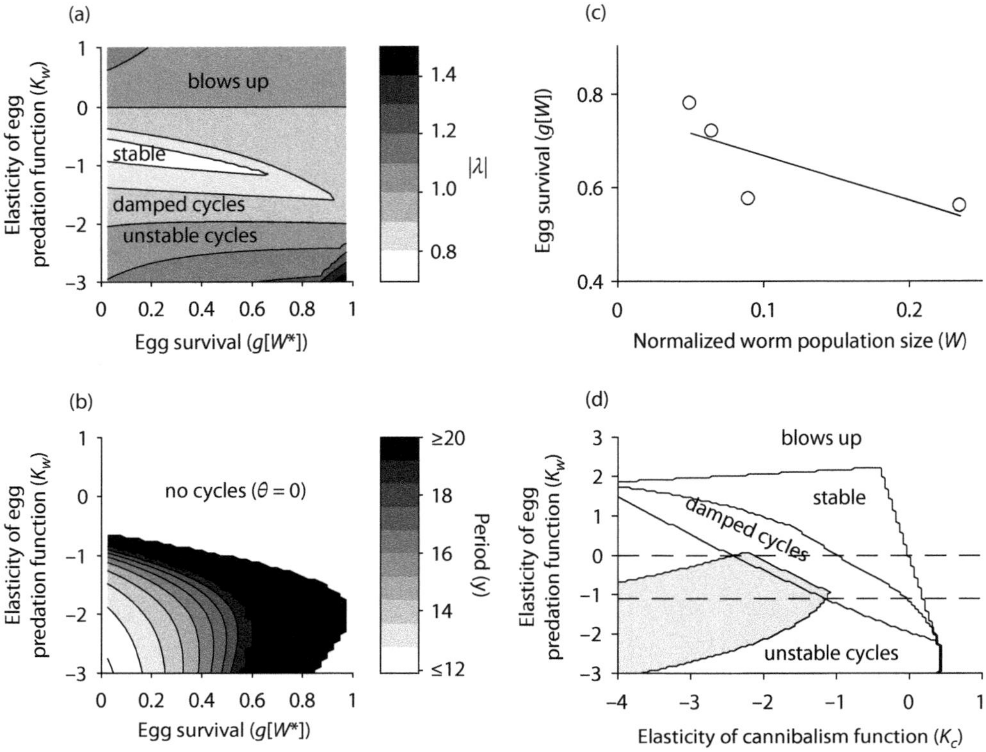

Fig. 7.5 Stability analysis of the Dungeness crab model with the egg-predator worm added to crab cannibalism in the density-dependent recruitment. Panel (a) shows how stability depends on the elasticity of worm-induced density dependence (K_W) for each value of the equilibrium value of egg survival ($g[W^*]$); the thick black line indicates the stability boundary, $|\lambda| = 1$. Panel (b) shows the corresponding period of locally unstable cyclic behavior. Panel (c) shows individual-level data on how egg survival varies with worm abundance. Panel (d) shows behavior of a model with both cannibalism and the egg-predator worm. The upper dashed line indicates the absence of a nonlinear response due to the worm ($K_W = 0$), and the lower dashed line indicates the value of $K_W = -1.1$, the value estimated from panel (c). The gray shading indicates values of $|\lambda| > 0.9$ and period between 9 and 12 years; that is, unstable or barely damped dynamics with close to the correct period. This became known as the "tongue of possible cycles." Adapted from data and results in Hobbs and Botsford (1989) and Botsford and Hobbs (1995).

the best estimate of that parameter; Fig. 7.5d). The combinations of values that would lead to unstable or nearly unstable cycles near the observed period of eight to ten years (i.e. $|\lambda| > 0.8$ with $8 < 2\pi/\theta < 10$, where θ is the angle of λ in the complex plane) are darkened in Fig. 7.5d. That area became known as the "tongue of possible cycles" because of its shape. The conclusion from this analysis is that neither worm predation nor cannibalism could—on their own—produce the type of cycles observed in natural populations. This is because the tongue in Fig. 7.5d does not intersect either the line $K_C = 0$ (no nonlinearity in cannibalism) and only barely touches the line $K_W = 0$ (no nonlinearity in worm predation; upper dashed line in Fig. 7.5d). However, there was a wide range of possible values of K_C that could produce the observed cycles if the worm effect had the empirically estimated value of 1.1 (lower dashed line in Fig. 7.5d).

Because of interest in the effects of coastal upwelling and other environmental variables on Dungeness crab catch, Botsford (1986) explored the frequency response (see Section 4.3.2) of the age-structured density-dependent model we described earlier for this species. Separate models of population dynamics and fishing showed: (a) a transfer function that showed a peak in sensitivity at frequencies near 0.14 y^{-1} (period of 7 y) and (b) a transfer function that declined with frequency for frequencies greater than 0.08 y^{-1} (periods shorter than 12.5 y). The practical conclusion from this analysis was that the stability conditions from the deterministic model were too strict; that the observed cycles could have been caused by a population that was deterministically slightly more stable, but that was also affected by random environmental perturbations. This type of environmentally driven cycling would be possible under conditions where the population was cyclic, but locally stable. Indeed, this is what Higgins et al. (1997) later found: the extremely cyclic historical catch record could be reproduced by a model that included age-structured density dependence but was deterministically stable, if small amounts of environmental variability were added.

The Dungeness crab fishery has continued since these modeling investigations during the 1970s–1990s, and catch (and abundance) continue to fluctuate, but the cyclic behavior analyzed here has not occurred since the early 1980s (Fig. 1.1). There was a decadal-scale shift in patterns of oceanographic variability in the northeast Pacific in the late 1970s, which likely led to this shift in population dynamics (Botsford and Lawrence, 2002). Empirical studies of the fluctuations continue, but there has not been a concerted effort to determine the cause of the large cyclic fluctuations seen in the left half of Fig. 1.1 (Rasmuson, 2013; Shanks, 2013).

7.4 An intertidal barnacle, *Balanus glandula*

Beginning in the late 1970s there was increasing interest on the U.S. west coast in the factors that influence the dynamics of rocky intertidal communities, in particular how and when patches of empty space opened up, and why some parts of the coastline consistently had more open space than others (Lubchenco and Menge, 1978; Paine and Levin, 1981; Gaines and Roughgarden, 1985; Connolly et al., 2001). One area of research was the role of the dispersing larval stages of the sessile species that occupy the rocky intertidal (Gaines and Roughgarden, 1985). As part of this effort, J. Roughgarden and colleagues developed a model for the barnacle *Balanus glandula*, which has a dispersing larval stage and a sessile adult stage. Because it was impossible to determine the sources of larvae settling out of the plankton onto any particular section of the intertidal habitat, they used an open population model to describe the number of recruits settling at time t, R_t:

$$R_t = sF_t, \tag{7.18}$$

where s is a constant reflecting the rate of arriving larvae ready to settle, and F_t is the amount of free space in the habitat. As previously noted, this open-population model assumes that there is some constant external source of larvae in the system, unlike the closed-population models we have dealt with so far in this chapter, in which all reproduction in the population originates locally. This assumption is clearly not strictly true, because the number of larvae available for settlement may vary considerably from day to day, based on ocean conditions, even within the reproductive season. Nonetheless it is a good starting point for examining how the larval settlement rate affects population stability in a system where most of the larvae were likely not produced in the study population.

The total amount of space in the habitat, A, is the sum of the free space and the occupied space,

$$A = F_t + \sum_{a=1}^{a_{max}} c_a n_{a,t}, \tag{7.19}$$

where c_a is the amount of space occupied by a barnacle of age a, and $n_{a,t}$ is the density of age-a barnacles. Adult barnacles have survival rate p. If we define recruits, $R_t = n_{0,t}$, as we have been doing in this chapter, then the full model is

$$R_t = s\left(A - \sum_{a=1}^{a_{max}} c_a n_{a,t}\right) \tag{7.20a}$$

$$n_{a+1,t+1} = p n_{a,t}. \tag{7.20b}$$

Although Roughgarden et al. (1985) did not analyze their model in this way, we can see that Eq. (7.14) could be rewritten in the same form as the other models we have been using in this chapter. From Eq. (7.2b), at equilibrium it would be

$$R^* = \mathrm{B}_C f\left(\Phi_C R^*\right), \tag{7.21}$$

where $B_C = s$ in their notation, Φ_C is the sum of $l_a c_a$ over all ages (l_a is the survival to age a, $^{p(a-1)}$), and $f\left(\Phi_C R^*\right) = A - \Phi_C R^*$. In this formulation, "effective population density," $\Phi_C R^*$, is the amount of occupied space. If we calculate the elasticity of f as we have done before (Eq. 7.3), we obtain

$$K = f'\left(\Phi_C R^*\right)\frac{\Phi_C R^*}{f\left(\Phi_C R^*\right)} = -\frac{\Phi_C R^*}{A - \Phi_C R^*}. \tag{7.22}$$

From Eq. (7.18), at equilibrium,

$$K = 1 - \left(\frac{A}{F^*}\right). \tag{7.23}$$

We know that equilibrium free space will be less than total space, $F^* < A$, so K will always be negative (i.e. there will always be density-dependent competition for space). We know from our previous analysis (Section 7.1) that over-compensation can begin to make the equilibrium unstable when K falls below a threshold K', whose value is less than -1 (for open populations). From Eq. (7.22), this stability threshold will occur when $F^* < 0.5A$; that is, when the free space is less than half the area. Roughgarden et al. (1985) derived this same result, terming it the "fifty percent free space rule," using essentially the same approach (analyzing the characteristic equation) even though they did not state the criterion in terms of the elasticity of recruitment survival. This finding led them to the conclusion that there is a "paradox of free space." That is, usually we think of limiting resources regulating population dynamics when they become scarce. In the case of barnacles, it might seem that space would not be a limiting factor regulating population dynamics when there is a lot of free space available, but actually the population is more stable when free space is abundant. Of course this is not actually a paradox, but rather an inevitable consequence of strong, over-compensatory density dependence.

The behavior of the open population model can be represented in a way similar to the representation of the closed population model in Fig. 7.2. Here we do so varying the rate at which settlers are provided (Fig. 7.6), rather than the strength of density dependence (as in Fig. 7.2). In the first column of Fig. 7.6 one can see that as larval settlement increases (moving down the rows of Fig. 7.6), the equilibrium amount of free space decreases and so the equilibrium recruitment decreases as well, making K more negative. In the second column, the magnitude of the influence function increases in proportion to the value of K determined in the first column. Because this open population model does not include an influence of age distribution on reproduction, the graphs in the second column contain only one influence function, rather than the difference between two influence functions as in Fig. 7.2. The resulting time series of the fraction of free space are stable in the top two rows, then become locally unstable as the value of s increases from 25 to 35 larvae/cm^2/wk.

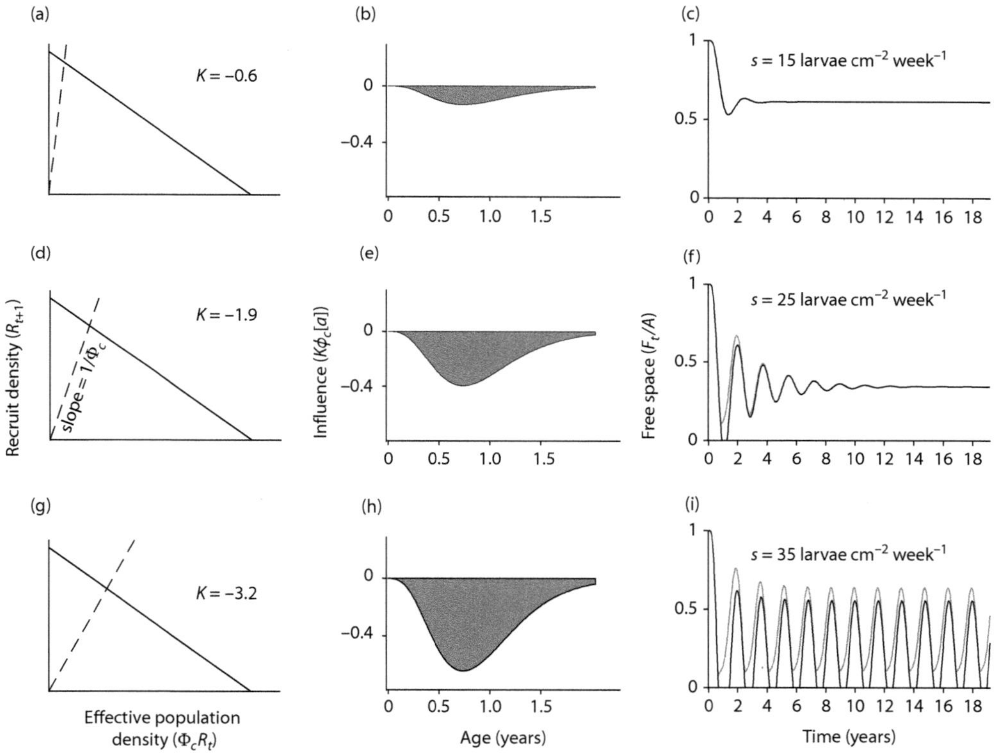

Fig. 7.6 Effect of different inputs of pre-recruits on the stability of the age-structured model of an open population with density-dependent recruitment: the barnacle, *Balanus glandula*, as an example. Pre-recruit larval settler rates vary from $s = 5$ to $s = 15$, to $s = 25$ larvae cm^{-2} week^{-1} from top row to bottom row. The left-hand column is plots of recruitment survival for the free space mechanism (solid line). The dashed line indicates the graphical solution for the equilibrium (filled circle). The central column plots the influence functions in Eq. (7.7b), which here is the product of space occupied at age and survival to age (Roughgarden et al., 1985), all multiplied by the value of K determined from the corresponding plot in the first column. The right-hand column is simulations of this model of age-structured space competition in barnacles. Simulations were initiated with an empty habitat ($F = A$). Results are shown for the basic model described in the text (black curves) and including additional density-dependent adult mortality as free space decreases (gray curves).

The basic source of instability in this system is the post-settlement growth of the barnacles. After a cohort colonizes an empty habitat patch, all of the settlers begin to grow, occupying more free space. The habitat will gradually fill in (with F near zero) to where no more recruitment is possible. Eventually the adults of that cohort die off, immediately opening up more free space to be colonized by a new cohort, and possibly producing a cycle in the availability of free space.

Roughgarden et al. (1985) explored two conditions that could stabilize these cycles. First, if the product $l_a c_a$ declined with age, the dynamics were always stable. This condition would mean that the space occupied by a cohort was always greatest right at the moment of settlement, because mortality outstripped growth as the cohort aged. This is equivalent to saying that the influence function $\varphi_C = l_a c_a$ has a peak at age 1 and declines towards a_{max}. In other words, loosely speaking, φ_C would then have a small mean, a large standard deviation, and thus a large coefficient of variation, which are the conditions favoring stability, as described earlier for the general case (Eq. 7.6). One can see how the results of these age-structured models of crabs and barnacles become very similar when expressed in terms of influence functions. Indeed, Roughgarden et al. termed φ_C the "net area function" in an analogy to the lifetime reproduction function.

The second factor affecting stability that Roughgarden et al. (1985) explored was the settlement rate, s (in their notation, B_C here); they showed how the population would transition from stable to damped to unstable (limit cycle) dynamics as s increased and F^* decreased accordingly (i.e. effective population density increases in Fig. 7.6). We can see how that holds in general for open population models by noting that from the equilibrium condition (Eq. 7.21), $B_C = R^*/f(\Phi_C R^*)$, which appears in the definition of K (Eq. 7.22). As a consequence, any increase in B_C would magnify the (negative) value of K. In the version of the model we have described here, once a cohort grows old enough that those barnacles occupy all of the free space, the habitat remains full until that cohort begins to die off, opening up space. To add a little more realism, Roughgarden et al. (1985) made the adult mortality increase by the factor $e^{-mF/A}$, so that there was increasing density-dependent adult mortality as the habitat got full (the parameter m determines how quickly that effect manifests). This approximates the phenomenon of "hummocking," in which large clumps of adult barnacles tend to pile up and become more vulnerable to being swept off the shore by waves. When we add the density-dependent mortality of adults (hummocking), free space never completely disappears in the cyclic population (gray lines in Fig. 7.6c, f, and i).

Gaines and Roughgarden (1985) empirically tested the model prediction that stability depended on the amount of free space and, in turn, the settlement rate by monitoring two rocky intertidal habitats near Hopkins Marine Station in Monterey Bay, California. They found that population dynamics were stable at the site with low settlement and consistently high free space (Fig. 7.7a; note how free space is at a relatively constant level, then drops to a slightly lower constant level after the small settlement pulse at around week 30), but cyclic behavior occurred at the site where the inflow of potentially settling larvae was greater (Fig. 7.7b; note that overall settlement is two orders of magnitude higher at this site). Unfortunately, this apparently elegant agreement of theory and data proved to be too good to be true; it turned out that the period of the cycles observed in the field did not match that predicted by the model. Possingham et al. (1994) later showed that an additional mechanism—attraction of a sea star predator to large aggregations of adult barnacles—was necessary to produce the observed cycles.

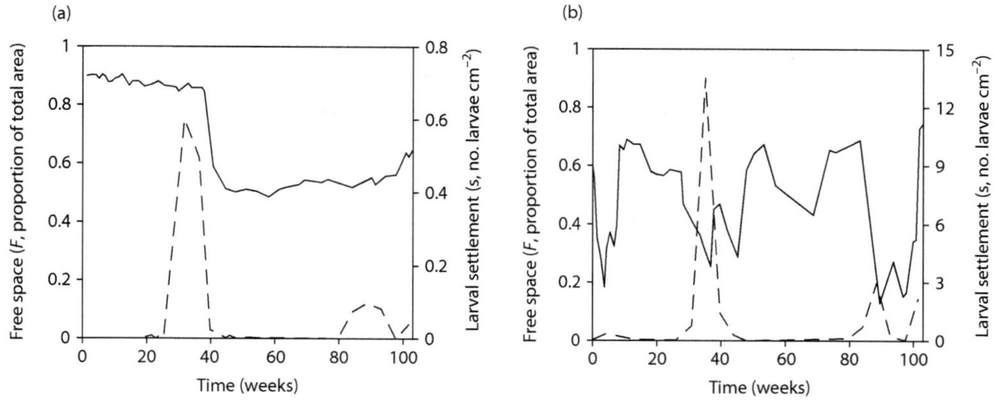

Fig. 7.7 Population dynamics of barnacles, *Balanus glandula*, at two sites in Monterey Bay California: (a) KLM and (b) Pete's Rock, monitored from August 1982 to August 1984. Solid curves indicate the proportion of free space in benthic quadrats; dashed curves indicate the density of larval settlement in those quadrats. Note that settlement to Pete's Rock is more than an order of magnitude greater than at KLM, and the amount of free space is both less and more cyclic. Redrawn from data in Gaines and Roughgarden (1985).

7.5 Cannibalism and the flour beetle, *Tribolium*

In the late 1970s there was an increasing ecological interest in cannibalism (Fox 1975, Polis 1981). This interest led to a "cottage industry" of mathematical modeling of cannibalism (see Claessen et al. (2004) for a review). There is not the space to cover that entire field here; rather we track one relevant thread with which we have direct experience: cannibalism in the flour beetle, *Tribolium*.

From the initial laboratory experiments with *Tribolium* in the 1920s, researchers noted that the relatively constant adult numbers stood in contrast to wild oscillations in larval abundance. They attributed these oscillations to cannibalism of eggs by larvae (Chapman, 1928; 1933), but the dynamics of this mechanism were not worked out until the late 1980s (Hastings, 1987; Hastings and Constantino, 1987: 1991).

Hastings and Costantino's (1987) model for *Tribolium* differed from the models we have developed so far in this chapter, but some aspects are similar. One difference is that the cannibalism occurred throughout two stages. The eggs could be preyed upon constantly throughout an egg stage of duration A_E by larvae of any age in the subsequent larval stage of duration A_L. The adult stage followed the larval stage, but because adults live an order of magnitude longer than either A_E or A_L, and they have been observed to have roughly constant density, that stage (and the pupal stage) were not modeled in detail so that the analysis could focus on the short-time-scale cycles in the egg and larval stages. A constant supply of eggs was assumed, B, making theirs an open model (as in the barnacle model in Section 7.4).

Hastings and Costantino (1987) used integro-difference equations to represent the egg–larval dynamics, but for consistency here we will describe the model as though it were in discrete time. Recruitment to the larval stage was assumed to result from the constant supply of eggs, which were subject to random encounters with larvae that had feeding rate c (i.e. a linear predator functional response) resulting in the expression

$$R_{t+1} = Be^{-c \sum_{x=t}^{t-A_E} \sum_{y=0}^{A_L} R_{x-y}}. \tag{7.24}$$

The terms in the exponent sum the number of larvae of each age that would have been present during each time step of the egg duration leading up to the larval cohort born at $t + 1$, times the cannibalism attack rate c (the death rate due to causes other than cannibalism is very small, so they assumed, for the sake of simplicity, that it could be ignored). This expression becomes much simpler at equilibrium, R^*, because the mortality due to cannibalism is just the attack rate times the duration of the egg stage, A_E, the duration of the larval stage A_L, and the equilibrium number of recruits,

$$R^* = Be^{-cA_EA_LR^*}. \tag{7.25a}$$

The elasticity, K, is then

$$K = -cA_EA_LR^*. \tag{7.25b}$$

As in the barnacle case, as the constant supply of eggs, B, increases, R^* increases, and K becomes more negative, tending to make the population less stable. This characteristic is apparent in the plot of stability boundaries (Fig. 7.8). For all combinations of durations of the larval and egg stages, increasing the egg input is destabilizing.

To evaluate the effect of the age structure on the stability threshold, we note that the width (with respect to age) of the effect of larvae on cannibalism is A_L, and the mean age is $(A_L + A_E)/2$, so that the width relative to the mean is $2A_L/(A_E + A_L)$. From our earlier finding that increasing the width of the age structure relative to its mean makes the critical value of K more negative, we would expect that increasing the duration of the egg stage would only increase the mean age, making the population less stable. This is also seen in Fig. 7.8; as A_E increases, the stability thresholds drop to lower values of B, making the thresholds more easily transgressed for a wider range of birth rates and larval durations. Increasing A_L increases the width relatively more than the mean, and hence would be expected to lead to a more stable population (i.e. thresholds move up to higher values of B), which is also consistent with Fig. 7.8. The expected period of the

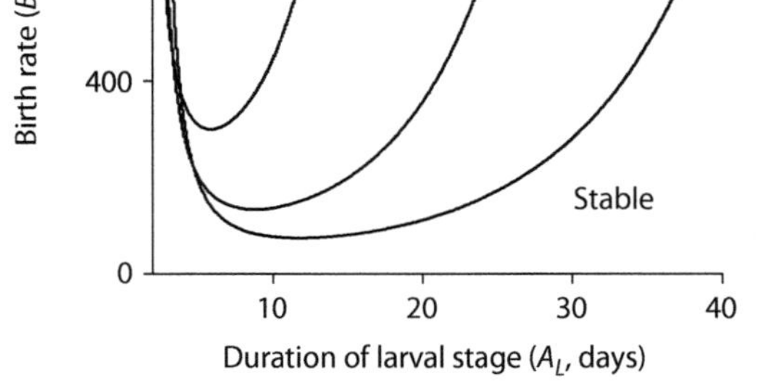

Fig. 7.8 Stability analysis for the Hastings and Costantino (1987) open population model of cannibalism in *Tribolium*. The contour lines show the boundary between stable ($|\lambda| < 1$) and unstable ($|\lambda| > 1$) population dynamics for different combinations of the constant rate of introduction of pre-recruits, B, and the duration of the larval stage, A_L. Curves are shown for three potential durations of the egg stage, A_E. Redrawn using model equations from Hastings and Costantino (1987).

cycles would be twice the mean lag in age, or $A_E + A_L$, as in Hastings and Costantino (1987). Hastings and Costantino (1991) later made their model more realistic by having cannibalistic tendencies increase with age. This diminished the relative width of the age-dependent effect on cannibalism, making the population less stable, as would be expected.

Hastings and Costantino (1987) also provided a simple heuristic explanation for why the period of the cycles is approximately $(A_L + A_E)$. Each cohort of larvae that is produced essentially consumes any eggs that enter the system until that cohort of larvae pupates. At that time, a new cohort of larvae is produced. For a discussion of a similar system, and an analysis by different means more analogous to that originally in Hastings and Costantino (1987), based on delay differential equations, see Gurney et al. (1983).

7.6 Effects of equilibrium conditions

An important consideration when linearizing nonlinear population models about an equilibrium is that equilibria can change due to exogenous factors, as well as changes in the focal population, that were not accounted for in the model. This is relevant to the models in the last half of Chapter 4 as well as to the analyses here in Chapter 7, because in both cases the population behavior predicted by the analyses described depend critically on equilibrium values of recruitment or abundance. Here we describe two aspects of equilibria that arose in our early analyses of Dungeness crab, but are relevant to other situations. One is the effect of single-sex hunting or fishing, which is a fisheries management approach applied to a number of species of crustaceans, as well as an approach to management of hunting in deer. The other is an effect that changes in individual growth rates can have on equilibria. Such effects will not be obvious in the age-structured models we have been using, but, importantly, they could lead to multiple equilibria.

7.6.1 *Single-sex harvest*

A common strategy in managing hunting for deer and harvest of marine crustaceans is to allow removals of males only. The underlying rationale is to maintain high reproductive rates in polygamous species. The drawback of this approach was first pointed out in a classic study of deer populations (Leopold et al., 1947). This kind of management works very well in maintaining populations at high abundance, but that ultimately is not desirable because it does not allow density to decline so that density-dependent increases in growth and fecundity rates can increase productivity.

We can illustrate this concept by writing the expression for equilibrium of an age-structured population (Eqs. 4.22 and 7.1) with density-dependent recruitment in terms of both sexes using additional subscripts m and f,

$$1 = \sum_{a=1}^{A} \varphi_{b,f}(a) f\left(R^* \sum_{a=1}^{A} \left[\varphi_{c,f}(a) + \varphi_{c,m}(a)\right]\right), \tag{7.26}$$

which becomes the following, written in terms of sums over influence functions:

$$1 = \Phi_{b,f} f\left[R^* \left(\Phi_{c,f} + \Phi_{c,m}\right)\right]. \tag{7.27}$$

If we harvest males only (and there are sufficient males to fertilize all females) we will not change the value of $\Phi_{b,f}$ Z. From Eq. (7.27) we see that if $\Phi_{b,f}$ does not change, the value of the function f cannot change, either. Thus, when we decrease $\Phi_{c,m}$ by fishing males only, the value of R^* must increase, which will maintain density at a high value. In other words, removing the males simply frees up resources for more females. Of course that result only describes what will happen at equilibrium; it is possible that there would indeed be transient increases in individual growth or fecundity rates immediately after a male-only harvest was initiated.

This explains the difference between the results of Botsford and Wickham (1978) and those of Ricker (1954): Ricker found that fishing an age-structured population with density-dependent recruitment made the population more stable, while Botsford and Wickham found that fishing the Dungeness crab population made the population less stable. The reason for the difference is that in Ricker's case both males and females were presumed to be fished at the same rate, and the density was reduced, thus moving it to a point where the elasticity at equilibrium was less negative. We can see in the first column of Fig. 7.2 that if females were harvested, reducing Φ_{b}, the equilibrium would move upward on the survival function f, towards a region with a less steep slope.

We should mention that it is not always safe to assume that there are enough males to fertilize all of the females in the population. For example, White et al. (2017) found that fertilization rates in the silverside *Menidia beryllina* decreased linearly with the proportion of males in the spawning population. Consequently the equilibrium population size depends on both the sex ratio and the shape of the relationship between sex ratio and reproductive output; in the context of the models developed in this chapter, $\Phi_{b,f}$ would become a nonlinear function of sex ratio. This would have to be accounted for in the management of sex-changing animals (e.g. many fish begin life as females then change sex to male, so males are at more risk from size-selective harvest; (Alonzo and Mangel, 2004; Easter and White, 2016) or in predicting the effects of sex-changing endocrine disrupting pollutants (White et al., 2017).

7.6.2 *Multiple equilibria*

In the 1970s, population biologists noticed that fishing of several species of fish over a number of decades had reduced populations to lower equilibria, but when fishing was diminished, these species remained at low abundance rather than returning to a level near their earlier equilibria (Holling, 1973). These included: (a) the population of Eurasian perch (*Perca fluviatilis*) in the north basin of Lake Windermere in the UK (Le Cren, 1958; Le Cren et al., 1977); (b) the population of Pacific Sardine (*Sardinops sagax caerulea*) off the west coast of the United States (Murphy, 1977); and (c) the central California population of Dungeness crab (*Metacarcinus magister*; Botsford, 1981). An explanation for this phenomenon is that if individual growth rates are density dependent, the size structure of the population could change substantially when it is harvested to low levels. If animals reached larger size at a younger age (as was observed in the Dungeness crab fishery, for example), then the influence functions Φ_b and Φ_c could shift to have higher values at lower ages, and be greater in overall magnitude. Together, the effect will be to change the equilibrium R^*. The expected direction of change can be seen from the solution for the equilibrium (Eq. 7.2),

$$1 = \Phi_b f\left(R^* \Phi_c\right).$$

An increase in Φ_b will require the value of f to decrease, which means an increase in R^* (if Φ_c remains unchanged). Similarly, an increase in Φ_c would require a decrease in R^* if Φ_b does not change. These results tell us what to expect: more reproduction increases the equilibrium, and more over-compensatory mortality reduces it. However, the actual combined effect on R^* depends on the relative changes in the shape of the two influence functions. In the case of Dungeness crab, Botsford (1981) showed that the effect of increased juvenile growth on compensation was greater than the effect on reproduction, and would generally lead to reduced R^*. Thus when fished to a low level, the crab population could have experienced a density-dependent increase in growth rate that changed the elements of the equilibrium condition in such a way as to "lock" the population at a new, lower equilibrium level, even if fishing pressure was later relaxed.

7.7 What have we learned in Chapter 7?

In this chapter, we have developed a framework for understanding how when older age classes have an over-compensatory density-dependent effect on the production and survival of younger age classes ("recruits", that influence can lead to cyclical fluctuations in the number of recruits over time with a period of $2T$ (i.e. two generations). We described how a number of investigators working on different species have independently reached the common conclusions that the occurrence of such behavior depends on a tension between two population characteristics: (1) the strength of the potentially destabilizing, over-compensatory density dependence in recruitment at equilibrium (as represented by the elasticity of the density dependence at equilibrium), and (2) the stabilizing effect of the "diversity" of ages over which juveniles and adults affected compensation (as represented by the ratio of the width divided by the mean of the age-dependent influence function for the operative density-dependent mechanism). We reviewed how different versions of essentially the same results were reached by a number of different investigators, independently, comparing modeling results to data from cycles and life histories from species such as crabs, barnacles, flour beetles, and more (Table 7.1).

The use of general expressions for density-dependent recruitment, and for influence functions, allowed us to write out population dynamics in terms of recruits only (the scalar R_t) instead of the entire age distribution (the vector of $n_{a,t}$ values). This, in turn, led to some simple conclusions. When there is over-compensatory density dependence, a population is more likely to exhibit unstable fluctuations (a) when the elasticity of the recruit survival function is more negative, (b) when there is a greater time lag in the over-compensation (i.e. when the mean age of the influence function is greater), and (c) when the effect of older individuals is concentrated into fewer age classes (i.e. when the width of the age function is narrower). Also, the unstable cycles tend to have a period of twice the generation time, unless (a) the population is very unstable and exhibits chaotic dynamics, or (b) the fluctuations occur on the ascending (compensatory, but not over compensatory) part of the stock–recruitment curve. In the latter case, the cycles of period $1T$ discussed in Chapter 4 could occur in the presence of environmental variability.

We illustrated the general framework with Figures 7.1–7.3. The first figure showed, in Figs. 7.1a and c, how each cohort would influence recruitment as it aged, giving both the familiar effect on reproduction (fecundity) and the effect on density-dependent recruitment (e.g. cannibalism). In the right-hand column, as examples of density dependence, we showed the recruitment survival functions from the Beverton and Holt (1957) model of stock recruitment and the model of Ricker (1954). We also showed at the top of that column that the expected Ricker and Beverton–Holt behavior would indeed be obtained using our age-structured model, and at the bottom, the values of recruitment survival elasticity K that would result. We learn later in the chapter that the possible values of K for the Beverton–Holt model will always be greater than -1, hence would never be negative enough to produce cycles with period $2T$. The threshold value of K below which cycles can occur (i.e. K') will always be less than -1.

In Fig. 7.2, we used those influence functions and a density-vague recruitment survival function to show how very steep declines in recruitment survival with density can lead to cycles of period $2T$. This was a closed population model and we showed how values of K varied with the slope of recruitment survival, and how that changed the population's age structure resulting from the influence functions. To show how adding fishing to this model would change stability we added fishing of males only (as in many crustacean fisheries and the hunting of cervids). We fished males only, because fishing females in this population will always move the equilibrium to lower values of density, hence in this population with over-compensatory density dependence it will always be stabilizing. The top of Fig. 7.3 showed how combinations of density dependence (i.e. K) and age structure determined stability. Combinations below K' (the black line) produce $2T$ cycles. Increasing fishing decreased stability by moving K' up to less negative values. The lower plot in Fig. 7.3 uses our prior understanding of eigenvalues to show us how the population becomes gradually less stable with increasing K (as one steps up in the triangles or bubbles), and crosses outside the unit circle. This happens differently for each value of fishing mortality, and we saw how the period of cycles depended on the point of crossing of the unit circle (from Section 4.1, it is 360/(the angle between the crossing point and the real axis)).

In the first application of that framework, we used the simplest possible age-structured model with density-dependent recruitment to demonstrate potential pitfalls in drawing general conclusions from population models, and to demonstrate that the approach with influence functions and the elasticity of recruitment survival does this well. It produces results that allow statements about the effects of adult survival and age of first capture that remain true over the full range of the other variable (Fig. 7.4).

Figure 7.5 describes $2T$ cycles in a model of Dungeness crab with both cannibalism and an egg-predator worm acting as density-dependent recruitment; with Fig. 7.5a showing how the elasticity of recruitment survival, and Fig. 7.5b showing the period of cycles for conditions under which $2T$ cycles would occur. This example introduced the notion of testing potential causal mechanisms using data from critical mechanisms over short time periods when long-term population data are not available (Fig 7.5c). While they are a tiny amount of data, they could be used to refine our ideas of what combinations of cannibalism and egg predation could be causing the observed cycles in Fig. 7.5d.

The next example was motivated by the often patchy distributions of the barnacle *Balanus glandula* along the U.S. west coast. Because the proposed mechanism involved an interaction between successful larval settlement and the density already settled at all ages, the reproductive source was not known, so an *open population* model was formulated. For an open population model, the presumed input of pre-density-dependence recruits has a critical effect on whether cyclic behavior occurs (the three rows in Fig. 7.6). Also, a different graphical solution is used to determine equilibrium, as shown in the left-hand column of Fig. 7.6. That equilibrium sets K, which multiplies the magnitude of the single influence function in the age structure (second column in Fig. 7.6), which can produce enough negative feedback to cause $2T$ cycles. The period of these cycles was short enough that this dependence of stability on the level of pre-settlement recruits could be tested in the field within a couple of years (Fig. 7.7).

The next example, cannibalism in the flour beetle, *Tribolium*, was mathematically a little different, but nonetheless showed similar characteristics. This model was an open population, assuming a constant supply of eggs from an approximately constant, large adult population. Figure 7.8, repeated from the analysis of Hastings and Costantino

(1987), clearly shows that, as in the open barnacle population, the population becomes less stable as the input supply increases. Increasing the duration of the egg stage increases the amount of variability in age, relative to the mean, as does increasing the duration of the larval stage. Thus both are stabilizing except at very small values of larval duration. Development of this early model of *Tribolium* has continued, and ongoing laboratory culture has enabled extensive comparison of models with data.

The examples in this chapter illustrate the advantages of "scaling up" from the individual level to the population level, as described in Section 1.5. In the examples, we saw how different types of age-structured density-dependent models can be formulated, using individual level rates (i.e. the influence functions) and the definition of the elasticity of recruitment survival, to draw conclusions regarding population level behavior (i.e. stability and cycles). These types of models would be useful, in conjunction with the appropriate data, in determining what specific mechanism is producing cyclic dynamics in populations. The generally shorter time scales required for observations at the individual level are an advantage compared with the difficulty in implementing a field experiment long enough to actually measure population level effects at the time scale of the cycle period. Scaling up from the individual level may be the best approach when attempting to evaluate the population level effects of "new" impacts on life histories, such as the effects of ocean acidification and climate change.

Another example of a population model that bears some similarity to the models described here is the model of canopy-forming kelp by Nisbet and Bence (1989). The basic density-dependent effect of adults on recruitment is the negative effect of shading by the adult canopy on recruitment, which they cast as a density-vague mechanism. As they note, the shading mechanism resembles the available space mechanism in barnacles in Section 7.4. They present an extensive comparison of model results with observations.

Finally, one should keep in mind that the dynamics we have described in this chapter all occur near the long-term equilibrium. Because real populations are not always near their equilibrium, they are also likely to exhibit transient dynamics (like those we described in Chapter 4) in addition to the long-term dynamics as we have described here.

CHAPTER 8

Age-structured models in a random environment

Because almost all biological populations are subject to random fluctuations in the environment, we turn to the effects of that random variability in this chapter. Our primary theoretical interest will be in how adding variability changes population behavior. We focus on age-structured populations at such a low abundance that any density-dependent effects are negligible (thus, populations that could be represented by a Leslie matrix). We have mentioned random environments at several points previously in this book, so we first place this chapter in that context. In Chapter 2, we described the risk of quasi-extinction of a simple population model (i.e. one with a single-state variable, abundance) in a random environment (Section 2.4.3). Here we describe how the addition of age structure affects the conclusions we drew there.

Chapter 4 generally addressed the deterministic, transient behavior of linear, age-structured populations before they reach the asymptotic state of geometric growth. Because it is also useful to view random variability as ongoing, transient buffeting away from the constant equilibrium of a density-dependent, age-structured population, we described random variability in age-structured populations with density-dependent recruitment in that chapter. In particular, we focused on the phenomenon of cohort resonance that emerges in age-structured populations.

It stands to reason that the material in this chapter would be related to the results regarding cohort resonance in Chapter 4, when density, and therefore density dependence, is near zero. Recall that cohort resonance involves sensitivity to environmental variability on two time scales: generational time scales (i.e. frequencies near 1/(generation time), hence the name) and very long time scales (i.e. frequencies near zero, corresponding to long-term trends in abundance), especially as populations decline in abundance (see Section 4.4). Those low-frequency trends are similar to the slow, random, geometric increases or decreases in abundance of linear age-structured models we will describe in this chapter. Nonetheless, the connections between density-dependent models with cohort resonance and the linear models in this chapter are not yet well understood.

Population biologists address three categories of population variability, but we will focus here on one of them, environmental variability. *Environmental variability* is random, environmentally induced variability in reproduction or mortality at a specific age (or ages) that affects all individuals at that age in the same way. This is in contrast to *demographic stochasticity* (described in Section 1.3.1), which leads to different random outcomes for each individual, and will not yield the same result each time at a given abundance. A typical example is variability in the sex ratio of offspring when population sizes are small; this will follow a binomial distribution (Box 1.4), so there could be very high or

Population Dynamics for Conservation. Louis W. Botsford, J. Wilson White, and Alan Hastings, Oxford University Press (2019).
DOI: 10.1093/oso/9780198758365.001.0001

very low sex ratios just by chance. A similar example with the stochasticity affecting individual survival is in Section 1.3.1. As demonstrated in those examples, demographic stochasticity generates variability endogenously, rather than being a simple response to a variable environment. *Catastrophic variability* refers to the environment causing a single large random change in state, that affects a major part of the population. Catastrophic variability could be addressed as a transient response to displacement to a state other than the stable age distribution, as described in Sections 4.1 and 4.2 on transient dynamics.

The primary practical concern regarding random environmental variability is the possibility that it could cause a population to cease to exist. As such, we will be concerned with the probability of that occurring (i.e. the probability of extinction), but for the reasons stated in Section 2.4.3, we will calculate the probability of quasi-extinction (i.e. declining to below a fixed threshold, as opposed to declining all the way to zero abundance). This process is the same as described in Section 2.4.3, except that here the model is age-structured. We can envision quasi-extinction, as we did in Fig. 2.13, as many time series of abundance projected into the future, each the result of a different sequence of random environments, some dropping below the quasi-extinction threshold before time T, others not.

The issue of extinction has a considerable history in ecology, including forays into some areas that we will not address in this chapter. For example, Willy Feller (Feller, 1939) developed what is known as the *birth–death process* model, which is essentially an unstructured linear model (like those we described in Chapter 2) with demographic stochasticity in the birth and death terms that comprise the growth rate λ. This type of model was later extended to include environmental stochasticity (Goodman, 1987; Leigh, 1981). The model by Lewontin and Cohen (1969), described in Section 2.4.3, also had a single variable (abundance), but it was structured differently.

At about the same time models addressing extinction developed a spatial view. One example consisted of birth–death populations, in which sub-populations occupying nodes on a lattice could exchange migrants (Bailey, 1968). Levins (1969) developed the first so-called metapopulation model, in which the state variable was simply the presence or absence of a sub-population (i.e. not sub-population abundance) at each spatial location. These sub-populations had a probability of extinction when extant and a probability of recolonization when extinct. MacArthur and Wilson's (1967) theory of island biogeography took a similar approach, describing the probability of population extinction on an island as a function of island area rather than population density. We will examine Levins' (1969) model and extensions of it in Chapter 9; for now we focus on cases in which the state variable is structured (primarily age structured) rather than simply abundance or presence.

This chapter lays out the population dynamic background for addressing practical problems involving potential extinction, but leaves discussion of the practical issues involved in application to Chapter 11. One important deferred topic is the important role of uncertainty in practical management of species at risk. We begin this chapter where we left off with unstructured models in Chapter 2 (Section 2.4.2), by describing the analytical *small fluctuation approximation* (*SFA*) for the growth rate of a structured population under environmental stochasticity (Section 8.1). Not surprisingly, that approximation depends on the sensitivities of linear models described in Section 3.5 for age-structured models, as well as the amounts of variability in each matrix component. We then show how we can use a diffusion equation to (approximately) translate the SFA into an estimate of the probability of extinction over some time span (Section 8.2). We are then able to express these results more generally by writing them in terms of the elasticities of the elements of the projection matrix, rather than the sensitivities (Section 8.3). We then use a case study, the *Totoaba* population from Section 3.5, to demonstrate how different levels of variability and correlations between life history stages influence the rate of increase in variance. We use the same species to see how well the SFA/diffusion approximation actually portrays the results of direct model simulations (Section 8.4). We learn there that that approximation does not always work, so we describe a method of avoiding the approximation (Section 8.5). Having assumed up to that point that environmental variability was uncorrelated over time, we then double back to describe what happens when that is not so (Section 8.6). We had learned earlier in Chapter 4 that the presence of such autocorrelation corresponded to "colored noise." We note several recent examples of surprising results regarding the effects of environmental spectra.

We then turn to several sets of examples of practical descriptions of random population behavior: (1) a regression model for determining random behavior from time series of abundance (Section 8.6) and (2) a review of 29 publications describing estimation of probabilities of extinction in various ways (Section 8.7). Because most of those examples have used stage-structured models, and we have expressed skepticism of stage models in Chapter 6, we then assess the use of stage-structured models to calculate probabilities of extinction (Section 8.8).

8.1 The small fluctuation approximation (SFA)

In this book we are primarily interested in the general nature of population behavior with different age-dependent parameters and under different conditions. Therefore, we will explore the probability of extinction using an analytical, mathematical approach, even though that may not necessarily be the best method to use for the practical application of estimating the risk of extinction of a specific population from available data. These results are due to Tuljapurkar and Orzack (1980) and Tuljapurkar (1989; 1990; 1982). They assume small fluctuations and primitive matrices (recall from Section 4.2.1 that the latter means that SFA methods will not necessarily work for obligate semelparous populations, i.e. populations spawning only once in their lifetime, all at the same age).

The basic results of these models deal with the scenario in which environmental variability causes some or all of the elements of the Leslie matrix to vary in each time step:

$$\boldsymbol{N}_{t+1} = \boldsymbol{A}_t \boldsymbol{N}_t. \tag{8.1}$$

Note that the subscript t on the Leslie matrix $\boldsymbol{A}_t$ indicates time variability. The annual values of environmental variability at each time are assumed to be statistically independent (i.e. there is no autocorrelation). Tuljapurkar and colleagues described population dynamics in terms of the distribution of the logarithm of the ratio of total population abundance (i.e. the sum over all age classes) at time t and at time 0, $x_t = \ln(N_t/N_0)$. After a certain amount of time, x_t will have a Gaussian distribution with mean $x_0 + \mu t$ and variance $\sigma^2 t$, where

$$\mu \approx \ln \bar{\lambda} - \frac{\sigma^2}{2} \tag{8.2a}$$

$$\sigma^2 \approx \frac{1}{\bar{\lambda}^2} \boldsymbol{\delta}' \boldsymbol{C} \boldsymbol{\delta}. \tag{8.2b}$$

$\bar{\lambda}$ is the average of the dominant eigenvalues, λ, of the $\boldsymbol{A}_t$ matrices, $\boldsymbol{\delta}$ is the vector of sensitivities of $\bar{\lambda}$ to each time-varying element of $\boldsymbol{A}_t$, and $\boldsymbol{C}$ is the covariance matrix for the elements of $\boldsymbol{A}_t$ that vary randomly. For example, if the elements b_3 and p_7 of a Leslie matrix varied randomly, the growth rate of the variance in x_t would be

$$\sigma^2 = \frac{1}{\bar{\lambda}^2}\begin{bmatrix}\frac{\partial\lambda}{\partial b_3} & \frac{\partial\lambda}{\partial p_7}\end{bmatrix}\begin{bmatrix}\sigma^2_{b_3} & \mathrm{cov}\,(b_3, p_7)\\ \mathrm{cov}\,(b_3, p_7) & \sigma^2_{p_7}\end{bmatrix}\begin{bmatrix}\frac{\partial\lambda}{\partial b_3}\\ \frac{\partial\lambda}{\partial p_7}\end{bmatrix}, \tag{8.2c}$$

where σ_i^2 denotes the variance of parameter i and cov(i, j) denotes the covariance of parameters x and y. The growth rate of the overall variance, σ^2, would in turn affect the growth rate of the mean of x_t.

The linear increase of the log of abundance in Eq. (8.2) indicates that when random variability is added to the Leslie matrix, population growth is still essentially geometric; that is not changed. However, here, as in the unstructured models in Section 2.4.3, the rate of increase of the mean is less than the deterministic response to the average model. In fact the expression for the growth rate of the mean of the logarithm of abundance, μ, resembles the earlier result for simple population models in terms of the approximation of the arithmetic mean (Eq. 2.24). Recall from Chapter 2 that the expectation (i.e. the mean) of ln λ, E(ln λ), is the logarithm of the geometric mean of λ, while the arithmetic mean of λ is simply E(λ). Rewriting Eq. (2.30), we obtain an expression for E(ln λ) that resembles the expression for μ in Eq. (8.2),

$$\mathrm{E}\,[\ln\lambda] = \ln\mathrm{E}\,[\lambda] - \frac{\sigma^2}{2\mathrm{E}[\lambda]^2}. \tag{8.3}$$

In both Eqs. (8.2a) and (8.3), the first term depends on the arithmetic mean of the growth rate, and the second term is a negative dependence on the variance of the random variability (we could have written Eq. (8.2a) with $\bar{\lambda}^2$ in the denominator instead of including it in Eq. (8.2b), and Eq. (8.2a) would be even more similar to Eq. (8.3), but writing it as we have becomes useful in Section 8.2). The important population dynamic consequence of these expressions is that increases in environmental variability shift the rate of temporal change in the distribution of population densities towards zero. We saw an example of the differences between the expected value of the growth rate and the expected value of the natural log of the growth rate in Fig. 2.13. Also, recall from Chapter 2 that this is a consequence of Jensen's inequality (Box 2.8).

In using these results, we need to keep in mind that they apply to a system where density-dependent effects are negligible; the effect of adding density dependence to the model would be that positive deviations in population density would be dampened (by whatever limiting factor produces the density dependence), but negative excursions would not be, so environmental variability would still shift the distribution of population densities downward (see Lande et al., 2003).

8.2 The first crossing solution

The equations describing the rate of change in the mean and variance of $\ln(N_t/N_0)$ (Eq. 8.2) can be used to predict the expected distribution of population abundances after T years, so the next step in describing quasi-extinction is to use that information to project the probability of extinction. As in the case of probability of extinction for simple models in Chapter 2, we cannot simply use the proportion of the distribution below the quasi-extinction threshold at time T as the estimate of the probability that the population will be extinct (actually "quasi-extinct") by time T. As mentioned in Chapter 2 (Section 2.4), for simple models that approach would not account for the possibility of some population trajectories dipping below the quasi-extinction threshold before t and returning to a higher level than the quasi-extinction threshold before t (Fig. 2.12). One way to solve this first crossing problem is to approximate the distribution of population trajectories as a diffusion process (Lande and Orzack, 1988). Such a process, also known as a Wiener process, essentially starts the population at state N_0 at time $t = 0$, then presumes that it behaves like a particle undergoing Brownian motion (with drift) in a fluid, with μ and σ^2 (Eq. 8.2) representing the mean and variance of the diffusive movement in each time step. The advantage of the diffusion approximation is that there is an analytical formula for the probability distribution of the time when a population that starts at N_0 will first encounter some arbitrary value N_{QE} (we do not provide the derivation here, but in Chapter 9 we will apply similar equations regarding physical diffusion to approximate the spread

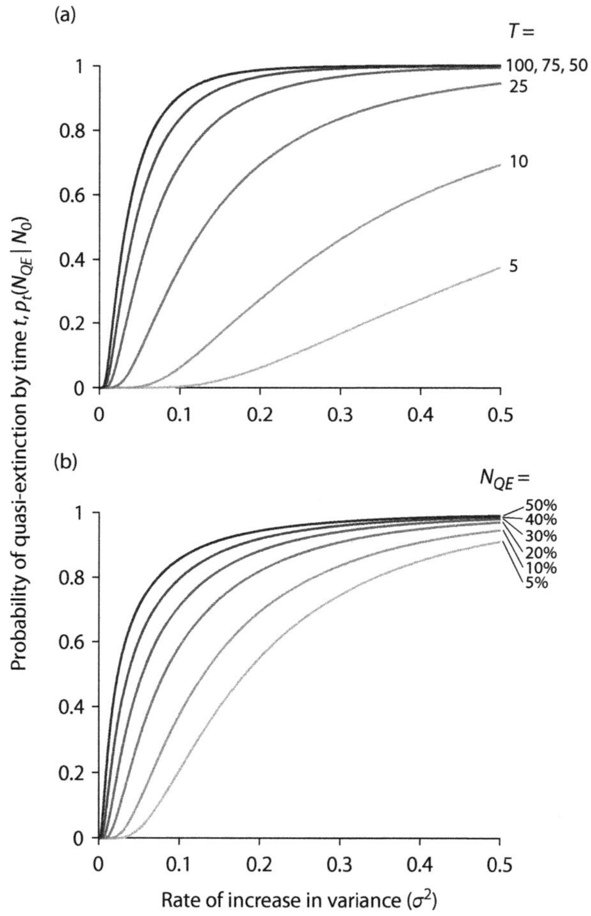

Fig. 8.1 Probabilities of quasi-extinction calculated using the diffusion approximation for first crossing. Each curve shows the probability of a population that has density N_0 at time $t = 0$ falling below the quasi-extinction threshold N_{QE} by time T, for different values of the growth rate of the variance of the logarithm of population density, σ^2. (a) The time horizon T was varied, holding the threshold N_{QE} constant at 10% of N_0. (b) The threshold N_{QE} was varied (values given as a percentage of N_0), holding the time horizon T constant at 25 y. Parameter values were from the *Totoaba* model (Cisneros-Mata et al., 1997).

of a population in space). If we take N_{QE} to be the quasi-extinction threshold, then we integrate that probability distribution from 0 to T to obtain the probability of abundance falling below N_{QE} before time T, given an initial abundance of N_0. Using the notation for a conditional probability, we refer to this probability as $p_T(N_{QE}|N_0)$:

$$p_T\left(N_{QE}\ |N_0\right) = \Phi\left[\frac{\ln\left(N_{QE}/N_0\right) - \mu T}{\sigma\sqrt{T}}\right] + e^{\frac{[2\mu \ln(N_{QE}/N_0)]}{\sigma^2}}\Phi\left[\frac{\ln\left(N_{QE}/N_0\right) + \mu T}{\sigma\sqrt{T}}\right], \quad (8.4)$$

where $\Phi(x)$ is the cumulative normal (i.e. Gaussian) probability function for value x, with mean 0 and variance 1 (note that Eq. (8.4) is rearranged slightly from the form given by Lande and Orzack (1988)). Recall that in the simple example in Section 2.4.2, we described only the probability of being below the threshold N_{QE} at time T, regardless of the past history. This is equivalent to the first term in Eq. (8.4). The second term essentially accounts for the probability that the population has crossed the threshold prior to T.

We can see from Eq. (8.4) how the probability of quasi-extinction will vary in general if we choose different values of the time horizon, T, and the quasi-extinction threshold, N_{QE}, for different values of the rate of increase of variance, σ^2, assuming $\ln\overline{\lambda} = 1.0$ (i.e. with very low variability, the population would not be increasing or decreasing). Figure 8.1 shows how probability of extinction, $p_T(N_{QE}|N_0)$, varies for different values of time period T (Fig. 8.1a) or quasi-extinction threshold N_{QE} (Fig. 8.1b). The key lesson from Fig. 8.1 is that extinction probabilities increase very rapidly with σ^2. Only a small amount of environmental variation is required for extinction probabilities to approach 1 if the time horizon is long enough, particularly if the population starts off close to the quasi-extinction threshold. In this example, only a combination of a very short time horizon ($T = 5$ y) and a low threshold ($N_{QE} = 0.1N_0$) causes extinction probabilities to increase slowly with σ^2 (lowest curve in Fig. 8.1a).

8.3 A more general version of the growth of variability

We can view the results regarding the spread of the distribution of population densities in Eq. (8.2b) in a less idiosyncratic way by writing them in terms of coefficients of variation and elasticities, rather than variances and sensitivities (Cisneros-Mata et al., 1997). These terms reflect the relative (or fractional) amount of variability and sensitivity, respectively. The rate of increase in variance, Eqs. (8.2b and c), becomes,

$$\sigma^2 = \boldsymbol{\varepsilon}'\,\boldsymbol{D}\boldsymbol{\varepsilon}, \quad (8.5)$$

where $\boldsymbol{\varepsilon}$ is a column vector of elasticities of the λ of the average matrix with respect to variation in each of the elements subject to random environmental variability. The general form of elasticity is given in Section 3.5, and Fig. 8.2a reminds us what the elasticities are for *Totoaba*. For matrix element a_i,

$$\varepsilon_i = \frac{\overline{a}_i}{\overline{\lambda}}\frac{\partial \lambda}{\partial a_i}. \quad (8.6)$$

$\boldsymbol{D}$ is a symmetric matrix with elements

$$D_{ij} = \mathrm{cv}_i \mathrm{cv}_j \rho_{ij}, \quad (8.7)$$

where cv_i and cv_j are the coefficients of variation in the ith and jth elements of the covariance matrix $\boldsymbol{C}$ in Eq. (8.2c), and ρ_{ij} is the correlation coefficient between those two elements. In other words, Eq. (8.5) is exactly the same as Eq. (8.2b), except that each

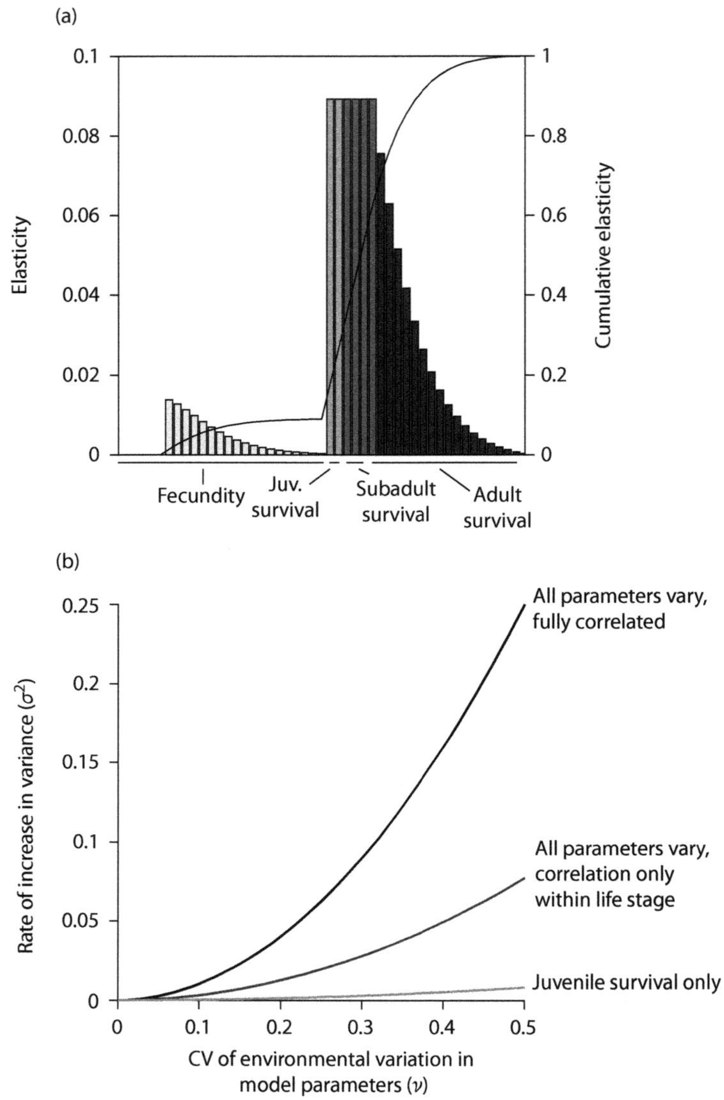

Fig. 8.2 Elasticities and the rate of growth in variance of *Totoaba*. (a) The elasticities (bars) of the asymptotic growth rate, λ, to each model parameter in the *Totoaba* Leslie matrix. The curve shows the cumulative elasticity. Model parameters m_a and p_a are ordered by age, a, along the horizontal axis, and the survivals corresponding to each life stage are labeled. (b) The growth rate of the variance in the logarithm of population size as a function of the coefficient of variation, v, of environmental fluctuations in model parameters. All parameters are assumed to have the same coefficient of variation, but the correlation among parameters varies among scenarios. Lines are labeled to indicate scenarios discussed in the text.

row and column of the covariance matrix $\boldsymbol{C}$ has been divided by the mean value of the corresponding model parameter, and the elements of vector $\boldsymbol{\delta}$ were multiplied by those means.

This expression (Eq. 8.5), along with the fact that the sum of all elasticities of a Leslie matrix is 1 (de Kroon et al., 1986), provides a benchmark for possible values of the rate of increase in variance, σ^2, in Eq. (8.2b): if all parameters vary by the same coefficient of variation (say v), and they are all correlated with correlation coefficient $\rho = 1$, the value of the rate of growth of the variance, σ^2, in Eqs. (8.2b) and (8.4) will equal v^2, and furthermore that will be the maximum possible rate of increase in variance (among all possible combinations of variability at that coefficient of variation; Cisneros-Mata, et al., 1997). This gives some sense of an upper limit on the rate at which the variance of $\ln(N_t/N_0)$ would be increasing.

Figure 8.2b gives us some idea of which probabilities of extinction in Fig. 8.1 are going to be important, at least in the context of *Totoaba* (Cisneros-Mata et al., 1997). It relates the rate of growth of the variance to more tangible features such as where in the life history the environmental variability occurs, and how variable it is. For example, taking the configuration with the most variability, we could assume that all elements of the Leslie matrix vary with a CV of $v = 0.3$, and that they are all completely correlated. From the upper curve in Fig. 8.2b that means that the growth rate of the variance will be 0.1. From Fig. 8.1, that means that with an extinction threshold of 0.1 N_0, probabilities of going extinct within 50–100 years will be substantial (>60%). It is also informative to note the effect of correlation among matrix elements on that result. From Fig. 8.2b, if the variability is only correlated within each life stage (middle curve in Fig. 8.2b), rather than across all life stages, the growth rate of the variance drops to about 1/4 of its value (relative to the upper curve), which corresponds to much less dire values of the extinction probabilities (in Fig. 8.1). Another example would be the case of a marine fish in which the predominant variability is in very early life, which we will take to be the juvenile survival. The CV would be expected to be much larger for an early life stage, perhaps 0.5, but this would lead to a growth rate of the variance of only 0.01 (lower curve in Fig. 8.2b).

8.4 Does the SFA/diffusion approximation work? *Totoaba* as an example

We have adopted the SFA and the diffusion approximation as a set of modeling results that tells us how the extinction of populations works, dynamically. One of the questions we need to answer is whether it actually represents extinction dynamics, and the associated risk. Obviously we cannot observe a hundred or so populations of grizzly bears in different natural environments to see what fraction drops below a quasi-extinction threshold in the next 50 years. We are confined to comparing the outcomes of the SFA to detailed simulations of a large number of populations in different random environments. An example is the analysis of Cisneros-Mata et al. (1997) for *Totoaba*. The goal of their study was to assess how correlations in environmental variability within the four ontogenetically different habitats of *Totoaba* affected population viability, while also trying to get a sense of how well the application of the SFA and diffusion approximation worked.

Cisneros-Mata et al. (1997) addressed the question of how well these models worked by comparing results from the SFA/diffusion approach to results from Monte Carlo simulations of coherent variability in the survival probabilities for each of the three *Totoaba* ontogenetic stages, and in fecundity. Quasi-extinction was defined to be total abundance dropping below half of the initial value within 100 y, and it was calculated over a range of coefficients of variation in each of the survivals and in fecundity. In all of their calculations they assumed that the value of $\bar{\lambda}$ for the mean matrix was 1.0, and that variability was perfectly correlated ($\rho = 1$) within each life stage (juvenile survival, subadult survival, adult survival, reproduction). They examined the effects of variability in each life stage separately, and compared numerical results to the SFA/diffusion results separately.

Their results showed that the SFA/diffusion model matched the numerical results for the three cases with variability in survival, but not for the case with variability in reproduction. Because diffusion equation approximations are known to fail when fluctuations are too large (Ewens, 1964; Grasman and Ludwig, 1983), they tried another approach in which abundance (and thus extinction) was defined in terms of adults only. With that change, the SFA/diffusion approximation model matched the numerical results.

We illustrate these relationships with a slightly different approach, using random numbers from a beta distribution (which is bounded between 0 and 1) for the survivals, and random picks from a lognormal distribution (which is bounded at 0) for the fecundities. Our results confirm the important result that the SFA did not come close to matching the numerical simulations when variability was in per capita reproduction (Fig. 8.3d). Apparently, variability in this parameter leads rapidly to excessive violation of the small fluctuations assumption. Because the parameter distribution is bounded at zero, variability tends to increase the mean and decrease the CV (relative to its intended value). Thus the analytical solution tends to (incorrectly) predict lower extinction rates as the CV increases. The nature of *Totoaba* life history is that reproductive output is very high (e.g. per capita reproduction of the largest age class is 4.6×10^6), so juvenile survival is very low (0.01). Thus fluctuations in reproductive parameters—and subsequently in age-1 recruits—

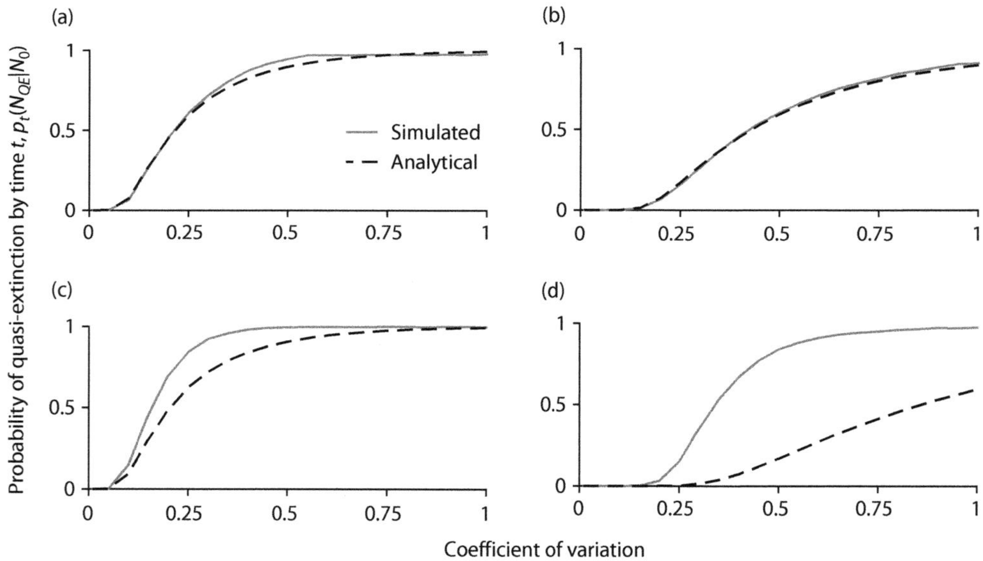

Fig. 8.3 Comparison of simulated *Totoaba* population dynamics to the analytical small-fluctuation approximation (SFA). For each value of the coefficient of variation of environmental fluctuations in model parameters, we made 10^4 simulations, starting at the stable age distribution, with $\bar{\lambda} = 1$. We recorded the proportion of simulations in which the population dropped below 50% of initial abundance within 100 y. Simulation results are represented by the gray curves, and the analytical SFA are the dashed curves. Variability was applied to (a) juvenile survival, (b) subadult survival, (c) adult survival, and (d) per capita fecundity (corresponding to variability in each of the four habitats that *Totoaba* occupy at different life stages); in each habitat random survival terms (or fecundities) were perfectly correlated across all of the survival terms (or fecundities) in that life stage.

become much larger than the diffusion approximation can accommodate. This indicates how extra care must be taken when applying the SFA to life histories with very low annual survivals at young ages (e.g. many species of fish).

Overall, we find that the SFA/diffusion approximation gives reasonable descriptions of the behavior of the population near the equilibrium, but does a very poor job of predicting dynamics, and therefore dynamics of extinction, when the population suddenly reaches very low levels. The heuristic explanation for the poor performance at small population levels is that in the underlying derivation of the diffusion approximation, the population is assumed to be a continuous, rather than discrete, variable, and a limit is taken as the number of changes to population abundance (i.e. individual births and deaths) goes off to infinity. This limiting procedure does not lead to quantitative problems if the population is large, but it does if the population is small. This problem can be overcome if one directly represents discrete changes in population abundance using an approach named the *master equation*. We describe the basic approach in Box 8.1, but given the difficulties involved in actually implementing the Master Equation, we do not describe it further.

Box 8.1 THE MASTER EQUATION

As discussed in Ovaskainen and Meerson (2010), there is an essentially exact description of population dynamics that can be used to understand population dynamics when numbers are small. This approach is known as the *master equation* (for the mathematically adventurous, Gardiner (2009) is the standard reference) for a system in continuous time. By taking explicit account of changes in the population size (i.e. up or down by one individual) we can obtain what is a precise, if somewhat unwieldy, description of the stochastic population dynamics.

It is easiest to describe this approach for a population that is unstructured, and this description makes clear the difficulties of using this approach for age-structured populations. The underlying variables in the model are the probabilities, $P_i(t)$, that the population is of size i at time t. The basic description is based on a set of equations for $dP_i(t)/dt$ describing the rate at which these probabilities change. In compact notation this can be written as $dP/dt = QP(t)$, where Q is a matrix with entries q_{ij} describing the rate of transition from a population of size i to a population of size j. If the population had m ages, and we were only considering a maximum of n individuals of each age, there would be $n \times m$ of these equations. Although numerical solutions of these equations are possible, various approximations have been used to find systems where analytic results are possible. However, as we have seen, these approximations do not necessarily work well for small population sizes. This is currently a very active area of research.

8.5 Color of the random environmental variability

From Eq. (8.2) up to this point, we have assumed that the environmental variabilities driving the changes in matrix elements are serially uncorrelated, by which we mean that the fluctuation in year t is statistically not related to the fluctuation in year $t + 1$. In other words, we assume the variability has a white noise spectrum, whereas actual environmental variability is likely to be autocorrelated. Autocorrelation indicates a pink, blue, red, or other spectrum (we discussed this distinction earlier, in Section 4.3.2). There has long been a suspicion that positively autocorrelated ("red") variability would lead to a greater risk of extinction, because the population is more likely to experience a series of "bad" years that collapse the population. This arose from early analyses of logistic models (Foley, 1994). However, more recent theoretical analyses have led to mixed results, showing both increases and decreases in extinction risk with autocorrelation (reviewed by Ovaskainen and Meerson, 2010). For the most part, these analyses have used simple, unstructured models of abundance, rather than age-structured models. For example, Schwager et al. (2006), using a simple model of abundance, pointed out that, while autocorrelated red noise increases the likelihood of a string of bad years, it also reduces the likelihood of any given year being catastrophically bad.

In addition to the fact that several different types of models have been used in these analyses, they also vary in the additional effects they include. For example, van de Pol et al. (2011) used several age- and stage-structured models to show the importance of (a) a nonlinear dependence of life history variables on the environment, (b) there being multiple environmental variables, and (c) the "whitening" effect of demographic stochasticity at low abundances. Another factor emphasized by Ferguson et al. (2016) is the degree that populations themselves contributed to the autocorrelation. This harkens back to an early fundamental question in ecology: whether population variability is due to the extrinsic variability in the random environment or to the intrinsic influence of population dynamics (Andrewatha and Birch 1954). Framing the question in terms of cohort resonance sheds light on this problem by separating the process into two elements: (1) the transfer function, which reflects population dynamics, and (2) the frequency content of the environment, which obviously represents the extrinsic environment.

One question that may arise in this era of a changing climate is what will be the effects on population persistence (probability of extinction) of changes in the frequency content of the environment? For example, as we noted in Section 4.3.2, paleontological records and climate projections indicate that the frequency of El Niño-Southern Oscillation (ENSO) events could change in the future. Reproduction in a marine bird, Brandt's cormorant, *Phalacrocorax penicillatus*, that breeds on the Farallon Islands off San Francisco, California, is diminished by warm water conditions that occur during positive ENSO anomalies. To test the effects of possible changes in ENSO frequency on the probablity of extinction, Schmidt et al. (2018) simulated cormorant populations were exposed to both more frequent ENSOs and less frequent ENSOs. Intuitively, one would expect that more frequent ENSOs would lead to greater variability and a greater probability of extinction. However, the result was the opposite. Being a relatively long-lived species (generation time of 5–7 y, roughly 10% survival to >15 years old) that maintains a consistent reproductive output over many years (unlike, say, a salmon), Brandt's cormorant's transfer function showed the greater sensitivity to low frequencies associated with cohort resonance, but did not exhibit the enhanced sensitivity to shorter, generation-time-scale frequencies that shorter-lived species exhibit. Hence, speeding up ENSOs to higher frequencies caused less population variability, and a lower probability of extinction.

In contrast to that example, climate projections for California's central valley show that the frequency content of expected future environmental variability is centered right in the cohort frequency band of greater environmental sensitivity of a shorter-lived species, spring run chinook salmon, *Oncorhynchus tshawytscha*. Since salmon are rather short lived, they have both of the bands of greater sensitivity associated with cohort resonance: one at cohort (generation time) frequencies and one at low frequencies, as in Fig. 4.15c. An environmental signal with almost all of its variance at the frequencies to which the population is most sensitive would be expected to produce high population variance, and result in high probabilities of extinction. A general investigation of the relative effects of having environmental variance at each of the two frequency bands to which salmon are sensitive (as compared to equal amounts of environmental variability in both bands, and white noise) showed that, by itself, environmental variability at cohort frequencies did cause high population variance, but would not cause high probability of extinction (Kilduff, et al. 2018). This is due to the effect mentioned in Section 8.4, that simple models do not represent dynamics at very low population abundances faithfully. By contrast, increased variability at the lower frequency associated with cohort resonance did produce the expected increase in extinction probability.

8.6 Application of SFA to population data

Another way of making use of the SFA/diffusion model is to use field data from species at risk to estimate μ and σ^2, then use that to estimate probabilities of extinction. This seems like an obvious thing to do, but there are some hidden dangers. First, any actual time series of population abundance is equivalent to just one possible iteration of a stochastic model, so that time series could deviate greatly from the mean abundance predicted by the model and still fall within the range of possible predicted outcomes. Second, and more subtle, is the concern that data for such an exercise are more likely to be available for populations that have been near extinction for a long time (generating a long data set), rather than ones that have gone extinct quickly (before much data could be collected). Thus the analysis will necessarily be limited to a particular type of population (cf. Boettiger and Hastings, 2012). The third cautionary note is that the derivation of the SFA assumes that the population can be described by a primitive Leslie matrix, hence it may not hold for populations that are semelparous, or close to semelparous, and thus have an imprimitive Leslie matrix (see Section 4.2).

Nonetheless, Dennis et al. (1991) developed a regression-based approach to obtain maximum likelihood estimates of μ and σ^2 for the SFA model (Eq. 8.1) from a time series of population abundance. If one assumes the distribution of population trajectories follows a diffusion-like process (a Wiener process, Section 8.2), the state transitions, N_t to N_{t+1}, should be independent, normally distributed variables (Wiener increments) suitable for linear regression. Then if one regresses the values of $\log(N_i/N_{i-1})$ versus the time increments, $\tau = t_i - t_{i-1}$, the slope of this relationship is μ and the error variance of the regression is σ^2. The benefit of the regression approach is that they could apply the standard set of regression diagnostic statistics to determine whether a data set fits the expected relationship well, and whether there were outlying data points. For five of the seven data sets that Dennis et al. tested (whooping crane *Grus americana*, Yellowstone grizzly bear *Ursus arctos horribilis*, Kirtland's warbler *Dendroica kirtlandii*, California condor *Gymogyps californianus*, and Puerto Rican parrot *Amazona vittata*), they estimated model parameters with low to modest uncertainty and statistically good fits to the data (Figure 8.4). For the remaining two species (Palila *Loxioides balleui* and Laysan finch *Telespyza cantans*; both Hawaiian honeycreepers), there were large fluctuations in the data that produced great uncertainty in parameter estimates. Dennis et al. interpreted this result, not as indicating that the SFA was inappropriate for these populations (the time series appear to describe a Wiener process), but that the model would not be useful for prediction because of the high uncertainty. They suggested that including density dependence could improve the parameter estimation. They also pointed out that for all of

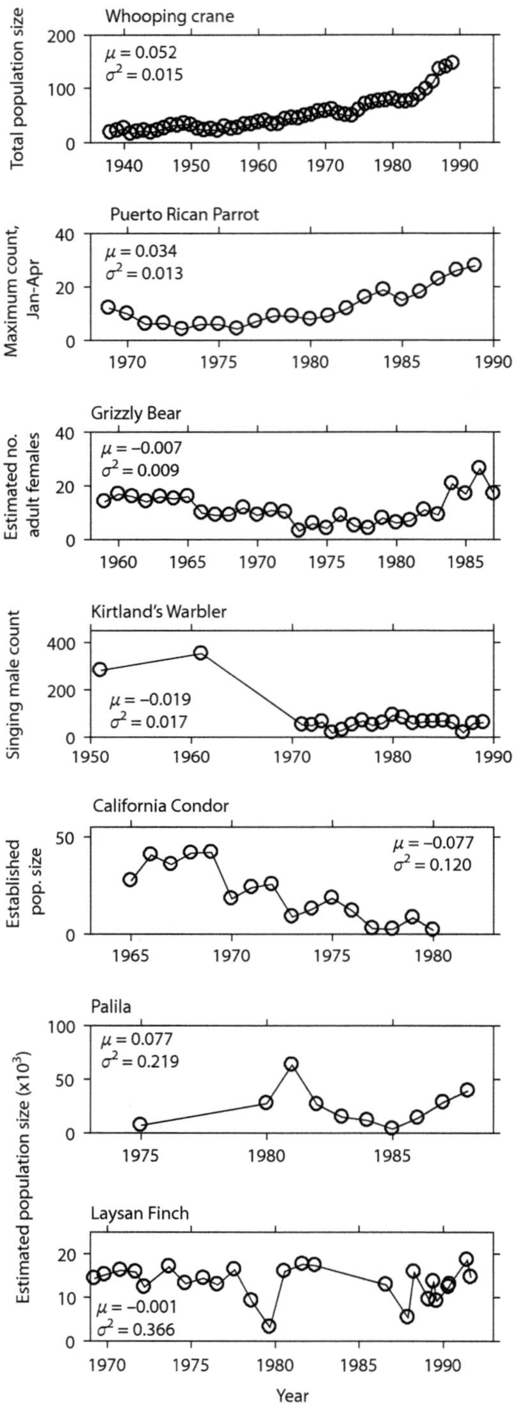

Fig. 8.4 Time series of population abundances used by Dennis et al. (1991) to estimate the rate of change in the mean (μ) and variance (σ^2) of $\ln(N_t/N_0)$. Estimated values of μ and σ^2 are given in each panel.

the species, the SFA model would not capture extinction risk due to catastrophic events, such as the sudden death of nearly half the population of Puerto Rican parrots in 1989 during Hurricane Hugo.

In summary, this model provides a means of fitting stochastic exponential growth to time series, and using that model to estimate probability of extinction. If the time series do not appear to exhibit exponential growth, they can be broken up into segments that do, or a constant maximum value of abundance can be assumed (see Dennis et al. (1991) for details of the examples in Fig. 8.4). A bothersome characteristic of the estimate of the mean of the growth in the log of abundance is that the estimate (Eq. in Dennis et al., 1991) contains the product of all of the sequential growth ratios, which ends up being merely the ratio of the final abundance to the initial abundance. The ratios of sequential abundances between the beginning and end of the series have no effect on the estimate (see Heyde and Cohen (1985) for more discussion on this).

One concern that Dennis et al. had for the longer-lived species in their analysis was that age-structured dynamics would lead to autocorrelation in the Weiner transitions. In the end, they did not detect that effect in the data, and dismissed that concern. However, in Chapter 10 we will revisit the calculation of extinction probabilities and consider an example in which the effect of the environment on the population age structure is represented differently (Botsford and Brittnacher, 1998).

8.7 State of the science quantifying extinction risk at the turn of the century

During the 1980s and 1990s, a variety of stochastic structured population models were developed to conduct *population viability analyses* (PVAs). The basic approach in PVA was to use some number of years worth of field data (the minimum was two) to estimate the transition probabilities between each age-, size-, or stage-class and to construct a projection matrix. That matrix was then analyzed to estimate the probability of quasi-extinction over some time horizon.

Fieberg and Ellner (2001) reviewed the application of structured population models to the estimation of growth rate of the mean, μ, in Eq. (8.2a) and probability of (quasi-) extinction, $p[E]$. They posed the problem as having collected demographic data over $t + 1$ years to estimate t projection matrices. If $t = 1$ then the analysis was deterministic and consisted of determining whether the dominant eigenvalue λ was less than 1. If there were multiple matrices available, then the analysis could include stochasticity, which is what Fieberg and Ellner focused on. The studies they reviewed used three distinct approaches to calculate μ and $p[E]$. The first was to conduct stochastic simulations in which one of the t matrices was chosen at random for each model year; this approach was termed the *random transition matrix* (RTM) method. The second was to use the t matrices to estimate the means and covariances of all the vital rates to parameterize a multivariate distribution. Stochastic simulations were then made; for each year in a simulation, the elements of the projection matrix were drawn at random from the multivariate distribution of vital rates. This approach was termed the *parametric matrix method* (PMM). The PMM category included many of the "canned programs" available for calculating $p[E]$ in PVA (see Chapter 10). Finally, the third method was to use the t matrices to estimate the variances and covariances of each vital rate, and use the SFA calculations we have described previously for *Totoaba*.

Fieberg and Ellner pointed out that there are advantages and disadvantages to each of the methods. Based on the results in Fig. 8.2b, we know that the degree of correlation among model parameters has a large effect on the predicted time to extinction. Thus estimation of that correlation is an important consideration. The RTM approach avoids the problem, because the inherent within-year correlation structure of the projection matrices is preserved within the simulations; thus it is simpler to use. It may also be helpful if the time series includes a rare but extreme event, such as a hurricane. The downside of that approach is that because it does not directly estimate the correlation structure, it is impossible to conduct a sensitivity analysis to understand how strongly the estimated extinction risk depends on that correlation structure (this is particularly important if the correlations are being represented by only a few years worth of data). The PMM and SFA do permit that sort of sensitivity analysis.

To illustrate the relative performance of the three methods, we performed an analysis similar to one that Fieberg and Ellner provided. First, we used the *Totoaba* model, with fecundity terms rescaled so that $\lambda = 1$, and simulated population trajectories for 1000 years with variation within each life stage (juvenile survival, subadult survival, adult survival, fecundity) set to CV = 0.01, 0.1, or 0.5. We calculated the rate of change in $\ln(N_t/N_0)$ for each of those three trajectories, and took those to be the "true" values of μ under each level of variability. We then applied SFA, RTM, and PMM to estimate μ for each of the trajectories, using only 2, 5, or 10 years worth of data, randomly drawn from the 1000 year trajectory. We repeated that process 1000 times to get distributions of the estimates of μ for each level of variability and amount of data (Fig. 8.5). In all cases the distribution of estimates of μ straddled the true value (even with only two years worth of data!), but the precision of the estimates improved with additional years, particularly when variability was higher. Additionally, there was virtually no difference in the estimates among the three methods (Fig. 8.5), so one could pick the method that best suited one's needs and data.

In addition to the wealth of information on the 29 applications in Fieberg and Ellner's review, the frequencies of the types of calculations made and the types of matrix models used are of the most interest here (Table 8.1). The majority of studies used the RTM approach (19), while PMM was second (9), and the SFA (2) was rarely used for PVA. However, Fieberg and Ellner suggested that modelers should favor PMM and SFA over RTM, because the former allow sensitivity analysis of the correlation structure while the latter does not. Because correlated variations among model elements can have a big effect (see Fig. 8.2), this is an important consideration.

More importantly, given our discussion of stage-based models in Chapter 6, most (18) of the models were stage structured of the SS/SG type (including those in which each stage was a range of sizes, denoted "size-stage" in Table 8.1), with about half as many age-structured models, and only two size-and-age models. In Chapter 6 we explained why stage-based models could not be expected to faithfully represent short-term transient dynamics, even if they faithfully described dynamics as the asymptotic stable stage distribution. Given that models with stochastic perturbations are never at the stable stage distribution, the fact that stage-structured models are commonly used for PVA is very concerning.

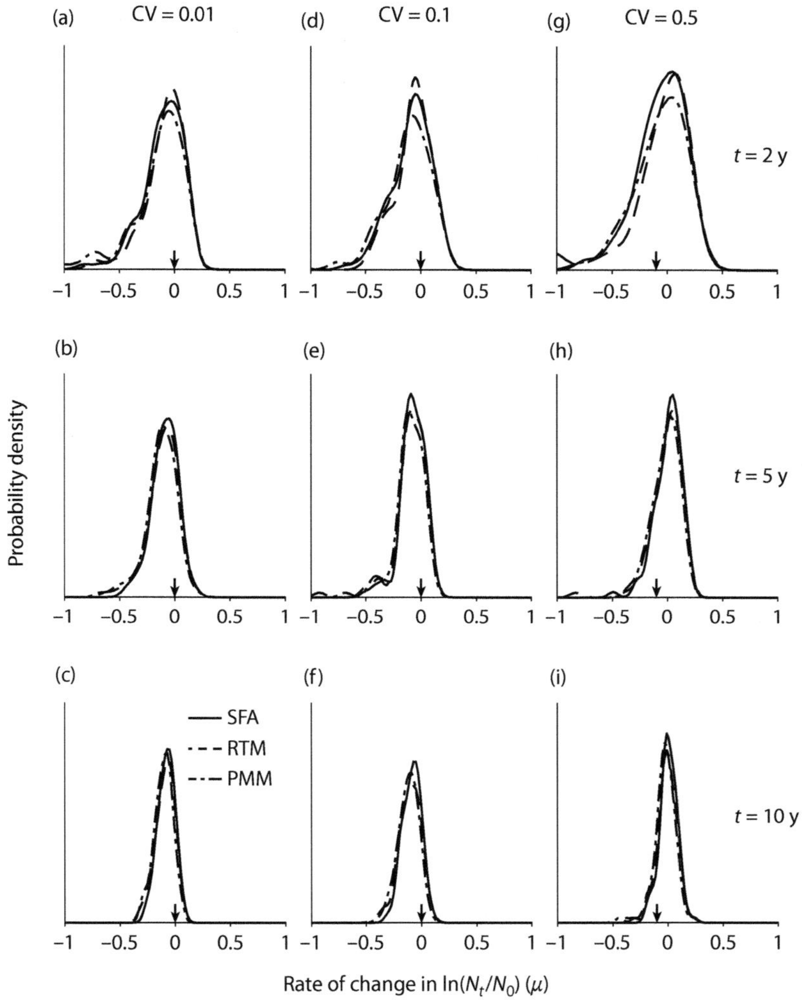

Fig. 8.5 Comparison of approaches in Fieberg and Ellner (2001) to simulating stochastic population dynamics to estimate μ, the growth rate of the mean of the logarithm of population size. In each scenario, t annual transition matrices were sampled from a simulated "true" sampled population trajectory for *Totoaba*. These matrices were used to estimate μ using the small fluctuation approximation (SFA, solid curves), the random transition matrix method (RTM, dashed curves), and the parametric matrix method (PMM, dot-dash curves). Population dynamics were simulated for 10^3 years for RTM and PMM. This estimation was repeated for 10^3 random simulations for each scenario. Scenarios varied in the number of matrices sampled for estimation: $t = 2$ (top row), 5 (middle row), or 10 (bottom row), and in the coefficient of variation in environmental fluctuations: 0.01 (left column), 0.1 (middle column), or 0.5 (right column). In each panel, the "true" value of μ is indicated with an arrow. In the model, variation in parameters for each life stage (juvenile survival, subadult survival, adult survival, reproduction) was perfectly correlated within stages, but uncorrelated across stages.

Table 8.1 Summary of stochastic matrix models reviewed by Fieberg and Ellner (2001). "Size-stage" indicates SS/SG models in which the stages correspond to size ranges; "size/stage" indicates SS/SG models in which the stages are a mix of size ranges and developmental/reproductive stages.

Species	Type of matrix	Method used
Barnacle (*Semibalanus balanoides*)	Age	RTM
Plant (*Plantago lanceolata*)	Size-stage and age	RTM
Plant (*Pedicularis furbishae*)	Size/stage	RTM
Marine bivalve (*Yoldia notabilis*)	Size-stage	PMM
Tortoise (*Gopherus agassizii*)	Size-stage	PMM
Subtidal snail (*Umbonium costatum*)	Age	RTM
Gorgonian coral (*Leptogorgia virgulata*)	Size-stage	RTM
Eagle (*Haliaeetus leucocephalus*)	Age	PMM
Herb (*Arisaema tyiphyllum*)	Size-stage	RTM
Fish (*Morone saxatilis*)	Age	RTM
Plant (*Panax quinquefolium*)	Size-stage	RTM
Plant (*Allium* sp.)	Size-stage	RTM
Brown alga (*Ascophyllum nodosum*)	Size-stage	RTM
Plant (*Haplopappusradiatus*)	Size-stage	RTM, PMM
Grass (*Andropogon breviffolius*)	Age	RTM
Grass (*Andropogon semiberbis*)	Size-stage	RTM
Plant (*Hudsonia montana*)	Siz-stage e	PMM
Deer (*Cervus elaphus*)	Age	RTM, SFA
Sedge (*Carex bigelow*)	Size/stage and age	RTM
Hawk (*Rostrhamus sociabilis*)	Stag-stage e	RTM, PMM
Five savanna plants	Size-stage	RTM
Bird (*Calidris pusilla*)	Age	RTM, PMM
Thistle (*Carduus nuans*)	Size-stage	RTM
Fish (*Totoaba macdonaldi*)	Stage	SFA, PMM
Bird (*Picoides borealis*)	Age	RTM
Herb (*Asarum canadense*)	Size-stage and stage	RTM
Plant (*Silene regia*)	Size-stage	RTM
Bird (*Vireo latimeri*)	Age	PMM
Bird (*Centrocercus urophasiansus*)	Stage	PMM

8.8 Perils of using stage models to characterize extinction risk

To illustrate the potential pitfalls of using stage-structured models in PVA, we revisit the loggerhead turtle model that we described in Chapter 6. Recall that Crowder et al. (1994) developed both a stage-structured and age-structured version of their model, In Fig. 6.4b we showed how the two models made alarmingly different predictions about the population trajectory after the introduction of turtle excluder devices (TEDs) on shrimp trawls increased post-juvenile survival; and in Fig. 6.6 the two models made dramatically different predictions regarding the stage distributions in cohorts. Here we explore this further by comparing extinction predictions for the two models. To examine the effect of variability alone, separate from the deterministic trajectory, we rescaled the mean matrix for both models to have $\overline{\lambda} = 1$. We simulated 10^4 80-year trajectories for both models, applying variability with a range of CVs to each stage in turn: stage I (eggs), stage II (small juveniles), and stage III (large juveniles; keep in mind that in most marine populations, most of the variability is in the younger age classes). For the stage model, we applied the same variability to both the "stay" and "go" terms for each stage. For the age model we applied the variability to the survival parameter for each age class in that stage (in each year, all of the survival terms of the age model that would fall in a stage received the same random value). We applied the same 10^4 sets of 80-year time series of random impulses to both models. We also started each simulation with an initial age (or stage) distribution that was very slightly perturbed from the asymptotic distribution.

The results of this exercise follow what one might expect from our discussion of the limitations of stage models. The egg stage is only a single-age class (the stage-structured model has only a "go" term, not a "stay" term), so the two models produce nearly identical results, and the 80-year extinction rate is very low (Fig. 8.6, left panels). However, for the

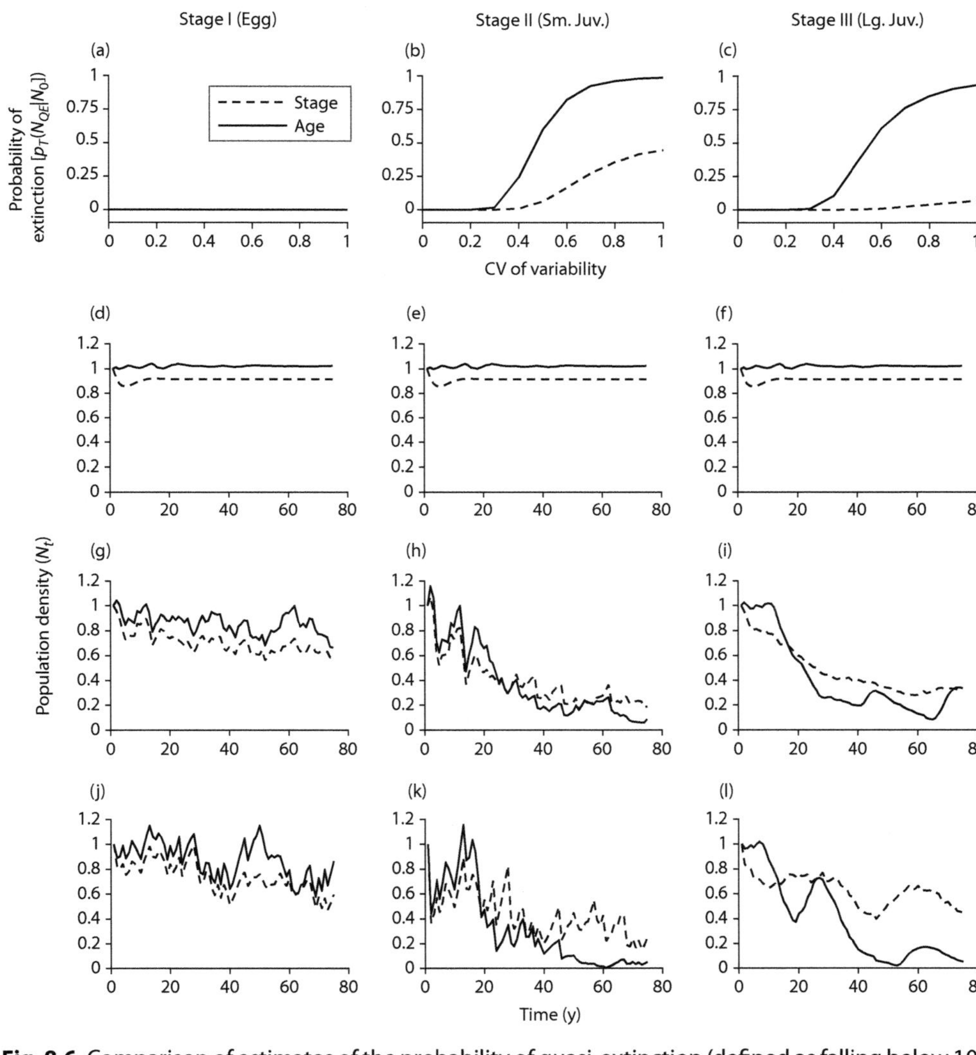

Fig. 8.6 Comparison of estimates of the probability of quasi-extinction (defined as falling below 10% of initial abundance within 100 years) using the stage- and age-structured versions of the loggerhead turtle model. Top row (panels a–c) shows the probability of extinction as a function of the CV of random variability, when that variability is applied to different life stages. The lower panels (d–l) show a representative trajectory for each model type, with variability applied to the same life stage as in the top panel in each column and the indicated CV.

small juvenile and large juvenile stages, each stage corresponds to multiple ages, and the age-structured model consistently predicts higher extinction probabilities than the stage-structured model, for the same level of variability (Fig. 8.6, center and right panels). This occurs because of the difference in how cohorts travel through the age or stage distribution in each model. In the stage-structured model, in each time step, a proportion of each stage moves on to the next stage, and a proportion stays behind. This rapidly spreads out the effects of any given disturbance through the stage structure, effectively dampening "bad" years faster (Fig. 6.6). By contrast, the age-structured model begins to experience cohort resonance, as negative deviations in one cohort echo in the following generation, producing cycles of period approximately 1/generation time (27 years in this model). This represents random (white noise) variability exciting a particular frequency response in the population, and creating a redder, cyclic pattern of variability, as we described in Chapter 4 (Section 4.3). This pattern is particularly evident when the variability is in Stage III (Fig. 8.6i,l). The low-frequency variation characteristic of cohort resonance is also evident in the age-structured trajectory shown in Fig. 8.6l, which is both cycling and declining more rapidly than the stage-structured model exposed to the same environmental variability. Note that there are also period 1/generation time cycles in the simulations without environmental variability, because the simulations did not begin exactly at the asymptotic age or stage distribution, and there are transient oscillations as the population converges on the asymptotic distribution (Fig. 8.6d–f). These cycles are deterministic (as we described in Chapter 4, Section 4.2), and unlike the cycles due to environmental variability and cohort resonance, do not lead to a decrease in the long-term growth rate. Keep in mind that this is a fairly cursory analysis that does not explore variability in multiple stages or covariance in variation, but nonetheless indicates potentially serious problems with using stage-structured models to estimate extinction probabilities.

8.9 What have we learned in Chapter 8?

Here we have expanded Chapter 2's brief foray into understanding the effects of a random environment on behavior of a simple, linear population model in a random environment, by addressing the same issue for linear, age-structured models. We used linear models because we are interested in extinction, which occurs at low abundance where presumably there would be no density dependence.

Initially, using the small fluctuations approximation (SFA), we found that it was useful to describe the distribution of possible future population densities, N_t, in terms of the probability distribution of the variable $\ln(N_t/N_0)$. The mean of that distribution increased linearly with time at a rate equal to $\overline{\lambda}$ (essentially the population growth rate in the absence of environmental variation) minus half the rate of increase of the variance. The rate of increase of the variance depends on the details of how variability affects the Leslie matrix. Thus, adding random variability to a linear, age-structured model leads to essentially geometric growth, with a growth rate that is less when there is more variability. This dependence parallels the result for a simple model in Chapter 2, and follows from Jensen's inequality (Box 2.8).

In order to be able to calculate probabilities of quasi-extinction using the SFA, we saw that we could parameterize a diffusion equation to approximate the distribution of $\ln(N_t/N_0)$. This allowed us to solve the "first crossing" problem, that is, estimate the probability of a population falling below the quasi-extinction threshold any time before time T. This allowed us to get some sense of how probabilities of quasi-extinction would increase with increasing environmental variability (Fig. 8.1).

We then rewrote the SFA in terms of coefficients of variation, elasticities, and correlations (rather than variances, sensitivities, and covariances). Since we know that the sum of all elasticities of a Leslie matrix is 1, this allowed us to connect the rate of increase in variance of $\ln(N_t/N_0)$ to the relative amount of variability in matrix elements (i.e. the coefficient of variation) for different correlation structures among the variable elements of the matrix (e.g. no correlation, versus all correlated, versus correlated only within a specific life history stage) (Fig. 8.2b). Differences in environmental variability (i.e. in CV) have a much greater effect with higher correlations of variability among life history stages.

Then, because the SFA and the diffusion approximation are both approximations, we asked whether they worked under various conditions, using *Totoaba* as an example. Comparing the SFA solutions to numerical simulations, we saw how the approximation broke down when variability was solely in the reproductive rate (Fig. 8.3). We concluded that this occurred because variability led to a highly skewed distribution of reproductive rates, which violated the assumptions of the SFA. We showed how this problem could be avoided by redefining probability in extinction, and in Box 8.1 we described an alternative. One can write what is called a master equation that expresses the dynamics of an age-structured model in a random environment more exactly. While this model is more exact, the numerical solution is somewhat unwieldy, so it would be useful only for fairly small populations.

Another shortcoming of the SFA/diffusion equation approach is that it assumed that the environmental variability was uncorrelated over time (i.e. it is white noise, as noted in Chapter 4), whereas most real environmental signals have some autocorrelation (red noise). The consequences of autocorrelation for extinction risk calculations are not well understood; simple models suggests that autocorrelation tends to lead to strings of "bad" years, which increases extinction risk. This does not always occur, however, and this issue needs further analysis. Recent results for age-structured models using a cohort analysis approach (Section 4.2) showed that differences in both the environmental spectrum and transfer function (frequency response) of the species could have unexpected effects on probabilities of extinction.

We then turned to examples for actual populations, the first being an application of SFA by Dennis et al. (1991). They used a regression approach to estimate the growth rates of the mean and the variance of the distribution of $\ln(N_t/N_0)$ for seven endangered species (Fig. 8.5). In five of the examples they estimated those parameters with moderate to low uncertainty, but—as in the *Totoaba* example—they found that the SFA did not work well when variability was extremely high.

The second series of examples was Fieberg and Ellner's (2001) review of 29 studies that had used age- or stage-based models to estimate probabilities of extinction. These examples used three distinct approaches for simulating variability in the projection matrix: SFA, RTM, and PMM. using one of three methods. All three methods produce similar estimates of the stochastic population growth rate (Fig. 8.5), but only PMM and SFA permit sensitivity analyses, while RTM may better accommodate rare, extreme events.

The important aspect of the Fieberg and Ellner review, as far as population dynamics is concerned, was that 18 of the 29 applications used projection matrices of the some go/some stay (SG/SS) stage-structured configuration. As we saw in our exploration of the transient dynamics of such models in Chapter 6, they may introduce bias in computed probabilities of extinction. Indeed, simulations using both the age-structured and stage-structured version of the loggerhead turtle model showed that as environmental variability increased, the stage-structured model generally produced much lower probabilities of extinction than did the age-structured model, meaning that the former is undesirably optimistic from a conservation standpoint.

CHAPTER 9

Spatial population dynamics

In this chapter we will consider population structure in a different dimension: space. Some of the first evidence that it is important to account for space in describing population dynamics came from fisheries (Secor, 2014; 2015). For example, Chong (1814) described how herring abundance in two different regions of modern-day Korea exhibited cycles that were out of phase with one another, suggesting that the two regions had distinct populations unconnected by migration. Later, Gilbert (1916) showed that sockeye salmon (*Oncorhynchus nerka*) spawning in different tributaries of the Fraser River, British Columbia, had distinctive coloration, size, growth rates, and migration timing. Additionally, Fraser River sockeye populations exhibit extreme, four-year cycles of abundance, and populations spawning in different tributaries have cycles of different phases (Ricker, 1950; White et al., 2014). These were the first indications that populations of the same species might be found nearby in space, yet experience distinct population dynamics or disturbances, as long as migration between them was somewhat limited. Around the same time, Hjort and Lea (1914) illustrated the converse point: Atlantic herring (*Clupea harengus*) sampled at different locations along the Norwegian coast exhibited dramatic, *identical* cycles in abundance over time. This (along with other evidence from tagging studies) indicated that the various locations were linked by dispersal.

Another aspect of spatial population dynamics was also developing in terrestrial ecology at the turn of the twentieth century. In *The Origin of the British Flora*, botanist Clement Reid (Reid, 1899) considered an apparent paradox in the recolonization of the British Isles by oak forests following the last glacial maximum: oak trees take >50 years to mature, and produce heavy acorns that do not fall more than a few meters from the parent tree. Thus oaks should recolonize empty space left behind by retreating glaciers at a scale of tens of meters per century. Yet they had fully covered Britian by Roman times, only 15,000 years after the glacial retreat. Skellam (1951) reframed this problem mathematically by developing a model of invasive spread. He calculated that the rate of spread must have been on the order of 500 m per year, and thus was surely assisted by animals transporting acorns as Reid had originally proposed (although he also noted that the fast spread could also be explained by the existence of refuge oak populations in valleys without glaciers that recolonized the island).

The development of both experiments and theory demonstrating the importance of space accelerated in the second half of the twentieth century. Huffaker's (1958) lab experiment with predatory mites, tracking their prey on a habitat array of paraffin-covered oranges, demonstrated that spatial heterogeneity and dispersal could promote stable coexistence. This confirmed a suspicion by Nicholson and Bailey (1935) that spatial heterogeneity could explain why predator–prey populations persisted despite the notorious instability of the model of a coupled host–parasitoid system that bears their

Population Dynamics for Conservation. Louis W. Botsford, J. Wilson White, and Alan Hastings, Oxford University Press (2019).
DOI: 10.1093/oso/9780198758365.001.0001

name (Hastings, 1977). Similarly, den Boer (1968) suggested that spatially distributed populations benefitted by spreading the risk of extinction over several populations with different environmental variability (termed "risk spreading"), allowing them to persist despite small-scale disturbances. MacArthur and Wilson (1967) developed the highly influential theory of island biogeography, predicting how the interaction of distance (limiting dispersal) and habitat area (promoting population persistence) should predict patterns of species diversity on islands. Simberloff and Wilson (Wilson and Simberloff, 1969; Simberloff and Wilson, 1969; 1970) successfully tested those predictions experimentally by completely defaunating mangrove islands and monitoring their recolonization (one of those experiments it is difficult to imagine obtaining a permit for in the modern day).

Levins, 1969; 1970) integrated many of these earlier observations into the *metapopulation* concept, describing the dynamics of a network of *subpopulations* all linked by dispersal. Levins' original, relatively simple model was greatly expanded upon in the late twentieth century, particularly by Ilkka Hanski and colleagues, who developed both theory and empirical observations to predict the dynamics of a metapopulation of Granville fritillary butterflies (*Melitaea cinxia*) in the Åland Islands, Finland (Hanski, 1991; 1994; Hanski and Gilpin, 1997; Hanski, 1998, Hanski and Ovaskainen, 2000; Hanski, 2001; Hanski and Ovaskainen, 2003). Notably, both island biogeography theory (MacArthur and Wilson, 1967) and these original efforts at metapopulation modeling (Levins, 1969; 1970; Hanski, 1994), focus only on the presence or absence of a species in a patch. That is, they are concerned only with rates of patch colonization and extinction, and do not explicitly consider within-patch dynamics, including age structure. As we shall see in this chapter, that is not always a useful approximation.

These late-twentieth-century lines of thought have now converged on our present understanding of the importance of space and dispersal for population dynamics (e.g. Levin, 1992; Tilman and Kareiva, 1997; Hanski and Gaggiotti, 2004; Kritzer and Sale, 2006). In a conservation context, this has led to a focus on the creation and management of reserves and parks; that is, conservation and management in a spatially explicit framework (Moilanen et al., 2008).

Here, we walk through the development of spatial descriptions of population dynamics. Once again, we use a density function, this time over space, not age or size, to represent the state of the population. These models are considered to be *spatially explicit*, meaning that the model operates on a one-, two-, or three-dimensional landscape, and keeps track of the distance between different locations in the landscape. Typically, individuals are only able to move between adjacent locations. We begin by examining the spread of invading populations, originally modeled by Skellam (1951) and Fisher (1937). Using the same framework we can also ask questions about the persistence of populations in fluid environments (e.g. streams), where the population may be spreading out by diffusion or advected downstream.

We then move on to considering metapopulation models, which were first developed primarily using *implicit* descriptions of space. In contrast to the spatially explicit approach for spatial spread we have just described, in the implicit approach we do not keep track of the relative position of different locations, but instead ask only what fraction of the habitat is occupied. We also do not restrict movement to adjacent locations. Additionally, these models typically assume stochastic extinctions and colonizations at the local (in space) level. You can probably already think of different species and biological situations that make one or the other of these two approaches to representing space (explicit versus implicit) more useful. We examine the original Levins extinction/colonization metapopulation model and the later developments of that type of extinction/colonization model by Hanski and others. Many of these more recent models have been spatially explicit, or even *spatially realistic*, which refers to a spatially explicit model in which the model landscape corresponds to a real geographical location. Finally, we consider metapopulation models that do more than simply track the presence or absence of organisms in a habitat patch, and actually represent within-patch dynamics (possibly age or size structured). We point out how this is necessary to represent the dynamics of metapopulations in which variation in patch abundance is important and actual patch extinctions are relatively rare.

9.1 Modeling the spread of a population

To begin exploring spatial dynamics, we start with a basic equation describing movement and local dynamics. At this point we are only considering space to be one-dimensional (i.e. along a straight line), and we ignore any differences among individuals other than their spatial location.

As in the M'Kendrick/von Foerster continuity equation (Eq. 3.2), the population dynamics are described using a dependent variable, $n(x,t)$, that is a density function, in this case on space (x indicates spatial location). Changes to $n(x,t)$ will depend on population dynamics (i.e. birth and death) at location x and on the movement of individuals past location x. We describe that movement as flux, J, which for a one-dimensional model has units of individuals per unit time (i.e. the number of individuals moving past a given point in a given amount of time; in two dimensions one would consider the flux crossing a line of a given length, and the units would be individuals per unit time per unit space; in three dimensions one considers flux through a two-dimensional box and the units include space squared).

We can understand the change in abundance that results from movement in terms of the difference in the number of individuals flowing in and the number flowing out of a small box on our one-dimensional landscape (as in Section 3.1 and Figure 3.2, in the context of a distribution over age rather than space). If we then take a limit as the size of that small box goes to zero, this difference becomes a rate of change in flux ($\partial J/\partial x$). Now, if there is a steady flow (flux) of individuals through the small box (J entering equals J exiting, so that $\partial J/\partial x = 0$), then the population density in the box will not change as a result of movement. Thus the contribution of movement to a change in density at a given location depends on the *change* in the movement rate ($\partial J/\partial x$), rather than the movement rate itself.

The total change in numbers at any location, $\partial n/\partial t$, is given by the sum of two terms: the change in the flux at location x plus any contribution from population dynamics (i.e. birth and death) at that location (see Okubo and Levin (2001)):

$$\frac{\partial n(x,t)}{\partial t} = -\frac{\partial J}{\partial x} + f[n(x,t)]. \tag{9.1a}$$

We use the function $f(n)$ to represent generically the birth and death processes going on at each point in space; these could be linear or nonlinear, and we will see later that there are several different options for what $f(n)$ could be.

Now the next step in the model development is to specify what the flux, J, is. If individuals move randomly, then they will tend to move from areas of higher concentration to lower concentration, which can be represented as

$$J = -D\frac{\partial n(x,t)}{\partial x}, \tag{9.1b}$$

where D is a measure of the rate of random movement and is termed the *diffusivity (or diffusion coefficient)*. It has units of distance squared per unit time. In chemistry, Eq. (9.1b) is known as Fick's first law, which describes the diffusion of molecules or particles down a concentration gradient. Hence, Eq. (9.1b) is simply the gradient in population density, $\partial n/\partial x$, times the movement rate, D. In a fluid dynamics context, D is determined by fluid viscosity, temperature, and other factors. In an ecological context, we could imagine it being determined by walking speed, the ease of moving across a landscape, and so forth. Substituting the flux (Eq. (9.1b)) into Eq. (9.1a), we get

$$\frac{\partial n(x,t)}{\partial t} = \frac{\partial}{\partial x}\left[D\frac{\partial n(x,t)}{\partial x}\right] + f\left[n(x,t)\right]. \tag{9.1c}$$

Equation (9.1) represents only movement in one dimension, x. This describes movement along a straight line, such as organisms moving through a habitat corridor or coastal organisms moving along a linear coastline. To add in additional spatial dimensions y (for a two-dimensional landscape) or even z (for three dimensions, such as plankton moving in a body of water), we use the "del" operator, ∇, to describe the partial derivatives in all of those dimensions,

$$\nabla = \frac{\partial}{\partial x} + \frac{\partial}{\partial y} + \frac{\partial}{\partial z}, \tag{9.2}$$

so the multi-dimensional spatial model is

$$\frac{\partial n(x,t)}{\partial t} = -\nabla J + f\left[n(x,y,z,t)\right]. \tag{9.3}$$

However, for now we will focus on the one-dimensional version; the results for additional dimensions are similar, but are harder to display in figures.

Like the other partial differential equation models we have analyzed, we must specify boundary conditions in order to solve the model. These include defining an initial density at $t = 0$ (which may vary over space), $n(x,0) = n_0(x)$, and defining the density at the edges of the spatial domain (unless the domain is assumed to be infinite). If the spatial domain is finite, one might presume that the habitat beyond the domain is uninhabitable, so it would make sense for the population density to be zero at the boundaries. So, for example, if x ranges from 0 to L, boundary conditions could be $n(0,t) = n(L,t) = 0$. These are known as homogenous Dirichlet boundary conditions. Another possibility would be to assume that individuals do not leave or enter the domain, so the flux at the boundaries would be 0. That is, $\partial n(0,t)/\partial x = \partial n(L,t)/\partial x = 0$. Kot (2001) provides several other possible boundary conditions and explains their consequences.

9.1.1 *The reaction–diffusion model*

As we proceed we will assume that the diffusion coefficient (i.e. random movement) is constant over space, which, from Eq. (9.1c), leads to the equation

$$\frac{\partial n(x,t)}{\partial t} = D\frac{\partial^2 n}{\partial x^2} + f\left[n(x,t)\right]. \tag{9.4}$$

This is the equation classically known as the reaction–diffusion model, originally proposed by Fisher (1937), who was interested in the movement of genes through a population, and also used by Skellam (1951). The name is from chemistry, and refers to the idea that there is spatial diffusion (the first term on the right-hand side) but also a "reaction" (the second term on the right-hand side) occurring that can increase the overall population density. With diffusuion alone, the density, or concentration, would gradually dissipate in all directions. If diffusion varies over space, so that it is a function of x, i.e. $D(x)$, then a more complicated expression would be required (see Okubo and Levin, 2001) for more information).

Notice that the rate of change in the population over space in Eq. (9.4) is proportional to the *second* derivative of n with respect to x. We illustrate this in Fig. 9.1 (in which we assume that there is no birth or death happening for the time being). Having the second derivative of n in Eq. (9.4) means that first, if there is no difference in density over space ($\partial n/\partial x = 0$), there is no net movement (top row in Fig. 9.1). This is ecologically sensible (organisms still move, but movement is equal in each direction). However, if there is a constant, linear gradient in population density over space (i.e. middle row in Fig. 9.1, where $\partial n/\partial x$ is constant but $\neq 0$), the second derivative $\partial^2 n/\partial x^2$ will be zero, and there will be no net diffusion of individuals in either direction. Diffusion only arises when the

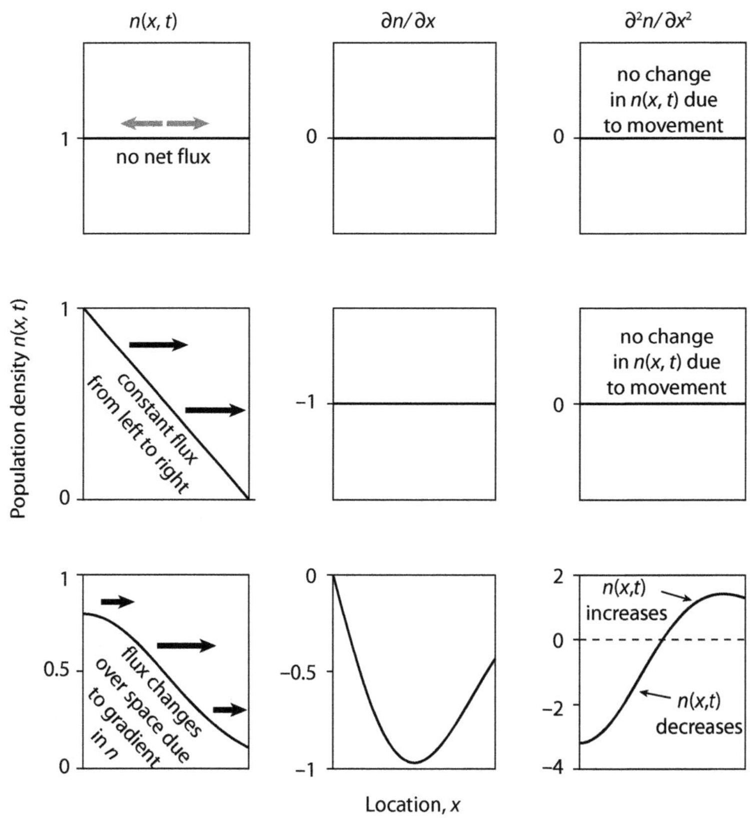

Fig. 9.1 Relationship between the shape of the distribution, $n(x, t)$ and flux (Eq. (9.1), assuming no "reaction" term). The first column shows three possible spatial distributions of population density, $n(x,t)$: no gradient, a linear gradient, and a nonlinear gradient. The second column shows the corresponding first partial derivatives with respect to space, $\partial n/\partial x$, and the third column shows the second partial derivatives, $\partial^2 n/\partial x^2$. The rate of change in population density due to movement (not population dynamics) is proportional to the second partial derivative (third column). Arrows in the first column indicate the relative flux in population density, assuming movement is random and moves down density gradients.

gradient is nonlinear (bottom row in Fig. 9.1), because in that case individuals are moving out of a location faster than they are moving in, or vice versa (note the arrows of different lengths in the bottom left panel of Fig. 9.1). It is not necessarily intuitive ecologically when this would be the case. However, as we will see, nonlinear gradients tend to arise when a population is growing exponentially, for example, and this can produce realistic patterns of movement.

9.1.2 *The asymptotic rate of spread*

The simplest function we could use for the birth–death part of Eq. (9.3) is exponential growth, $f(n) = rn$. The best way to understand the consequences of that assumption is to consider what happens when an invading population starts out at $t = 0$, at low population density in the middle of the spatial domain (Fig. 9.2a). If $r > 0$, the population begins to grow exponentially at the spot of the invasion, and then also begins to spread outward. After a short time, several patterns emerge. First, the population is always largest at the point of the invasion, where initially it began to increase exponentially. Second, the exponential growth produces a nonlinear spatial gradient in population density (notice the spatial distribution is bell-shaped, like a normal distribution), which as we noticed earlier, would be necessary for net movement. Finally, the outward spread of the population is not exponential, but very quickly becomes a wave traveling at a constant rate. This is termed the asymptotic rate of spread (ARS).

To find an expression for the ARS, we rely on the fact that particles diffusing according to Fick's equation (Eq. 9.1b) will eventually follow a normal distribution with standard deviation $2\sqrt{D}$ (Okubo and Levin, 2001). In other words, if organisms do not die or reproduce (i.e. $r = 0$), and there are initially n_0 individuals at $x = 0$ and $t = 0$, then the solution to Eq. (9.1) is

$$n(x,t) = \frac{n_0}{\sqrt{4\pi Dt}} e^{-\frac{x^2}{4Dt}}. \tag{9.5}$$

Kot (2001) shows how this expression can be derived by applying separation of variables to Eq. (9.4). If the population is also growing exponentially at rate r ($r > 0$), then the solution is slightly modified to

$$n(x,t) = \frac{n_0}{\sqrt{4\pi Dt}} e^{rt-\frac{x^2}{4Dt}}. \tag{9.6}$$

To find the ARS, we imagine that the organisms can only be detected by observers when they reach some very small threshold density, n_c. The "edge" of the spreading population is the point at which the population density is at that threshold, so we set the density equal to n_c and solve for the velocity, x/t,

$$\frac{x}{t} = \pm\sqrt{4Dr - \frac{4D}{t}\log\left[\frac{n_c}{n_0}\sqrt{4\pi Dt}\right]}. \tag{9.7}$$

As t gets larger, the second term under the square root symbol goes to zero, since t (in the denominator) is larger than the square root of t in the numerator. Once t is sufficiently large that we can ignore that term, we obtain a remarkably simple expression for the ARS,

$$\frac{x}{t} = 2\sqrt{rD}. \tag{9.8}$$

This approximation holds true for diffusion in one, two, or three dimensions. In Fig. 9.2b we have plotted this solution along with the actual rate of spread in the model simulation

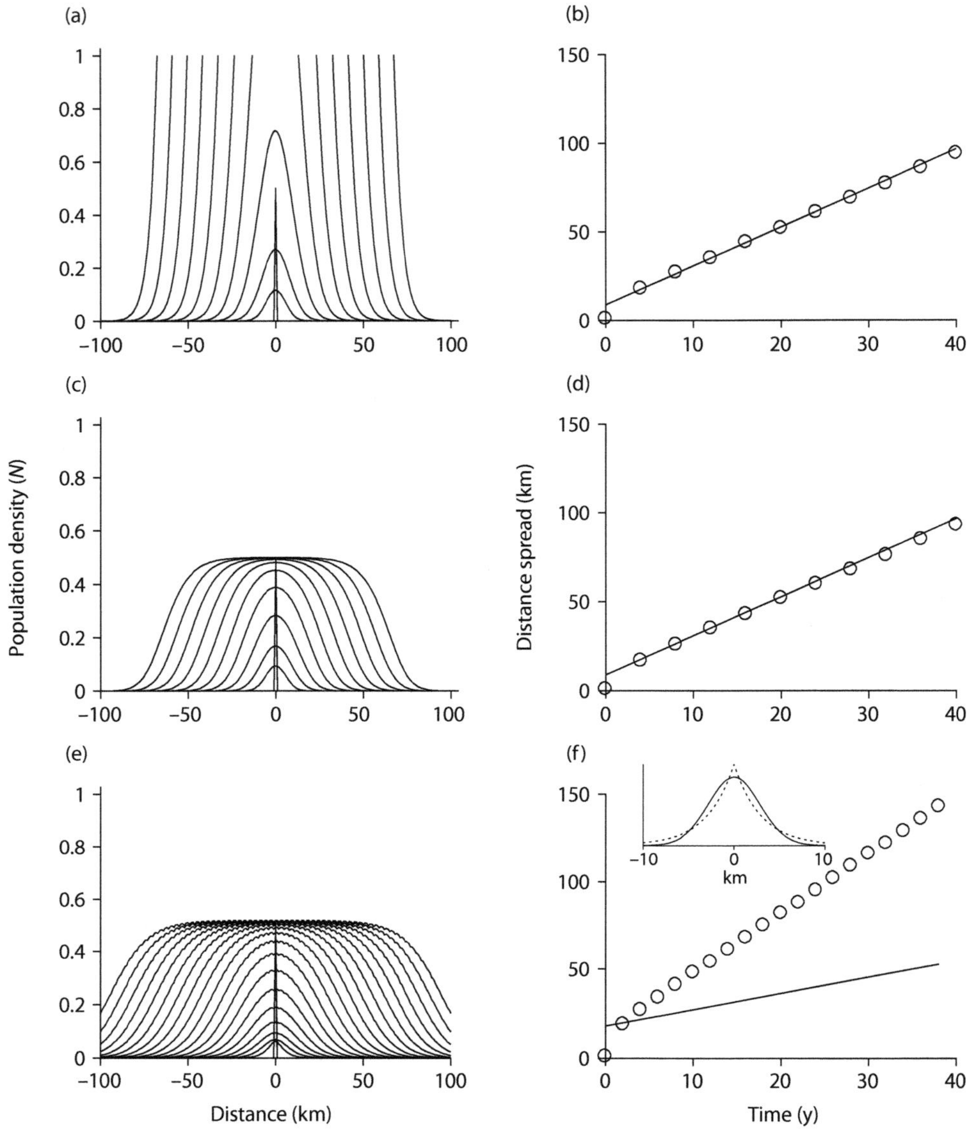

Fig. 9.2 Spread of invading populations. In (a) the population grows exponentially; in (c) and (e) the population grows logistically to a carrying capacity. Panels (b, d, f) show the expansion of the leading edge of the invasion wave (open symbols), along with the analytical estimate of the asymptotic rate of spread (solid line). Movement in (a–d) is diffusive and normal (Gaussian), whereas movement in (e,f) follows a leptokurtic dispersal kernel (see inset in f).

to show the correspondence. Skellam (1951) arrived at this same solution using a slightly different approach; he was interested in diffusion in two dimensions, so he integrated Eq. (9.4) to find the radius of the circle that contained all but a very small proportion of the entire population at time *t*.

The solution in Eq. (9.8) also holds if we use a slightly more realistic model for $f(n)$, such as a logistic model (Fig. 9.2c,d). This is actually the version used by Fisher (1937), and for an invading population the outcome is that population density reaches the carrying capacity in the center of the invasive range, but the population is growing exponentially at the edges, where population density is low (Fig. 9.2c). This produces a nonlinear gradient in population density at the edge, and the population expands at the same rate as in exponential growth, $2\sqrt{rD}$ (Fig. 9.2d). In fact, this result (Eq. (9.8)) will generally hold whenever the population grows exponentially at the edge of its range; i.e. whenever there are no Allee effects and $f(n)$ is greatest at low n (Lewis, 1997). Kot (2001) provides a library of other forms that $f(n)$ could take, including a model with an Allee effect. The reason the same answer arises here as in the density-independent model is that the population density is always small at the edge of the range, so density dependence does not enter.

The reaction–diffusion model has had great success in describing and predicting the spread of invasive species in a number of case studies. A major difference among studies using this model has been the way in which D was estimated. Ideally, one would use mark–recapture data, and estimate D based on the rate of increase in variance of recapture location over time (recall that D has units of distance^2 per unit time; Okubo and Levin, 2001). However, this type of data has not always been available. For example, Skellam (1951) did not have an independent estimate of D for oak trees, but used the model to estimate a lower bound for D given an estimate of r and the minimum time taken for the population to spread across Britain. Lubina and Levin (1988) studied the expansion of the population of sea otters (*Enhydra lutris*) along the central California coastline. Sea otters had been driven nearly extinct by the early twentieth century due to the fur trade, but a relict population at remote Pt. Sur began to expand after otters were protected by treaty in 1911. The California coast is relatively linear, so Lubina and Levin (1988) used the linear version of the reaction–diffusion model to approximate otter population dynamics. They estimated D based on the variance of the spatial distribution of otter population density from aerial survey data (rather than individual mark–recapture estimates), and also estimated r from the slope of the logarithm of population density versus time from the first few years of surveys (when densities were low) in order to calculate the ARS. They obtained the best results from estimating D and the ARS separately for the northern and southern range boundaries, and obtained good agreement between model and data for the period 1938 to 1972 (Fig. 9.3). The ARS was slower in the north, apparently because of a smaller D. The same effect could be attributed to higher mortality in the north, and thus smaller r, but Lubina and Levin (1988) rejected that hypothesis. In 1972, both edges of the population reached large gaps of no habitat (Morro Bay in the south and Monterey Bay in the north), and there was a dramatic jump in the range size before the population resumed expansion at the ARS. In the north, the ARS increased from below to above the model prediction after 1973, apparently either because movement became easier (higher D) or mortality rates decreased (higher r).

The sea otter range expansion/reinvasion continued after Lubina and Levin's 1988 publication; in 2008, Tinker et al. published an updated analysis of the spread (Tinker et al., 2008). They used a more complex model than the simple diffusion assumption, allowing them to account for spatial variation in demography (with a stage model) and other factors, but nonetheless predicted that the population would continue to expand at a linear rate not very different from the original ARS calculation.

Andow et al. (1990) modeled the spread of three invasive species, muskrat (*Ondratra zibethica*, which was also originally studied by Skellam, 1951), the small cabbage white butterfly (*Pieris rapae*), and the cereal leaf beetle (*Oulema melanopus*). They were able to use individual mark–recapture data to estimate D, and found that the reaction–diffusion model correctly predicted the ARS for the first two species. The cereal leaf beetle, however, was spreading much faster than predicted by the model; Andow et al. (1990)

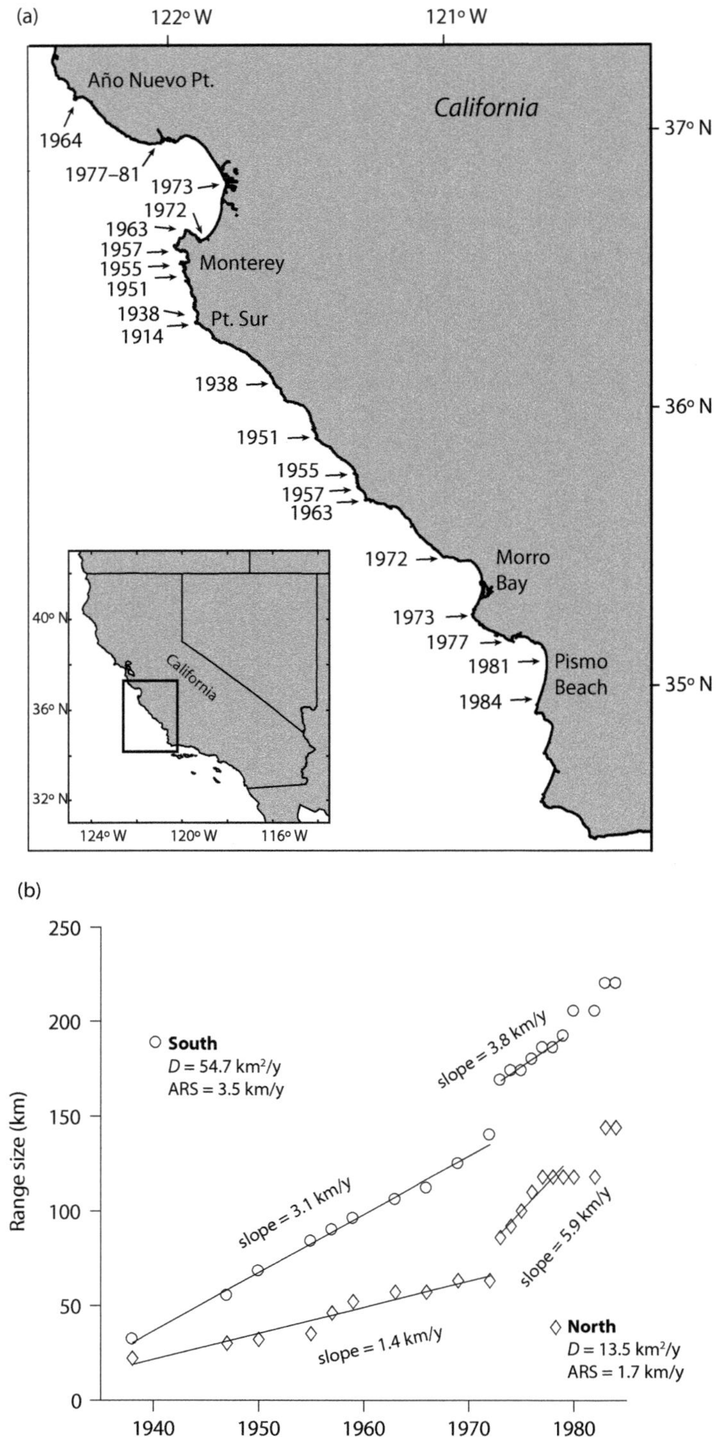

Fig. 9.3 Spread of re-invading sea otter (*Enhydra lutris*) populations in central California. (a) Map showing the historical progression of the reinvasion along the coast. Pt. Sur is taken to be the dividing line between southern and northern halves of the range. (b) Symbols show observations of the distance to the leading edge of the invasion in the south (circles) and north (diamonds). Solid lines are linear regressions fitted to those points, estimating the asymptotic rate of spread (ARS). Independent estimates of the ARS based on calculations of diffusive movement (D) from aerial surveys of otters are shown for both south and north. Redrawn from data in Lubina and Levin (1988).

speculated that this could be due to occasional long-range movements on air currents or by hitchhiking on humans.

9.1.3 *Leptokurtic dispersal*

The observation that some invasions move faster than the ARS predicted by the reaction–diffusion model has led to the development of other dispersal models that allow for more frequent long-distance movement. If we consider the spatial distribution of a single cohort of offspring released from one point in space, diffusion alone will eventually lead to a normal distribution. In order to have more frequent long-distance movement, the tails of that distribution would have to be "fatter", i.e. a greater proportion of individuals move longer distances. This leads to the consideration of "fat-tailed", or leptokurtic, dispersal patterns. As we hinted previously, the derivation of the reaction–diffusion model essentially ignores the possibility of substantial long-range movement so we need to use another approach.

This effect can be seen if we switch to discrete time and use an integro-difference equation, which is also a type of model that is commonly used to represent dispersal (e.g. Kot et al., 1996). A very simple integro-difference model would be

$$N_{t+1}(x) = \int_{\Omega} k\,(x - z)\, b\,[N_t(z)]\, dz, \tag{9.9}$$

where $N_t(x)$ is the population density at time t and location x, $b[N]$ describes density-dependent fecundity, and $k(x - z)$ is the dispersal function, which describes the probability of offspring born at point z dispersing to point x. This dispersal probability depends on the distance between x and z. The Ω indicates that integration is over the entire landscape (i.e. all possible locations z). Thus the model describes a population with non-overlapping generations; the offspring all disperse according to the dispersal function, and the new population density at point x in time $t + 1$ is obtained by integrating over the number of offspring produced at every other location times the probability of dispersing from that location to x. The function k is also called the dispersal *kernel*. Because we are now operating in discrete time, this model is better suited to describing species that reproduce and disperse in discrete time intervals, perhaps seasonally.

You may recognize that Eq. (9.8) has a very similar form to an integro-difference equation we used previously (Eq. 5.15) when we introduced integral projection models (Section 5.4). The models are very similar; in IPMs the kernel describes the probability of individuals growing from every size to every other possible size, while here it describes spatial movement. As with Eq. (9.1), we are simply substituting space for size as the structuring dimension.

To analyze Eq. (9.9), we must specify what the kernel, k, is. One option is a normal distribution,

$$k(x) = \frac{1}{\sqrt{4\pi D}} e^{-\frac{x^2}{4D}}. \tag{9.10}$$

We have written this so that the standard deviation of the distribution is $2\sqrt{D}$. Recall that simple diffusion produces a normal distribution of dispersing particles (Okubo and Levin, 2001); it turns out that this model is analogous to the reaction–diffusion model (Eq. 9.4) and if $b[N]$ is a discrete-time formulation of the logistic model (Eq. (2.6)), then this model will have an ARS of $2\sqrt{rD}$, just like the reaction–diffusion model.

To represent greater longer-distance dispersal, we must replace Eq. (9.10) with a more leptokurtic dispersal kernel. One example is the negative exponential function, which is also termed the Laplacian function,

$$k(x) = \frac{1}{2a} e^{-a|x|}, \tag{9.11}$$

where a is a parameter that determines how quickly the dispersal probability declines with distance, x. This kernel is much more peaked in the center with longer, fatter tails than the normal distribution (Fig. 9.1f inset). Simulating the same population invasion as we did earlier (Fig. 9.2c), but using this kernel, produces a much faster-spreading population (Fig. 9.2e). Indeed, the ARS is substantially faster than one would observe for a normal dispersal kernel with the same standard deviation; i.e. $2\sqrt{rD}$ (Fig. 9.2f). If two kernels have the same standard deviation, it means that the average distance dispersed by offspring in a cohort is the same. Thus the difference in invasion speed must be due to the shape of the tail of the distribution: more individuals make extremely long jumps using the Laplacian kernel. Kot et al. (1996) provide an example of using leptokurtic kernels such as these to describe the spread of dispersing fruit flies, *Drosophila pseudoobscura*.

9.1.4 *When diffusion is not a good representation of movement*

While we can learn a lot about how populations move in space using the reaction–diffusion approach, it is important to recognize that not all types of animal movement are well represented by Fickian diffusion (Eq. 9.3) or by dispersal kernels like those used in Section 9.1.3. For example, many animals exist in territories or home ranges; although they may move great distances, there may be no net movement in any direction, regardless of gradients in population density.

One example in which this nuance becomes very important is in spatial models that include harvested areas and reserves. For example, Walters et al. (2007) developed a model of rocky reef fish populations along the coastline of central California to evaluate the potential costs and benefits of placing a network of marine protected areas (MPAs). An important consideration in MPA design is the movement of adult fish (Grüss et al., 2011). Fish cannot detect MPA boundaries, so highly mobile species may cross the boundary frequently and spend a lot of time in areas where they are vulnerable to harvest, even if the center of their home range lies within the MPA. As a consequence, MPAs of a given size are generally less effective for protecting more mobile species.

Walters et al. (2007) modeled this phenomenon using a reaction–diffusion approach, so adult fish moved according to the discrete-space approximation of Eq. (9.3). The problem with this approach is that at the boundary between a heavily fished area and a no-take MPA, the gradient $\partial n/\partial x$ will be very steep, leading to rapid diffusion out of the MPA. In fact, as harvest is increased and the gradient steepens, the flux of fish out of the MPA will increase; in other words, the MPA becomes less effective as harvest increases, simply because fish are being sucked out of the MPA to their doom by the gradient. Indeed, Walters et al. (2007) concluded that MPAs would not be effective for most of the species they modeled. This is contrary to the observation that many adult fishes maintain stable home ranges in the vicinity of MPA boundaries and remain inside the MPA (or near the boundary) rather than diffusing outwards (e.g. Lowe et al., 2003).

Walters et al. (2007) also proposed what turns out to be a better way of representing the movement of animals in home ranges, although they did not actually implement this approach in their model. The idea is to assume that adult fish do not disperse, but experience fishing mortality in proportion to the proportion of their home range that is outside the MPA. This requires making an assumption about the distribution of time spent in different parts of the home range (e.g. more time spent near the center, or near the edges?), but avoids the artifact of steady diffusion across the MPA boundary (see Moffitt et al., 2009; 2011 for examples of this approach). Maciel and Lutscher (2013) have also proposed several alternative modifications to the diffusion framework that account for more realistic behaviors of individuals at patch or habitat edges. The lesson of all of this work is that mathematical approximations of diffusive movement can sometimes introduce undesirable artifacts that can alter model results.

9.2 Population persistence in aquatic habitats

9.2.1 *The KISS model: persistence of a patch of plankton*

Models similar to those we have used to describe the spread of invading populations (i.e. transient dynamics) have also been widely used to investigate questions about the persistence of existing populations (i.e. steady-state dynamics). For example, Kierstead and Slobodkin (1953) were interested in blooms of harmful phytoplankton (red tides). They used a version of Eq. (9.4) with exponential growth to describe the rapid reproduction of the phytoplankton as well as their diffusive movement within a mass of water,

$$\frac{\partial n(x,t)}{\partial t} = D\frac{\partial^2 n}{\partial x^2} + rn. \tag{9.12}$$

The nature of the blooms in question is that they are contained within a water mass that has conditions favorable for growth (e.g. low salinity), but if the phytoplankton cells diffuse out of the water mass, they die. A bloom will occur if exponential growth in the center of the water mass can offset the diffusion of plankton out of the patch. Therefore, they modeled a spatial domain of size L with homogeneous Dirichlet boundary conditions (i.e. population density is zero at the boundaries). They found that the population would have net positive growth (i.e. a bloom would occur) if the patch was bigger than

$$L > \pi\sqrt{D/r}. \tag{9.13}$$

This model became known as the KISS model, after Kierstead, Slobodkin, and Skellam (the latter having also addressed this topic), and the critical patch size is the KISS size (Skellam, 1951; Kierstead and Slobodkin, 1953; Slobodkin, 1953; Okubo, 1978). The derivation of the solution in Eq. (9.13) is too lengthy to provide here (see Kot, 2001), but the basic form is intuitive because it includes the ratio of diffusivity to the rate of exponential growth. You can also see that the units are correct, as D has units of space squared per time. Here we also see that larger L (or smaller D) improves the chance of persistence, since either change would make the rate at which individuals reach the boundary smaller; the boundary is the only death term in the model.

It turns out, of course, that the actual physics of water masses and blooms are somewhat more complex than this; for example, the presence of fronts or converging currents at the patch edge would require different boundary conditions (Okubo and Levin, 2001). The KISS model is also notable for being, perhaps, the first ecological model to be named with a somewhat strained acronym (one has to take the first *two* letters in Kierstead to make it work) in order to be a little cute.

9.2.2 *The drift paradox*

Another type of persistence problem concerns populations living in a flowing environment, such as a stream. So far we have only considered movements that are equal in every direction, such as diffusion. But in a flowing environment there is net movement in one direction. We can rewrite Eq. (9.3) to include the effect of a flow that has velocity v:

$$\frac{\partial n(x,t)}{\partial t} = v\frac{\partial n}{\partial x} + D\frac{\partial^2 n}{\partial x^2} + f\left[n(x,t)\right]. \tag{9.14}$$

If we keep the same boundary conditions we used before, with $n(x,t) = 0$ at both the upstream and downstream edges of the habitat domain, we encounter a problem. If there is net flow of organisms in one dimension, it seems inevitable that eventually the population will "wash out" off the downstream edge of the habitat and go extinct. This is the so-called "drift paradox" (Hershey et al., 1993; Speirs and Gurney, 2001).

The drift paradox has been investigated for several decades, particularly in the context of insects in streams (Müller, 1954). A few different resolutions have been proposed, such as the idea that the downstream drift of larval insects is balanced out by the upstream flight of winged adults, which then oviposit in the upper stream reaches to replenish the upstream edge of the population (Müller, 1954). The problem with this resolution is that not all populations subject to drift have a flying adult stage (Waters, 1972). Fortunately, a simpler resolution to the paradox can be found simply by considering the role of diffusive movements in addition to the net downward drift.

To determine whether a population can persist in the face of drift, we take the same approach as in non-spatial models (e.g. Chapter 2) and ask whether a population at very low density will increase away from zero or decline to extinction. Thus we are ignoring any nonlinear density dependence in births or deaths (assuming there are no Allee effects), and set $f(n) = rn$ in Eq. (9.14). This model is linear with respect to n, so we try to find a solution that has the usual form for solutions to linear differential equations,

$$n(x,t) = e^{\lambda t}g(x), \tag{9.15}$$

where $g(x)$ is an as yet unknown function of x. In other words, the population at position x will eventually grow exponentially, with initial conditions that depend on location, as long as $\lambda > 0$. Speirs and Gurney (2001) provide the lengthy set of algebraic steps necessary to find the conditions under which $\lambda > 0$, and we will not detail them here. Essentially, they found that persistence was more likely if velocity, v, was slower or if the length of the habitat, L, is longer. This makes sense; the reproduction is more likely to exceed washout if washout is a slow process (because of slow v or long L). It is also possible to calculate the critical length, L_c, that is the minimum length required for persistence. It turns out that L_c is only a real number if the following relationship holds:

$$v < 2\sqrt{rD}. \tag{9.16}$$

This means that population persistence is only possible if the combination of reproduction (r) and diffusive processes (D) exceeds the downstream velocity (v). Notice that this threshold value is the same as the asymptotic rate of spread we derived earlier for expanding populations (Eq. 9.8).

A different population model involving a similar kind of interaction between advection and diffusion is the model that depicted how nearshore ocean circulation affects the larval phase of the barnacle, *Balanus glandula* (Roughgarden et al., 1988; Possingham and

Roughgarden, 1990). Recall that we discussed the cycles arising from density-dependent competition for space in this species in Section 7.4. As in many marine organisms, barnacle larvae are planktonic and can be transported by ocean currents prior to settling back onto rocky shores and metamorphosing into adult barnacles. The initial version of Roughgarden and colleagues' spatial model was a two-dimensional advection/diffusion model representing the movement of larvae in the offshore/onshore and alongshore directions, coupled to a simple model of adults at the shoreline boundary. The primary result of this model was similar to the previous "drift paradox": populations could persist when alongshore advection was non-existent or low, but as advection increased, larval losses from populations along the shore eventually became too great for the populations to maintain sufficient replacement for persistence.

Another version of this two-dimensional larval model added a reflecting boundary offshore (Alexander and Roughgarden, 1996). This represented a prominent feature of oceanography along the California coast, *upwelling*, by which equatorward winds along the coast cause cold water to upwell to the surface, then move offshore in a surface layer. That mass of upwelled surface water forms an offshore boundary with non-upwelled water known as an upwelling front. It was thought at the time that barnacle larvae would reside primarily in that upwelled surface water. In this model, even with no offshore advection, diffusion of larvae away from the shore could reduce replacement in the benthic adult population to the point that the population could not persist. The upwelling front had to be within a certain distance of the shore for a sufficient number of larvae to return to settle onshore and allow the population to persist. Alexander and Roughgarden also examined the effect of temporal variability in the offshore position of the upwelling front, representing the actual movement of such fronts in response to weekly-scale variability in upwelling winds. The temporal variability in this cross shelf mechanism interacted with the "drift paradox" mechanism, with specific combinations of them leading to extinction. Another application of this type of model served to show how circulation features (e.g. the convergence of two alongshore currents) could interrupt alongshore larval transport enough to be responsible for some of the species range limits commonly seen along the world's coastlines (Gaylord and Gaines, 2000).

The predictions of these models by Roughgarden and colleagues have led to a long series of empirical and theoretical investigations into the causes of variability in larval recruitment and the mechanisms by which intertidal populations persist in highly advective environments (reviewed by Menge and Menge, 2018; Shanks and Morgan, 2018). One important detail that has emerged is that larvae are not passive particles, but exhibit strong vertical swimming behaviors that help avoid being swept offshore (Morgan et al., 2009). Accounting for that behavior—along with key details of the way currents behave very close to shore—requires more complex modeling approaches (e.g. Nickols et al., 2015), though the basic rule of thumb expressed by Eq. (9.16) remains a useful first approximation.

At the beginning of this chapter we noted that the models we used to describe movement through space were not unlike the age- and size-structured models we had considered in Chapters 3 and 5. As we have worked through these different examples, you may have noticed parallel structures in these spatial PDEs (e.g. Eqs. 9.12 and 9.14) and the earlier M'Kendrick/von Foerster PDEs. There is one key mathematical difference, which is that the spatial models here are *second-order* PDEs and the age- and size-structured models were *first-order* PDEs. For the reader who is so inclined, Box 9.1 delves deeper into the relationships among these models.

Box 9.1 RELATIONSHIPS AMONG PDE MODELS

The partial differential equations in this chapter differ mathematically from the M'Kendrick/von Foerster partial differential equations in Chapters 3 and 5, and it is helpful to understand their differences. Basically, adding the diffusion term with the second partial derivative makes this a *second-order PDE*, whereas both the age-structured and the size-structured M'Kendrick/von Foerster equations are *first-order PDEs*.

We can briefly review our mechanistic explanations of how each term worked in the PDEs in earlier chapters to gain some understanding of the important differences in how they work among these equations. It turns out there are only a couple of types of terms in these equations.

In Chapter 3, the first term on the right-hand side of the *age-structured* M'Kendrick/von Foerster equation (Eq. 3.2, here Eq. B9.1-1) was the derivative of the product of the density function, $n(a,t)$ times the rate of growth (in age, not size) of an individual, da/dt,

$$\frac{\partial n(a,t)}{\partial t} = -\frac{\partial}{\partial a}\left[n(a,t)\frac{da}{dt}\right] - D(a,t)\,n(a,t). \tag{B9.1-1}$$

In that case the latter term, da/dt, was equal to 1, so did not appear in the final equation. We used Fig. 3.3 to explain the function of that first term on the right-hand side in this model: the rate of increase in the age distribution with time (left-hand side of the equation) was equal to -1 times the slope of the age distribution in the age direction times the rate at which that distribution was moving in the age direction. We will refer to this type of term here as an *advection term*, because it describes how the age distribution is moving along the age axis.

For the *size-structured* M'Kendrick/von Foerster equation in Chapter 5, again the first term on the right-hand side was minus the rate of change in the size direction of the product of the size density, $n(m,t)$ times the growth rate (in size) of an individual, $g(m,t)$ (Eq. B5.1-4, here B9.1-2),

$$\frac{\partial}{\partial t}n(m,t) = -\frac{\partial}{\partial m}\left[n(m,t)\,g(m,t)\right] - D(m,t)\,n(m,t). \tag{B9.1-2}$$

Using the product rule of differentiation on the first right-hand side term led to two terms (Eqn 5.2, here B9.1-3),

$$-\frac{\partial}{\partial m}\left[n(m,t)\,g(m,t)\right] = -\frac{\partial n}{\partial m}g(m,t) - \frac{\partial g}{\partial m}n(m,t). \tag{B9.1-3}$$

To explain the effect of the first term, we used Fig. 5.1a to provide the same explanation as we used for the advection term in Chapter 3, except that this time the rate of movement was not equal to 1, but rather actually depended on the growth rate of individuals. This size dependence would distort the size distribution because, in general, growth rate varies with size, m. The second term in Eq. (5.2) was a new kind of term with different behavior. In the second term in Eq. (5.2), the rate of change in the growth rate of individuals, $\partial g/\partial m$ (which we also called $g_m(m)$) multiplied by the size distribution caused *deterministic dispersion*, which amounted to "piling up" (i.e. an increase in $\partial n/\partial t$) if it was negative or "stretching out" (i.e. a decline in $\partial n/\partial t$) if it was positive (shown in Fig. 5.1b). We will refer to this type of term here as a *deterministic dispersion* term.

In the spatial models of this chapter, an advection term was not introduced until Eq. (9.14), where it is the first term on the right-hand side (i.e. the term describing flow downstream).

There the distribution passes a point x at a rate v, causing an increase or decrease in $n(x,t)$ according to the slope of the distribution in the x direction (i.e. $-\partial n/\partial x$).

Because the first PDE model in this chapter (Eq. 9.1) did not contain an advection term, we can use it to understand how the diffusion term works. First we note that the definition of flux (Eq. 9.1b) has the same form as the advection term in the size model, in the sense that $D(x)$ acts the same as $g(m)$ in moving the distribution in the x (or m) direction with time, causing an increase in n that depends on the local slope of the distribution in the x (or m) direction. This may help in understanding what flux is. Similar to the advection term in the size model, flux here describes the contribution to the rate of increase in the spatial distribution due to the fact that the distribution is moving across space at the rate $D(x)$.

The big difference in this chapter is that this flux term does not appear in the basic model of the rate of increase in the time direction: only its derivative with respect to space appears, which makes this a second-order PDE. We place flux into the first term of the spatial PDE model in Eq. (9.1c). Applying the product rule of differentiation to the first term on the right-hand side then leads to two terms, one with $\partial D/\partial x$, or D_x (similar to g_m), which is therefore like a deterministic dispersion term, except that it is multiplied by the derivative of the distribution with respect to x, rather than by the distribution itself (as in the size equation). The other term is another new type of term with the second derivative in the x direction. This term is similar to an advection term, except that it is advecting the spatial rate of change of the spatial distribution in the positive x direction, and therefore contains the second derivative of the spatial distribution in the x direction. The advection occurs at a rate $D(x)$, but in a number of models (e.g. Eq. 9.4, Eq. 9.13), D does not depend on x, so this term alone will represent diffusion as illustrated in Fig. 9.1. If D does depend on x, the first term, containing D_x, will lead to a piling up or smoothing out effect, as in the size model (except that it is multiplied by the spatial derivative of the distribution, rather than the distribution).

When formulating these PDE models (or reading someone else's formulation) it is important to consider how these term "work," and whether that is biologically reasonable. In particular, diffusion has the most potential for misrepresenting actual biological processes, as we saw in Section 9.1.4.

Another mathematical attribute that makes these equations useful for describing populations of individuals is that they guarantee "conservation of mass," where mass here is abundance. In moving through age, size, or space, individuals will not inadvertently be created or destroyed (other than through reproduction, mortality, or boundary conditions). This can be seen by writing the age-structured version as

$$\frac{\partial n(a,t)}{\partial t} + \frac{\partial n(a,t)}{\partial a} + D(a,t)\,n(a,t) = 0 \tag{B9.1-4}$$

and the size-structured version as

$$\frac{\partial n(m,t)}{\partial t} + \frac{\partial}{\partial m}\left[n(m,t)\,g(m,t)\right] + D(m,t)\,n(m,t) = 0. \tag{B9.1-5}$$

These can be interpreted as saying that every individual either gets older (gets bigger) or dies. Similarly, the diffusion equation guarantees the same behavior: without reproduction or death, individuals cannot be created or destroyed, so

$$\int_{-\infty}^{\infty} n(x,t)\,dx$$

will remain constant with time.

9.3 Metapopulations

9.3.1 *The Levins model*

So far we have considered models that represent a homogenous landscape across which organisms move freely and frequently. In the models of invading populations (Section 9.1) or those subject to drift in a stream (Section 9.2), the interior part of the population is at carrying capacity and well mixed, and the interesting part of the system is the leading edge that is either expanding into empty territory or retreating due to washout. Levins (1969) conceived of an entirely different type of system: a metapopulation, in which subpopulations occupied discrete habitat patches that were linked by infrequent dispersal, presumably across an uninhabitable landscape (the uninhabitable part is often referred to as the landscape "matrix," which can be a little confusing given our reliance on mathematical matrices to analyze these models).

The original Levins (1969) model is quite simple; it is spatially implicit, meaning that it does not specify where in space each patch was, or how far apart any particular pair of patches were. Instead it assumes that during (rare) dispersal events, any patch could be reached equally well from any other patch. Additionally, the state variable is not population abundance; instead the model only keeps track of whether patches are occupied by an extant subpopulation, or are empty because the subpopulation in that patch has gone extinct. The state variable is P, the proportion X of patches that are occupied, and the only two demographic rates are c, the rate at which empty patches are colonized by occupied patches, and e, the rate at which subpopulations in occupied patches go extinct. The differential equation describing this process is

$$\frac{dP}{dt} = cP(t)\,[1 - P(t)] - eP(t). \tag{9.17}$$

This says that the rate of change of the proportion of occupied patches is equal to the colonization rate times the proportion of occupied patches (which will send out the colonizers) times the proportion of empty patches available for colonization, minus the current proportion of occupied patches times the probability of them going extinct.

To find the equilibrium solution to the Levins model we take the usual step of setting $dP/dt = 0$, and solving for P^*,

$$P^* = 1 - e/c. \tag{9.18}$$

There is only a biologically meaningful equilibrium ($P^* > 0$) if the colonization rate c exceeds the extinction rate e, and the smaller the ratio e/c is, the closer the metapopulation will be to full occupancy at equilibrium.

The Levins model has strong intellectual ties to island biogeography theory (IBT; MacArthur and Wilson, 1967), in that it is concerned with the colonization and extinction of habitat patches, but does not make predictions about the size of individual subpopulations (an occupied patch could have two individuals, or two thousand). Indeed, it is possible to recast Eq. (9.17) as a sort of single-species version of IBT by assuming that all colonists originate from a mainland that is always occupied (Hanski, 2001). The Levins model was very influential in advancing the metapopulation concept in conservation biology, but because it is spatially implicit and assumes all patches are identical and equidistant, it is difficult to apply to real systems.

9.3.2 *Incidence function models*

Hanski (1991, 1994, 1998, 2001) and his colleagues adopted the basic Levins focus on colonization and extinction, but improved the realism of the model by being spatially explicit and accounting for differences among patches and their locations relative to one another. Specifically, one might expect that extinction rate would depend on patch size, because smaller patches support smaller populations that are more vulnerable to stochastic disturbances. Also, colonization of an empty patch is likely to depend on the distance to nearby occupied patches. Rewriting Eq. (9.17) to describe the dynamics of just patch i, we have a version of the model used by Hanski and Ovaskainen (2000),

$$\frac{dP_i}{dt} = C_i(t)\,[1 - P_i(t)] - E_i P_i(t) \tag{9.19a}$$

$$E_i = \varepsilon / A_i \tag{9.19b}$$

$$C_i(t) = c\sum\nolimits_{i \neq j} e^{-\alpha d_{ij}} A_j P_j(t). \tag{9.19c}$$

In this model, E_i and C_i are the extinction and colonization rates, respectively, of patch i, and $P_i(t)$ is the probability of occupancy state of i at time t. Extinction depends on patch area, A_i and colonization depends on the distance d_{ij} from i to every other patch j, the area of that patch (bigger patches have larger populations and thus produce more migrants), and whether that patch is occupied at time t (hence C_i is time varying). The rate of migration between patches is a negative exponential function of distance, with the average distance traveled equal to α. The parameters ε and c are constants.

This is known as the *incidence function model*, named after the concept of incidence of occupancy developed by Diamond (1975), who, working within the IBT framework, was investigating why some habitat patches had been colonized by particular species and others had not. In this concept, the incidence for patch i, J_i, is found by finding the equilibrium of Eq. (9.19a) (and assuming for the time being that C_i is constant with time),

$$J_i = C_i / \left(C_i + E_i\right). \tag{9.20}$$

Writing the expression for incidence only gets us so far, because as we suggested earlier, the colonization rate of patch i will depend on whether the other patches around it are occupied. Thus the incidences of all the patches are interconnected, and we need to analyze the entire model to learn which patches tend to be occupied. Unfortunately, the model in Eq. (9.19) is somewhat cumbersome, because now there is a separate equation for each patch. Analysis can proceed in two ways. First, the model can be simulated in a stochastic manner, with each patch going extinct or being colonized in each time step according to the probabilities E_i and C_i. Multiple simulations of the model can be calculated to estimate whether the metapopulation persists, and what proportion of patches are occupied at equilibrium (e.g. Hanski, 1998). This approach has led to the name *stochastic patch occupancy model*, or SPOM (Hanski and Ovaskainen, 2003).

As we have argued before (Chapter 1), it is difficult to derive a general understanding of a system from stochastic simulations alone. Helpfully, there is a second approach to analyzing incidence function models that allows us to make use of the mathematical tools already developed for structured population models. Ovaskainen and Hanski (Ovaskainen and Hanski, 2001; Hanski and Ovaskainen, 2003), proposed a deterministic approach to analyzing Eq. (9.19) by assembling the *landscape matrix*, $\boldsymbol{M}$. This matrix has elements m_{ij} as follows:

$$m_{ij} = ce^{-\alpha d_{ij}} A_j \frac{A_i}{\varepsilon} \tag{9.21}$$

and elements m_{ii} along the diagonal are zero. Thus each off-diagonal element is the contribution of patch j to colonization of patch i (from Eq. 9.19c) times the inverse of the extinction rate of patch i (from Eq. 9.19b). The inverse of the extinction rate is the average time a population will survive before going extinct. Hanski and Ovaskainen (2003) offer the explanation that m_{ij} measures the proportion of time patch i would be occupied if patch j were the only source of potential colonizers.

The landscape matrix $\boldsymbol{M}$ is in some way analogous to the projection matrix used in nonspatial structured population models, such as the Leslie matrix (Chapter 3). First, the dominant eigenvalue of $\boldsymbol{M}$, λ_M, determines whether the metapopulation persists. If $\lambda_M > 1$, the metapopulation will not have a stable equilibrium at zero, just as in the nonspatial models (of course, unlike the Leslie matrix model and other linear models, the state variable in this model, P_i, is bounded at 1 and cannot increase without bound). We should note that Hanski and Ovaskainen typically did not include the parameters c and ε in the expression for m_{ij}, so the persistence criterion was $\lambda_M > \varepsilon/c$. We adjusted this in Eq. (9.21) to make the analogy to nonspatial models more clear.

Unlike, say, the Leslie matrix, the landscape matrix $\boldsymbol{M}$ is not typically used to project model dynamics. However, one could rewrite Eq. (9.19a) in terms of $\boldsymbol{M}$ (and in discrete time) if we define vectors $\boldsymbol{P_t}$ and $\boldsymbol{E}$, where $\boldsymbol{P_t}$ is the vector of occupancy probabilities for each patch, and $\boldsymbol{E}$ is the vector of corresponding extinction rates (Eq. 9.19b). We will also use the version of $\boldsymbol{M}$ that does not include parameters c and ε in the expression for m_{ij}. Then the model is

$$\boldsymbol{P}_{t+1} = \boldsymbol{P}_t\,(1 - \boldsymbol{E}) + (1 - \boldsymbol{P}_t)\,\boldsymbol{cMP}_t. \tag{9.22}$$

Then we can use the dominant eigenvalue of $\boldsymbol{M}$ to write a simpler expression for the long-term, asymptotic dynamics of the model,

$$\hat{P}_{t+1} = \hat{P}_t\left(1 - \hat{E}\right) + \left(1 - \hat{P}_t\right)\hat{C}\hat{P}_t, \tag{9.23}$$

Where $\hat{P}_t$ is a weighted average of patch occupancy, $\hat{E} = \varepsilon/\omega$ and $\hat{C} = c\lambda_M/\omega$, where ω is the weighted average of patch areas. The weights in ω are given by the dominant eigenvector of $\boldsymbol{M}$. Essentially this reduces the system of equations for each of the patches in Eq. (9.19a) to a single equation, and it can be used to calculate the equilibrium patch occupancy, $\hat{P}^*$:

$$\hat{P}^* = 1 - \frac{e}{c\lambda_M}. \tag{9.24}$$

Hanski and Ovaskainen (2000) called λ_M the "metapopulation capacity" of the landscape, because it describes how connectivity among patches determines the equilibrium occupancy of the metapopulation.

9.3.3 *Patch value in the incidence function model*

Because the incidence function model and its derivatives are spatially explicit and account for patch differences, this lends itself to asking questions beyond simply whether the metapopulation will persist. Now we are able to investigate the role each patch plays in the metapopulation, and determine which patches contribute more (or less) to metapopulation persistence. This would clearly be valuable information in conservation planning, particularly for identifying patches that should be protected from harvest or habitat degradation.

The idea that different patches make different contributions is linked to the idea of source–sink dynamics (Pulliam, 1988). Pulliam's (1988) original definition of a source was quite simple, that it was a patch in which the birth rate exceeded the death rate; the opposite was true in a sink. Thus sinks depend on immigration from sources in order to persist, while sources can persist independently. Like the Levins metapopulation model, this has proven to be a thought-provoking concept, but there are difficulties in implementation. For one, it is difficult to apply this definition to species with obligate dispersive stages, such as pond insects with flying adults or benthic marine species with planktonic larvae. If potentially all local reproduction emigrates, and each new cohort immigrates, how does one calculate the within-patch reproductive rate (Figueira and Crowder, 2006)? More broadly, a patch within a metapopulation may make important contributions to metapopulation persistence even if it is not a "source" under Pulliam's definition; for example it could serve as a stepping stone linking disjoint subnetworks of the metapopulation. Therefore it can be more useful to consider the "value" of the patch to the metapopulation (Ovaskainen and Hanski, 2003).

Ovaskainen and Hanski (2003) catalogued a variety of metrics for patch value in the incidence function model. They also identified two main ways to measure value. The first using *perturbation* metrics, in which the value of a patch is estimated by simulating its removal from the metapopulation. Essentially this measures how much the patch would be missed if it were gone. If we consider the value of a patch to metapopulation persistence (or metapopulation capacity, in Hanski and Ovaskainen's terminology), then the perturbation measure of patch i's value, V_i, can be defined to be

$$V_i = \lambda_M - \lambda_{M-i}, \tag{9.25}$$

where the subscript $M-i$ indicates that the eigenvalue is calculated from the landscape matrix M with the column and row corresponding to patch i removed. Ovaskainen (2003) expanded this concept to describe patch value in the context of both habitat destruction and habitat restoration.

The second approach to measuring value is what Ovaskainen and Hanski (2003) termed *metapopulation dynamic* metrics, in which the value is calculated based on the dynamics of the intact metapopulation, without perturbation. The metapopulation dynamic equivalent of V_i (Eq. (9.25)) can also be approximated by calculating the dominant right and left eigenvectors of $\boldsymbol{M}$, $\boldsymbol{w}$, and $\boldsymbol{v}$, respectively,

$$V_i \approx \lambda_M w_i v_i, \tag{9.26}$$

where w_i and v_i are the ith elements of $\boldsymbol{w}$ and $\boldsymbol{v}$ (Ovaskainen and Hanski, 2001; 2003). To interpret this formula, recall that in the age-structured Leslie matrix model (Chapter 3), the right eigenvector is the stable age distribution, and the left eigenvector is the future reproductive contribution of each age. In the metapopulation context, this means that patch value depends on the proportion of time patch i is occupied (w_i) times the contribution of patch i to colonization (v_i). Also, the sensitivity of λ to the i,j element of the Leslie matrix was proportional to $w_i v_j$ (Eq. (3.35)) Thus the patch value here is the metapopulation capacity λ_M times the proportional reduction in λ_M if the contribution from patch i were reduced. The two ways of estimating V_i are not exactly equal, because the metapopulation dynamic approach (Eq. 9.26), in analogy to the eigenvector sensitivity calculations, is only considering a small change to patch i in the landscape matrix, while the perturbation metric (Eq. 9.25) is simulating a total removal.

We illustrate both calculations for a hypothetical metapopulation in Fig. 9.4. In Fig. 9.4a we show the spatial arrangement and size of the patches, and Fig. 9.4b displays the two measures of patch value. One can see that both patch size and patch location contribute to patch value; the large, centralized patches, 5 and 6, have much higher value than smaller

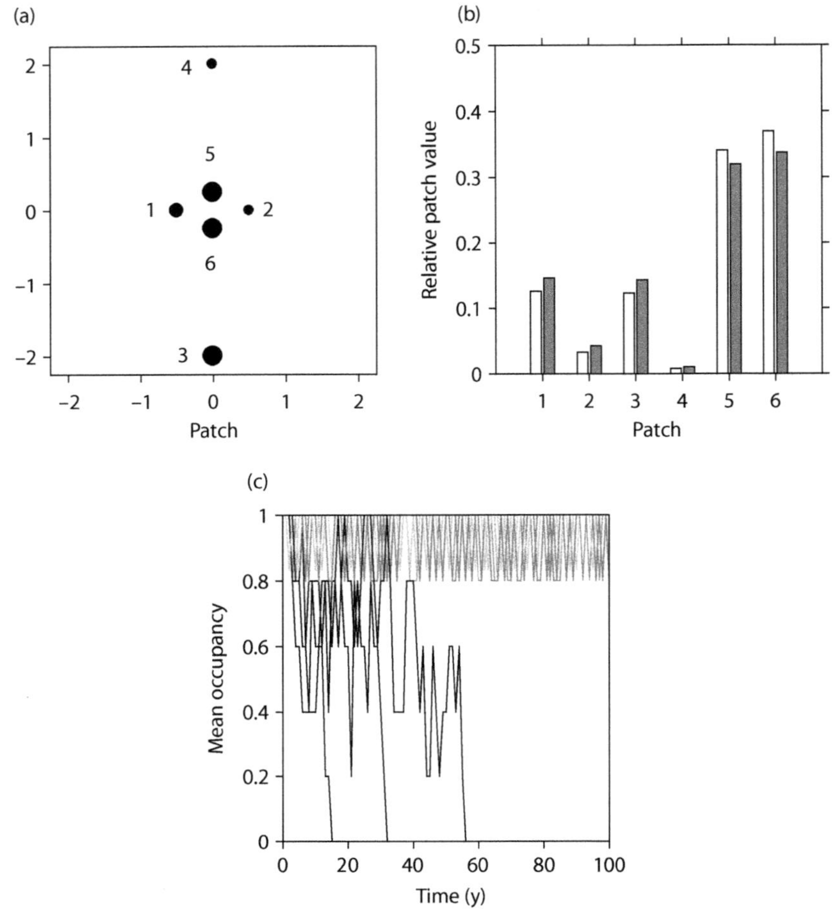

Fig. 9.4 Incidence function model of a metapopulation. (a) Spatial map of the six patches in the metapopulation; circle area corresponds to patch area. (b) Value of each patch, calculated using the perturbation metric (white bars) or metapopulation dynamic (eigenvector) metric (gray bars). (c) Simulated metapopulation dynamics for the entire metapopulation (light gray curves), metapopulation with patch 2 removed (medium gray curves), and with patch 6 removed (black curves). Three stochastic simulations are shown for each scenario. Parameter values are $a = 0.5$, $c = 0.5$, and $a = 0.5$. Adapted from Ovaskainen and Hanski (2003).

or more distant patches. The exact contribution of each patch depends on the values of the model parameters ε , c, and α. Then, in Fig. 9.4c we show simulations of the stochastic dynamics for the overall metapopulation and for the metapopulation with a low-value patch (2) or a high-value patch (6) removed. This example illustrates that removing a high-value patch leads to metapopulation extinction much more quickly than removing a low-value patch; in this example removing the low-value patch did not even reduce average metapopulation occupancy.

The example in Fig. 9.4 also highlights an important distinction between deterministic calculations, such as λ_M, and stochastic simulations. In Fig. 9.4c, all of the simulations with patch 6 removed go extinct, despite the fact that λ_{M-6} is still greater than 1 ($\lambda_{M-6} = 2.7$, while $\lambda_M = 5.4$). Thus the extinctions are not due to a failure to meet the persistence threshold (though having a smaller eigenvalue certainly pushed the metapopulation closer to deterministic extinction), but are instead due to demographic stochasticity—the chance extinction of all of the occupied subpopulations in a single time step. That sort of stochastic event would have been less likely in a metapopulation with more patches. This raises another caveat to the interpretation of either version of V_i: patch value does not account for the current occupancy state of the metapopulation, and if the metapopulation is currently very close to extinction, patch value may have little predictive value because chance events (or habitat destruction) may cause extinction before the long-term patch values become important (Ovaskainen and Hanski, 2003).

The incidence function/SPOM approach to modeling metapopulation dynamics has been applied to a large number of study systems, including the American pika *Ochotona princeps*, a lagomorph; (Smith and Gilpin, 1997), Netherlands tree frogs *Hyla arborea*; (Etienne et al., 2004), and perhaps most famously, a variety of butterfly species (Thomas and Hanski, 1997). While this modeling approach has been very successful, it remains somewhat limited by the use of patch occupancy as a state variable. Because occupancy is binary, these models cannot account for potentially important within-patch dynamics, which we address in Section 9.4.

9.4 Models with internal patch dynamics: structure in space and age

In many cases it may be important to account for within-patch population dynamics and population structure when modeling metapopulation dynamics (Levin and Paine, 1974; Hastings and Wolin, 1989; Gyllenberg et al., 1997). For example, for a metapopulation in which some patches are harvested and some patches are unharvested reserves, harvested patches are likely to have different population abundances (and possibly age structures), and thus different reproductive output. In some cases, the possibility of extinction still dominates the dynamics, and there are possibilities of other stochastic events as well. A full, but rather abstract, framework for these dynamics using the implicit space approach is given by Gyllenberg and Hanski (1992, 1997), with a recent readable overview in Ovaskainen (2017).

However, in many marine systems, a different approach is more appropriate. Reproduction includes a planktonic larval stage, and larvae may disperse broadly on ocean currents (Roughgarden and Iwasa, 1986; Botsford et al., 1994; Cowen and Sponaugle 2009). These metapopulations violate the basic assumption of the incidence approach: that local extinctions are common and dispersal/colonization is uncommon. Subpopulations are seeded with propagules after every reproductive event, and local extinction of patches is exceedingly rare (Kritzer and Sale, 2004; Sale et al., 2006). This actually led some authors to argue that these systems were not actually metapopulations at all (e.g. Grimm et al., 2003). Kritzer and Sale (2004) responded by suggesting that the metapopulation framework was useful in marine systems as long as dispersal was sufficiently limited such that among-patch variability in demographics led to variability in local subpopulation dynamics. Setting that debate aside, the point remains that if extinction is rare, then patch occupancy is not a useful state variable; there are likely to be important differences among patches that are not captured simply by its occupancy state, but will be reflected in the different local population abundances or age structures.

For the dynamics of marine metapopulation with dispersing larvae, the question of population persistence was approached in two different ways. The first approach we will consider took advantage of matrix mathematics by dealing with different areas of space as patches (as in many of the spatial models described in this chapter), without specific reference to their size and spacing (i.e. it was spatially implicit). The second approach was motivated by the spatial heterogeneity created by marine reserves (i.e. areas with no fishing). Because the size of reserves and the spacing between them were important, this approach explicitly included a spatial dimension. We consider each of these approaches in turn.

9.4.1 *Metapopulation persistence: replacement over space*

One approach to determining persistence of marine metapopulations with dispersing larvae begins with a simple metapopulation composed of a number of patches, each containing a semelparous population with non-overlapping generations. We ignore age structure within each patch for this explanation, although it can easily be included. Since we are concerned for now with persistence, we can also ignore density dependence, because we only want to know whether a population at very low abundance will increase away from zero.

The vector of abundance (or density) within each of n subpopulations at time t is the $1 \times n$ vector $\boldsymbol{N}_t$. Each time step the individuals in patch i spawn r_i propagules, which can then disperse to any of the other patches (including back to patch i). The probability of dispersal and survival to patch j is p_{ji}. Then the discrete-time update equation is

$$\boldsymbol{N}_{t+1} = \boldsymbol{C}\,\boldsymbol{N}_t, \tag{9.27}$$

where $n \times n$ matrix $\boldsymbol{C}$ has entries $c_{ij} = r_i p_{ji}$. $\boldsymbol{C}$ is called the connectivity matrix, but it could also be called the metapopulation projection matrix. It is analogous to the Leslie matrix in nonspatial models. To understand the different ways that metapopulations can persist, first imagine that there are only two patches (Fig. 9.5b). If $\boldsymbol{C}$ has the form

$$\boldsymbol{C} = \begin{bmatrix} 1.1 & 0 \\ 0 & 1.1 \end{bmatrix}, \tag{9.28}$$

(as it would if $r_{11} = r_{22} = 1.1$, $p_{11} = p_{22} = 1$, and $p_{12} = p_{21} = 0$) we can tell two things: first, the two patches do not actually exchange propagules, because the off-diagonal elements are 0 (so this is not actually much of a metapopulation). Second, each of the two populations is persistent, because they will increase in abundance each time step (we can tell this because the values on the diagonal are >1). We would say that each of the subpopulations is *self-persistent*, because it can persist without contributions from other subpopulations. As a consequence, the metapopulation itself is also persistent.

Now consider the opposite scenario,

$$\boldsymbol{C} = \begin{bmatrix} 0 & 1.1 \\ 1.1 & 0 \end{bmatrix}, \tag{9.29}$$

(as it would be if $r_{11} = r_{22} = 1.1$, $p_{11} = p_{22} = 0$, and $p_{12} = p_{21} = 1$). In this case, neither patch is self-persistent, because none of the propagules spawned in a patch stay in that patch. Instead, the propagules spawned in patch 1 all disperse to patch 2, where they mature and spawn offspring that all then return to patch 1. The consequence of this is that the individuals in patch 1 are not replaced by their own offspring, but they are replaced by their grandchildren. Thus replacement occurs over multiple generations, and both patches are persistent. We say that this metapopulation is *network-persistent*, because persistence depends entirely on connectivity among patches rather than any one patch being self-persistent.

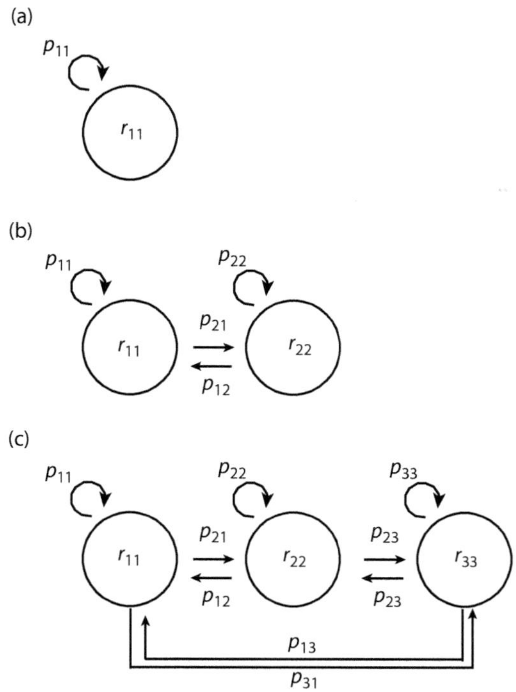

Fig. 9.5 Connectivity diagram for (a) one-patch (not spatial), (b) two-patch, and (c) three-patch metapopulations. The parameters p_{ij} give the fraction of propagules dispersing from patch j to patch i in each model time step, and the parameters r_{ii} give the per capita reproduction in patch i.

To express the persistence conditions more generally, we can say that by definition the metapopulation will be persistent if any subpopulation is self-persistent, that is if either $r_1p_{11} > 1$ or $r_2p_{22} > 1$. If no patch is self-persistent, the metapopulation can be network-persistent if the following condition is met (Hastings and Botsford, 2006a):

$$\frac{(r_1p_{21})\,(r_2p_{12})}{|(1-r_1p_{11})\,(1-r_2p_{22})|} > 1. \tag{9.30}$$

The numerator is the gain in replacement from the loop through both populations, and the denominator is the product of the shortfalls in self-replacement loops in each of the two patches. The interpretation is that replacement via network connectivity (numerator) must exceed the shortfalls in self-replacement (denominator) for the metapopulation to persist. This same idea can be used to determine persistence in metapopulations with more than two patches. For example, for three patches (Fig. 9.5c), the condition for persistence is the sum of the persistence conditions for each closed loop in the metapopulation,

$$\frac{(r_2p_{12})\,(r_3p_{23})\,(r_1p_{31})}{|\,(1-r_1p_{11})\,(1-r_2p_{22})\,(1-r_3p_{33})\,|} + \frac{(r_3p_{31})\,(r_2p_{32})\,(r_2p_{21})}{|\,(r_1p_{11}-1)\,(r_2p_{22}-1)\,(r_3p_{33}-1)\,|} +$$
$$\frac{(r_1p_{21})\,(r_2p_{12})}{|\,(1-r_1p_{11})\,(1-r_2p_{22})\,|} + \frac{(r_1p_{31})\,(r_3p_{13})}{|\,(1-r_1p_{11})\,(1-r_3p_{33})\,|} + \frac{(r_3p_{23})\,(r_2p_{32})}{|\,(1-r_2p_{22})\,(1-r_3p_{33})\,|} > 1. \tag{9.31}$$

The two terms in the first row represent connectivity through the loops $1 \rightarrow 3 \rightarrow 2$ and $1 \rightarrow 2 \rightarrow 3$, respectively, and the terms in the second row represent the loops between each of the three pairs of patches. For a metapopulation of four patches and greater, the conditions follow the same logic, but are more complex to write out (see Hastings and Botsford, 2006a). You may recognize this approach to replacement and persistence as being essentially the same as the one we described for stage-structured populations in Chapter 6 (Section 6.5). Indeed, Hastings and Botsford (2006b) developed that stage-structured calculation after noting that the concept of replacement loops applied equally well to spatial structure and age (or stage) structure. This also means that the alternative approach to calculating replacement proposed by de-Camino-Beck and Lewis (2008) for age/stage models could also apply to metapopulation persistence in space.

Calculating persistence conditions for larger metapopulations is not practical using this approach, but the general results of this analysis are illuminating. First, this analysis tells us the importance of inter-patch connectivity; it must be strong enough to make up for the shortfalls in self-persistence: the less self-persistent component populations are, the more inter-patch connectivity will be required for persistence. This links metapopulation persistence to local subpopulation persistence (i.e. the $r_i p_{ii}$ terms, Fig. 9.5a). Second, these results tell us that persistence depends on closed *loops* of replacement, not on the one-way flows described by the source–sink concept. When discussing sources and sinks (even when calculated using the approach of Figueira and Crowder (2006)), it is important to keep in mind that, by themselves, they do not determine metapopulation persistence. Finally, these conditions also show that it is the fractions of propagules leaving one patch that arrive at another patch (the p_{ii}s and p_{ij}s) and the per capita reproduction within patches (the r_is) that are important. This point is not appreciated by many. For example, the most commonly measured and reported variable in empirical studies of marine larval connectivity is "self recruitment," a ratio which is calculated as the number of locally produced larvae settling in a patch divided by the total number of larvae settling to that patch from all sources. This quantity is easier to measure than p_{ii}, which is a ratio with the same numerator as self-recruitment, but the denominator is the total larval production in that local patch, not total settlement from all patches. Unfortunately p_{ii} is relevant to population persistence but self-recruitment is not (Botsford et al., 2009; Burgess et al., 2014).

9.4.2 *Population persistence in heterogeneous space*

Another approach to examining persistence in marine metapopulations arose against the backdrop of the rising popularity of marine reserves in the 1990s (Lubchenco et al., 2003). The question of how best to design spatial networks of reserves led to the need for a way to understand, quantify, and maintain population persistence in those reserves.

Understanding how this persistence condition for metapopulations related to the persistence of single populations requires a brief diversion to mention how fisheries management viewed population persistence at that time (a topic covered more fully in Chapter 11). For non-spatial populations, resource scientists had taken advantage of the graphical interpretation of the equilibrium condition for age-structured models with density-dependent recruitment, described in Chapter 4 (Eq. 4.24, Fig. 4.11). Using that relationship, they concluded that they could maintain populations at a positive equilibrium by maintaining lifetime reproduction above a certain level (Sissenwine and Shepherd, 1987). For reasons that will be explained in Chapter 11, they assumed this level could be described generally as a fraction of the lifetime egg production expected in an unharvested population. Here we will call this the *critical replacement threshold* (CRT). This approach to persistence of a single population would also describe persistence in the spatial case of a metapopulation distributed along a coastline, as long as population parameters and fishing were the same all along the coast. It would not hold, however, if survival rates varied over space, as they would if a network of marine reserves caused some patches to have high survival (reserves) and others to have low survival (fished areas).

To determine the effects of spatial heterogeneity in survival and reproduction on metapopulation persistence, Botsford et al. (2001) characterized that heterogeneity with a simple one-dimensional spatial model with areas of no fishing (i.e. marine reserves) of specified width and spacing along an arbitrarily long coastline. The subpopulations at each point along the coast were age structured and for the initial analysis, fishing was assumed to remove all reproduction at fished locations (the so-called "scorched earth" assumption, which is the worst-case scenario in terms of population persistence). In the original formulation larval dispersal was represented using a Laplacian function (Eq. 9.11), but here we use a normal dispersal kernel, as in Section 9.1.3 (Eq. 9.10). We refer to the standard deviation of the kernel as L_{diff}, the diffusive length scale of dispersal, and the mean of the kernel (i.e. the average displacement along the coastline) as L_{adv}, the advective length scale. However, we initially assume that $L_{adv} = 0$, so dispersal is equal in both directions alongshore.

The results (Fig. 9.6) show how populations would persist for certain combinations of the width of individual reserves (relative to the length scale of the dispersal function, L_{diff}) and the fraction of the coastline covered in marine reserves (Botsford et al. 2001). The metapopulation will persist for any combination of those variables to the upper right of the solid lines in Fig. 9.6. First consider the line labeled "CRT = 0.35" in Fig. 9.6a. This corresponds to the scenario in which a single, non-spatial population would require lifetime egg production to be 35 percent of the unfished level in order to persist (a commonly assumed "safe" value, Chapter 11). That solid line shows that populations could persist in two different ways. One way can be seen by moving from the origin to the right along the reserve width axis: if a reserve is a small fraction of the coast (essentially isolated from other reserves, i.e. a single reserve), as reserve width increases from zero, a population can persist within the reserve once its width becomes equal to or greater than L_{diff} (this is the average distance a propagule will disperse in either direction (in the absence of advection), and so is often referred to as the "mean dispersal distance"). Because the larval origin and destination are in the same reserve, this is self-persistence, just as in Section 9.4.1. The other way populations can persist can be seen by moving upward along the vertical axis, the fraction of coastline in reserves. Even when reserves are vanishingly small (i.e. near zero on the horizontal axis), populations will persist when the fraction of the coastline covered becomes larger than 0.35 (i.e. the CRT). This is termed "network persistence." As in Section 9.4.1, the general meaning of the modifier "network" is that it is an effect on population behavior that depends on many reserves working in concert, whereas a single one would not cause that effect. That modifier applies here because, while the reserves cover 35 percent of the coastline, they need not individually be larger than L_{diff}, and be able to support persistence on their own.

We can also illustrate the consequences of the two different types of persistence by plotting the equilibrium distribution of larval recruitment along the coastline (Fig. 9.7). As an example, we show a linearized representation of the pattern of rocky habitat and marine reserves along the central California coast (Kaplan et al., 2009; Botsford et al., 2014). We plot the equilibrium recruitment for two hypothetical species that occupy rocky habitat; one has a short dispersal distance ($L_{diff} = 2$ km) and the other has a longer dispersal distance ($L_{diff} = 15$ km). We assume that the CRT = 0.25, and there is no reproduction outside of the reserves (i.e. the scorched earth assumption). At equilibrium, the

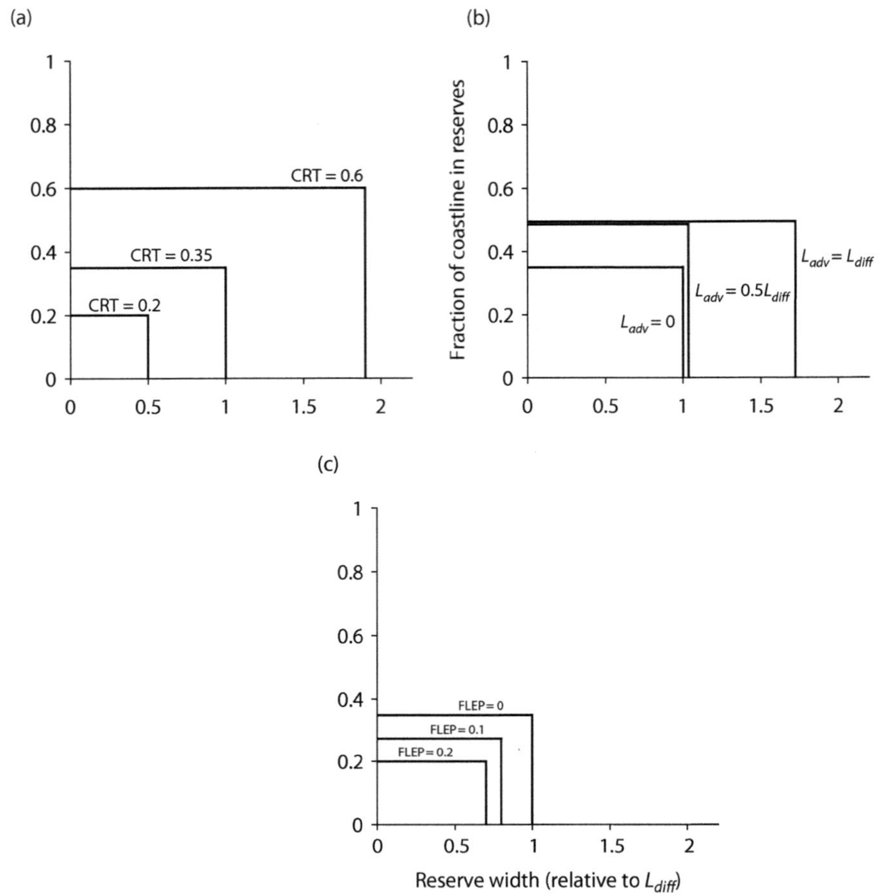

Fig. 9.6 Persistence thresholds for marine reserves of specified width and spacing along a linear coastline. The larval dispersal function has a Gaussian shape with a constant, arbitrary standard deviation (the length scale of diffusion, L_{diff}). Populations persist to the above and to the right of each of the boxes, and are not persistent inside each box. In the null case, panels (a–c), fishing between reserves reduces larval production to zero (FLEP = 0), the non-spatial *critical replacement threshold* CRT = 0.35, and there is no alongshore advection, $L_{adv} = 0$. (a) Persistence thresholds for different values of the non-spatial (CRT). (b) Persistence thresholds for different levels of alongshore advection, L_{adv}, expressed relative to L_{diff}. (c) Persistence thresholds when fishing between reserves reduces FLEP to values greater than zero.

short-dispersal species has nonzero recruitment in virtually every area in reserves that cover good habitat, with very little recruitment outside these areas. Each of those reserves is self-persistent. The long-dispersal species has a broader distribution of recruitment, covering rocky habitat both inside and outside of the reserves. It also has the greatest recruitment near the center of the habitat where reserve areas covering good habitat are spaced close enough together to cover more than 25 percent of the coastline locally. The areas that are more sparsely covered with reserve/habitat combinations have very little recruitment. This illustrates the "network" aspect of persistence in this species; no single reserve is self-persistent for the longer distance disperser.

Returning to Fig. 9.6, the rest of that figure evaluates the consequences of several of the assumptions we made initially. In panel (a), we see that if the fraction of lifetime egg production required for persistence (i.e. the CRT) was actually less than 0.35, fewer

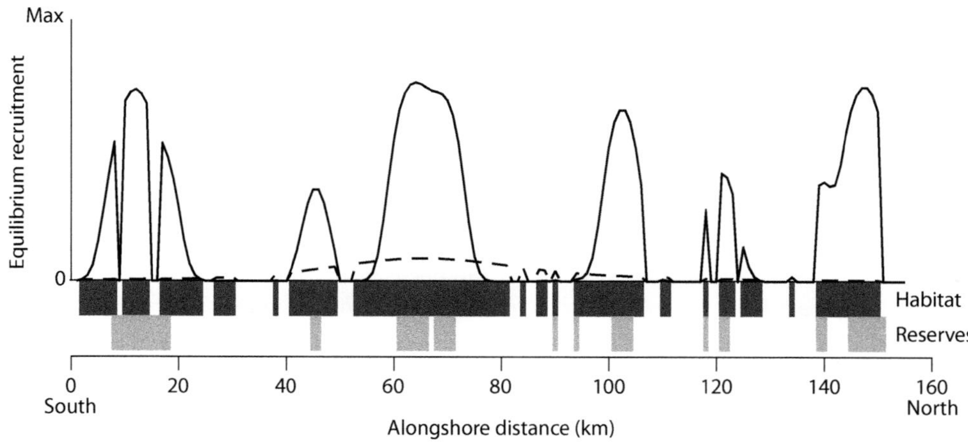

Fig. 9.7 Example MPA configurations along an approximate one-dimensional representation of the central California coastline. The presence of rocky reef habitat suitable for kelp forest fishes and invertebrates is shown in dark bars, and the locations of marine reserves are shown in light gray bars. The solid and dashed curves show relative larval recruitment at equilibrium for a species with relatively short dispersal distance (diffusive length scale of dispersal $L_{diff} = 2$ km, solid line) and a species with longer dispersal distance ($L_{diff} = 15$ km, dashed line). Locations where the equilibrium recruitment is greater than zero will sustain adult populations. The CRTs for both species are 0.25, and the shapes of the dispersal kernels are Gaussian. Fishing reduces FLEP to 0 outside reserves (FLEP = 1 inside reserves). Based on Kaplan et al. (2009).

and smaller reserves would be required (and the opposite is true if CRT > 0.35). This was an important finding because, as we will see in Chapter 10, the CRT is a major source of uncertainty in fishery management, and few realized that the performance of reserves also depended on that uncertain value.

This linking of metapopulation persistence to single population persistence was an important result. Notice that along the vertical axis, the network persistence threshold is simply equal to the CRT. The interpretation is that if there are many reserves, each very small relative to the dispersal distance, then the total amount of larval production along the coastline (relative to the unfished condition) is equal to the proportion of the coastline in reserves, and widespread dispersal spreads those larvae equally among all the reserves. Thus if the fraction of coastline in reserves > CRT, replacement (and persistence) is possible (note that the CRT is the threshold in the limit as the coastline becomes very long relative to individual reserve widths; for shorter, more realistic coastlines bounded by unsuitable habitat, some larvae will be "lost" when they disperse beyond the edge, and as a result the persistence threshold will be slightly greater than the CRT). Along the horizontal axis, enough locally produced larvae must remain in the reserve to exceed the replacement threshold. In this case, the reserve width thresholds correspond approximately to the proportion of the dispersal kernel—centered at one edge of the reserve—that would lie within the reserve boundaries.

Panel (b) in Fig. 9.6 shows that if the larval dispersal kernel is subject to various amounts of alongshore advection, more and larger reserves would be required. Notably, oceanographers have developed a single number, the Peclet number, that summarizes the balance between the two processes. The Peclet number is simply $\text{Pe} = L_{adv}/L_{diff}$ (Largier, 2003); the larger the value, the more difficult it is to achieve persistence (White et al., 2010a). This is analogous to the findings related to the drift paradox in Section 9.2.2.

Finally, panel (c) shows that if the scorched earth assumption were not true (i.e. if fishing did not reduce reproduction in areas outside reserves all the way to zero), fewer and smaller reserves would be required. This was important because, of course, although the scorched earth condition demonstrates the requirements for persistence quite clearly, they are not likely to exist in reality. Moreover, it means that predicting population dynamics inside reserves requires knowing something about fishery management and fishing mortality *outside* reserves, which puts something of a damper on the idea that reserves can be an independent "insurance policy" for poor fishery management (White et al., 2010b). However, this result is good news for population persistence, if one considers how unlikely it would be to actually set aside 35 percent of a coastline in reserves, or to designate a coastal reserve that is wider than the length scale of larval dispersal, which is likely dozens of kilometers for most marine species (though there may be some cases in which it is only on the order of 10 km, and thus more feasible; Hameed et al., 2016).

An interesting—and not previously appreciated—aspect of the curves in Fig. 9.6 is that they are rectangular, rather than rounded, at the elbow. That is, persistence requires the metapopulation to either satisfy the self-persistence or the network-persistence requirements. There is no "synergy" between the two types that would allow a metapopulation to persist if it is pretty close to both persistence thresholds, but does not satisfy either. Reconsidering the persistence condition in Eq. (9.30) suggests why this is the case: persistence is more likely if the p_{ii} terms are greater (shrinking the denominator), but as the p_{ii}s increase, the p_{ji} terms would necessarily shrink (because all of the dispersal probabilities sum to 1), which decreases the numerator. Thus the two types of persistence work in opposition to each other (e.g. if $r_2 p_{22}$ approached 1.0, $r_2 p_{12}$ would go to zero).

The example in Fig. 9.7 illustrates a generality of the results from this model: that they apply not just to reserves, but to the combination of the spatial distribution of the habitat and the reserves. Moreover, one could characterize the differences in adult survival between "reserve" and "no reserve" simply as a difference between "good" and "bad" habitat. With that view, one can see how this understanding of metapopulation persistence over space applies to any landscape (or seascape) with heterogenous habitat, even without reserves.

Another general view that arises from this deterministic model is that, in the presence of spatial heterogeneity, shorter-distance dispersers will be more likely than longer-distance dispersers to persist when good habitat is sparse, up to the point that the fraction of good habitat exceeds the network persistence threshold (e.g. 35 percent), after which species dispersing any distance will persist. This would not necessarily hold in a random environment (Williams and Hastings, 2013).

Another important result from this formulation (though not shown here) is that the shape of the dispersal kernel (e.g. leptokurtic versus normal) does not affect the persistence thresholds, as it does in Section 9.1 for invasion (Lockwood et al., 2002). The reason for this is that invasion speed is very sensitive to the small probabilities at the leading edge of the invasion front, whereas persistence depends primarily on the mean (or standard deviation) dispersal distance. Thus kernels with different shapes but the same mean and standard deviation will produce the same persistence results (unlike in Fig. 9.2).

9.5 Spatial variability across populations

Adding a spatial dimension as we have in this chapter raises the issue of how populations at various locations might covary. There is a substantial literature on the topic of covariability in ecology; however, this being a book about population dynamics, we confine ourselves to variation over space within single species. This topic is often referred to as *synchrony* among populations, but it rarely refers to complete synchrony, rather to a certain degree of synchrony. The level of synchrony is usually empirically examined in a stochastic context, but there are also efforts to explain how it might be produced by deterministic mechanisms.

Spatial variability across populations has become a conservation concern because it is a form of the classical diversity–stability relationship (Elton, 1958; Kinzig et al., 2001). This concern began with the realization that a group of randomly varying ecological assets will display lower aggregate variability and greater stability if they vary in an uncorrelated way, an effect termed the *portfolio effect* (Tilman, 1999), after the parallel effect in financial markets (Markowitz, 1952). (Note that the definition of "stability" here is low variability, one of the many ecological definitions of stability mentioned in Chapter 1.) There are many different kinds of diverse assets in ecology (e.g. genetic diversity, species diversity), but here we are concerned with population diversity, the degree of statistical independence among time series of populations of the same species at different locations (Luck et al., 2003). Conservation managers sometimes focus on this portfolio effect based on the notion that lower aggregate variability of a number of populations leads to better ecosystem performance and greater conservation of resources (Hilborn et al., 2003; Luck et al., 2003; Liebhold et al., 2004a). We will describe causal mechanisms underlying this phenomenon in this chapter, deferring actual management issues to Chapter 10.

The degree of similarity (or synchrony) between populations is most commonly expressed in terms of the Pearson product–moment correlation between abundances. This can be calculated as correlations between sequential differences in abundances ($N_t - N_{t-1}$) or sequential logarithms of abundances ($\ln(N_t) - \ln(N_{t-1})$). These calculations may describe correlation between population growth rates or the influence of a large-scale environmental variable. Another approach is to calculate correlations between measured survival rates (Kilduff et al., 2015). The measure of stability is usually the aggregate coefficient of variation, or CV (i.e. the (aggregate standard deviation)/(aggregate mean)). For a set of n populations, this can be written as

$$\mathrm{CV} = \frac{\sqrt{\sum_{i=1}^{n} \sigma_i^2 + \sum_{i=1}^{n} \sum_{j \neq i}^{n} j = 1 \sigma_i \sigma_j \rho_{ij}}}{\sum_{i=1}^{n} \mu_i}, \qquad (9.32)$$

where μ_i is the mean of the time series for population i, σ_i is the standard deviation of series i, and ρ_{ij} is the correlation coefficient between series i and series j. This can be approximated in a number of different ways, as explained in Chapter 10.

There are many examples of intraspecific covariation over space, primarily in vertebrates and insects (Table 9.1). Three categories of causal mechanisms are commonly proposed to underlie high levels of synchrony: (a) dispersal among subpopulations (Goldwyn and Hastings, 2008), (b) a regional scale (or larger) common environmental driver influencing a number of otherwise independent subpopulations (termed the Moran effect, after Moran, 1953), and (c) a common trophic or other ecological interaction with another species (Liebhold et al., 2004a). Species with a high level of intraspecific covariation are commonly, but not always, cyclic. Another characteristic of interest is that the level of covariation often declines with distance between populations (Liebhold et al., 2004a). The large-scale synchrony of hare and lynx (a coupled predator and prey system; Krebs et al., 2013; 2018) in different regions in Canada, and synchronous fruiting of trees

Table 9.1 Summary of synchronous population dynamics observed in a range of taxa. Adapted from Liebhold et al. (2004a), where all relevant literature citations are provided.

Taxon	Geographical extent of synchrony
Protista: ciliophora	10–500 cm (microcosm)
Fungal plant pathogen	0.5–3 km
Viral human pathogen	1–1000 km
Insect detritivores	5–20 m
Insect herbivores	1–1000 km
Insect predators and parasitoids	10 m–400 km
Fish	10–500 km
Amphibians	0.2–100 km
Birds	5–2000 km
Mammals	10–1000 km
Mollusks	2–30 km

(masting, likely an example of the Moran effect; Liebhold et al., 2004b) are prime examples of this phenomenon.

Models to describe how the synchrony across space could arise typically start with a model that has oscillatory local dynamics due to a negative feedback (e.g. overcompensatory density dependence in discrete time or predator–prey dynamics) and then include coupling between populations as well as stochasticity (e.g. Goldwyn and Hastings, 2011). Although the question of determining the cause of synchrony is often posed as which of the three factors we listed above is responsible, it is more reasonable to determine the relative role of each of the three potential causes.

When synchrony arises due to cyclic population dynamics, the form of the local (in space) oscillation plays a large role in determining the importance of different factors that could lead to synchrony. Oscillations that include multiple time scales, known as relaxation oscillations, are more easily synchronized by dispersal. An example of this kind of oscillation is predator–prey dynamics in which both species are rare and growing slowly for a significant time during each cycle, interspersed with times of relatively rapid population increase or decrease (as opposed to a harmonic oscillator, like a sine wave, that is always increasing or decreasing on the same time scale). Without writing out a detailed model, one could imagine that small numbers of dispersers among patches when the populations are low will have a strong synchronizing effect. In contrast, if the oscillations lack this feature, the time scale over which spatially distinct populations will become synchronous scales as the inverse of the dispersal rate. This is likely to be too slow to lead to synchrony over ecological time scales (Goldwyn and Hastings, 2008; 2011).

9.6 What have we learned in Chapter 9?

In this chapter, as in Chapters 3 and 4 on age structure, Chapter 5 on size structure, and Chapter 6 on stage structure, we have included a new dimension—space—to our descriptions of the distribution of abundance. Spatial population models can include space either explicitly, where distance between populations is actually included, or implicitly, where populations are presumed to be in different locations, without specifying where they are or how large they are; each type of representation can be useful in certain situations.

We began by modeling the spread of a population over space. The first model stated that the rate of change in the spatial distribution with time depended on the rate of change in the movement rate of individuals at each location, a quantity called flux, plus the change produced by population dynamics (birth and death). By assuming that movement was the same as the movement of randomly moving molecules, Fick's Law in chemistry (which may not be the way individuals organisms move over space), we obtained an equation that involved diffusion of individuals over space. Assuming that the diffusion coefficient did not vary over space gave us the reaction–diffusion model, in which the "reaction" is the population dynamics at each location. The consequence of this for an invading population with simple exponential growth and a population with logistic growth are shown in the first and second rows, respectively, of Fig. 9.2. The populations eventually spread outward at a constant rate, termed the asymptotic rate of spread (ARS). These two examples are controlled by a Gaussian shaped dispersal function, whereas many populations have relatively more individuals moving large distances, i.e. the leptokurtic distribution as shown in the third row of Fig. 9.2.

These ideas and models can be used to characterize the spread of populations. The model of the transient northward spread of the sea otter (*Enhydra lutris*) northward along the California coast following protection (Fig. 9.3) allowed explanation of the observed changes in the distribution, in terms of the population dynamics at each location and rates of individual movement that produced the changes.

Next we addressed the question of population persistence in these models: if a population is increasing exponentially at a point, but is constantly being diminished by diffusion away from that location, under what conditions can it avoid declining to zero individuals? This question was addressed by the KISS model developed by Kierstead and Slobodkin (1953) for blooms of harmful phytoplankton (i.e. red tides). They showed that a bloom would occur when the size of the bloom exceeded a function of the diffusivity and the inherent growth rate of the population (r).

We also addressed another persistence problem involving movement over space, the case in which populations occur in a one-dimensional habitat, such as a stream or a coastline, and there is substantial flow of individuals in one direction (e.g. downstream). Such advective movement could have a substantial negative effect on population persistence (i.e. the drift paradox). However, if there is also substantial diffusion, it can counteract the advection, so that under certain conditions populations will persist. Similar advection–diffusion models along coastlines contributed to the understanding of persistence of the barnacle populations studied in Chapter 7.

We then turned to the dynamics of *metapopulations*, collections of individual subpopulations, connected by dispersal among them. This concept was initiated by Levins (1969) who formulated a simple model with implicit spatial structures consisting of individual patches.

That concept led to the development of two kinds of spatially explicit metapopulation models. The first, incidence function models, was most closely related to Levin's concept, resulting in matrix equations that resembled Levin's simple population model. These models consist of a number of separate subpopulations, each of which can colonize empty patches at specified rates, and can also go extinct at certain rates. The stochastic version of this type of model is a stochastic patch occupancy model, or SPOM. We demonstrate the behavior of such a model in Fig. 9.4.

The second kind of spatially explicit metapopulation model consisted of subpopulations with actual internal population dynamics (e.g. logistic or age-structured models), connected by dispersal among populations. Since these were developed primarily in the context of marine metapopulations, the links between subpopulations are most frequently through a dispersing larval stage, although less often fish swimming among subpopulations is included.

One type of metapopulation model with explicit population structure in the subpopulations was based on patches (implicit space), rather than explicit space. It revealed a view of persistence of spatial populations that was based on replacement, and was similar to the way we calculated R_0 for stage-structured models by calculating replacement across several loops in Section 6.5. In that model, by definition, the metapopulation would persist if any patch was self-persistent (i.e. R_0 including the larval stage was greater than 1.0). If no patch was self-persistent, persistence required that for all possible replacement loops in the metapopulation, the product of replacements through the loop, divided by the product of the shortfalls in self-replacement for each patch in the loop, be greater than the product of the shortfalls from self-persistence. Calculation of this through the many loops as in cases with many subpopulations would be very unwieldy, but the general conclusion that spatial persistence depends on replacement in the way indicated in Fig. 9.5 was a valuable increase in the understanding of network persistence. For example, it told us that whether subpopulations are so-called sinks or sources is not a complete description of persistence, as is often presumed.

A spatially explicit version of this model was used to show other aspects of how populations could persist in spatial models. We knew from Chapter 4 that in order to persist, even populations with density-dependent recruitment had to reproduce enough in their lifetime to replace themselves. This threshold, which we called the CRT (critical replacement threshold), and expressed as the proportion of lifetime egg production in an unfished population, can be estimated from the stock–recruitment relationships observed in harvested populations, as we will discuss further in Chapter 10. The answer to the spatial persistence question for metapopulations was that it was just a more complicated version of persistence in non-spatial models. If larval dispersal distances are short enough that a large fraction of larvae settle within their patch of origin, a population could persist in that patch through the usual rules of replacement (self-persistence). However, if larval dispersal distances are too great for self-persistence, there is a second way that populations could persist, and it also involves replacement. If habitat is heterogeneous but the population can have a large fraction (greater than the CRT) of its larvae settle in favorable habitat, that population will persist. This rule was plotted in Fig. 9.6, and an example is shown in Fig. 9.7, in which the habitat heterogeneity is produced by a coastline that has both fished patches and unfished patches (i.e. marine reserves). This was merely an example of how metapopulation persistence works at this point; we will examine more thoroughly how metapopulation theory can inform marine reserve design in Chapter 10.

Finally, we also learned that variability in populations can have varying degrees of synchrony or covariation over space, either because of external, large-scale environmental forcing, dispersal among subpopulation, or because internal population dynamic mechanisms lead to synchronized fluctuations. We will describe the practical consequences of this synchrony in Chapter 10, when we discuss the portfolio effect.

CHAPTER 10

Applications to conservation biology

This chapter on the application of population modeling to conservation biology, and the next, Chapter 11 on the management of marine populations, differ from the earlier chapters in that we deal with the problem of uncertain knowledge. In order to engage in the management of populations one must confront the question of how much can we know about the current state and future behavior of a population. Chapters 2 through 9 describe the behavior of various populations without any concern over how we would know that a population is behaving in a certain way. In these two chapters the application of population dynamics to management involves determining the state of the focal population. We do not go so far as to cover methods of estimation, but rather refer generally to the observations that are made and how the information gleaned is used—in concert with population models – to manage populations. For readers interested in the estimation side of these topics, we recommend Hilborn and Mangel (1997), Caughley and Gunn (1996), Hilborn and Walters (1992), Walters and Martell (2004), Clark (2007), and Ovaskainen et al. (2016).

We begin the chapter with a perusal of relevant parts of the book for information on extinction and population equilibria that we might use to guard against populations being driven to low levels. We conclude that keeping track of two metrics, abundance and replacement, and trying to keep them above a safe level, would guard against population collapse. This extends the conventional approach beyond simply trying to maintain adequate abundance in a way that our conventional calculations of probabilities of extinction are still valuable. We describe a few examples in which modern assessments of population viability have addressed the existing problem of parameter uncertainty, and the effects of the frequency content of environmental variability (concerning an elephant population and a fox population). We then demonstrate the idea of tracking both abundance and replacement in an example that includes several types of error (a salmon population).

Finally, to end this chapter we sought an example of how methods of assessing and controlling population viability differed when we controlled as much as possible for taxon (i.e. different approaches to the same species or genus). Although we recognize that many of the examples already used in this book are from anadromous Pacific salmon, the only example we could find of an organized program of many different analytical groups addressing PVA of the same taxonomic group was focused on salmon. It is probably not a coincidence that salmon have been so well studied in the context of PVA, as the problem of salmon conservation has motivated major conceptual developments in the implementation of the U.S. Endangered Species Act (ESA; Waples, 1991). We describe

Population Dynamics for Conservation. Louis W. Botsford, J. Wilson White, and Alan Hastings, Oxford University Press (2019).
DOI: 10.1093/oso/9780198758365.001.0001

the evolution of that effort over a decade or so, including the conclusion that measures of probability of extinction were not a consistent as managers might prefer.

10.1 Lessons from earlier chapters

We begin this chapter by asking what our understanding of population dynamics from Chapters 3 through 8 suggests about how we could achieve specific practical management and conservation goals. A goal in nearly all applications (except problems involving reduction of pest populations) is to avoid population extinction. Achieving this goal is frequently made more complicated by the fact that there is a second goal that is at cross purposes to the goal of preventing extinction. In fisheries management (Chapter 11), this other goal is clear: harvest as many fish as possible. In other problems there is almost always another "use" of the population or its habitat (otherwise why would there be a concern regarding extinction?), but this use may be less clearly defined and controlled.

For population problems in conservation the situation is typically one in which a population has recently declined, drawing attention to increasing risk. There may be time series of estimated abundance, and possibly some observations of life history parameters. There are typically more data from recent times than from the past. Because of the recent decline in abundance, there is uncertainty regarding how much density dependence has changed the recent values of vital rates (density-dependent survival, reproduction, or growth rates of individuals). Moreover, in conservation decisions there is usually little appetite for waiting for more data to accumulate, a notion codified in U.S. environmental laws under a "*best available science*" rule. The basic idea is that it is not worthwhile to put off making a decision until later based on the current high uncertainty, rather such decisions need to be made immediately on the basis of the best available science (National Research Council, 2004). So our problem is then how to make management decisions given the data up to the current date.

To manage such populations to achieve a specific goal, we need to identify population quantities related to that goal that we can observe and control. These quantities will allow us to track the state of the managed population, and make adjustments when needed. To establish which indicators are useful to monitor for the prevention of extinction, we will examine our expression for the probability of extinction of a linear population near collapse (i.e. with very little density dependence) from Chapter 8. Keep in mind that we do not know the future parameter values for the population, as it continues to decline (and density dependence possibly changes them). In spite of that we can use the basic description of the probability of extinction in Chapter 8 to indicate how we might guard against extinction. Specifically, we look back to Eq. (8.4), which used the first crossing solution to estimate the probability of quasi-extinction between $t = 0$ and $t = T$, p_T, given the initial population abundance N_0 (recall that we defined *quasi-extinction* in Section 2.4.3, as falling below some threshold abundance N_{QE} below which the population is unlikely to recover). From Eq. (8.4) we can see that p_T depends on N_0, and because N_0 appears only in the denominator of a ratio with N_{QE}, we can make the trivial observation that the probability of extinction increases as N_0 declines. This is also seen in Fig. 8.1b, which shows that p_T is higher when N_{QE} is a larger fraction of N_0. From this we can conclude that monitoring abundance and keeping it high, say above a certain pre-agreed level, would guard against collapse. However, there is not a level that would guarantee no extinctions, and even if there were, it would be different for different species and for different populations of the same species at different locations. But at least we know that in general, higher N is safer.

We also know from the definition of μ (the linear growth rate of the ratio $\ln[N_t/N_0]$) in Eq. (8.1a) that the probability of extinction would be less when $\overline{\lambda}$ (the average of the dominant eigenvalues of the population projection matrix) is larger. It would not make sense to consider a population to be at a safe level if $\overline{\lambda}$ were less than 1. However, estimating $\overline{\lambda}$ requires observations of adults and young over (recent) time. This is often difficult because observations might not be started until a population is thought to be in jeopardy. Additionally, keep in mind that the expected value of μ increases when variability in λ is lower; but of course large sample sizes are also necessary to estimate the variability in λ, and as we shall see in this chapter, measurement error associated with estimating $\overline{\lambda}$ also causes problems.

Alternatively, the lifetime reproductive rate, R_0, varies monotonically with λ, and, while estimating it also requires monitoring reproduction over a certain time, it is a more straightforward calculation. Thus, a second way of keeping the probability of extinction low would be by keeping this measure of replacement at low abundance high. This conclusion is reinforced by the condition for equilibrium in an age-structured model with density-dependent recruitment from Chapter 4 (Fig. 4.11). That also suggests that keeping replacement (LEP) above a certain level would tend to keep the equilibrium abundance high, thus further guarding against collapse. Again, the problem with this suggestion also is that we do not know with much certainty what the required threshold levels for long-term persistence should be. As we saw in Chapter 4, we know that technically the threshold level for equilibrium remaining above extinction can be estimated as the inverse of the slope of the adult–recruit curve at low abundance. But that would (typically) be logistically difficult to estimate because it requires reducing the population to low abundance, a level managers, agencies and stake holders usually want to avoid in conservation problems.

In summary, the accumulated knowledge of population dynamics tells us that management can guard against extinction by keeping track of one or both of two quantities: abundance (the state of the system) and replacement (an indicator of the rate of change). For both we know that we should keep them high, but we do not know how high they need to be to be safe. The former is a direct indicator of abundance, and the latter indicates how much the natural age structure has been changed by additional anthropogenic mortality, thus reducing replacement (and, in the long run, abundance, as in the equilibrium condition in Fig. 4.11).

Because these two variables are related (i.e. a certain level of replacement is required to maintain a desired level of abundance, as specified by the equilibrium condition in Chapter 4), it would make sense that they be managed jointly. This is similar to control systems in which both the state of the regulated variable and the value of its derivative are monitored, and form the basis of feedback control. This allows such strategies as using the value of the derivative as a predictor of the future state of the regulated variable.

A common step (and one taken in fisheries management) is to assume that the safe levels of both abundance and replacement are a fraction of the natural level of each. For replacement, it is reasonable to expect that as more examples are examined, we may find that the critical threshold amount of replacement is a similar fraction of the natural, unperturbed level for taxonomically similar species in similar environments (there is some indication of this in fisheries; see Section 11.3). For abundance, the fraction could be set to a level that would not allow extinction to occur before management could realize the danger and respond. Indeed, this is essentially what we have been doing when calculating probabilities of quasi-extinction (Section 8.8). That is, we have been using a calculation of the probability of quasi-extinction to choose a level of abundance that would provide an acceptable risk, given current abundance.

10.2 Probabilities of extinction: the problem of measurement uncertainty

The question of a population's risk of extinction has been the centerpiece of modeling efforts to manage endangered species over the past several decades (e.g. the examples of PVA in Table 8.1). The purpose of estimates of the probability of extinction has been to describe the consequences of the uncertainty in the future random environment. Sometimes the environmental effects in the calculations were simply a random rearrangement of past environments, but more frequently the population models in extinction calculations were driven by white noise.

These probabilities of extinction have played other roles in the management of populations at risk, such as those species listed in the United States under the Endangered Species Act. For example, the probability of extinction has been an important tool for triage, as management agencies choose among various endangered species which has the greatest need of additional resources for greater protection. But mainly they have served as a goal on which to base "delisting" criteria, i.e. quantitatively formulating the question of what is a safe level of abundance that would allow the species to be removed from the endangered or threatened list.

Conservation biologists realized early in the history of conservation biology that uncertainty in unknown parameters would interfere with the calculation of the probability of extinction. Taylor (1995) pointed out that the uncertainty in parameter estimation led to a broader spread of predicted extinction timing, which biased estimates of probability of extinction. Ludwig (1999) made a stronger statement by asking the question in the title of his publication, "Is it meaningful to estimate a probability of extinction?" He included the estimation of the two parameters from a simple lag-1 Ricker model in his PVA to show that the results could end up being "meaningless," because of a substantial number of outcomes in which the probability was either 1.0 or 0.0. As examples, he applied his model to data from three species of birds: Palila (*Loxiodes balleui*), Laysan finch (*Telespyza cantans*) (the time series for these first two were shown in Fig. 8.4), and snow goose (*Anser caerulescens*). Ludwig showed that if one uses the upper and lower 95% confidence intervals on the estimates of the intrinsic growth rate, r, one obtains extinction probabilities close to either 0 or 1. The problem gets worse once one accounts for the possibility of measurement error, which Ludwig represented as φ, a ratio of the magnitude of observation error to the variance in population abundance. His results for those three species are shown in Fig. 10.1a; essentially, as φ increases, the range of possible values of r increases (often spanning zero), and as a result the range of possible extinction probabilities widens rapidly, suggesting that we actually have no predictive capacity whatsoever.

Fieberg and Ellner (2000) replied to Ludwig (1999) by asking "When is it meaningful to estimate an extinction probability?" in their title. They used the simple random geometric model from Lewontin and Cohen (1969; Eq. 2.25) to show what happens when measurement error is accounted for. They defined the annual multiplier (the L_i's in Eq. 2.25) in terms of their logarithms ($r_i = \ln(L_i)$), and assumed they were distributed randomly, with some value of the mean r and a standard deviation of 0.2. In Fig. 10.1b, we show that as the mean value of r declines, the probability of extinction, $p(E)$ increases, as we would expect (we show this calculation using both simulations and the small

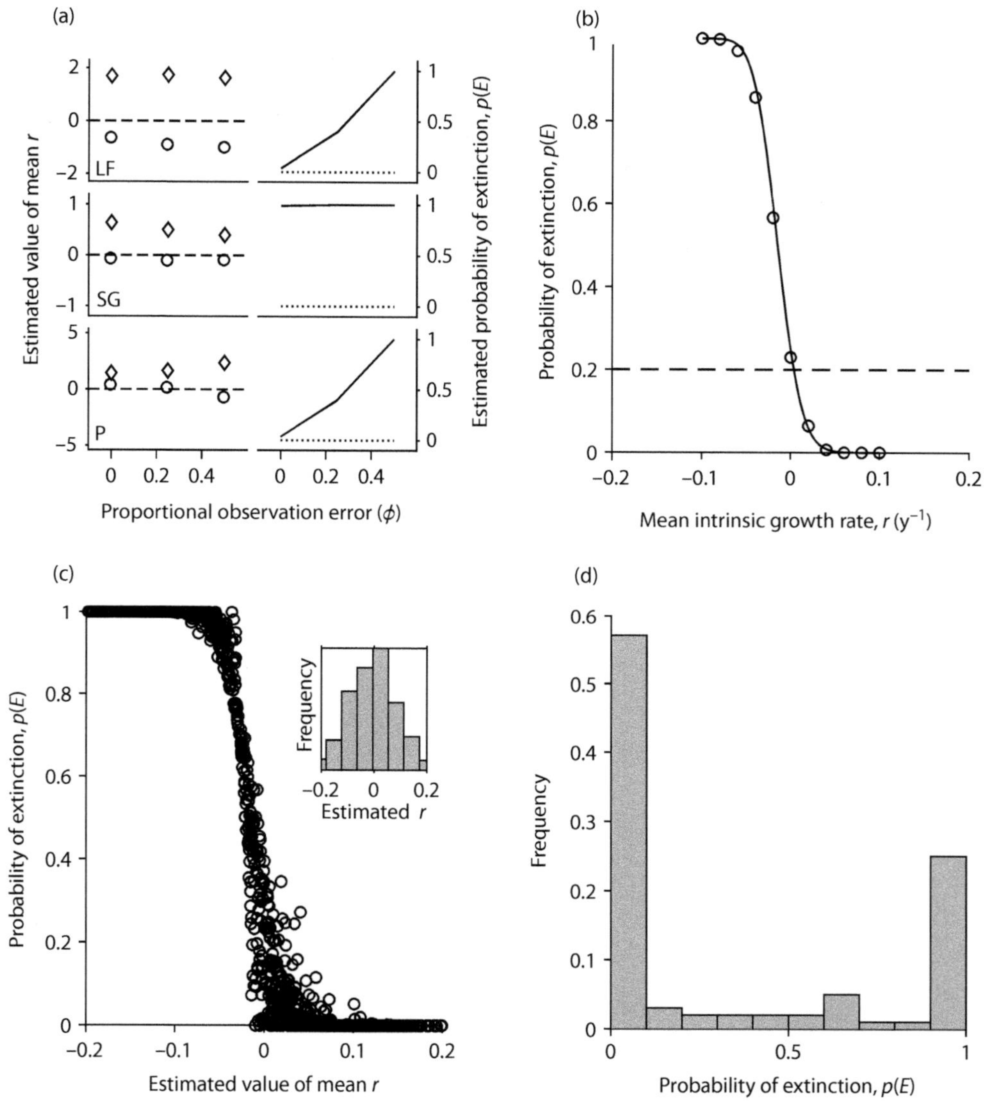

Fig. 10.1 Effects of measurement error on the estimation of extinction probability. (a) Ludwig's (1999) estimates of the upper (diamonds, dotted lines) and lower (circles, solid lines) bounds on *r* (left panels) and *p*(*E*) (right panels) as a function of φ, the proportional observation error in the time series. Dashed line indicates $r = 0$. Results shown for Laysan finch (LF), snow goose (SG), and palila (P). (b) Relationship between mean *r* (randomly varied with standard deviation = 0.2) and probability of extinction in $T = 100$ y, *p*(*E*), both via simulation (circles; $n = 10^3$) and using the SFA (line). (c) Relationship between the estimated mean *r* (based on five values sampled from the simulations in panel (a) and the resulting estimate of *p*(*E*). Inset: the sampled distribution of estimates of *r*. (d) Histogram of the *p*(*E*) estimates from panel (b). Panels (b–d) adapted from Fieberg and Ellner (2000).

fluctuation approximation). With that amount of variability in *r*, $p(E) = 0.2$ even when $r = 0$ (corresponding to $L = 1$). Now we consider the problem we would face if we did not know the mean *r*, and had to estimate it from a series of observations of the population. Those estimates of a mean would have a Student's *t*-distribution, with a standard deviation equal to the estimated standard deviation divided by the square root of the sample size.

The inset in Fig. 10.1c shows an example of such a distribution for an actual value of mean r equal to 0 and a sample size of five observations. Figure 10.1c shows that this process ends up producing a large number of estimated mean r values for which the probability of extinction is either 0.0 (if mean r is estimated to be slightly greater than zero) or 1.0 (if mean r is estimated to be slightly less than zero). The result is a U-shaped distribution of $p(E)$, with peaks at 0.0 and 1.0 (Fig. 10.1d). This error is minimized if a longer time series is available to estimate r (assuming r is constant over time, which of course it may not be), and if the time horizon for estimating $p(E)$, T, is shorter, so that small errors in estimated r do not propagate as far. Based on these results, Fieberg and Ellner concluded that for a meaningful estimate of $p(E)$, T must be $<$10–20 percent of the length of that series used to estimate r.

These publications alerted population biologists to a fundamental limitation in calculating extinction probabilities. However, they did not quantify the level of error in actual PVAs of real species, rather they simply identified the problem using very simple models. For example, Ludwig (1999) used lag-1 Ricker models to represent the dynamics of birds with generation times much greater than one year, so his results may not be good representations of actual population dynamics. We revisit the problems of both measurement uncertainty (are the parameters correct?) and structural uncertainty (is it the right type of model?) in examples using Pacific salmon and the island fox later in this chapter.

10.3 Probabilities of extinction: the importance of environmental spectra

While the calculations of probability of extinction as conventionally calculated would still be part of an abundance-plus-replacement metric, they could be improved by attending to the problem raised in Section 8.5: there is an increasing realization that the probability of extinction depends on the color of the environmental variability, beyond simple reddening or blueing, but in the complex ways that cohort resonance influences calculations of the probabilities of extinction. Probabilities of extinction are as much a statement about the expected future environment as they are about the population dynamic responses to it. In the future we will have to pay as much attention to that dependence as we do to parameter uncertainty (e.g. Schmidt et al. 2018).

An early analysis of the probabilities of extinction of African elephant (*Loxodonta africana*) populations under various conditions incorporated assessment of the effects of environmental frequency spectra (Armbruster and Lande, 1993). This model also included the effects of density-dependent recruitment, rather than assuming linear behavior at low levels; the age at first reproduction and the calving interval (the average number of years between reproduction for females) increased with the number of adult elephants per square kilometer, representing resource competition. This allowed the consideration of how extinction probability depended on the size of a protected reserve (assuming habitat area = reserve area). The model was age structured with an elephant lifetime composed of twelve five-year increments, but that does not incur the problems associated with stage structure outlined in Section 6.3, because five-year time steps were used in the calculations. Reproduction depended on the age of first reproduction, the calving interval, and the presence of at least one adult male.

Armbruster and Lande modeled the effects of the time scale of environmental variability using different survivorship schedules depending on whether a given five-year time

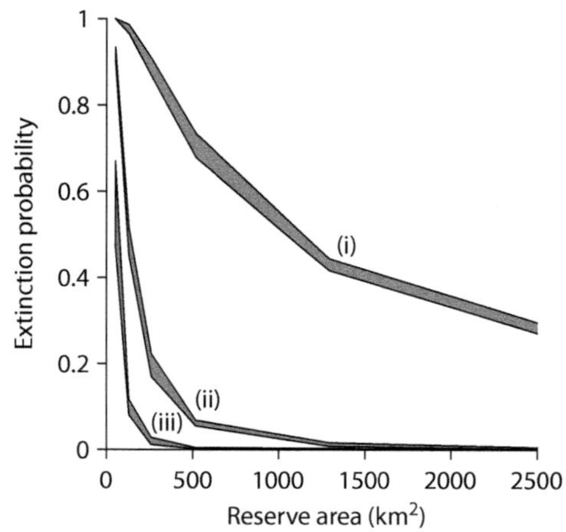

Fig. 10.2 Probability of extinction over T years as a function of reserve area for elephants. Shaded areas represent 95 percent confidence intervals for $p(E)$ for three different possible future mixtures of drought frequencies, expressed as the mean periods between droughts: (i) 10/25/125, (ii) 10/50/200, (iii) 10/100/500 drought patterns. Based on data in Armbruster and Lande (1993).

interval experienced a 10-year, a 50-year, or a 250-year drought, with lower survivals for the less frequent, more severe droughts. Given possible uncertainty in estimates of the true frequency of less frequent droughts, they also examined a higher-frequency regime, with mild, medium, and severe drought conditions occurring every 10, 25, and 125 years, and a lower-frequency regime in which they occurred every 10, 100, and 500 years.

Their results indicated probabilities of extinction were lower with larger reserves, and much greater for more frequent droughts (Fig. 10.2). The latter is seemingly at odds with the results of an increased frequency of warm-water ENSO signals on Brandt's cormorant populations that we described in Section 8.5 (Schmidt et al., 2018). The difference between these lies in how speeding up and slowing down the environmental variable are defined. In Schmidt et al. (2018), "speeding up" the ENSO signal increased the frequency of the signal, which made both ENSO durations and non-ENSO durations shorter, without increasing the total duration of warm water periods. When periods of poor conditions are described in terms of a return time, as in descriptions of weather (e.g. a 100-year storm), decreasing the return time increases the total amount of time in poor conditions confounding those two effects, as in the elephant example.

10.4 Replacement as an extinction metric

As noted in Section 10.1, specifying abundance without monitoring replacement to make sure it is adequate leaves us vulnerable to further decline. In existing practices in conservation biology, calculated probabilities of extinction answer the question we posed in Section 10.1 regarding how high the abundance threshold should be to maintain an acceptable level of risk. However, additional information on how anthropogenic activities have changed population structure, hence replacement, and consequently the ultimate equilibrium level, is also quite informative. While the current practice of calculating probabilities of extinction, of course, contains information regarding population growth rate, it does not directly address the question of how anthropogenic effects on the focal population have reduced replacement. In other words, a PVA is likely based on estimates of the population growth rate over time, but does not necessarily include information on what the present-day replacement is, and thus how the population growth rate will change in the near future. Information relative to this question may have arisen in the course of studying the population to elucidate the causes of the observed decline, but there has not been a focus on replacement rate as a general goal, or control variable.

The inclusion of replacement as a monitored metric in management of population persistence is a new idea in conservation biology, and does not have the history that the use of abundance alone has. However, we can say that it seems possible and has promise. It would involve tracking changes in age structure of a population, and there is a long-standing interest in the age distributions of wildlife populations, e.g. (Caughley, 1977; Hutchinson, 1978). There are various methods for aging wildlife, more so for mammals than birds. Among plants, trees stand out in terms of their age-ability. The ability to age individuals becomes less necessary as the technology for tagging individuals continues to increase, with an accompanying increase in the ability to analyze tagging data.

An increase in efforts to observe age structures would address the problems with stage-structured models we noted in Chapters 6 and 8. The inability to age individuals was the major driver in the rise in popularity of stage-structured models. The example below in Section 10.6 illustrates the value of tag–recapture methods in establishing better age-structured models.

10.5 An example with abundance, replacement, and measurement error

The original PVA for the first Pacific salmon listed under the U.S Endangered Species Act (Sacramento River winter-run chinook salmon, *Oncorhynchus tschawytsha*; the time series of decline of this stock is shown in Fig. 2.4) illustrates the roles of several types of error and of including a measure of replacement in a two-dimensional metric of persistence (Botsford and Brittnacher, 1998). This salmon population was modeled using a linear, discrete-time renewal equation with recruitment each year being the weighted sum of recruitments two, three, and four years earlier, all multiplied by a random variable termed a cohort replacement rate (CRR), that essentially represented the lifetime reproductive output of a cohort (the weightings in the sum reflected the proportion of a recruit cohort that would return to spawn at age 2, 3 (the most common age), or 4). This was a realistic portrayal of the effect of the random environment, since the dominant effect of the random environment in salmon is in the conditions that a cohort of salmon encounter when they first enter the ocean (Pearcy, 1992). Hence variation in cohort survival and reproductive output can be described by a single variable representing the effects in the year of ocean entry. The criteria for delisting this stock (i.e. removing it from the list of endangered species under the Endangered Species Act) included a minimum value of the CRR (actually the mean of the natural logarithm of the CRR, reflecting the importance of geometric means of rates, from Chapter 2), in addition to the conventional condition of a specified level of abundance. The combined effects of these show (trivially) how, for any specific desired level of probability of extinction, having a greater replacement rate allows one to maintain the same $p(E)$ with a lower overall initial abundance (Fig. 10.3a). In this example, initial abundance is actually a variable that can be manipulated somewhat by increasing hatchery production (hatcheries are facilities in which salmon are artificially spawned and juveniles are reared until being released back into rivers). This is why it makes sense to plot it as an independent variable in Fig. 10.3, as opposed to just plotting probability of extinction versus abundance, as in most endangered species.

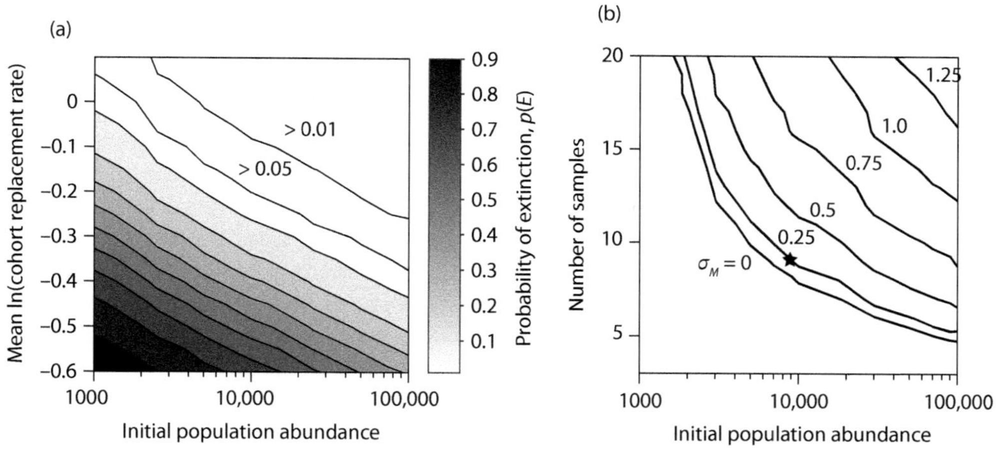

Fig. 10.3 Population viability of Sacramento River winter-run chinook salmon. (a) Effect of cohort replacement rate and initial population size on the probability of quasi-extinction (abundance < 100) over 50 years. (b) Number of independent estimates of cohort replacement rate (i.e. years worth of data) required to estimate a probability of extinction ≥ 0.1, as a function of the initial population abundance and the standard deviation of measurement error, σ_M. Based on model results in Botsford and Brittnacher (1998).

A counting station at a dam below the spawning grounds of the salmon allowed estimation of the spawning run abundance each year with a specified standard error. If at least five consecutive years worth of data are available, it is possible to obtain a point estimate of the CRR (based on the known contribution of spawning two, three, and four years in the past to the current spawning run). The delisting criterion based on CRR was chosen to be a minimum level of the average CRR estimated over n annual estimates. The effect of computing a mean CRR from n noisy samples was included in the model, and used to calculate the consequent variability in the estimates of the probability of extinction. In Fig. 10.3b we show an example of the effects of this variability on the minimum initial abundance needed to have $p(E) > 0.1$ over 50 years, assuming the true mean CRR $= 1$ (so $\ln(\text{CRR}) = 0$), for a range of values of n (realize that n is the number of estimates of CRR, so the total number of years worth of data would be $n + 4$, because five years worth of data are required to get the first estimate) and of the measurement error in estimating CRR, expressed as a standard deviation σ_M. This plot adds to the information in Fig. 10.3a (which assumes we have perfect information about both abundance and replacement) by showing how many samples of spawning abundance would be required in order to be confident that one would maintain the desired probability of extinction. In this example, measurement error σ_M was thought to be 0.25, so Botsford and Brittnacher (1998) proposed that delisting could occur if $n = 9$ estimates of CRR were obtained (and had a geometric mean of 0.0) after the population was restored to an abundance of at least 10,000, i.e. the point indicated by a star on Fig. 10.3b.

10.6 Comparative studies: Pacific salmon

Comparing PVAs of multiple Pacific salmon populations listed under the U.S. Endangered Species Act provides insight into the range of approaches to PVA that have been employed over a variety of ecological conditions and groups of analysts. Specifically, we focus on two meta-analyses, separated by a decade, allowing a view of how approaches to PVA have evolved (Busch et al., 2013; McClure et al., 2003). Of course, because these PVAs dealt with anadromous salmon, they differ from "typical" PVAs in three ways. First, reliable counts of all spawners were often made for multiple years, so long that high-quality data sets are available. Second, some populations have been supplemented by juveniles raised in hatcheries. These hatchery fish supplement the fishery, and some may contribute some reproduction, but do not "count" as part of the wild population that is being assessed. Finally, these species are semelparous, with a few being obligate semelparous (all spawning at the same age). The ages at which most spawning occurs range from 3 y to 5 y, depending on species and population, with nearly all spawning occurring within one year of the dominant age (Hill et al., 2002). This implies that the populations would be described by imprimitive, or close to being imprimitive, projection matrices (see Section 4.2.1; note that in this discussion we are leaving out steelhead, *Oncorhyncus mykiss*, which are sometimes iteroparous).

McClure et al.'s (2003) initial analysis of 152 listed salmon stocks and 24 nonlisted stocks in the Columbia River drainage (these were stocks of chinook salmon, steelhead, coho salmon *O. kisutch*, and chum salmon *O. keta*) essentially followed the diffusion approach that Dennis et al. (1991) had taken (see Section 8.6). They used time series of population abundance to estimate λ and the rates of change in the expected mean and variance of $\ln(N_t/N_0)$ (μ and σ^2, respectively; see Eq. 8.2) in order to estimate probabilities of extinction. They made several modifications to that approach in order to accommodate the semelparous salmon life history. First, they replaced the total population counts used by Dennis et al. (1991) with a four-year running sum, to reduce the variability in the overall abundance trend introduced by interannual variability in the abundance of individual cohorts. Using a running sum presumably forms a variable that approximates the total (incompletely counted) abundance. However, this also removes the information on the distribution of annual recruitment variability (i.e. the variability sought in Botsford and Brittnacher (1998)). The estimate of μ based on this running sum of abundance contains the same problem as mentioned in Chapter 8 regarding the Dennis et al. (1991), method, that calculating the mean of the product of a number of ratios of sequential observations in a time series simply results in the ratio of the first observation to the last observation, essentially excluding information from all of the intervening time points. With regard to σ^2, McClure et al.'s estimation of the variance term, σ^2, differed from that in Dennis et al. (1991), and instead used the "slope" method derived by Holmes (2001).

The second problem to be dealt with was the potential reproduction by hatchery fish. The usual (ideal) practice is that all adults of hatchery origin that return to the spawning river are culled and their gametes are used to spawn the next generation of hatchery fish, i.e. they do not contribute to natural recruitment (hatchery fish are visibly marked as juveniles, so it is possible to remove them at a weir or fish ladder as they make their way back upstream). Some of those fish almost certainly do escape and arrive at the spawning grounds, though they typically have lower reproductive success and contribute less to the next generation.

To deal with the uncertainty associated with hatchery reproduction, McClure et al. estimated the values of λ in two ways, first assuming hatchery fish were not reproducing at all, and then assuming that they reproduced at the same rate as wild fish. Thus they obtained two estimates representing the best-case and worst-case scenarios that bracketed the likely true value. The best-case estimates of λ for most of the listed stocks was <1.0, and in the worst case most were <0.9 or even <0.8 (for comparison, the average for unlisted stocks was 1.02). Thus most of these stocks were headed towards deterministic extinction. The possible effects of environmental variability were tested by comparing values of λ from decadal periods with known poor ocean conditions to periods favorable to salmon survival. They then calculated the increase in λ that could be achieved by each of two possible management actions: reducing harvest to zero and reducing mortality due to passage through hydropower dams to zero. To do this they expressed the effects of reduced mortality from either source in terms of its expected increase in replacement, R_0, then used the definition of generation time as the time required for a population growing geometrically at rate λ to increase by an amount R_0 in order to translate changes in R_0 to changes in λ (cf. definition T_3 in Box 4.2, Eq. 6.21).

For some populations in the data set, time series were of actual abundance (rather than an index of abundance), making it possible to calculate actual probabilities of extinction within $T = 50$ years using the first crossing solution in Section 8.2. These estimates tended to follow bimodal distributions, with $p(E)$ estimated to be either 0 or near 1 for each stock. This level of divergence should not be surprising given the possible sources of error in these calculations that we discussed in Section 10.2. It is possible that more consistent (and perhaps more accurate) estimates would be possible over a shorter time horizon, as Fieberg and Ellner (2000) suggested.

Our description of McClure et al.'s (2003) results gives a sense of the state of PVA science at the turn of the last century: the basic SFA framework was still being used (with some improvements over the summary we gave in Chapter 8), despite the possible sources of bias or uncertainty. A decade later, Busch et al. (2013) summarized the progress that had been made since then in developing a wide variety of approaches to describing the risk status of Pacific salmon in the contiguous United States. They selected 34 populations of chinook and coho salmon with appropriate data and applied the different approaches to them, so that the various groups doing these analyses (recovery teams) could be viewed as quasi-replicate groups working on a common viability problem. We do not describe the various approaches here (see the appendix of Busch et al., 2013), which primarily vary in whether and how the methods included density dependence, and the way different recovery teams estimated population parameters and included (or not) possible environmental covariates (Busch et al. termed this "local tailoring"). Rather, we focus on the results of the comparison, and mention a couple of examples of interest.

From the point of view of managing extinction risk, managers would like to think that there was consistency among the different methods. That is, whatever adjustments were required by recovery teams to tailor methods to local requirements would not dominate outcomes. Indeed, a possible positive outcome would be that multiple models could be applied in an ensemble approach and the ensemble average would provide a more accurate estimate of the true state (much like ensemble approaches to atmospheric modeling; Anderson, 1996). However, it turned out that this was not the case. The methods produced very different distributions of extinction, so that individual populations had a broad range of probabilities of extinction from the different methods, and particular methods seemed to consistently estimate either higher or lower probabilities of extinction, regardless of the data set used. We illustrate this in Fig. 10.4, where the gray shading indicates the estimate of $p(E)$ for each population using each method. As one scans from more pessimistic models on the left, to models that consistently estimate low $p(E)$ on the right, there is no apparent correlation between columns. This suggests it may be worth more research on the overall reliability of estimated probabilities of extinction from PVAs.

One of the approaches included in Busch et al.'s analysis provides a good example of presenting the consequences for exinction risk of both measures of abundance and replacement for a population, and providing a probabilistic description of the current

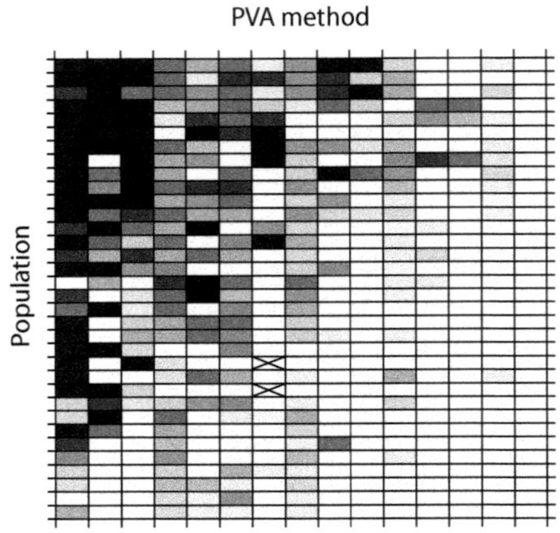

Fig. 10.4 Variability in estimated probability of extinction in different PVA methods. Shading indicates the probability of extinction (white is 0.0, black is 1.0) estimated by each of 15 different PVA approaches applied to data from each of 34 salmon populations. Methods are ordered from pessimistic to optimistic from left to right. Redrawn from Busch et al. (2013), where details on the PVA methods and populations can be found.

status of the population in that context. McElhany et al. (2006) focused on salmonids spawning in the Willamette River and the Columbia River (chinook, coho, chum, and steelhead). They assembled a number of viability criteria, each of which is associated specific metrics and threshold levels of those criteria for particular levels of extinction risk. The first two metrics were population abundance and productivity (lifetime reproduction), as we discussed in Section 8.1. McElhany et al. portrayed the combined effects of abundance and productivity on probability of extinction in what they called viability curves (Fig. 10.5). These curves are similar to the iso-probability curves in Fig. 10.3a. The difference is that instead of depending only on reproductive age structure (as in Botsford and Brittnacher's (1998) analyses), these curves account for a density-dependent stock–recruit relationship, and they define abundance as the equilibrium abundance, rather than the initial abundance. The "replacement" measure in this case is the number of recruits produced per spawner at low density (essentially the initial slope of the stock–recruit curve) so it can be much greater than 1 without implying a long-term geometric increase (because per capita productivity slows as density increases).

With the quasi-parabolic viability curves in hand, what remained to be done to express population status was to determine the current levels of abundance and productivity of the population. McElhany et al. estimated these from available time series, and added in the effects of both estimation error (due to limited samples in the time series), and measurement error (due to errors in individual annual counts). These estimates are indicated on Fig. 10.5 as a contour of probability density, indicating the confidence level associated with the point estimate.

In addition to abundance and replacement, McElhany et al. (2006) accounted for two other facets of viability, which we will not cover here: diversity and spatial structure. With abundance and productivities that is a four-dimensional description of viability. Their account of the effects of spatial distribution is similar to our description of spatial population diversity in Section 10.8, but we do not address the issues of genetic diversity that they used in their fourth criterion.

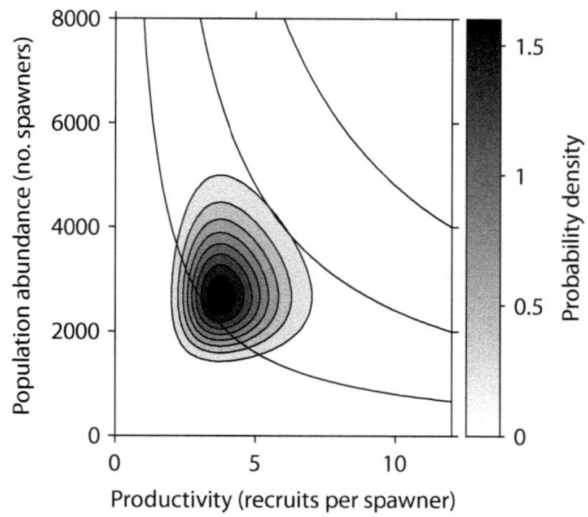

Fig. 10.5 Hypothetical population status evaluation. Black contour lines indicate probabilities of extinction as a function of abundance and productivity (a measure of replacement at low density). Shading indicates the probability density of estimates of current stock status. Adapted from McEhany et al. (2006).

10.7 Addressing exogenous variability: drivers and errors

There has been a growing awareness of how uncertainty in life history parameters (and the true model structure) degrades the quality of PVA predictions, as we have seen in previous sections. Fortunately, we have also seen a steadily improving ability to estimate parameter values via mark and recapture methods. For example, Bakker et al. (2009) conducted PVAs for the island fox (*Urocyon littoralis*), a species endemic to the Channel Islands off the coast of California. There are six subspecies, each occupying one of six islands. In 2004 four of these subspecies were listed as endangered under the U.S. ESA, after declining to extremely low levels due to predation by golden eagles (*Aquila chrysaetos*) on three islands (the birds were attracted to the islands by human-introduced pigs and deer) and disease on another.

Bakker et al. formulated a sex-specific version of the simplest density-dependent age-structured model (Section 7.2), using four categories: males and females of age 1 (juveniles) and age >1 (adults). Live trapping and tagging over various multi-year periods on various islands allowed for aging and estimation of both age-specific survival (age 1 and >1) and annual reproduction rates. These life history parameters were related to three "ecological drivers": fox density, predation (eagle numbers), and annual rainfall. The rainfall effect was attributed to ENSO as a categorical variable, distinguishing two years worth of high, ENSO-driven rainfall in their data set from other normal years. For all of the mark–recapture time series the estimation program MARK (White and Burnham 1999) was used to separate the temporal variance in survival and reproduction into sampling variance and process variance.

Analyses of the resulting simple age-structured model proceeded by first using the deterministic version of the 4×4 projection matrix (two ages + two sexes) to determine λ and various sensitivities and elasticities, as though it were a linear model (i.e. without density dependence), though they accounted for density effects by making the calculations at three separate effective densities. The behavior of the *deterministic skeleton* of this model was determined by the hump-shaped dependence of both juvenile and adult survival on adult density (i.e. Fig. 10.6a), which gets lower as eagle abundance increases. Because of

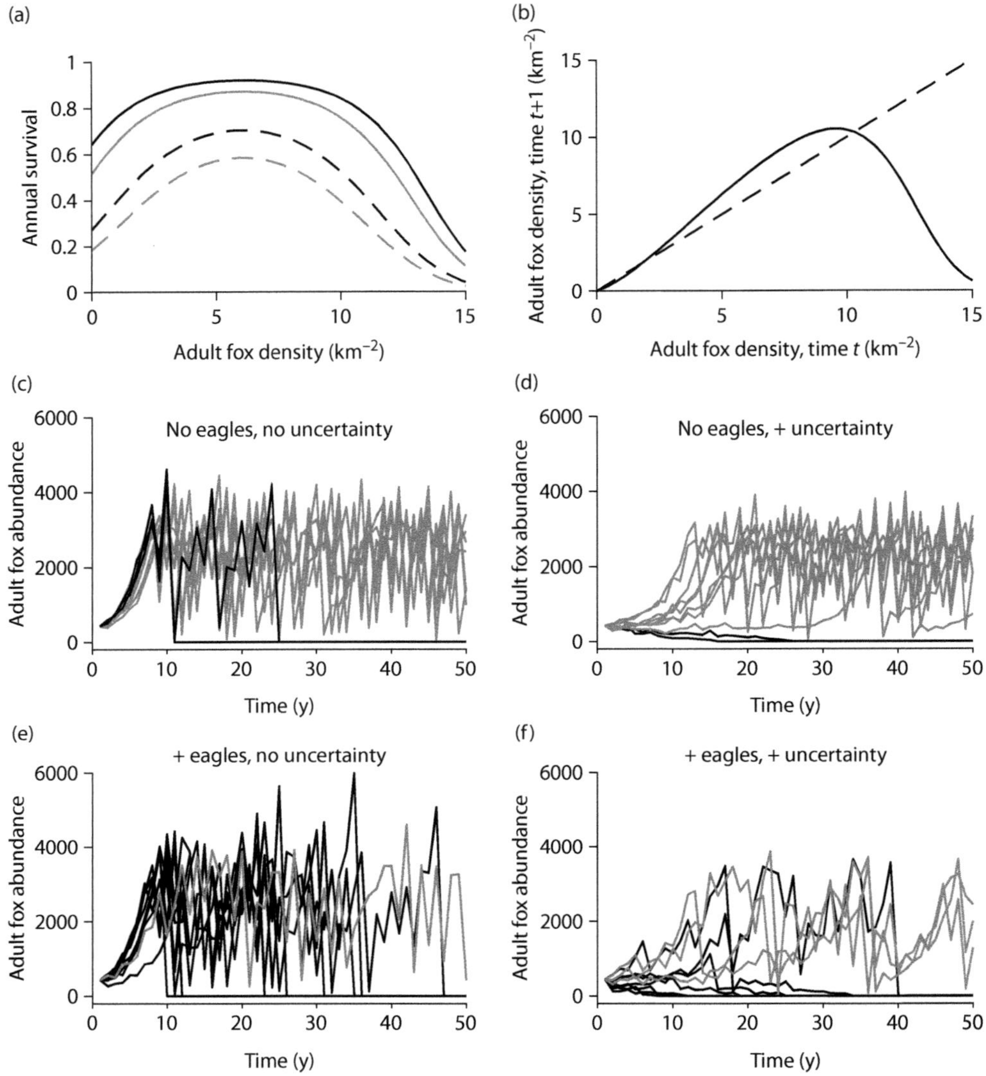

Fig. 10.6 PVA for the island fox population on Santa Cruz Island. (a) Survival of adults (black) and juveniles (gray) as a function of adult abundance, in the absence of eagles (solid curves) or with additional mortality equivalent to that caused by 13 eagles (dashed curves). (b) 'Stock–recruit' relationship in the absence of eagles (solid curve), the dashed curve is the 1:1 replacement line. (c–f) Example simulations of the model without (c, d) or with (e, f) mortality equivalent to six eagles and without (c, e) or with (d, f) including measurement error in model parameters. Gray curves are simulated populations that did not fall below the quasi-extinction threshold (30 foxes), black curves are those that did. Redrawn using model parameters in Bakker et al. (2009).

this shape, a plot of adult density at time $t+1$ versus adult density at time t has the hump-shaped appearance of an over-compensatory Ricker stock recruitment function (Fig. 10.6b, cf. Fig. 2.8). Thus we can already predict, based on the form of density dependence and the two-stage structure of the model, that it will tend to exhibit period-2 oscillations. The tendency to oscillate would be less if adult survival were higher, effectively broadening the reproductive age distribution (as we saw in Section 7.2).

Bakker et al. then analyzed simulated model projections, including both eagle predation (or not) and parameter uncertainty (or not), or both factors (the simulations also included environmental variability (lower survival in ENSO years, which had a specified probability of occurring in any given year) and demographic stochasticity in offspring sex ratio). This served to convey the range of possible outcomes, and the degree to which management actions (e.g. removing eagles or their primary prey) could outweigh observational uncertainty in terms of estimating the probability of extinction. As one might expect based on the lessons of Chapter 8, adding uncertainty in the population growth rate increased the probability of extinction (compare Figs. 10.6c and 10.6d), as did increasing the mortality rate due to eagles (compare Figs. 10.6c and 10.6e). Interestingly, adding both eagle mortality and observational uncertainty reduced the extinction rate (relative to eagle predation without uncertainty; Fig. 10.6e,f). This is because the added variation in the population growth rate, which had a mean <1 due to the eagles, allowed some populations to have a positive growth rate and persist.

The island fox example underscores some lessons about PVA regarding: (1) including density dependence, (2) including environmental drivers, and (3) including uncertainty. With regard to density dependence, we have noted that for age-structured models with density-dependent recruitment, the equilibrium ultimately depends on the (presumably greater) offspring survival at low densities, but that would be less important in a population that moved frequently between low values and a high equilibrium, as in the Ricker-like cycles of this population. With regard to environmental drivers, we now understand better how the autocorrelation structure (hence the frequency content) of the environmental signal actually influences probability of extinction (see Section 8.5). Moreover, it is remarkable that they found no autocorrelation in their rainfall time series and thus chose to model ENSO effects as a white-noise process, because the ENSO spectrum is definitely not white (Schmidt et al. 2018). On the other hand, the relationship between rainfall along the U.S. west coast and ENSO is complex, and varies with latitude. There is clearly value in including different kinds of uncertainty in extinction calculations, but it is possible that more could be done than just lumping observation uncertainty in with the environmental variability. For example, managers might be interested in knowing how extinction predictions might be changed by reducing the uncertainty in a key variable with additional data collection, or as more years worth of monitoring data become available. Finally, ecologists are increasingly facing the problem of attempting to predict the effects of environmental variability, when both the mean and the variance of that variability may be shifting due to climate change. Thus our estimates of "environmental variability" could be biased because the data from which we are estimating variance is changing monotonically rather than being *stationary*. This is an ongoing research challenge (Wolkovich et al., 2014).

10.8 Population diversity

In Chapter 9 we introduced the idea that variability in abundance in a number of populations of the same species would have greater or lesser variance in the aggregate total, depending on the levels of correlation among time series of their abundance. The general effect of having lower aggregate variability with increasing statistical independence (lower correlation) among time series was referred to as the *portfolio effect*, a term borrowed from investment theory (Markowitz, 1952; Tilman, 1999). This is therefore another ecological example of diversity leading to greater stability (Loreau, 2010; here greater stability refers to lower variability of a quantity, relative to its mean, i.e. the coefficient of variability).

Resource managers are interested in the benefits of managing to maintain a high level of this kind of population diversity. For example, Hilborn et al. (2003) pointed out the historical advantage of having maintained so many subpopulations of sockeye salmon in the north Pacific, an effect further quantified by Schindler et al. (2010). Other examples include the contribution of variability in small-scale habitat to persistence of a rare plant in an alpine ecosystem (Abbott et al., 2017), the maintenance of high algal grazing pressure on coral reefs due to asynchronous fluctuations in herbivorous reef fish populations (Thibaut et al., 2012), and increased persistence of a semelparous plant due to intra-population diversity in the age at reproduction (Acker et al., 2013).

As in Section 10.1, here again we are interested in a metric that we can use to track the state variable of interest—in this case diversity – in order to manage to obtain a high level of stability. Looking back to Chapter 9 for suggestions, we note that Eq. (9.32) expresses the desired coefficient of variation (CV) in terms of operative correlations between populations (the ρ_{ij} terms). We reprint this equation here:

$$\mathrm{CV} = \frac{\sqrt{\sum_{i=1}^{n} \sigma_i^2 + \sum_{i=1}^{n} \sum_{j \neq i}^{n} j = 1 \sigma_i \sigma_j \rho_{ij}}}{\sum_{i=1}^{n} \mu_i}. \tag{10.1}$$

This expression contains both the effects of variability (the σ_i terms) and covariability (the ρ_i terms). Here we are interested in the effects of the latter. The portfolio effect depends on multiple independently fluctuating populations reducing overall variability; if populations have a correlated response to environmental fluctuations (i.e. positive ρ_i terms), this will increase the overall CV.

Nonetheless, a number of different approaches have been used to represent the level of correlation among populations. Many authors compare the aggregate CV to the average of the CVs of individual subpopulations, following the lead of Doak et al. (1998) (e.g. Schindler, et al. 2010; Anderson et al., 2013). A difficulty with that approach is the latter quantity, which does not appear in the expression for the aggregate CV (Eq. 10.1) and thus does not precisely quantify the contribution of among-population correlations. Alternatively, Loreau and de Mazancourt (2013) developed a *synchrony index* that is the ratio of the numerator of Eq. (10.1) to its value if all entities were completely synchronous. Another widely used metric derived from financial markets is the Sharpe ratio, θ, usually expressed as (Koellner and Schmitz, 2006)

$$\theta = \frac{\mu_P - R}{\sigma_P}, \tag{10.2}$$

where μ_P is the portfolio mean, σ_P is the portfolio standard deviation, and R is the "risk-free" rate of return. There is no such thing as a risk-free rate of return for a natural system, so $R = 0$ (Koellner and Schmitz, 2006) and the ratio simply becomes the inverse of the aggregate CV. This metric is seeing increased usage in comparing the relative portfolio benefits of different salmon fisheries (Griffiths et al., 2014; Moore et al., 2010).

A different metric focuses directly on the specific effect of the correlations in Eq. (10.1) by subtracting the value of the CV with variance terms only (without any correlation) to form a *diversity deficit* (Yamane et al., 2017). This reflects changes in the CV due to changes in the ρ_i terms (covariance among populations) as opposed to the σ_i terms (overall

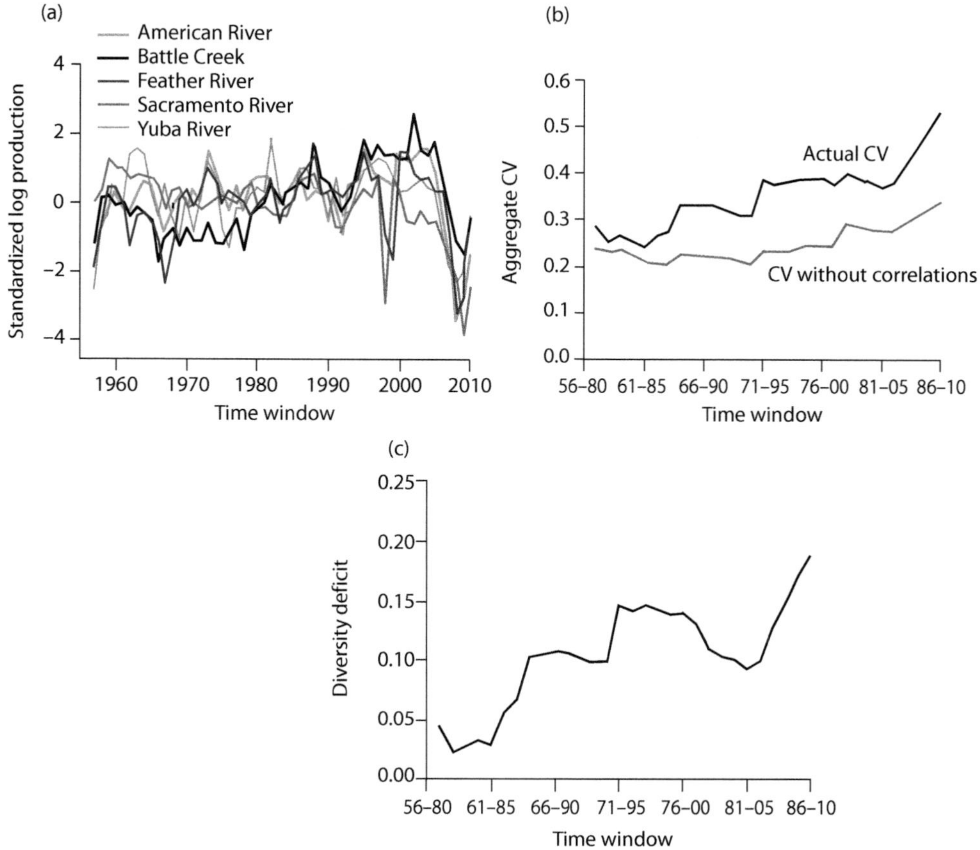

Fig. 10.7 Population diversity in Sacramento River fall-run chinook salmon. (a) Standardized log abundance of populations spawning in five tributaries (and the mainstem) of the Sacramento River. Raw data were log-transformed, the mean subtracted, and divided by the standard deviation. Abundance is measured in terms of "production," which is the number of returning spawners plus the number of fish harvested. (b) The aggregate coefficient of variation (CV) across all five populations (black curve), calculated over 25-year moving windows. The gray curve is the CV calculated assuming no correlation among populations. (c) The diversity deficit, based on the CVs in (b). Redrawn from Yamane et al. (2017).

variance). As an example, Fig. 10.7a shows the standardized (mean = 0, variance = 1) log-transformed abundance of five independent populations of Sacramento River fall-run chinook salmon (each population occupies one of the four main tributaries of the river, plus the mainstem). Note that this is a totally different fishery stock from the Sacramento River winter-run chinook salmon that we discussed in Section 10.3 (different "runs" of the same species ascend the same river at different times and often spawn in different habitats, so they cannot interbreed freely and are thus truly separate populations). Figure 10.7b illustrates the long-term increase in aggregate CV in the system, as well as the CVs due to variances alone, and Fig. 10.7c shows the consequent diversity deficit. Historically, these populations varied independently, but have been synchronized somewhat, likely due to loss of some spawning habitats (due to dams) and due to supplementation of the stock with hatchery fish. Note in Fig. 10.7c that there were sharp increases in the diversity deficit in the 1980s and then again in the 2000s; the latter being especially apparent with the concurrent decline of all five populations during that period (Fig. 10.7a), which led to a fishery closure in 2008–9. Notice that those periods of increased population synchrony are reflected by increases in the aggregate CV, but not in the CV calculated without correlations (Fig. 10.7b). Hence the diversity deficit (the difference between those two values) directly accounts for the loss of portfolio diversity in a way that other metrics do not. A metric such as the diversity deficit could be used by managers as an additional "state" reference point, in addition to abundance (as McElhany et al. (2006) had suggested). Yamane et al. (2017) showed that if portfolio variance were reduced by management interventions that decreased the diversity deficit, the probability of the aggregate stock falling below the level that would trigger a fishery closure would be substantially reduced (Fig. 10.7c). Note that the latter analysis is not unlike a PVA, but with the target being a fishery reference point (we discuss these further in Chapter 11) rather than a quasi-extinction threshold.

10.9 What have we learned in Chapter 10?

In conservation biology, the primary use of population dynamics has been to calculate probabilities of extinction and use them to learn how different management actions could be used to maintain a low risk. Here we began afresh by asking how our analyses of population dynamics in the earlier chapters had indicated we could manage to guard against extinction. We decided that continuing to use abundance as a metric to track was a good idea, but that it made more sense to also track replacement or population growth rate, as evidenced by current population structure (i.e. not the estimate of population growth rate one obtains from fitting a model to the time series of past abundance). We noted that while keeping these two variables high was a good idea, we did not know what a critical persistence level would be (other than keeping replacement greater than 1.0, if we had the survival and reproductive estimates to calculate it).

Because calculation of the probability of extinction was still a vital part of this scheme, we then briefly reviewed existing questions regarding that calculation. A primary limitation attracting attention over the past several decades has been that a rigorous accounting of the effects of errors in estimating parameter values or annual growth in abundance, showed that distributions of probabilities of extinction would tend to be U-shaped with peaks at 0.0 and 1.0 (Fig 10.1d) due to estimation error, particularly over longer time horizons. Another problem, identified more recently through analyses of the cohort resonance effect, was the fact that probabilities of extinction depend on the spectrum of environmental variability. This, of course, is not an error, rather an effect that has to be accounted for. In other words, the convention of calculating probabilities of extinction using white noise as the environmental driver can be misleading.

We then turned to several examples of calculating probabilities of extinction that address some of the identified shortcomings in the conventional approach to managing against extinction. The first example was a viability analysis of an elephant population that included the effects of density dependence and different time scales of environmental variability. Arguably the most valuable lesson from this example was the difference between expressing the time scale of environmental variability in terms of frequency, as we have thus far in this book, and describing it as meteorologists often do, in terms of the return time between occurrences (e.g. a 100-year storm). Basically, the latter increases the variance as the frequency increases, making the elephant/drought result seem different from the results with Brandt's cormorant and ENSO frequency (Section 8.5).

The second example analyzed viability of the first Pacific salmon listed under the U.S. ESA, the Sacramento River winter-run chinook. In this analysis the basic definition of probability of extinction depended on both abundance and replacement, mainly because abundance can be artificially controlled with hatcheries. This probability of extinction was used to explore how delisting criteria would depend on measurement error and estimation error. Adding error in estimating mean abundance and the uncertainty in annual counts increased the number of years worth of sampling required to indicate a threshold value of spawning abundance (10,000 spawners) while maintaining replacement greater than 1.0.

Then, staying with the topic of extinction of salmon, we took advantage of the fact that the listing of Pacific salmon provided multiple examples of the state of the science in both 2003, and later in 2013. We saw that in 2003, the best available method was a modified version of the approach used by Dennis et al. (1991; see Section 8.2), producing values of λ between 0.85 and 1.0. In part due to the effects of estimation error, calculations of the probability of ending up below 10% of their initial value after 50 years ranged from 0.0 to 1.0. The later study reported in Busch et al. (2013) was actually organized as a comparison of a number of research teams analyzing the same data sets. We saw that many of the groups had moved beyond the use of a modified version of the approach of Dennis et al. (1991), including density dependence and other factors, and some used methods that included both abundance and replacement metrics. It is notable that results by the different recovery teams did not seem to produce broadly consistent results. This raises questions about the use of probability of extinction as a management metric, a topic that seems to be worth investigating further. Of course, as Brook et al. (2002) wrote at the time that Ludwig et al. were debating whether it made sense to calculate probabilities of extinction, many people are aware of the problems associated with PVA, but as yet no one has proposed a better alternative. Perhaps the best way forward is to take care in ensuring that model structures faithfully represent the intended ecological process, a theme we have sounded throughout this book (i.e. realism).

The next example, viability analyses of the island fox, was notable for its estimation of actual survival and reproductive rates from mark–recapture data. This closer scrutiny of life history rates produced a model whose deterministic skeleton produced the nonlinear behavior of a Ricker stock–recruitment function. The strong effect of that functional form upon dynamics persistence seems to be worth investigating further.

Our final topic in Chapter 10 was an introduction to a relatively new idea in population viability brought about by the existence of populations distributed over space. In Chapter 9 we introduced the idea that the overall aggregate variability in a group of populations was less if the populations were more statistically independent. In this chapter we focused on how to manage that effect. That is, if resource managers wanted to diminish overall variability by maintaining greater statistical independence among them, how would they go about tracking progress? There are many proposed ways of doing this, but we described only the one—the diversity deficit—which focuses directly on the specific effect of the correlations among populations on the aggregate CV.

CHAPTER 11

Population dynamics in marine conservation

The predominant use of population modeling in marine conservation is in the analysis of the effects of fishing, a concern that arose in the early part of the twentieth century. Consequently, we focus here on how population models have been used in the process of managing fisheries, and more recently marine protected areas. Fisheries management has not been, shall we say, universally successful (Worm et al., 2006; 2009), so here we describe how models contributed to the success or failure of fisheries management, as well as how models have improved with time. Of course there is not enough space here for a deep discussion of all of the factors contributing to the success of fisheries management (e.g. Gutiérrez et al., 2011). In particular, we do not emphasize bioeconomic aspects (see Clark (2010) for a comprehensive summary of these approaches, and Sethi (2019) for a general introduction to optimal control theory, which underpins most of those economic analyses); we limit our focus here to the role of population modeling.

Our description of the use of population models in fisheries begins with the three famous models of the 1950s, omitting much interesting, related, history dating from the late 1800s. That early history involved concerns about overfishing, and the related questions of how much abundance should decline with fishing, how much fishing should be allowed, and the causes of fluctuations in abundance and catch. To address these, the concept of logistic growth (Chapter 2) emerged in fisheries, promulgated by early fishery scientists, such as J. Hjort, in parallel with its development among general ecologists, such as A. J. Lotka. Fishery scientists came to understand that the optimum catch would be obtained not at the highest abundance, nor the lowest, but rather at a middling level (as in Section 2.3.1). The interested reader can read about these developments in Smith (1994), and can find a bibliography with reference to this period and later in Botsford (2013). By the 1950s two other models, with greater realism (i.e. age structure), had also developed. Relationships between these three models were understood, but further development was limited by computing speed. By the 1970s, some fisheries had declined to low levels, and fishery biologists (and others) began to appreciate the importance of maintaining adequate replacement in populations. By the 1990s, there was a deeper appreciation for the role of the large uncertainty in fisheries management, which led to international development of a precautionary approach to management that changed the ways that population dynamics was used. Fisheries management developed in the direction described in Chapter 10: avoiding fishery collapse by tracking estimates of abundance and replacement (termed reference points), and developing rules for reacting to observed values (termed control rules). Concerns about uncertainty and the collapse of many conventionally managed fishery stocks also led to interest in a completely

Population Dynamics for Conservation. Louis W. Botsford, J. Wilson White, and Alan Hastings, Oxford University Press (2019).
DOI: 10.1093/oso/9780198758365.001.0001

different way of controlling fishing: spatial fishery management, particularly the use of marine protected areas (MPA). In tracing these developments in the science of fishery management, we will draw on the science of structured population dynamics, including material from Chapters 2, 3, 4, 7, 8, and 9.

11.1 Three models from the 1950s

Three predominant fishery models were developed in the years after World War II: the logistic fisheries model (Schaefer, 1954), the single cohort model (Beverton and Holt, 1957), and the stock–recruitment models (Ricker, 1954). We make use of several plots from Schaefer (1967a) to illustrate how the relationships between these three models were viewed.

11.1.1 *The logistic fishery model*

The logistic model typically used biomass as the state variable, but occasionally it was abundance. For biomass, B, it is often written as

$$\frac{dB}{dt} = kB\left(1 - \frac{B}{B_\infty}\right) - Y, \tag{11.1}$$

where k and B_∞ are constants (the growth rate and maximum biomass (i.e. carrying capacity), respectively, the variables r and K in Section 2.3.1), and Y is fishery yield (the catch rate). Assuming the fish populations were at an equilibrium, so $dB/dt = 0$, the maximum of the resulting expression for yield can be found as we did in Chapter 2 (Section 2.4.4): set dB/dt to zero and solve for the value of Y at equilibrium, then differentiate that equation with respect to B. As noted in Chapter 2, this approach produces the result that the biomass at which yield is a maximum is half the carrying capacity,

$$B_{Y_{max}} = \frac{B_\infty}{2}. \tag{11.2}$$

To relate these results to the available data, primarily fishery yield from catch data, fishery biologists assumed that catch (yield) per unit fishing effort (i.e. the amount of fishing gear times the duration deployed, E) was proportional to biomass with proportionality constant q,

$$Y = qEB. \tag{11.3}$$

We assume that B is at equilibrium so dB/dt in Eq. (11.1) is zero. We solve Eq. (11.3) for B, so $B{=}Y/(qE)$. Substituting that expression for B into Eq. (11.1) at equilibrium, and solving the resulting equation for the yield, Y, produces an expression for yield as a function of effort,

$$Y = B_\infty qE - \frac{q^2 B_\infty}{k} E^2. \tag{11.4a}$$

Plotting this equation indicates the amount of effort that should be prescribed to obtain the maximum yield, which Schaeffer plotted as Fig. 11.1a. This is a hump-shaped relationship: yield is zero if there is no effort or if effort is so high that the population is driven extinct, and the maximum yield is in the middle (at the level of yield that corresponds to an equilibrium biomass of $B_\infty/2$, as given in Eq. 11.2).

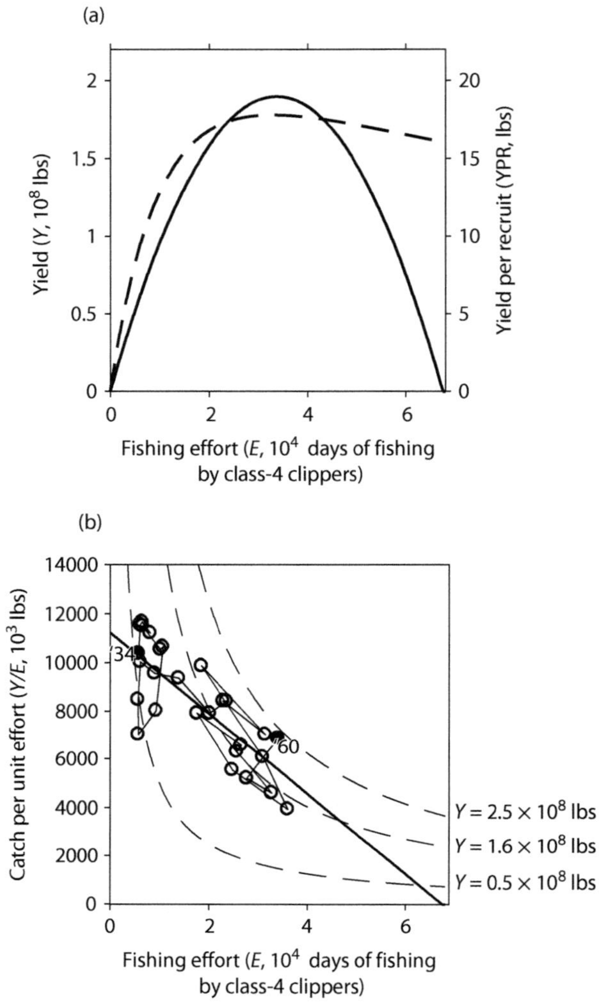

Fig. 11.1 An example of the logistic fishery model applied to Eastern Pacific yellowfin tuna, *Thunnus albacares*. (a) Yield as a function of harvest effort (Eq, 11.4a, solid curve). Yield-per-recruit, calculated from the single cohort model, is shown for comparison (dashed curve). (b) Catch per unit effort as a function of effort, illustrating how two of the parameters of the logistic model can be estimated from a regression line (solid line) fitted to catch and effort data (open symbols). Lines connect the catch data indicating the progression of time from 1934 to 1960 (the starting and ending years are shown as filled symbols). Yield isopleths (dashed curves) show the combinations of effort and catch per unit effort associated with the indicated equilibrium values of yield. Redrawn using model parameters from Schaefer (1967): $B_\infty = 2.9548 \times 10^8$lbs, $k = 2.5681\ y^{-1}, q = 3.806 \times 10^{-5}$ y. The yield-per-recruit curve was calculated assuming 2.5×10^4 clipper days produced $F = 0.69\ y^{-1}$.

Dividing Eq. (11.4a) by E gives a plot of Y/E versus E, that is a straight line,

$$\frac{Y}{E} = B_\infty q - \frac{q^2 B_\infty}{k} E. \tag{11.4b}$$

This equation was originally the basis for determining two of the three constants in this model (k, B_∞, and q). In Fig. 11.1b we depict the yield and effort data that Schaefer (1967a) had for Pacific yellowfin tuna (*Thunnus albacares*); he fitted a regression line to the Y/E

versus E data and used the slope and intercept of that line to estimate k and B_∞. The third parameter, q, had to be estimated independently from tagging experiments. This basic approach was refined by later authors, as we describe later.

In Fig. 11.1b we have also plotted isopleths of yield; that is, lines showing the yields that would result from the different combinations of E and Y/E. From these and the associated data, one can see that from 1934 to 1960, adding additional effort to the fishery resulted in an overall higher yield (the later data points are on higher-yield isopleths) even as CPUE declined.

This model, in particular Eq. (11.2), was a statement of the enduring concept of maximum sustained yield (MSY). This concept from this simple model has been widely used to manage regional to national fisheries, and as the basis for international treaties. It was preferred over other models because the data required—past catch and fishing effort (assuming the value of q could be obtained elsewhere)—were more commonly available than the more biological data required for other models. In the 1970s this approach began to fall out of favor for a number of reasons. Larkin (1977) catalogued the litany of problems with MSY-based management, from the argument that it was anti-precautionary (because inter-annual variability in recruitment means that populations are often far from equilibrium, so the MSY level of harvest is too high), to the economic point that the level of catch that maximizes monetary profit is not likely to be the same as MSY. Larkin closed his article, entitled "An epitaph for the concept of maximum sustained yield" with an actual epitaph:

M.S.Y.
1930s–1970s
Here lies the concept, MSY
It advocated yields too high
And didn't spell out how to slice the pie
We bury it with the best of wishes,
Especially on behalf of fishes.
We don't know yet what will take its place,
But hope it's as good for the human race

Yet this epitaph proved to be a bit premature, as the general concept continued to be used, and is still used today, albeit more commonly as a precautionary reference point for the upper limit of yield, rather than as a target for the desired level of yield (Smith and Punt, 2001; Finley, 2011; Botsford, 2013).

A reader of this book (particularly of Chapters 1 and 2) will realize that this population model (Eq. 11.1) suffers from the structural uncertainty of not having a *bona fide* state variable. For example, as fishing increases, the age structure of the population will change, and that will not be represented in abundance or biomass. The importance of this limitation was not universally appreciated among early fishery biologists. For example, Rothschild and Suda (1977) favored this model, pointing out that while it may have lacked realism, a more complex model might have been too difficult to understand. However, the authors of one of the other models developed in the 1950s, Beverton and Holt (1957) appreciated the structural limitations of this simple model, saying (on their p. 330) that "The simple logistic interpreted in terms of biomass implies that the rate of increase of the population depends solely on its total weight, and is independent of the age- and size-composition of the individuals comprising it: the population is to be regarded, in effect, as a continuous, homogeneous mass of living matter."

Other problems with this model include the assumption that catch-per-unit-effort is proportional to abundance, the assumption that the population is at equilibrium, and the assumption that there is no variability so a purely deterministic approach can be used. The first is contained in the assumption in Eq. (11.3) that $Y = qEB$, and is subject to a plethora of problems dealt with in the literature on population estimation (Maunder et al., 2006; Maunder and Punt, 2013). One important example is the use of $B = Y/(qE)$ to depict long, declining time series of abundance of fished stocks, warning of severe reductions in biomass, (e.g. Myers and Worm, 2003; Worm et al., 2006). Cautionary rebuttals to that sort of analysis often criticize use of CPUE as an index of abundance, primarily because catchability, q, is unlikely to be constant over the entire history of exploitation of a fishery, as the fishing fleet becomes more efficient or switches to other target species, or if catchability itself depends on abundance (e.g. Hampton et al., 2005; Maunder et al., 2006, Sibert et al., 2006; Worm et al., 2009). The assumption of equilibrium also complicates this estimation because fish populations are rarely at equilibrium. Schaefer and others realized this problem, and developed methods to attempt to overcome it (Schaefer, 1954; Pella and Tomlinson, 1969; Fox, 1970; Schnute, 1977).

Fundamentally, MSY provided a false sense of security from its beginning, because use of the word "sustainable" was chosen on the basis that the population was assumed to be at an equilibrium (Schaefer, 1967b). As we saw in Chapter 2, not all equilibria are stable. Also, assuming that the population is at an equilibrium in the derivation of Eq. (11.2) does not guarantee a persistent population.

11.1.2 *The single cohort model (also known as the dynamic pool model, yield-per-recruit model)*

The single cohort model tracks the decline in abundance with age of a cohort of fish, along with the increase in size (length or weight) of each fish with age. It also includes fishing at a constant rate, for all ages greater than a specified lower age limit on the fishery. It therefore allows one to evaluate the consequences of the tradeoff between allowing the fish to become larger before harvesting them and the fact that some will die before they get larger. This classical model is included in Section 2 of 20 in the 530-page tome, *On the Dynamics of Exploited Fish Populations* by Ray Beverton and Sidney Holt (Beverton and Holt, 1957). Beverton and Holt called this the "dynamic pool model" because it conceives the fished population as a "pool" of individuals, with older ones constantly being removed by death or fishing and new recruits constantly entering. This is in contrast to the logistic model in which no distinction is made between "old" and "young" biomass; essentially the dynamic pool concept is synonymous with age structure. The remaining 18 sections of Beverton and Holt's book greatly expand and apply this basic view of age-structured dynamics to a very broad range of fishery questions. Notably, the advice frequently given in fisheries that, "If you think that you have discovered something new and original in the analyses of fisheries management, you had better check Beverton and Holt to see whether they have already done it," is often worth following.

In this model, survival is simple exponential decay due to natural mortality after the age of recruitment to the population, a_r, up to the age of first capture a_c, after which the cohort is fished. Recall that the term "recruitment" generally refers to the entry of individuals into a certain stage in the population (e.g. the larval stage, the adult population). Here a_r is defined as the age at which the fish arrive on the fishing ground and are potentially available to the fishery (the time between a_r and a_c is typically determined by gear type, e.g. the mesh size of trawl nets determines how large fish have to be before they are captured). We define n_a as a density function on age, a (we may use the shorthand that n_a is the number of individuals of age a, which is not precise—because age is a continuous variable in this formulation, we should speak of the number of individuals in some age interval $a + \Delta a$). Writing cohort abundance surviving to an age in terms of recruitment and mortality rates leads to

$$n_a = Re^{-M(a-a_r)} \qquad a_r < a < a_c \tag{11.5a}$$

$$n_a = Re^{-M(a_c-a_r)}e^{-F(a-a_c)} \qquad a_c < a, \tag{11.5b}$$

where R is initial abundance in a cohort at age a_r (units: individuals), M = instantaneous natural mortality rate (units: y^{-1}), and F = instantaneous fishing mortality rate (units: y^{-1}). Weight-at-age was assumed to be weight cubed, with proportionality constant c,

$$w_a = cl_a^3 \tag{11.6}$$

and length at age was assumed to have the von Bertalanffy form (Box 3.8),

$$l_a = L_\infty\left(1 - e^{-k(a-a_0)}\right), \tag{11.7}$$

where a_0 is the effective "age" (often negative) at which length would be zero. Lifetime yield for the cohort is then the sum of yield over all ages being fished, which is the fishing mortality rate times biomass at each discrete age, summed over all fished age classes,

$$\text{YPR}\,(a_c, F) = \sum_{a=a_c}^{a=A} Fn_a w_a. \tag{11.8a}$$

Note that this would be the same as the annual catch if recruitment were constant and the population were at the stable age distribution. If we substitute in expressions for n_a and w_a, we obtain the expression that Beverton and Holt developed,

$$\text{YPR} = R'\,FW_\infty \sum_{i=0}^{3}\left[u_i e^{-ika_c}\frac{1}{M+F+ik}\right], \tag{11.8b}$$

where $R\prime$ is the recruitment to the fishery, W_∞ is the asymptotic maximum biomass, k is the von Bertalanffy growth rate, and $u_i = 1, -3, 3, -1$ for $i = 0, 1, 2, 3$. In Box 11.1 we show the derivation of that expression (which does not require summing over every age class).

Box 11.1 DERIVATION OF THE YIELD PER RECRUIT CALCULATION

To derive the expression for yield per recruit, we will first consider the rate at which fish are caught in terms of abundance (i.e. number of fish) and then in terms of biomass. To do this it is useful to think about dynamics in continuous time, and note that (a density function on age a for) the rate at which individuals of age a are caught, $Y_A(a)$, i.e. the number of fish of that age harvested, will be the fishing rate F times the abundance, n_a,

$$Y_A(a) = Fn(a). \tag{B11.1-1}$$

The subscript A indicates that we are dealing with abundance. If both F and recruitment, R, are constant over time and F is constant over all ages $a > a_c$, one can calculate the total yield

(Continued)

Box 11.1 CONTINUED

(in abundance) for a cohort of recruits over its lifetime, Y_A, as the integral of the age-specific catch rate over all ages that are subject to harvest,

$$Y_A = \int_{a_c}^{\infty} Y_A(a)da, \tag{B11.1-2}$$

by integrating Eq. (B11.1-1) from a_c to ∞ . In doing this we substitute in the expression for n_a from Eq. (11.5) and factor out the term with $-M(a_c - a_r)$ in the exponent so we can find an explicit expression for the definite integral,

$$Y_A = Re^{-M(a_c-a_r)} \int_{a_c}^{\infty} Fe^{-(M+F)(a-a_c)}da. \tag{B11.1-3}$$

To simplify, we substitute $R' = Re^{-M(a_c-a_r)}$ to represent the number of recruits entering the fishery. Then evaluating the definite integral, we obtain

$$Y_A = R'\frac{F}{M+F}. \tag{B11.1-4}$$

The interpretation of this equation is that all of the recruits R' will die, and of those, a fraction $F/(M+F)$ will die from being caught by the fishery. As an aside, if we were interested in catch over some finite age range (rather than the entire life of a cohort), then Eq. (B11.1-2) would be integrated over some time interval Δt and the solution would be

$$C = R'\frac{F}{M+F}\left(1 - e^{-(M+F)\Delta a}\right), \tag{B11.1-5}$$

which is a famous expression in fisheries science known as the Baranov catch equation (Baranov, 1918), where Δa is usually 1, indicating annual catch.

The calculation for yield in units of biomass, Y_B, is similar to that of catch, with the addition that the right-hand side of Eq. (B11.1-1) is also multiplied by weight-at-age, $w(a)$ to produce an expression for (the density function on age for) the age specific yield,

$$Y_B(a) = Fn(a)w(a). \tag{B11.1-6}$$

To calculate total yield per recruit, Y_B, from this expression, we combine the expression for length-at-age (Eq. 11.7) and biomass as a function of length (Eq. 11.8). We define the asymptotic maximum biomass to be $W_\infty = cL_\infty^3$, and for simplicity we also assume that $a_0 = 0$. This leads to an expression for biomass-at-age,

$$w(a) = W_\infty\left(1 - e^{-ka}\right)^3. \tag{B11.1-7}$$

We expand the polynomial as

$$w(a) = W_\infty \sum\nolimits_{i=0}^{3} u_i e^{-ika}, \tag{B11.1-8}$$

where $u_i = 1, -3, 3, -1$ for $i = 0, 1, 2, 3$. Then to calculate total yield per recruit, we substitute Eq. (B11.1-7) into the analog of Eq. (B11.1-2) and integrate, just as for catch,

$$\text{YPR} = R' \int_{a_c}^{\infty} Fe^{-(M+F)(a-a_c)} W_\infty \sum\nolimits_{i=0}^{3} u_i e^{-ika} da. \tag{B11.1-9}$$

Evaluating the definite integral produces

$$\text{YPR} = R'FW_\infty \sum\nolimits_{i=0}^{3} \left[u_i e^{-ika_c} \frac{1}{M+F+ik}\right]. \tag{B11.1-10}$$

This equation is equivalent to the version given (in discrete time, not continuous time) as Eq. (11.8a) in the text: the product of the fishing rate, abundance, and biomass, summed over all fished ages, with the benefit of being simpler to calculate using a hand calculator. Additionally, the polynomial expansion of the expression for $w(a)$ will also be useful later when describing the transient dynamics of biomass inside marine protected areas (Box 11.4).

Beverton and Holt (1957) used North Sea plaice (*Pleuronectes platessa*) for an example of YPR as a function of the values of fishing mortality rate, F, and age of first capture, a_c (they also helpfully suggested that the calculations (using the formula we give in Box 11.1) could take as little as three hours if one is quick with a hand calculating machine; see Fig. 11.2). In Fig. 11.3 we show how this calculation of YPR works, for a single value of F and three values of a_C: ages 5, 10, and 14. In Fig. 11.3a, survival declines exponentially with age until age a_c, when it declines more steeply due to the addition of fishing mortality; meanwhile biomass increases with age as length cubed. In Fig. 11.3b we show the resulting curves for yield-at-age; yield per recruit is F times the integral (or sum, if we are dealing with discrete ages) of each of these, and since F is the same for all three, in this example YPR increases as a_C moves from five to ten years, because the fish are allowed to get bigger before being caught, and it declines again by $a_c = 14$ years, because so few fish survive to that age.

In Fig. 11.4a, we have redrawn Beverton and Holt's original plot of YPR isopleths for the full range of combinations of a_C and F. Plots of YPR for other species of course vary with life histories, but have the same basic shape of the yield isopleth "cupped" about the

Fig. 11.2 Ray Beverton and Sidney Holt in 1949 at the Fisheries Laboratory, Lowestoft, UK. Holt is operating a hand calculator, and next to Beverton is a hand-made yield-per-recruit isopleth surface, like that in Fig. 11.4a.

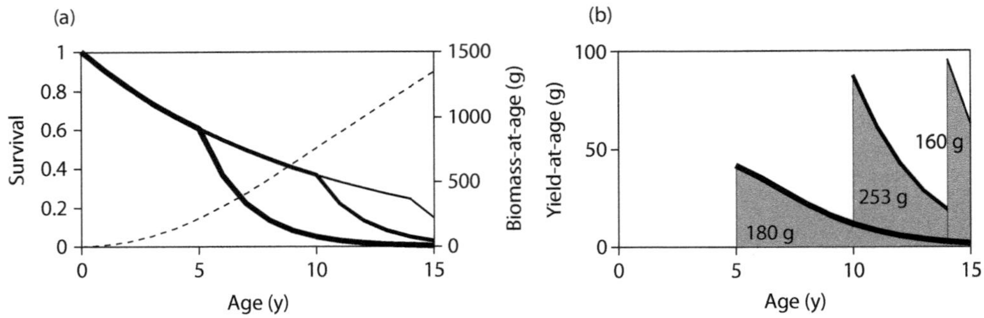

Fig. 11.3 Yield-per-recruit analysis for North Sea plaice. (a) Survival to age (solid curves, left axis) for fishing rate $F = 0.4\,y^{-1}$ and age of entry to the fishery (a_C) equal to 5 y (thickest curve), 10 y (medium curve), or 14 y (thinnest curve). Biomass (individual weight) at age (dashed curve, right axis) is also shown. (b) Yield-at-age (black curves) for the values of a_C shown in panel (a). The integral (or sum) of each curve (gray shading) is the total yield per recruit, which is indicated in Eq. (11.8).

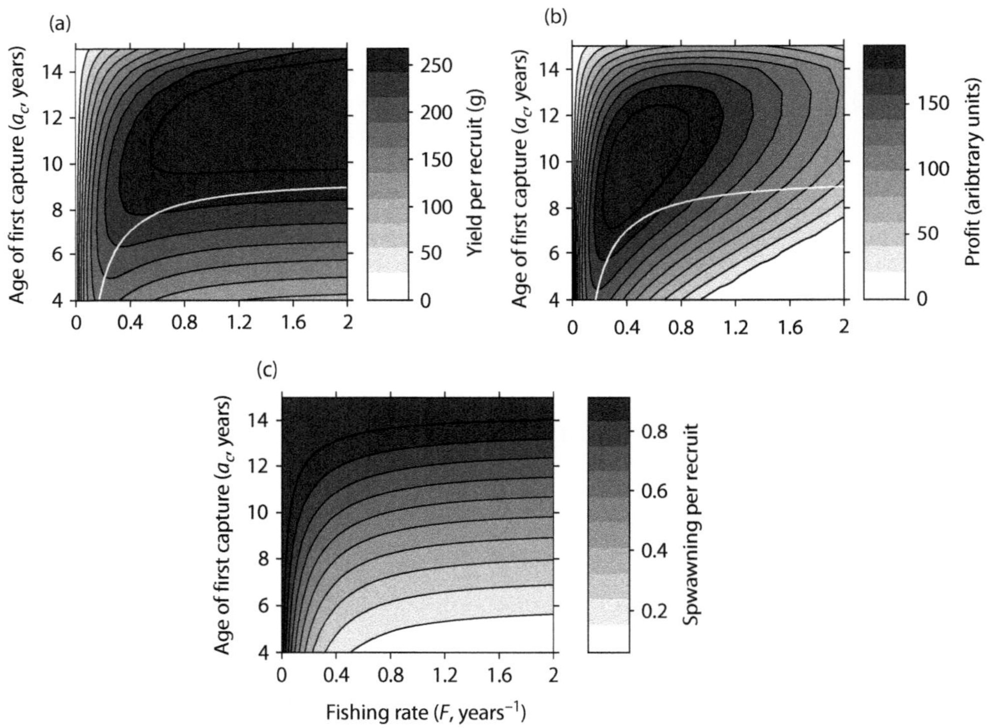

Fig. 11.4 Yield-per-recruit analysis for North Sea plaice, as a function of the harvest rate (F) and the age of entry to the fishery (a_C). (a) Yield-per-recruit isopleths. (b) Profit isopleths, assuming profit was equal to yield minus $100 \times F$. (c) Fraction of unfished spawning per recruit (SPR) isopleths. The white curves in (a) and (b) are the 40% SPR isopleth from panel (c). Redrawn using model parameters from Beverton and Holt (1957).

maximum YPR which occurs at a_c equal to the age of maximum unfished cohort biomass and $F = \infty$. That age was termed the critical age (Ricker, 1945), and on this plot it would have had a value of 13 y. Because $F = \infty$ cannot actually be achieved, one could use a plot such as this to find the maximum YPR, for some fixed values of either F or a_c.

The basic lesson from this plot is that if harvest begins when fish are too small (small a_c), they are captured before they can grow large, and biomass yield suffers (this is termed "growth overfishing"). In actual management, setting the age of first capture, a_c, could only approximately be realized by adjusting the mesh size of the net in a trawl fishery, or a lower size limit in a hook-and-line fishery. In the case of plaice, the value of a_c corresponded to a particular size of the net mesh in the trawl nets. Beverton and Holt noted that the pre-World War II F for plaice ($F = 0.73$) was already well past the optimal value for the value of a_c at that time, approximately 4 y.

The impression from the isopleths in Fig. 11.4a, that it would be desirable to fish at $F = \infty$ (for a particular value of a_c) to obtain the maximum yield, arises because there is no consideration of the cost associated with fishing, and thus of the net profit (rather than the gross yield). If the cost of fishing were proportional to F, subtracting it from the yield would cause the yield surface to rotate about the $F = 0$ axis, so that the maximum would shift downward and to the left, moving it away from infinity. We illustrate this in Fig. 11.4b by plotting the isopleths of "profit," represented (to a crude first approximation) as YPR $- 100 \times F$ (the constant of proportionality for F was chosen arbitrarily to show the effect).

The single cohort model was (and still is) used to manage fisheries when there were data available to determine growth and mortality rates. However, it cannot be used to evaluate a global maximum yield since it lacks the mechanism linking reproduction to the age structure. Instead of attempting to evaluate maximum sustainable yield, one merely evaluates relative YPR. An example of the relationship between the logistic model and the single cohort model was shown by Schaefer (1967) as the dashed line in Fig. 11.1a. That line is based on estimates of von Bertlanffy growth parameters, M, and fishery characteristics (a_c, a_r) from yellowfin tuna. It is surprisingly conceptually consistent with the yield curve from Schaefer's logistic model of the same species, with a maximum at about the same place (although the exact location of the peak in YPR depends on the relationship between effort, E, and fishing mortality rate, F; Schaefer (1967b) showed a few different possibilities). The YPR curve has greater yield than the logistic curve to the right of the maximum. This can be explained by the fact that the single cohort model only accounts for growth overfishing, and does not account for the decline in reproductive output with increased fishing (*recruitment overfishing*). The logistic model does account for this, though not explicitly in terms of the reduced reproductive output of older, fished age classes (as the *complete* models in Sections 11.1.4 and 11.2 do).

To summarize, the single cohort model achieved greater realism than the logistic model, having explicit growth and survival processes, but it was missing a vital population process, reproduction. It was therefore useful only under the assumption of constant recruitment. Nonetheless it continues to serve a useful purpose in applications where fisheries scientists want to investigate the relative effects of different factors on yield (or reproductive output), without having to explicitly account for the cost of fishing or the effects of fishing on reproduction, e.g. (Pikitch, 1987; Goodyear, 1993).

11.1.3 *The stock and recruitment model*

Ricker's (1954) analysis of stock and recruitment was motivated by his realization that the evolving "modern theory of fishing" was missing a vital part of the story: how fishing affected future recruitment through the dependence of recruitment on adult abundance at age. He referred to the concept of *population regulation* being developed by the Australian entomologist A. J. Nicholson (Nicholson, 1933) and others, as well as the various kinds of "density dependence" (compensatory and depensatory) discussed by Smith (1935), Neave (1953), and others. Importantly, he noted that to be realistic, accounting for population regulation and density dependence required separating the effects of immature and mature individuals on each other, an effect not afforded by the representation of density dependence in the logistic fishery models in continuous time (for example, in some species the adults cannibalize immature juveniles). On that basis he conducted a study of the population dynamic consequences of stock and recruitment in the semelparous case using the graphical cobwebbing concept from Chapter 2. We described his findings regarding how over-compensatory density dependence could lead to period $2T$ cycles in Chapter 7, though he achieved them with the graphical solution in Section 2.3.2 and simple numerical simulations, rather than the mathematics and computational capability used here.

11.1.4 *Complete age-structured models: linking cohorts with a stock–recruit curve*

The obvious next step in the development of fisheries population dynamics was to add a means of accounting for the effects of fishing on future recruitment to the combination of the cohort growth and survival rates from the single cohort model. Schaefer envisioned this but did not develop it rigorously, and Beverton and Holt (1957, Section 6 of 20) actually added fecundity-at-age and a stock–recruitment function to achieve a complete "self-regenerating" model. However, with calculations limited by the computing power of the 1950s, there was little appetite for taking advantage of these more complete realistic models. As vacuum tubes were replaced by transistors, then integrated circuits through the 1960s and 1970s, these models became more widely used, especially in countries with highly developed fisheries management infrastructure. We examine this type of model next, in Section 11.2.

11.2 Replacement in fully age-structured fishery models

With two parallel trends in the late 1970s and early 1980s, (1) the increasingly common use of complete age-structured models and (2) the increasing awareness of, and dissatisfaction with, fisheries collapsing to low levels, the focus in fisheries management shifted to greater attention on the goal of avoiding low abundance (i.e. the goal addressed in Chapter 10). A major development came from fish biologists addressing the effects of power plants' once-through cooling systems on population persistence. These cooling intakes can entrain and kill vast numbers of fish, particularly larvae, which is concerning if a population's reproductive output is also reduced by fishing. Goodyear (1977) quantified the way that the change in age structure due to fishing would affect population persistence. He did this by defining a *compensation ratio* (CR). The idea behind the CR is that if a population with some type of compensatory density dependence experiences a reduction in lifetime egg production (e.g. from fishing), then some other vital rate must increase in order for the population to remain at equilibrium. In most fisheries, the most important source of density dependence is in juvenile survival, so reductions in egg production are compensated for by increases in per-capita juvenile survival; the CR indicates how much compensation is needed to maintain complete replacement (see review in Goodyear, 1993). This is based on the idea that the product of lifetime egg production (LEP) and juvenile survival must equal 1 if the population is at a stable, nonzero equilibrium. Looking back at the graphical method for determining the equilibrium of an age-structured population with density-dependent recruitment (Fig. 4.11), one can see that as egg production decreases, equilibrium recruitment initially changes very little, implying that per capita juvenile survival has increased (essentially, fewer eggs reduces competition among juveniles). In that graphical solution, juvenile survival is the ratio of recruit abundance to the abundance of eggs (i.e. the slope of the diagonal lines in Fig. 4.11). At equilibrium, the number of recruits times LEP times juvenile survival must equal the number of recruits, so the condition that LEP times juvenile survival equals 1 is produced by the graphical solution. This relationship eventually appeared in the Sissenwine and Shepherd (1987) reference we have used throughout this book to refer to an equilibrium condition for age-structured models with density-dependent recruitment.

In Goodyear's (1977, 1993) formulation, the CR is the ratio of LEP in an unfished stock to that in a fished stock, for a given level of fishing. That ratio gives the factor by which juvenile survival (or some other factor) would have to increase in order for the fished population to still persist. For example, CR = 100 implies that LEP is reduced to one percent of its unfished value, and for the population to persist, compensatory density dependence must increase juvenile survival by a factor of 100. An important difficulty in applying the CR concept is that there is a limit to how much compensation can occur: there will be some threshold beyond which the stock cannot compensate for additional increases in harvest, and it goes extinct. This *replacement threshold* will be determined by the shape of the density-dependent function; specifically the slope at the origin (e.g. Fig. 4.11): steeper slopes lead to greater compensation. Unfortunately, that threshold CR was essentially unknown at the time of Goodyear's original publication, and remains unknown for many fishery stocks today. We address this point in Section 11.2.1.

11.2.1 *Stock–recruit curves, lifetime egg production (LEP), and spawning per recruit (SPR)*

The approach to fisheries management emerging from these considerations about CR was to choose a value of lifetime egg production that was safely above the replacement threshold, and attempt to set F and a_c to maintain that level of egg production. Throughout this book we have referred to these replacement concepts in terms of lifetime egg production (LEP) or eggs per recruit (EPR). Both of these require knowing fecundity as a function of age. Frequently, fecundity (i.e. the number of eggs produced) is not actually known, so biomass-at-age is used as a proxy, and termed *spawning stock biomass per recruit* (SSBR). Two additional, related quantities both have the acronym SPR. The first is spawning potential ratio (Goodyear, 1993), which is the ratio of fished SSBR to unfished SSBR (thus it is the inverse of the compensation ratio). The second is spawning per recruit (Mace and Sissenwine, 1993; in the same book as Goodyear, 1993), which can either be in terms of eggs or biomass (hereafter we will use "SPR" to refer to the first definition, unless we say otherwise). For each of these quantities there is a minimum value required for persistence (Section 4.3.1), commonly referred to as the critical replacement threshold (CRT), which is determined by the slope at the origin of the stock–recruit curve. That slope is commonly described in terms of *steepness*, which is the proportion of unfished recruitment obtained at 20 percent of unfished SSB (this is a nondimensional value, facilitating comparison of relative compensation across different stocks). In Table 11.1 we summarize the relationships of all of these different quantities.

By the 1990s, fishery assessments in the USA and Europe were predominantly managed using SPR as a percentage of the unfished maximum, with a goal of maintaining a

Table 11.1 Terms and acronyms used in the context of biological reference points in fisheries. Entries sharing the same number of stars (*, **, or ***) are mathematically related and convey equivalent demographic information.

Acronym	Name	Definition
CRT *	Critical replacement threshold	Minimum value of lifetime reproduction required for replacement and population persistence. Inverse of the slope at the origin of the egg–recruit relationship
EPR **	Eggs per recruit	Average lifetime reproductive effort of a new recruit. Equivalent to LEP, but usually calculated in a different way (Hilborn and Walters, 1992)
FLEP ***	Fraction of lifetime egg production	Fraction of unfished LEP realized by a fished population. Conceptually identical to SPR, but calculated in a manner that is less sensitive to estimates of the natural mortality rate (O'Farrell and Botsford, 2005)
LEP **	Lifetime egg production	Average lifetime reproductive effort of a new recruit. Equivalent to EPR, but usually calculated in a different way (O'Farrell and Botsford, 2005)
SPR ***	Spawning potential ratio	The ratio of spawning stock biomass in a fished population to that of an unfished population (Goodyear, 1993)
SPR **	Spawning per recruit	Average lifetime reproductive effort of a new recruit. Equivalent to EPR and LEP, but can be expressed either in units of eggs or spawning biomass (Mace and Sissenwine, 1993)
YPR	Yield per recruit	Average biomass yield resulting from a single new recruit (Beverton and Holt, 1957)
Other definitions		
***	Compensation ratio	A measure of the level of density-dependent compensation required to offset a reduction in lifetime reproduction. Equal to 1/SPR (first definition) (Goodyear, 1980)
*	Slope at the origin	The slope at the origin of the egg–recruit or stock–recruit relationship
*	Steepness	A measure of the slope of the stock–recruit relationship. Defined as the ratio of recruitment at SPR $= 0.2$ and recruitment at SPR $= 1.0$. Equal to $1/(1 + 4 \times \text{CRT})$ (Hilborn and Walters, 1992)

minimum "safe" percentage of SPR that would ensure population persistence (Goodyear, 1993; Mace and Sissenwine, 1993; ICES, 1983; 1984). However, to manage for replacement using SPR, one must know what the replacement threshold is, i.e. what is the initial slope of the stock–recruit curve. This slope is highly uncertain because fishery managers try to avoid fishing populations to low abundance; thus we would not likely have any data near the origin of the stock–recruit curve and thus would not have any information on the slope there. Of course, some populations have been fished to low abundances, hence we have some idea of what that slope is for those populations. For example, Mace and Sissenwine (1993) used life history information and data on recruitment and spawning

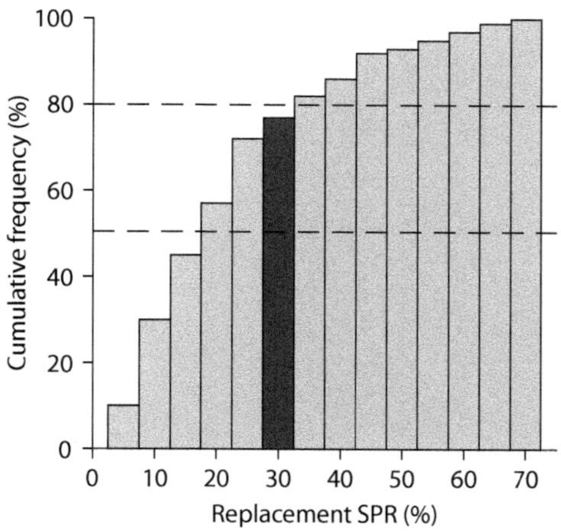

Fig. 11.5 Cumulative frequency histogram of estimates of the level of SPR needed for persistence, for 91 fishery stocks, based on stock and recruitment data. Dashed lines indicate 50% (the mean) and 80% (the level chosen to be conservative); the shaded bar is 30% SPR. Redrawn from Mace and Sissenwine (1993).

biomass to estimate the slope of the stock–recruitment curve for 91 species, primarily from the North Atlantic. They found that the average replacement level of SPR was 20 percent (Fig. 11.5); some taxa were more resilient, with steeper stock–recruit curves (e.g. Atlantic cod, *Gadus morhua* and flatfish in the families Bothidae, Pleuronectidae, and Soleidae), and some were less resilient (e.g. small pelagic fishes such as the Atlantic herring, *Clupea harengus*, and sardine, *Sardina pilchardis*).

As an example of differences in available data for estimating replacement thresholds, Fig. 11.6 shows stock–recruitment data for two rockfish populations from the U.S. west coast. Yellowtail rockfish, *Sebastes flavidus*, (Fig. 11.6a) have not been fished heavily, so the stock–recruitment data do not offer much guidance on what the slope near the origin should be. As examples, we show four hypothetical curves with very different initial slopes, each of which provides a reasonable fit to those data. Pacific Ocean perch, *Sebastes alutus* (Fig. 11.6b), on the other hand, have been harvested down to low abundance, and there are plenty of data points near the origin, providing a good estimate of the slope. Extrapolations from species like Pacific Ocean perch for which we have good estimates of the replacement threshold (CRT) can be used for species with less data near the origin. For example, Dorn (2002) used a hierarchical model to jointly estimate stock–recruitment curves for a suite of west coast rockfishes; his estimates for yellowtail rockfish and Pacific Ocean perch are shown as the dark curves in Fig. 11.6. This illustrates how information from other stocks can inform the management of species such as yellowtail rockfish, in order to set minimum SPR limits. We discuss more about how these limits are chosen in Section 11.3.

Examining the stock–recruit data in Fig. 11.6 as a time series, focusing on the progression of data from the first data point to the last, illustrates a fundamental conundrum of fisheries management that we refer to as the "blind chicken problem." In the classic 1955 film *Rebel Without a Cause*, the protagonist (portrayed by James Dean) is challenged to a game of "chicken" by a rival, in which each drives a car towards a cliff, and they compete to see who will be the last one to jump out of the car before reaching the precipice. At the outset of a new fishery, there is typically pressure to keep increasing F (and thus decreasing SPR) to increase fishery yield. Amplifying this pressure is (a) as the fishery grows, fishers make capital investments in gear, and landings must remain high to pay off those investments; and (b) the initial "windfall" yield from a new fishery is very high, as the oldest, largest individuals are removed from the population, giving the misleading

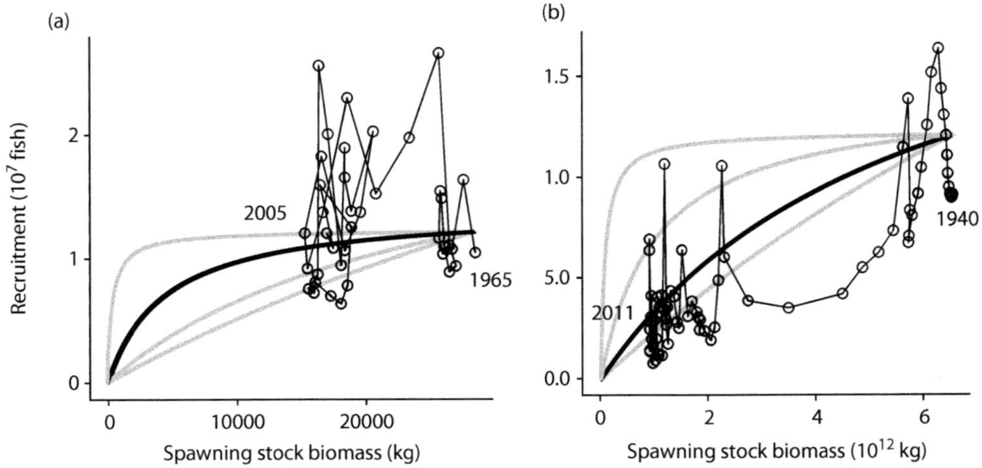

Fig. 11.6 Stock–recruitment relationships for (a) yellowtail rockfish, *Sebastes flavidus*, and (b) Pacific Ocean perch, *S. alutus*. Curves are a range of possible Beverton–Holt functions that could be fitted to the data. Each point represents one year; the black line connects years and the first and last year in the data set are labeled. In each panel, the black curve is the curve estimated for that stock using a hierarchical Bayesian model incorporating information from 11 *Sebastes* stocks in the northeastern Pacific Ocean (Dorn, 2002). Stock–recruit data obtained from the RAM legacy stock assessment database (ramlegacy.org; Ricard et al., 2011).

impression that very high yields could be sustainable even after those large individuals are no longer present. Thus as the fishery matures, SPR moves from right to left on the stock–recruitment plot. The goal is to avoid "falling off the cliff" by reducing SPR below the replacement threshold; the problem is that this is a blind game of chicken because we do not know where the cliff is until the stock begins to collapse, revealing the slope at the origin of the stock–recruit relationship (as in Pacific Ocean perch). A desire to avoid the blind chicken problem led to a focus on precautionary management, as we describe in Section 11.3.

11.2.2 *Replacement and optimal fishery yield*

Adding the requirement that one must maintain a certain level of SPR complicates the question of what is the combination of lower size limit and fishing mortality rate that produces the highest yield (Fig. 11.4a). Adding an SPR requirement turns the YPR analysis into a constrained maximization problem; one wants to find the highest YPR that permits some minimum SPR. To illustrate what this problem looks like, we plotted the isopleths of SPR for North Sea plaice in Fig. 11.4c. To visualize the effect of the constraint, one would overlay the contour corresponding to the target SPR on the YPR isopleths (e.g. SPR = 40 percent, the white curve on Fig. 11.4a,b), and only choose solutions above that contour. Interestingly, the constraints tend to run parallel to some of the yield isopleths in Fig. 11.4a, so a range of possible *F* values could satisfy the constraint and achieve a particular level of yield. Botsford and Hobbs (1986) provided an analytical solution to this problem, including the additional issue of choosing the level of *F* (along the solution surface) that also accounted for the cost of fishing (so inefficient, high levels of *F* would be avoided). As a practical matter, at this stage, the goal of "optimization" probably puts too fine a point on the question to be of practical use, but it is informative to view the playing field and see how the different components interact.

Given the similarity in the shape of the dashed and solid curves in Fig. 11.4a, it is perhaps not surprising that given the usual two fishery management goals of (a) protecting stocks by maintaining a certain level of replacement and (b) seeking maximum yield, choosing a_c and F to seek one goal will often come close to achieving the other (the constrained optimum would be where these lines cross, which, given the uncertainty, could occur anywhere). Indeed, several publications reviewed in Clark (1991) showed that for a number of species with reasonable choices of stock recruitment functions, adjusting age limits and fishing mortality rates to yield 35–40 percent replacement also yielded something close to MSY.

11.3 The precautionary approach and modern fishery management

As both disenchantment with management using MSY as a management target (Larkin, 1977) and the number of overfished stocks continued to grow in the 1980s and 1990s, a precautionary approach to fishing was developed and promoted at the national and international levels of fisheries management (Garcia, 1994; Mace, 2001). The precautionary approach was motivated by a request in 1991 that the Food and Agriculture Organization (FAO) of the United Nations develop an International Code of Conduct for Fisheries. It was discussed at the International Conference on Responsible Fishing in Cancun, Mexico in May 1992, then it was presented at the UN Conference on Environment and Development (UNCED) the following month in Rio de Janeiro. Principle 15 of the Rio Declaration states that "in order to protect the environment, the precautionary approach shall be widely applied by states according to their capabilities."

11.3.1 *Precautionary management and reference points*

Precaution is also part of United States fishery law (the authors of this book are all U.S. scientists, so we confess to being most familiar with that jurisdiction, as one might guess from the examples we have used throughout this book). In the USA, the federal fishery law (the Magnuson Stevens Fisheries Conservation and Management Act, MSFCMA; originally passed in 1976, amended in 1996, and reauthorized in 2006) has ten National Standards, which are interpreted by the National Standard Guidelines, and these call for the precautionary approach in their specification of optimum yield in fisheries.

This precautionary approach involves a number of ideas that influence fisheries management globally (see Garcia (1994), Hilborn et al. (2001) for reviews). The first is the reversal of the burden of proof to favor the environment, which for fisheries means that the management agencies are not required to prove that a fishery is overfished, rather those that think otherwise should prove it is not. A second part of the precautionary approach is the implementation of *reference points*, which are specific population states that could be one of two types: target reference points and limit reference points. *Target reference points* are population states that are goals, such as MSY has historically been. *Limit reference points* were new, and they are pre-agreed lower thresholds for populations which—if breached—trigger specific, draconian management measures (e.g. a fishery severely restricted, or even closed). These have led to a number of different proposed targets and limits using various kinds of available data (Mace, 1994; Caddy and Mahon,

1995; Caddy and McGarvey, 1996). A key aspect of limit reference points is that they are "pre-agreed," and thus avoid the common pitfall of successively reanalyzing fisheries each time they reach a new low level, then declaring them fixed, with no real changes in management.

Another aspect of the precautionary approach is that management impacts on the environment should be reduced below uncertain calculated safe values, as a measure of precaution. An example of this was that the calculated value of fishing that would lead to MSY, F_{MSY}, was demoted from being a target reference point to being a limit reference point, i.e. an upper limit on fishing (this is somewhat arbitrary, since the MSY concept and the replacement concept depend on different modeling constructs). For example, F_{MSY} could be set as the upper limit reference point, and the target reference point could then be set as some fraction of F_{MSY} (e.g. ten percent lower). That fraction would represent a degree of precaution: more uncertainty in the estimate of F or F_{MSY}, or more natural variability in the fished population, would presumably call for larger fractions and thus lower target reference points (Caddy and McGarvey, 1996) (the reference points could also be expressed in terms of, say, the biomass of the fishery at MSY, rather than in terms of F). A key functional advantage of reference points was that they implied observations of state (and thus uncertainty in those observations) and feedback control, which provides greater assurance of sustainability than simply calling a yield "sustainable," because it satisfies an equilibrium condition under certain restrictive assumptions, as was done for MSY under the logistic model.

As an example of how the science of reference points evolved in the early 1990s, Clark (1991) used an estimate of the stock–recruit curves and life history parameters for a range of U.S. west coast rockfish (*Sebastes* spp.) and found that reducing SPR to 35 percent of the unfished value should produce something close to MSY for most species. This led to a management target of $F_{35\%}$ (later updated to $F_{40\%}$ after the potential for serially autocorrelated recruitment was accounted for). After a decade of management under that framework, it was recognized that that level of harvest was depleting stock biomasses to alarmingly low levels. A decade after the original analysis, Clark (2002) expressed concern that management targets of 35 percent or 40 percent SPR led to values of F that were actually greater than F_{MSY} for a range of possible stock–recruit curve shapes (he also admitted that the original selection of stock–recruit curve parameters may have been "sketchy"). Indeed, Mace and Sissenwine (1993) suggested that $F_{30\%}$ could be a conservative management limit (rather than a target) for data-poor stocks, as 80 percent of the stocks in their analysis have replacement LEP below that level (Fig. 11.5); they suggested that Clark's $F_{35\%}$ might be too close to that level to use as a target. Clark (2002) then suggested an alternative, more cautious management approach of estimating F_{MSY} from harvest data, then fishing at a fraction of that value. More recently, Hilborn (2010) pointed out that if we forget about MSY and instead try to obtain "pretty good yield" (a term originally coined by Alec MacCall), defined as 80 percent of MSY, then one can obtain satisfactory yield for a wide range of possible stock–recruit curve shapes, while also maintaining SPR (or total biomass) at what should be conservative levels. Modern management of west coast rockfish now uses both biomass and SPR reference points, as we describe in Section 11.3.2.

11.3.2 *Managing to avoid overfishing*

In the 1990s, fishery managers in the USA began to seek definitions of overfishing and ways to manage to avoid it (Rosenberg et al., 1993; Restrepo and Powers, 1999). Their indicators were organized around the two variables identified in Chapter 10 as being useful in avoiding extinction, (1) current abundance, relative to virgin (i.e. unfished) abundance, and (2) current replacement (relative to unfished replacement). Note that (1) is a state (the current abundance or biomass of fish in the water) and (2) is essentially a rate (the rate at which that state will grow into the future, given current harvest rates). If the estimated current abundance is less than a certain fraction of the virgin abundance (often 0.4), the population is considered to be *overfished*. If the estimated current replacement (LEP or SPR of incoming recruits) is below a certain fraction of the virgin replacement, typically also 0.4 (based on the analyses of replacement LEP discussed in Section 11.3.1), the population is considered to be *undergoing overfishing* (to relate this to our use of the term "overfishing" earlier, note that this would be recruitment overfishing, not merely growth overfishing).

These two definitions of overfishing, along with the associated target and limit reference points, are depicted in a two-dimensional diagram called a control rule (Fig. 11.7). The horizontal axis determines control of abundance (state), and the vertical axis determines control of replacement (rate). We have been discussing replacement in terms of SPR, i.e. will the current cohort of new recruits have SPR that is greater than the replacement threshold? In practice this is often expressed in terms of F rather than SPR, both because F is what is actually adjusted in order to change SPR (via the effects of F on the age structure), and because the SPR limit rule is often defined as F_{MSY} rather than in terms of SPR *per se*. At equilibrium, when the population is at its stable age structure, the proportional spawning biomass per recruit (relative to the unfished maximum) will be the same as the proportional population biomass (relative to the unfished maximum). In practice, an actively managed population in which F is being repeatedly adjusted will rarely be near equilibrium. For example, if a population became overfished and F was reduced so that it was no longer undergoing overfishing, the SPR of new recruits might be quite

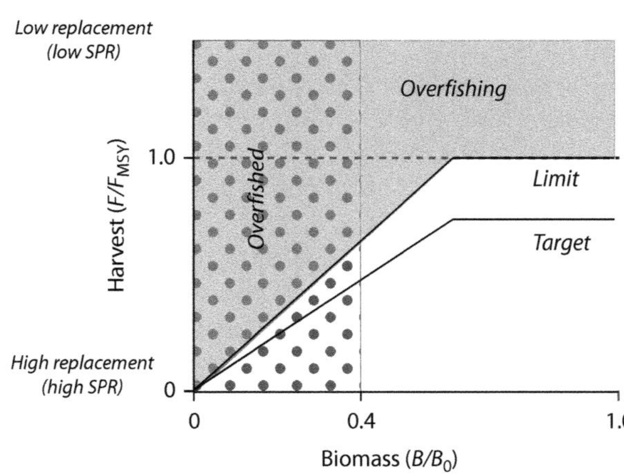

Fig. 11.7 Hypothetical example of a control rule like those used in U.S. fisheries. The horizontal axis is biomass, expressed relative to the unfished state (B_0); if biomass falls below the limit reference point (often 40%, as shown here) the stock is considered overfished (dotted region). The vertical axis represents current replacement, as represented by the fishing mortality rate (relative to the MSY rate). If catch exceeds the limit rate, there is too little replacement in the fishery, and the stock is undergoing overfishing (gray shaded region). In this example the replacement limit declines as biomass decreases, reflecting the management goal of slowing the approach towards the biomass limit (and having a faster rebuilding if that limit is crossed). Adapted from Restrepo and Powers (1999).

high (because those new recruits will experience a low F), but the population biomass may remain quite low until those recruits fill in the age and size distribution of the overharvested population. This filling-in process would follow the patterns we described for transient dynamics in age-structured populations in Chapter 4.

In Fig. 11.8 we show an example of this type of sequence of such changes in a fishery, this time for another west coast rockfish, the chilipepper, *Sebastes goodei*. Figure 11.8a shows the state variable, spawning stock biomass, and Fig. 11.8b shows the rate variable, SPR. This stock is managed with a target reference point of 40 percent depletion of SSB (i.e. 40 percent of the unfished value) and a limit reference point of 25 percent depletion of SSB; SPR is managed with a target (and limit) of 0.5 (Field et al., 2016). As the west coast fishing fleet grew and modernized in the mid-twentieth century, fishing effort (primarily commercial trawl fishing) on chilipepper gradually increased (indicated by decreasing SPR), which decreased SSB. SPR declined past the MSY level in the 1970s, which eventually led to SSB declining past both the target and limit reference points by the 1990s. This led to a major reduction in effort (and thus a drastic increase in SPR), which eventually led the stock to rebuild SSB above the target reference point. This example illustrates how a stock can experience all possible combinations of overfished (or not) and undergoing overfishing (or not); in Fig. 11.8c, we plot both the state and rate control variables in a phase plot (the latter is plotted as $1 - \text{SPR}$, a measure of the level of exploitation). In this way one can see the looping progression of the stock from healthy (lower right) to overexploited (upper left) and back again (this type of phase plot is also known as a Kobe plot, because this approach to visualizing fishery status was adopted by the joint tuna Regional Fisheries Management Organization process; also known as the Kobe process because the first group meeting was held in Kobe, Japan in 2007).

The present-day management of chilipepper—and most other fished species in developed nations—is driven by an integrated *stock assessment* process (Maunder and Punt, 2013), in an age-structured population model that is statistically confronted (using likelihood-based or Bayesian approaches) with multiple data streams (e.g. fishing effort, numbers of fish caught at each size or age, growth and aging data) to estimate unknown parameters (e.g. annual recruitment rates, or the natural mortality rate). The fitted model is then used to estimate SSB, SPR, or other quantities for comparison to reference points. The stock synthesis approach is a commonly used modeling framework for this type of assessment (Maunder and Punt, 2013; Methot and Wetzel, 2013).

As we have alluded to several times already, there is some uncertainty with the choice of management reference points. The threshold fraction of abundance that determines overfished status is somewhat arbitrarily chosen, and calculation of the threshold is subject to considerable uncertainty in estimating the original, unfished abundance (which had often been measured decades earlier, with different gear, environmental conditions, etc.). The required fraction of unfished replacement is, in some cases, more closely based on empirical information. As in the rockfish example (Fig. 11.6), we have estimates of the slope at the origin of the stock–recruit curve (and thus the minimum sustainable value of SPR) for those stocks that have been harvested down to low abundance. For those stocks we can also calculate the SPR at which MSY occurs, using an age-structured population model. On the basis of these values, reasonable fractions of required replacement are chosen for different taxa (Mace, 1994; Clark, 2002). Because the legal goal of fishery management in the USA is to ensure sustainable yield, the fractions of SPR used as limit reference points are typically chosen to be somewhat greater than the SPR for which the equilibrium population abundance would be zero.

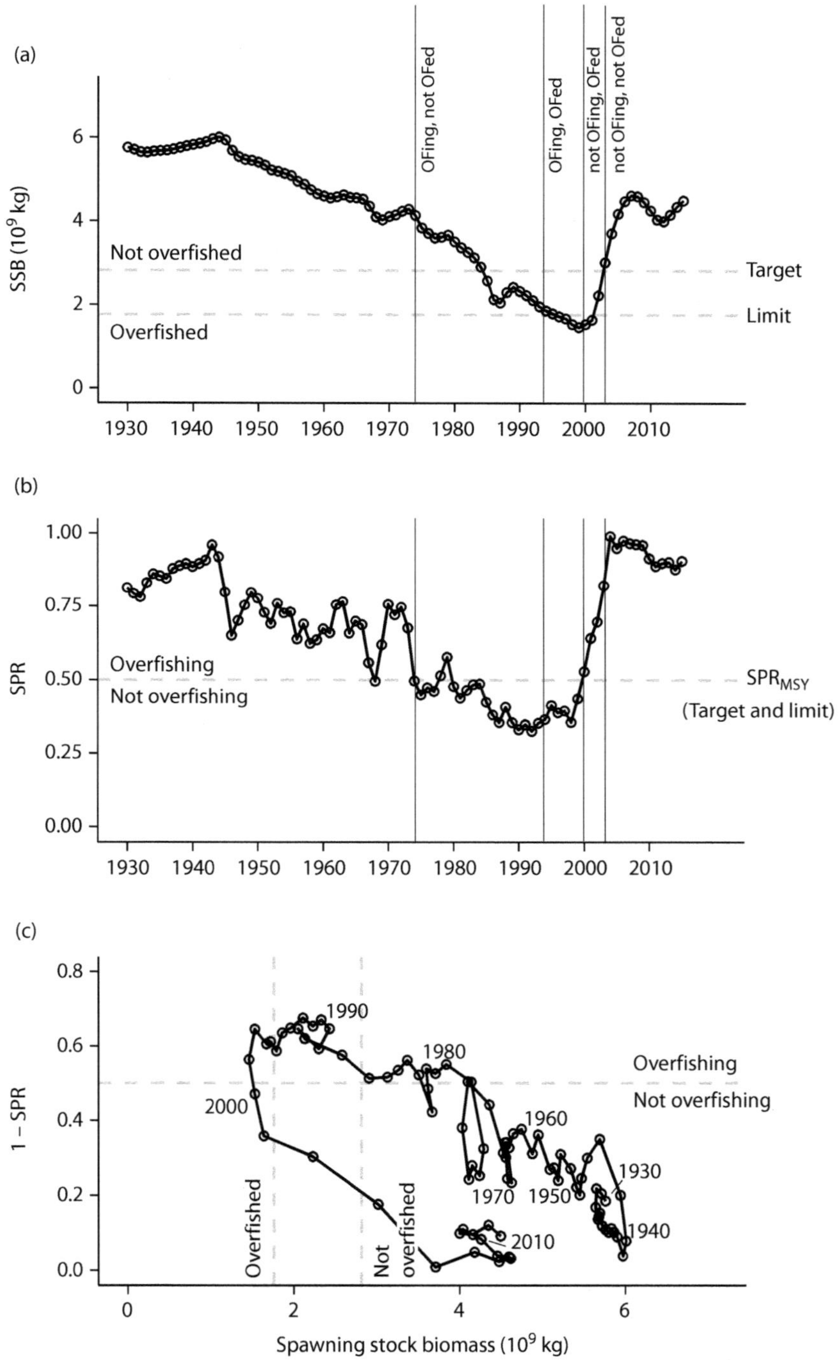

Fig. 11.8 Historical trends in the U.S. west coast stock of chilipepper, *Sebastes goodei*. (a) Spawning stock biomass time series (SSB; points and line) with target and limit reference points (dashed lines). (b) Time series of spawning biomass per recruit (SPR) relative to the unfished level (points and line), with MSY reference point (dashed line). (c) Kobe plot showing 1 − SPR and SSB in phase plane. Data obtained from the RAM legacy stock assessment database (ramlegacy.org; Ricard et al. (2013)).

11.4 Spatial management: marine protected areas

In the 1990s, management of marine populations for biodiversity and fishery yield by implementing areas with no fishing (or limited fishing), called marine protected areas (MPAs), started to become popular (Murray et al., 1999; Lubchenco et al., 2003). Of course, there had been a long history of restricting harvest in particular locations in traditional fishery management in Oceania (Johannes, 1978; McClenachan and Kittinger, 2013) and species- and gear-specific closures have been a common part of conventional, western fisheries management for many years (Hilborn, 2014). What changed in the later part of the twentieth century was a shift towards protected areas that protected all species, not just particular fishery species, and a global increase in the number and size of those protected areas (Wood et al., 2008). In part this rise was associated with the United Nations-led goal of "protecting" ten percent of the ocean by 2020 (Woodley et al., 2012). The goals of these protected areas were typically to protect habitats and the associated populations. Additionally, there were typically expectations that they would buffer against the uncertainties associated with conventional fisheries management (several important global fisheries had collapsed or were collapsing in the 1990s) and perhaps even allow biomass to "spill over" outside of the protected area and supplement fisheries (Lubchenco et al., 2003).

In response, in the late 1990s there was a growing interest in the science of protected areas, and how an understanding of population dynamics could inform their design and management. At the time, there was little understanding of how marine populations with a dispersing larval stage would persist when the landscape (seascape, actually) became heterogeneous, rather than presumably equally fished everywhere. Additionally, there were untested assumptions that populations inside protected areas would not be affected by fishing outside, but that the former could benefit the latter. Because these were poorly understood problems with important general questions, early modeling began with a strategic approach and somewhat simple, strategic models.

The umbrella term for the types of areas we discuss here is marine protected areas (MPA), which can include areas with some degree of fishing for some species. The term "marine reserve" is typically reserved for fully no-take areas. Because the focus of our discussion is primarily on the effects of harvest restrictions on individual species in MPAs, without regard for whether harvest is also banned for other species (or whether any other activities are prohibited) we will use the broader term "MPA" throughout this chapter.

11.4.1 *Strategic models of MPAs*

Strategic models of how MPAs might function were originally focused on the same two metrics we have addressed previously for conventional (i.e. non-spatial) management: population persistence (not going extinct) and fishery yield. As the popularity of MPAs grew, a number of claims regarding their benefits involved these two issues, but the patterns and dynamics of persistence and yield in a heterogeneous seascape were not well understood. We described the results of strategic models assessing persistence in MPAs in Sections 9.4.1 and 9.4.2, and we describe their consequences for the practical implementation of MPAs here. Results from the strategic models for yield in a spatially variable setting were not described in Chapter 9, so we will describe them here, along with their practical consequences.

In Chapter 9 we described how the persistence of spatially distributed populations with a dispersing larval stage had been addressed using two different strategic models: a patch-based model (Section 9.4.1) and a spatially explicit model of a linear coastline (Section 9.4.2), and we showed that they produced complementary results regarding the behavior to expect from a system of reserves. The patch model indicated that we could expect two kinds of persistence: self-persistence and network persistence. The former was a single patch (or a single MPA encompassing several patches) that had sufficient locally produced larvae returning to provide replacement. The latter occurred when no patch was self-persistent, but persistence was possible because of additional replacement arising from connectivity paths through the other MPAs, over multiple generations. Thus some species might persist in a single MPA, if enough of their larvae settled in that MPA. But if that self-connectivity were insufficient, they could still persist by depending on replacement paths through other nearby MPAs if connectivity and fishing were appropriate. For the purposes of illustration, we make these statements assuming all replacement went through MPAs, but of course, replacement would be provided through settlement in fished areas, it would just be less.

The practical bottom line of this complex result would be that single MPAs could provide for persistence of species with enough self-replacement (these would be species with short-distance dispersal). But longer-distance dispersing species could also persist, depending on how much total area was protected in MPAs (or, equivalently, how far apart MPAs were spaced), and how much fishing diminished persistence between reserves.

The spatially explicit strategic model of spatial persistence (Section 9.4.2) produced results similar to those from the patch model, but also provided more explicit information on the spatial scale. When there were many MPAs, persistence depended explicitly on the critical replacement threshold (CRT) from our previous description of conventional (non-spatial) management of single populations: the minimum fraction of the coast in reserves had to be greater than the CRT (or less, depending on the level of fishing outside reserves). A species in a single isolated MPA would persist if the alongshore width of the MPA was greater than the mean dispersal distance (though the exact minimum MPA width also depended on the CRT). This unanticipated dependence of spatial management on one of the fundamental uncertainties of conventional, nonspatial management meant that MPAs would not be producing the independence from uncertainty that some advocates were touting. It can be most clearly explained for the situation in which there is extremely heavy fishing so that we can assume no reproduction outside the MPAs. In that situation, if the fraction of coastline covered in MPAs is less than the CRT, MPAs will only protect species with mean dispersal distances less than their width. However, when the fraction of the coastline covered is greater than the CRT, MPAs will cause species with all mean dispersal distances to persist. If we relax the assumption that there is no reproduction from outside the MPAs, the threshold for this network effect becomes increasingly less than the CRT.

These basic concepts about MPA persistence were first developed by Botsford et al. (2001), who used a CRT of 35 percent as their baseline scenario (though they explored other values). That choice has led to some enduring "rules of thumb" for MPA design among readers who did not appreciate that the CRT may not be 35 percent for all species. We explore this potential misconception in Box 11.2.

Box 11.2 RULES OF THUMB FOR MPA DESIGN—IS 35 PERCENT A MAGIC NUMBER?

A recurring question in the science of MPA design is, what fraction of a coastline (or the global ocean, for that matter) should be protected? This is, at its heart, a policy question, not

(Continued)

Box 11.2 CONTINUED

a scientific question. Over the past decades there have been calls by scientists to protect 10–30 percent of the global ocean (O'Leary et al., 2016); these were codified to some degree in the 2010 addendum to the United Nations Convention on Biological Diversity that, in part, committed signatories to protecting ten percent of coastal oceans by 2020. Without delving into the scientific and political checks and balances that lead different scientists to advocate particular round-number percentages of ocean to protect, suffice it to say that the types of models presented in this book cannot provide the answer, though they can provide guidance for smaller-scale management scenarios of specific ecological populations or fisheries.

For example, if the desire is to ensure that fished species have persistent populations, regardless of harvesting happening outside MPAs, we showed in Chapter 9 (Fig. 9.6) that the proportion of the coastline required for persistence varies widely depending on species parameters, oceanography, and other factors. Thus it defies belief that it is possible to pick a single number that would apply globally or even to all of the species present in an MPA. That said, it is common for papers describing MPA models to use a value of approximately 35 percent as the minimum (O'Leary et al., 2016). This seems to derive from the analysis of Botsford et al. (2001), one of the first spatially explicit analyses of population persistence in MPA networks. In that paper, the authors used a value of 35 percent for their main results, because that was a commonly used estimate of the inverse of the slope of the stock–recruit function for many west coast groundfishes at the time, e.g. Clark (2002). Although Botsford et al. explored other possible values and showed their consequences (as in Fig. 9.6), the 35 percent result has become a commonly used baseline value, e.g. (White et al., 2010a; Pelc et al., 2010). It is important to remember that in most of these modeling exercises, the shape of the stock–recruitment function is assumed to be unknown, so it is prudent to assume a fairly shallow slope (e.g. 1/0.35, based on the precautionary limit reference point for replacement SPR estimated by Mace and Sissenwine (1993)), and that choice is easily justified by referring to the prior examples that did so. Similarly, the oft-repeated rule of thumb from Botsford et al. (2001) that MPAs must be approximately as wide as the average larval dispersal distance also depends on that critical 35 percent value (Fig. 9.6a). As always it is important to be vigilant about understanding which results of a model are due primarily to the choice of key parameters.

The second question addressed with strategic models was how fishery yields with MPAs would compare to fishery yields with conventional methods. There was great interest in the answer to this question in the 1990s and 2000s because meta-analyses of MPAs were showing that in some cases MPAs were producing two- to ten-fold changes in abundance and biomass relative to fished areas (Roberts and Polunin, 1991; Halpern, 2003; Lester et al., 2009). Without further analysis, it appeared that fishery yields could then be expected to increase similarly. The question was answered by both Mangel (1998) and Hastings and Botsford (1999) in different, but related ways. Using a logistic model, Mangel (1998) identified a critical invariant, which was the product of the harvest rate and one minus the proportion of the coastline in MPAs. An *invariant* is a number that is the mathematical combination of two or more parameters; one obtains the same overall results from a model as long as the invariant has the same value, regardless of what the values of the component parameters are. In this case, one could obtain the same fishery yield with any number of MPAs, as long as the harvest rate was increased proportionally to the fraction of the coastline in MPAs. This meant that catches obtained by either management approach (conventional or MPAs) could be equal.

Hastings and Botsford (1999) took a different approach. Using the simplest possible model of a fishery (simple age structure, implicit space, i.e. the model in Eq. 7.8), they showed that choosing the level of fishing effort that maximizes yield in a conventional fishery and choosing the proportion of habitat in MPAs that maximizes yield are equivalent choices (we detail their calculations in Box 11.3). Thus, again, either management approach could produce the same yield, or if both management approaches were used in concert they would produce that same total yield. This provides some context for interpreting the results of the meta-analyses we mentioned earlier: in most cases, the reason for high percentage increases in abundance and biomass in MPAs was that abundance and biomass had been fished down to very low levels prior to MPA implementation. The increase in abundance and biomass after MPAs were put in place did not mean that those populations would produce a sustainable yield any higher than the conventional MSY.

Box 11.3 EQUIVALENCE IN YIELD BETWEEN MARINE RESERVES AND CONVENTIONAL MANAGEMENT

To examine how marine reserves and conventional management would differ in terms of expected fishery yields, Hastings and Botsford (1999) used the simplest possible model of a fishery, with no age structure and implicit space. The state variable is the population density of adults in time t, N_t, adults reach maturity at age j, each adult produces b juveniles, juveniles experience density-dependent survival according to nonlinear function $f(bn_t)$ (f could be something like the Ricker or Beverton–Holt function; the exact form does not matter), and adults have annual survival s. They first considered a conventional fishery in which a fraction H of all adults are harvested each year. In that case, the dynamics of the adult population are

$$N_{t+1} = (1 - H)\left[f\left(mN_{t-j}\right) + sN_t\right]. \tag{B11.3-1}$$

This system has a steady-state equilibrium, N^*, that satisfies

$$N^* = (1 - H)\left[f\left(mN^*\right) + sN^*\right]. \tag{B11.3-2}$$

The total fishery yield is the term $H\left[f\left(mN^*\right) + sN^*\right]$, so rearranging Eq. (B11.3-2) to have that term on one side, one has

$$H\left[f\left(mN^*\right) + sN^*\right] = \left[f\left(mN^*\right) + sN^*\right] - N^*. \tag{B11.3-3}$$

If one wishes to find MSY, one must find the maximum of this expression,

$$MSY_{conventional} = \max\left\{\left[f\left(mN^*\right) + sN^*\right] - N^*\right\}, \tag{B11.3-4}$$

where the variable N^* (equilibrium adult population density) is what has to be adjusted to find the maximum (this is similar to the MSY example in Chapter 2, where we found the equilibrium population abundance that produced the highest yield). Then, to represent management using reserves, they took a spatially implicit approach and assumed that if a proportion c of the coastline were inside reserves, and larval dispersal were widespread (and adults are sedentary, so movement between reserve and fished areas is minimal), then the proportion of larvae spawned anywhere that settled into reserves would be c, and the proportion settling into fished areas would be $1 - c$. So the dynamics of population density

(Continued)

Box 11.3 CONTINUED

inside the reserve, $N_{r,t+1}$ would be

$$N_{r,t+1} = f\left(cmN_{r,t-j}\right) + sN_{r,t}, \qquad \text{(B11.3-5)}$$

with equilibrium

$$N_r^* = f\left(cmN_r^*\right) + sN_r^*. \qquad \text{(B11.3-6)}$$

If the fishery outside the reserve simply harvests all adults each year, maximizing yield requires maximizing the number of larvae that are spawned in the reserve and disperse to the fished area, $(1-c)\,f\left(cmN_{r,t-j}\right)$. With some algebra, it is possible to rearrange Eq. (B11.3-6) so that that expression is on one side and $\left[f\left(cmN_r^*\right) + csN_r^*\right] - cN_r^*$ is on the other (to do this, start by multiplying both sides of Eq. (B11.3-6) by c). Then the optimization problem has identical form to the conventional case (Eq. (B11.3-4)),

$$MSY_{MPA} = \max\left\{\left[f\left(cmN_r^*\right) + csN_r^*\right] - cN_r^*\right\}. \qquad \text{(B11.3-7)}$$

This means that choosing the level of fishing effort that maximizes yield in a conventional fishery and choosing the proportion of habitat in MPAs that maximizes yield are equivalent choices.

The implications of these two publications are best illustrated by numerical results from simulations by White et al. (2010b; Fig. 11.9), which mirrored earlier results by Holland and Brazee (1996). White et al. (2010b) used a spatially explicit model like the tactical models described in Section 11.4.2, but with an idealized linear coastline with evenly spaced MPAs. The top panel (Fig. 11.9a) shows how yield changes as the fishing rate increases (and lifetime egg production decreases) under conventional management (no MPAs): yield rises to a maximum, then decreases towards population collapse (in this example, the CRT = 35 percent). In this figure, the effects of fishing are expressed in terms of the values of fraction of lifetime egg production produced, FLEP (i.e. essentially SPR; Table 11.1). The bottom panel (Fig. 11.9b) shows management with MPAs (and larval dispersal that is very broad relative to the size of MPAs). If the fishing rate were low enough (i.e. FLEP high enough; the solid curve in Fig. 11.9b) for the population to be persistent without MPAs, adding MPAs (moving from right to left on the horizontal axis) would cause fishery yield to decline, whereas at higher values of fishing rate (lower FLEP) adding MPAs would cause it to increase initially (because yield is low or nonexistent from a population that is near collapse) then decline again for very high MPA coverage. Notice that the curve corresponding to FLEP = 0 (*scorched earth*) is nearly the same as the curve in Fig. 11.9a, which reflects the equivalency of the conventional and spatial optimization problems identified by Hastings and Botsford (1999). That is, adjusting the harvest rate in the conventional fishery is equivalent to adjusting the proportion of habitat in MPAs (if the harvest rate is 100 percent outside of MPAs). In this case the MSY under MPAs is slightly lower than the conventional MSY, because the model accounts for the larger biomass of older fish, and FLEP = 0 represents growth overfishing (the Hastings and Botsford (1999) model did not account for this effect).

Keep in mind that the very general result of equivalence in yield between MPAs and conventional management depends on the assumption that fishing is the same in all locations, that habitat quality and larval connectivity are also spatially homogeneous,

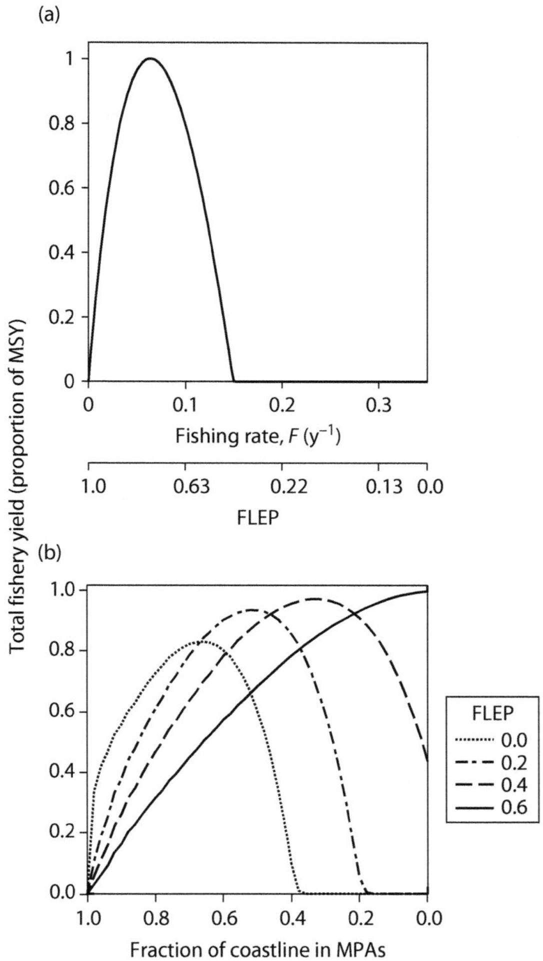

Fig. 11.9 Example of the relationship between equilibrium fishery yield and fishing effort, with and without MPAs. In this spatially explicit, age-structured model of a population along an effectively infinite coastline, the species had a symmetrical dispersal kernel with mean of 0, and the critical replacement threshold was a fraction of unfished lifetime egg prodution (FLEP) = 0.35. In (a), there are no MPAs; the horizontal axis is labeled both with the fishing rate, *F*, and the equivalent FLEP. In (b), MPAs were evenly spaced along the coastline, and the width of each MPA was 10% of the standard deviation of the dispersal kernel (i.e. network persistence conditions). The fishing rate was set to produce the indicated value of FLEP in non-MPA parts of the coastline. In both panels, the vertical axes show the total fishing yield at equilibrium (relative to MSY yield if the coastline were managed without MPAs). Redrawn from White et al. (2010b).

and that density dependence occurs primarily via competition among juveniles (i.e. it is intra-cohort density dependence). This equivalence begins to break down as we relax some of those assumptions. Gaylord et al. (2005) analyzed a spatially explicit, stage-structured model with inter-cohort density dependence (i.e. adults prey on or compete with juveniles), and found that managing with MPAs would lead to greater yields than in conventional management. They attributed this to accounting for size (stage) structure, because older, larger individuals could accumulate in MPAs and export larvae, and to the harvest of adults outside of MPAs improving the survival of juveniles there, increasing fishery productivity. Later, White and Kendall (2007) used a spatially implicit model to show that intercohort density dependence in which the abundance of adults diminishes recruitment (i.e. the Ricker form) was sufficient to produce greater yields with MPAs. More recently, a global meta-analysis (Barneche et al., 2018) revealed that fish reproductive output may not be isometric (proportional to biomass) but rather hyperallometric (increases nonlinearly with biomass), meaning that large fish in MPAs may be disproportionately valuable for replacement, relative to the same biomass of small fish outside MPAs. This also suggests that greater yields may be possible with MPA management (De Leo and Micheli, 2015) than using conventional management. Of course, none of these analyses accounts for other potential costs of MPA management, such as additional travel costs to new fishing grounds or the cost of enforcement (reviewed by Gaines et al., 2010).

11.4.2 *Tactical models of marine protected area design*

By the mid-2000s, strategic models of population dynamics inside MPAs had begun to be replaced with direct simulation of age- or size-structured models to act as tactical tools for actual MPA planning actions. For example, a 1999 California law mandated the implementation of a network of new MPAs along that state's coastline, which led to a planning process (2006–2013) that involved a series of spatial metapopulation models of increasing complexity over time (White et al., 2013a; Botsford et al., 2014). The presumption was that tactical models of specific species to evaluate proposed configurations of MPAs would produce results that largely matched the earlier strategic models.

As an example, one of the models used in the California MPA design process (White et al., 2013a) had age-structured dynamics (similar to those we described in Chapters 3–5) within each patch, with the addition of harvest. The model was intended to represent a relatively sedentary fish or invertebrate species that would not move out of a habitat patch as an adult, but which could disperse between patches during the planktonic larval stage, so it had the spatial structure of a metapopulation, like the models we described in Section 9.4. The abundance of individuals of age $a+1$ at time $t+1$ in patch i depended on the natural mortality rate M and the local fishing rate, F_i,

$$N_{i,a+1,t+1} = N_{i,a,t}e^{-(M+\varphi_a F_i)}, \tag{11.9}$$

where φ_a is the fishery selectivity (the proportion of age-a individuals that are vulnerable to the fishing gear). Length-at-age followed the von Bertalanffy equations (Box 3.8), and egg production was proportional to biomass. Then, the larvae spawned in each patch, $L_{i,t}$, were dispersed among patches according to the dispersal matrix $\boldsymbol{D}$ (with elements $D_{i,j}$ giving the probability of a larva dispersing from patch i to patch j),

$$\boldsymbol{S}_t = \boldsymbol{D}\boldsymbol{L}_t, \tag{11.10}$$

where $\boldsymbol{L}_t$ is the vector of $L_{i,t}$s and $\boldsymbol{S}_t$ is the vector of settling larvae. Settling larvae experience density-dependent mortality in each patch, following the Beverton–Holt function, before becoming the new first age class,

$$N_{i,1,t+1} = \frac{aS_{i,t}}{1 + aS_{i,t}/b}. \tag{11.11}$$

The a parameter in the Beverton–Holt function represents survival at low density within each patch, and b represents the maximum settler density in a cell.

In its initial formulation (Kaplan et al., 2009), the model operated on a habitat domain that was a linearized version of the California coast (i.e. each habitat patch represented one alongshore segment of the coast, extending from the beach out to the 3 nautical mile boundary of ocean waters under California's jurisdiction). Larval dispersal was represented as a one-dimensional kernel that was the same for all patches (as in Section 9.4), with a standard deviation set to approximate the estimated larval dispersal distance of each species being modeled. In later iterations the model represented space at a finer grain, with 1 km^2 cells representing habitat in two dimensions. The proportion of each cell that contained suitable habitat for the species being modeled (e.g. rocky benthos, for most kelp forest fish species) determined the maximum density of settling larvae, b. Larval dispersal was represented more realistically by obtaining the dispersal probabilities D_{ij} from the movement of simulated larval particles in a three-dimensional flow field from a numerical ocean circulation model (Mitarai et al., 2009; Drake et al., 2011). Finally, fisher behavior was made somewhat more realistic by allowing fishing effort to move around in space in a manner akin to the ideal free distribution (Fretwell, 1969), such that more effort was allocated to patches with more fish biomass, and at equilibrium, catch-per-unit-effort was the same in all non-MPA patches. Details of all of these model components are described by White et al. (2013a).

One simplification employed by most of the strategic and tactical models we have described so far is the assumption that adult fish remain within a single habitat patch, so that all movement and connectivity occurs in the larval stage. However, few fish remain in one place throughout their life; many move within a home range which can be observed and quantified (Secor, 2015). This will diminish persistence (because fish will move outside MPA boundaries and be vulnerable to catch), but have a positive effect on yield. Moffitt et al. (2009) extended the early version of the tactical models used in the decision making for California MPAs to include home range movement. The response of yield and abundance to increasing home range size depended on the type of persistence conferred by the larval stage: for self-persistent species (i.e. short mean larval dispersal distance), persistence and abundance declined gradually as home range size increased from zero, but for network-persistent species (i.e. longer mean dispersal distances), as home range size increased from zero, persistence gradually declined up to a certain home range size at which it dropped to zero, and yield gradually increased up to the same threshold range size, then also dropped to zero. This indicates the expected tradeoff between yield and persistence one would expect from fish swimming outside the MPAs, with the constraint that persistence is required to have any long-term yield at all. This type of adult movement was included in later tactical models (White et al., 2013a).

The results of these tactical models, with their much greater realism, largely followed the broad-brush patterns obtained from the strategic MPA models described in Section 11.4.1, thus instilling some confidence in the strategic conclusions. For example, in Fig. 11.10, we show results for red abalone (*Haliotis rufescens*), a species with short-distance larval dispersal and negligible adult movement, and two fishes with longer-distance larval dispersal, cabezon (*Scorpaenichthys marmorata*), with little adult movement, and black rockfish (*Sebastes melanops*) with a substantial home range. Each point on the figure corresponds to their calculated response to a proposed MPA configuration. We show both the response in ecological value (biomass) and economic value (yield). Based on earlier findings from the strategic models, results were calculated under two possible fishing scenarios: the fishery outside MPAs is overharvested (FLEP $<$ CRT) or sustainably harvested (FLEP $>$ CRT); the CRT used here was 0.35. Comparing the species in the overharvested scenario, we see that the pattern of increase with MPAs follows the order expected on the basis of the strategic persistence result: less movement leads to greater persistence (Fig. 11.10a). Abalone are self-persistent, and they persist inside an MPA of any size (regardless of total MPA coverage along the coast; cf. Fig. 9.7), whereas the black rockfish,

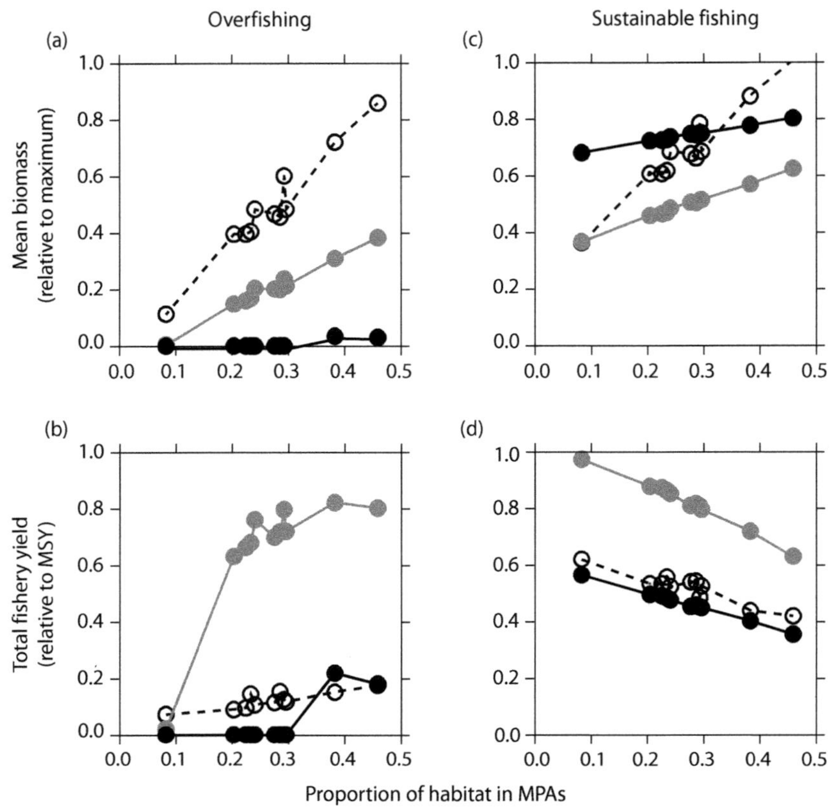

Fig. 11.10 Effect of eleven alternative marine protected area (MPA) networks on equilibrium population biomass, and fishery yield in the cases where the populations are overfished and sustainably fished, as predicted by a model used for tactical decision making in the North Central Coast region of California, USA. Panels (a, c) show mean biomass (relative to the unfished maximum), and panels (b, d) show total fishery yield (relative to the maximum sustainable yield without MPAs) for three species: red abalone *Haliotis rufescens*, (open circles), cabezon *Scorpaenichthys marmorata*, (gray circles), and black rockfish *Sebastes melanops*, (black circles). Each point is the region-wide average value for one MPA network for one species. Adapted from White et al. (2010b).

with broad larval dispersal and a large home range, can only have network persistence, once the total proportional MPA coverage is greater than the CRT (0.35), consistent with the earlier strategic results (Botsford et al. 2001). Turning to the responses of fishery yield in the over-harvested case (Fig. 11.10b), we see that abalone yield increases much less with MPA area than cabezon does because, while abalone persist in every MPA, this has little effect on the fishery because the larvae do not disperse far from the edge of an MPA. Black rockfish of course cannot contribute to fishery yield until it the population is persistent (i.e. after 35 percent coverage). When the fishery is sustainably managed without MPAs, biomass of all three species increases with fraction of habitat covered (Fig. 11.10c), but starts from a greater level than the overfished case (Fig. 11.10a). With sustainable fishery management, fishery yield declines with increasing area in MPAs because the amount of area being sustainably fished is declining (Fig. 11.10d), just as the earlier strategic results predicted (Mangel, 1998; Hastings and Botsford, 1999).

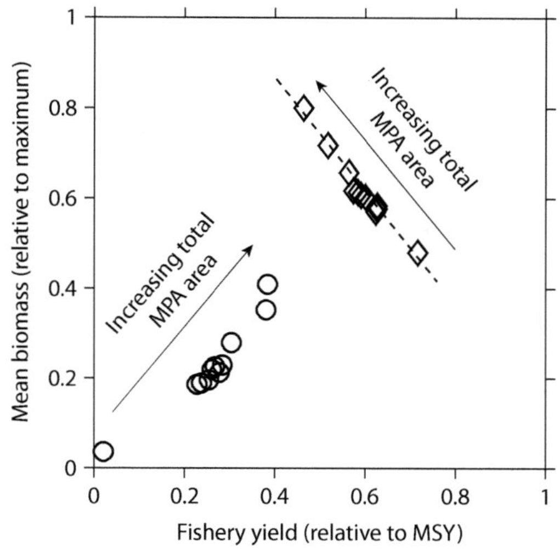

Fig. 11.11 Plots of the averages over species of biomass and yield from Fig. 11.10, to illustrate the tradeoffs (or not) between biomass and yield for the cases with sustainable management near MSY outside MPAs (diamonds) or overfishing outside MPAs (circles).

In the California MPA decision making process (White et al., 2010b; 2013a), the modeling results were typically presented to decision-makers in the form of the summary in Fig. 11.11, with the conservation benefits (biomass) and the economic benefits (fishery yield) of each MPA proposal plotted together (in this example, the results for the three species in Fig. 11.10 are averaged together for each proposal). This type of plot illustrates the basic tradeoff (or not) of MPA and fishery management in a single panel: if there is overfishing, both biomass and yield can increase with additional MPA area, but if the fishery is sustainable then there is a tradeoff between the two. Thus MPA planners must account for what is likely to happen with fishery management outside the MPA, and also choose where they think the system should be along the biomass–yield tradeoff frontier.

While the basic results in Figs. 11.10 and 11.11 seem to be consistent with the results of the earlier strategic models, the added value of the more realistic tactical approach is that it is possible to find some MPA proposals that are worse (or better) than expected for both biomass and yield, such as the few diamonds that are very slightly below and to the left of the tradeoff frontier defined by the dashed line in Fig. 11.11. Those proposals correspond to MPA configurations that may not include particular high-value patches, or perhaps lack particular connectivity pathways, e.g. Rassweiler et al. (2012). To identify such patches (or collections of patches) that would produce the best-functioning set of MPA networks (for either conservation or fishery goals), one can apply the same sort of metapopulation patch value metrics that we described in Section 9.3.3. In that section we described calculating patch value for the incidence function models, which only dealt in terms of patch colonization and extinction. However, the same sort of approach can be taken for models with internal patch dynamics, typically by constructing a connectivity matrix with elements c_{ji} that represent the average lifetime number of offspring per new recruit that will be spawned in patch i and disperse to patch j, assuming that densities are low and thus density dependence is minimal (similar to the matrices used to calculate metapopulation persistence in Section 9.4.1). One can then apply the eigenvalue-based metrics discussed in Section 9.3.3 to identify high-value patches, e.g. (Jacobi and Jonsson, 2011; Watson et al., 2011).

11.4.3 *Other types of models used in MPA design*

The focus of this book is on structured population models, so we have paid most attention to how that type of model can inform MPA design. Of course, there are other quantitative approaches to designing MPAs and MPA networks. For example, it is possible to describe the larval connectivity between habitat patches using spatial graphs, and apply the mathematics of graph theory to identify patches that are more or less well-connected to each other, e.g. (Treml et al., 2007; Thomas et al., 2014). In principle, this is little different from the type of patch-value calculations we described in Section 11.4.2, because that type of graph can be represented as a matrix with each element c_{ji} representing the edge length between nodes i and j. The danger in the way this method is typically applied is that it addresses larval connectivity probabilities only (i.e. the probability that a single larva will travel from patch j to patch i), without also considering how many larvae will be produced per individual settling in each patch, and whether particular patches have sufficient connectivity to persist over time, i.e. actual population dynamics (Burgess et al., 2014).

A second very common MPA design tool is the software Marxan (Ball and Possingham, 2000). Marxan is a descendant of an older method named SPEXAN, which is an abbreviation for "SPatially EXplicit ANnealing" (Marxan is just marine SPEXAN). In short, this method identifies the optimal configuration of reserves, given a set of planning units (e.g. habitat patches) and planning constraints (e.g. economic cost of protecting each patch, minimum reserve size). The optimization routine uses an annealing algorithm, meaning that it switches around a lot of different randomly chosen configurations at first, and then "cools" as the algorithm proceeds and gradually hones in on the optimal set of possible configurations. Marxan is widely used and can handle a wide variety of planning information and design constraints. The major critique of this approach is that it does not account for population dynamics, but rather assumes that protecting a particular habitat patch will automatically also protect the species that occupy that habitat. As we have seen, that is not always the case, particularly for widely dispersing species. However, it is possible to include population dynamic calculations into the Marxan routine (White et al., 2014b).

Finally, in the absence of sophisticated numerical methods for MPA network design, some have suggested designing networks of MPAs using simple guidelines for the minimum size of individual MPAs and the maximum desirable distance between adjacent MPAs along the coast: so-called "size and spacing" rules. These are applied to networks that contain many species. These are favored by global-scale planners because any specific combination of size and spacing will specify a fraction of coastline to be set aside in MPAs, which is the form in which international conservation recommendations are made.

In principle, size and spacing guidelines could be obtained from the results of strategic models like Botsford et al. (2001) for each species in the network: find the minimum MPA size needed for self-persistence (vertical axis of Fig. 9.6), then find the total of the coastline needed for network persistence (horizontal axis of Fig. 9.6), and divide the latter by the former to determine how many MPAs are required, and thus how far they should be spaced (in practice one would also want to account for potential adult movement range in the minimum size estimation, as well as accounting for differences among species if the MPAs are intended to protect more than one species). This is not a completely unreasonable approach to MPA design in low-information settings, but it would provide a different answer for each species. It is worth thinking about such guidelines from a population dynamics standpoint. If it is possible to have a minimum MPA size that allows self-persistence, presumably of a short-dispersing species, then adjusting spacing will potentially add a greater number of self-persistent MPAs, but do little to promote connectivity among reserves (because dispersal is so short). If instead a species has long dispersal distances, then network persistence is not possible until the total reserve area increases (i.e. spacing decreases) to the persistence threshold. That is, there is no marginal benefit to increasing size or decreasing spacing, unless spacing is close enough to allow persistence (as seen for the black rockfish in Fig. 11.7a). In other words, one should not expect smooth increases in MPA performance as size increases or spacing decreases, but rather big jumps when persistence thresholds are crossed. Moffitt et al. (2011) explored these ideas in more detail.

11.4.4 *Adaptive management of MPAs*

Ideally, MPAs are managed using adaptive management (Walters, 1986; Grafton and Kompas, 2005). In general, we are interested in whether placing an MPA (or MPAs) leads to an increase in the abundance or biomass of fished species (though there are other possible goals of MPAs which we do not discuss here, such as protecting unique habitat features). This means that adaptive management would entail monitoring those populations over time after MPA establishment and comparing those observations to the expected response, then adjusting MPA management accordingly (e.g. enlarge the MPA or perform more enforcement if poaching is expected). As we mentioned in Chapter 4, generating those expected initial transient population responses requires a different analysis from the long-term equilibrium results commonly used in the MPA design process (e.g. Fig. 11.10).

An important problem in the adaptive management of MPAs is how to represent the projections of "successful" MPA performance for comparison to the empirical results obtained by monitoring the populations inside the MPA. MPAs have been predicted to accomplish many things: increase abundance, increase biomass, increase biodiversity, etc.; which should we choose? The first thing that happens when fishing is removed from a population is that the age structure, which has been truncated by fishing, will begin to fill in. Because this step is a precursor to almost any proposed benefit of MPAs, and it is essentially the reversal of the effect of fishing, we have chosen it as the general indicator of the fact that the MPA is "working" for a species.

The form of the population response to MPAs depends quite a bit on larval connectivity. If there is little connectivity, so that most larvae settling into an MPA were spawned in that MPA, then the population will be relatively demographically closed, and population dynamics could be described by the transient response of an age-structured model to a perturbation (the removal of fishing is the perturbation) as it returns to the stable age distribution. We presented the calculations relevant to that scenario in Section 4.2.3. Alternatively, if most of the larvae settling in the MPA are spawned elsewhere, then the MPA will be demographically open and a different analysis is required, which we present here. White et al. (2013a) described both scenarios and evaluated what relative amount of external larval supply is required for dynamics to track the open versus closed population scenarios.

Immediately following the implementation of an MPA, the first consequence of reducing fishing to zero will be for the population age structure to begin to "fill in" as the population returns to its unfished state. In Box 11.4 we derive an analytical expression describing how that process proceeds. If we express the abundance at some time t after MPA implementation, $N(t)$, relative to the initial, pre-MPA abundance, $N(0)$, we obtain a remarkably simple expression,

$$\frac{N(t)}{N(0)} = (M+F)\left[\frac{1}{M} - \left(\frac{1}{M} - \frac{1}{M+F}\right)e^{-Mt}\right]. \tag{11.12}$$

As t increases, this equation approaches the solution

$$\frac{N(t)}{N(0)} \approx \frac{M+F}{M} \tag{11.13}$$

asymptotically. So the abundance will ultimately change by a factor of $(M + F)/M$, and the difference between fished abundance and unfished abundance will decline as e^{-Mt} (i.e. the rate at which the second term in brackets in Eq. (11.12) approaches zero). Therefore, the population should reach 90 percent of the eventual value on a time scale of about $2/M$, and 95 percent of the value on a time scale of about $3/M$ (Fig. 11.12). The key implication of this result is that the responses to implementation of MPAs will vary with local levels of fishing, and that the factor of increase is independent of the amount of larval recruitment. Kaplan et al. (2019) extended these results by deriving an expression similar to Eq. (11.13) for biomass, which increases to a higher level—but more slowly—because older fish are also much larger (see Box 11.4). They also derived an expression for how variability in larval recruitment translates into variability in this response ratio.

Box 11.4 "FILLING IN" AGE STRUCTURE IN BIOMASS INSIDE MPAS

To see how the increase in the abundance of older fish inside a new MPA will unfold over time, note that the MPA implementation will separate the age structure into two parts: those individuals that were already in the population at the time of implementation, and those that recruit to the population after the implementation. Assuming there is some age a_c at which fish become available to the fishery, the effect of the MPA will only be felt by the part of the population that is age $\geq a_c$, so we only consider that part of the age structure. Thus the two parts of the population (already recruited and will recruit after the MPA) will be separated at $a_c + t$, where t is the time since MPA implementation (that is, the fish already in the population at $t = 0$ will keep getting older). The younger fish, N_{low}, will only experience natural mortality, M, so we can determine their abundance at time t from

$$N_{low}(t) = Re^{-Ma_c}\int_{a_c}^{a_c+t} e^{-Ma}da, \tag{B11.4-1}$$

where Re^{-Ma_c} describes a constant rate of external larval settlement from the plankton (R) and the survival of those settlers to age a_c. Evaluating the integral yields

$$N_{low}(t) = Re^{-Ma_c}\frac{1}{M}\left(e^{-Ma_c} - e^{-M(a_c+t)}\right). \tag{B11.4-2}$$

For the older age classes, N_{high}, they will have experienced both natural mortality and fishing mortality F prior to the MPA, and only M since, so the expression for their abundance involves terms for both $e^{-(M+F)}$ and e^{-M},

$$N_{high}(t) = Re^{-Ma_c}\int_{a_c+t}^{\infty} e^{-Ma}e^{-(M+F)(a-(a_c+t))}da = Re^{-Ma_c}\frac{e^{-M(a_c+t)}}{M+F}. \tag{B11.4-3}$$

It is also useful to calculate the initial abundance, just before implementation,

$$N(0) = N_{low}(0) + N_{high}(0) = Re^{-Ma_c}\frac{e^{-M(a_c)}}{M+F}. \tag{B11.4-4}$$

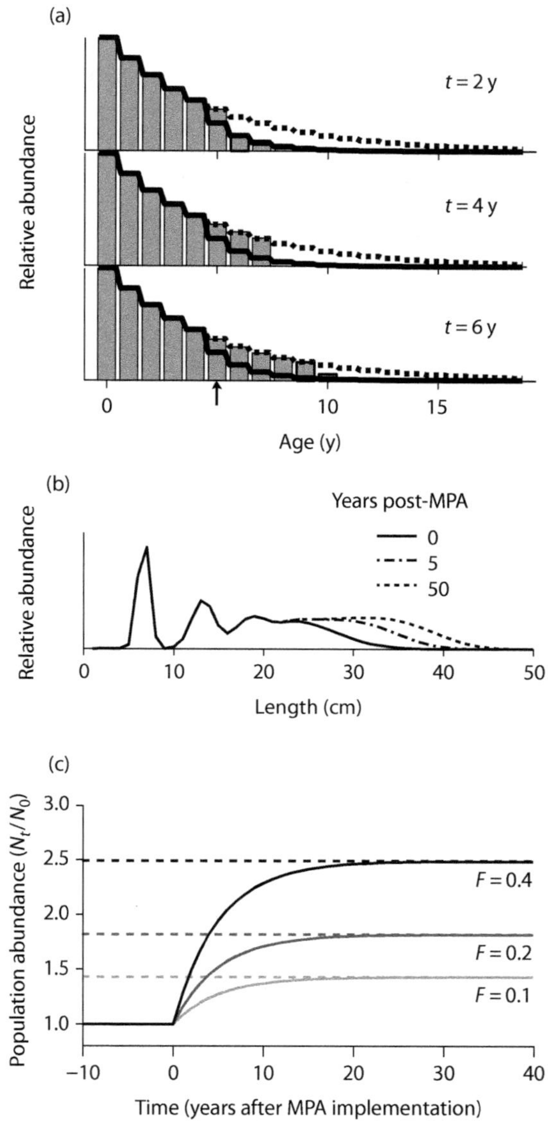

Fig. 11.12 Transient population dynamics following MPA implementation. (a) Filling in of the age structure at $t = 2$, 4, and 6 years after fishing is stopped, for the case with constant larval recruitment. The black curve indicates the unfished stable age distribution; the dotted curve indicates the age distribution under fishing ($F = 0.8\ y^{-1}$). The arrow indicates the age at which individuals enter the fishery (a_c). (b) Filling in of the size distribution for the same scenario depicted in (a). (c) The pattern of population increase (measured as the ratio of abundance at time *t*, relative to the starting abundance, for fished age classes only), for three different values of *F*. The dashed lines indicate the eventual asymptotic abundance, $(M + F)/M$. Adapted from White et al. (2013a), using parameters for kelp rockfish (*Sebastes atrovirens*; $M = 0.2\ y^{-1}$).

Box 11.4 CONTINUED

Then we can obtain a remarkably simple expression by normalizing the post-MPA abundance by the initial abundance,

$$\frac{N_{low}(t) + N_{high}(t)}{N(0)} = \frac{M+F}{M}\left(1 - e^{-Mt}\right) + e^{-Mt} = \\ (M+F)\left[\frac{1}{M} - \left(\frac{1}{M} - \frac{1}{M+F}\right)e^{-Mt}\right]. \tag{B11.4-5}$$

We discuss this equation in the text as Eq. (11.12) and in Eq. (11.13) show that as t becomes very large, the right-hand side approaches $(M + F)/M$.

One can follow a similar derivation to obtain the expression for increase in biomass, $B(t)$, rather than abundance. For this, we take advantage of the biomass-at-age expression originally derived by Beverton and Holt (1957), and which we derived in Box 11.1, Eq. (B11.1-8). Substituting that expression for biomass into Eq. (B11.4-2 and B11.4-3), then simplifying, we obtain a solution similar to Eq. (11.13), where once again k takes on values of 1, −3, 3, and −1 for $i = 0, 1, 2, 3$,

$$\frac{B(t)}{B(0)} = \frac{\sum_{i=0}^{3} K_i \frac{e^{-(M+ik)a_c}}{M+ik}}{\sum_{i=0}^{3} K_i \frac{e^{-(M+ik)a_c}}{M+F+ik}}. \tag{B11.4-6}$$

This expression is, of course, more complicated than the simple $(M + F)/M$ result, but does lend some insight. First, as in Eq. (11.13), there is a term similar to $1/M$ in the numerator and one similar to $1/(M + F)$ in the denominator, so the parallel to $(M + F)/M$ is there. The addition of the k and i terms adds complexity, but because they appear as a product in a negative exponential term, their influence diminishes with increasing magnitude. Thus the most influential term will be $i = 1$, $k = -3$, which adds a negative number to both denominators. This will serve to increase the overall ratio, meaning that the biomass ratio will always be greater than the abundance ratio, and that the biomass ratio will approach its asymptote more rapidly (Kaplan et al., 2019).

The problem with applying the result in Eq. (11.13) to MPA management is that it depends on the local, site-specific value of F, but the spatial scale at which fishing mortality rates are estimated in stock assessments is typically hundreds of kilometers. This has motivated development of a new *state space* method for estimating local values of fishing mortality rate from size distribution data using integral projection models (Chapter 5). That new method (White et al. 2016) also addressed one other limitation of the $(M + F)/M$ result and the types of projections in Fig. 11.12, which is that the population is not likely to be at an equilibrium age structure when the MPA is implemented (as Eq. 11.12 implicitly assumed), rather the age and size structure at $t = 0$ will reflect the recent history of recruitment variability (as we saw in other size-structured populations in Chapter 5). Therefore it is important to account for that variability and make projections starting from the actual initial age or size distribution of the population (Nickols et al., 2019).

11.5 What have we learned in Chapter 11?

Our historical description of the different ways of thinking about the population dynamics of fisheries began in the 1950s with two different points of view: (1) a logistic model linked to catch and effort data, and (2) an age-structured model with cohort survival and growth linked to biological data. There was a common awareness that the latter would need an additional description of reproduction and a density-dependent recruitment function in order to be a complete population model like the logistic model, but limited computing power and a paucity of the appropriate data originally prevented that. The cohort model indicated that the values of F and a_c that produce the maximum yield from a cohort (yield per recruit, YPR) would be obtained by not fishing until the biomass of the cohort reached a maximum, then catching all of the fish ($F = \infty$). The yield isopleths were "cupped" around the peak at infinity, but the maximum YPR was easily moved to a peak at lower values of F by including the cost of fishing in the model. The raw effect of fishing on reproduction could be seen by plotting eggs per recruit (or spawning biomass per recruit, SPR) on the same graph; this allowed one to choose the maximum yield that would maintain a certain minimum egg production.

In the 1980s greater attention began to be paid to preventing overfishing, leading fishery analysts to account explicitly for the loss of replacement, i.e. the reduction in lifetime egg production (instead of depending just on having the word "sustainable" as the S in MSY). This effort was aided by the development of the precautionary approach, which involved the use of reference points, indicators of fishery status that could be used to direct management responses through control rules. These could be target reference points (i.e. goals) or limit reference points (pre-agreed thresholds beyond which draconian reductions in harvest would be implemented).

While reference points add valuable formal structure to the decision making, they are still hampered by a fundamental uncertainty in the minimum values of abundance and replacement necessary for sustainable fisheries. Fishery management in situations with adequate infrastructure typically monitor two reference points: (1) estimated abundance or biomass and (2) replacement, calculated as a minimum value of LEP or SPR, but operationally represented by values of F.

In the 1990s fishing began to be controlled by setting aside areas with no fishing, marine protected areas (MPAs). This explicitly created fish populations distributed over heterogeneous space, which required a new approach to estimating persistence and yield. We were able to apply results from Chapter 9 to show that, depending on the spatial scale of larval dispersal, populations could have self-persistence in a few large MPAs, or network persistence in a network of many smaller MPAs. In practice, this means that species with different dispersal patterns and life histories will respond differently to any given MPA configuration. Unfortunately, determining the minimum persistence thresholds for MPAs requires knowing some of the same uncertain quantities as conventional fisheries management (e.g. the slope at the origin of the stock–recruit curve), which dampens the value of MPAs as a less uncertain alternative management approach.

Using both a simple strategic model and complex tactical models we saw that there were fundamental similarities between conventional fishery management and MPA management. The two afford similar maximum sustainable yields (though depending on the ecology of the species, MPAs might produce somewhat higher yield). Additionally, the management in fished areas affects MPA performance; MPAs will increase both biomass and fishery yield only if the fishery is overexploited. If it is sustainably managed, then MPAs only reduce catches (recognizing that there may be other, non-fishery goals for MPAs).

Finally, we examined transient population dynamics in MPAs, which is crucial for adaptive management. Unfortunately, we found that the pattern of population increase depends strongly on the level of fishing prior to MPA implementation, which is rarely known.

CHAPTER 12

Thinking about populations

In this book we have seen that population dynamics is a collection of ways of understanding different aspects of how populations function. Each of these has been encountered by researchers over the years, and the understanding of population dynamics has improved as a result, particularly in the past several decades. We hope this book demonstrates how the development of this field is at a point that it can be viewed as a unified, coherent whole, worthy of studying and mastering before embarking on addressing a population question. Perhaps the best example of this cohesiveness is Taylor's (1979) analysis of the duration of the cyclic transient phase of annually emerging insect populations by determining their first three eigenvalues from life histories. Decades later his results (redrawn in Fig. 4.7c) helped show that transient cycles will be stronger when reproductive age distributions are narrower. Later, in the study of cohort resonance, they helped us understand how sensitive different populations of fish would be to certain frequency bands of environmental variability. An example is sockeye salmon, which were shown to be more sensitive to frequencies near $1/T$ with narrower spawning age structure (Fig. 4.15c,d,e,f).

A benefit of population dynamics being a coherent whole is that studying the general field of population dynamics allows one to benefit from earlier population analyses by others, without having to "re-invent the wheel." A downside of the coherence of this field, on the other hand, is that unlike basic statistics, population dynamics cannot be compartmentalized into a toolbox format. This problem arises in teaching a course on population dynamics, when students more used to their statistics classes do not attend class until the lecture is on a topic in which they are interested; they often find that they cannot understand that lecture because they lack the material leading up to it. It can also arise among practicing ecologists, who might be tempted to pick a model "off the shelf" for a particular problem—perhaps because it is the type of model they are most familiar with—without considering the origins of the structure and the often hidden assumptions of that model and their consequences for the results they will obtain.

In this chapter we first recount how we arrived at the models and analyses that we have presented, to impart some notion of what should be expected of them. We begin with the initial philosophical considerations, and choice of mathematical approaches. Next we describe some of the principles that emerged from our analyses of these models. Because of the complex and uncertain nature of the ecosystems in which these populations exist, these principles take the form of general trends or rules of thumb. For example, we can say that as the value of some factor x increases, some quality of the population (e.g. persistence, stability, cyclicity) will increase or decrease. These principles are few and they are simply different, potentially useful ways of viewing and interpreting population dynamics as a whole. Last in this chapter, we describe how valuable this type of understanding is in management applications, because it can direct the focus of monitoring and data collection in order to measure or detect key quantities, or test specific

Population Dynamics for Conservation. Louis W. Botsford, J. Wilson White, and Alan Hastings, Oxford University Press (2019).
DOI: 10.1093/oso/9780198758365.001.0001

hypotheses. However, the principles we have explored in this book are not sufficient by themselves for actually managing population, because that requires specifying values of catch or areas to be protected, based on data. Essentially, we need the parameters to populate the models we have described in this book. From a societal point of view, this is the major challenge facing the field of population dynamics for conservation: we have largely developed a sensible understanding of the ways in which different types of populations operate, but we need more data to better quantify the actual breakpoints where population behavior changes, especially for the boundary between persistence and extinction.

12.1 Modeling philosophy and approach

The relevance of discussing modeling philosophy may not have been readily apparent when first encountered, but it can be a valuable guide that helps to identify needed characteristics of models and avoid nonsensical models. For example, here it led to the conclusion that whether one was using a model in a deductive scheme to answer a scientific question, or an inductive scheme to manage a population, models needed to possess what Levins (1966) called realism. This led to greater attention to carefully defining what a *state variable* is (Caswell et al., 1972; Caswell, 2001). That definition was the foundation of the organization of this book in terms of different chapters focused on models with different *i*-state variables (age, size, space), and different ways that we classified individuals.

Our choice about which model and which *p*-state variable to use was based in part on the different levels of ecological organization (Section 1.5). We realized that the variables on which many population models are based, total abundance (N) and total biomass (B), were not adequate descriptions of state because individuals of different ages and sizes contributed to reproduction and survival in different ways. Indeed, age is a natural choice as an *i*-state variable because the dynamics of any system, biological or physical, depend on the arrangement of different internal time-lagged effects, and age is in the same units as time. The first age-based population model we examined (the M'Kendrick/von Foerster model) was formulated with a strict conservation quality (i.e. it kept track of what happened to all individuals at each point in time), and was based on the rates at which individuals disappear and appear, that is die and are reproduced. More precisely, individuals that are at a given age at a given time either grow older with time or die. This view was based on our understanding of the levels of ecological integration, and that populations are made up of individuals. The equation with this conservation characteristic was the same as the continuity equation used in physics to describe the flow of incompressible fluids. The conservation characteristic showed up when we solved the model to follow a cohort as it ages, and observed that the only change in numbers is due to mortality.

The principle of continuity contained in this equation even predates the invention of calculus. Early in the sixteenth century, Leonardo da Vinci observed that water flowed faster through narrow parts of ditches than through wider parts, and described a verbal model to explain this (Klein, 2010). He pictured a group of men dancing a Polonaise through a narrow alley. The rule was that they had to maintain the same distance from each other. Where the alley opened into a wider spot their speed along the street had to be less. This concept is akin to the result that size distributions are lower where the growth rate is higher in the red sea urchin stand distribution, in Section 5.2.

Age-structured models, describing how the distribution of individuals over age, $n(a, t)$, changed with time, allow us to understand many aspects of population dynamics. However, age is not easily determined for many species, and the ecological interactions among individuals within a population, and with exogenous entities (resources, predators) are commonly determined by size, rather than age. This complicates our ability to describe population dynamics, which inherently depends on time lags, as caused by age differences. We got around this problem by formulating influence functions in terms of size-at-age. For example, in some models of cannibalism in Dungeness crab in Chapter 7, the amount of cannibalism occurring at each age was based on weight-at-age, based on the known growth curve of the crab.

However, many interesting ecological phenomena occur as a consequence of temporal changes in growth curves of individuals. The relationship between age and size (i.e. the individual growth curve) is often changed dramatically by changes in individual growth rates, which can change population equilibria and dynamics. For example, the proposed multiple equilibrium phenomenon in Botsford (1981); Section 7.6) was an example of a mechanism including a temporally varying growth rate. There, harvesting reduced population densities, leading to an increase in individual growth rates that then changed the equilibrium condition, which locked populations into a new, lower-density equilibrium. Also, in the cohort resonance study of the effects of environmentally induced temporal variability in growth rate (Worden, et al., 2010), we represented temporal variability in growth by varying the mean of a fixed distribution of spawning over age, whose mean varied (i.e. a cohort with a lower mean age of reproduction representing faster growth, and vice versa).

Other than these approximations, we did not pursue models in which changing growth rates change dynamics. There is at least one comprehensive other source of further information in this understudied area. In the first chapter of their book, de Roos and Persson (2013) lament the fact that "Basic ecological models thus ignore without much qualm the most prominent process in an individual's life history, one that is unique to biological systems and has no counterpart in physical and chemical systems." They go on to explore a variety of models of physiologically structured populations with variable growth and ontogeny.

The habit that ecologists sometimes have of ignoring ontological development has also influenced the development of models with stage as a state variable. The development of stage models was based on the question of which *i*-state best described reproduction, with little regard for how rapidly individuals grew to that reproductive state (i.e. their ontogeny). Also, as a practical matter, the wide use of stage-based models occurred primarily because the timing of ontogenetic development (i.e. their growth curve) was not known. One of the examples of more recent models that we have described, that of the island fox (Chapter 10), integrated into management the kind of tagging studies that could alleviate this problem.

12.2 Replacement, an organizing principle

In our mathematical search for (non-mathematical) principles of population dynamics, the role of replacement in persistence and other behaviors was foremost. Its simplest version, that populations will not persist unless each individual reproduces enough to replace themselves, is most important, and widely known, even to the general public, but in this book we have seen that the role of replacement goes much farther, governing population behavior through age-dependent ***influence functions*** in both static as well as dynamic ways. The "static role" was determining the equilibrium, and that role was played by the area under an influence function, lifetime reproduction (the integral over age of reproduction at age). In linear age-structured models that value was compared to 1, as the persistence threshold. This simple idea portrays population persistence as a population being unstable about the equilibrium of the zero state ($N = 0$). The "dynamic role" in linear age-structured models determined initial behavior before populations reached their asymptotic geometric behavior. That role was played by the shape of the same influence function (reproduction at age), as determined by the way that shape influenced the magnitude of the second and third eigenvalues compared to the magnitude of the dominant eigenvalue (Figs. 4.2 and 4.3; the damping ratio in Eq. 4.10). In biological terms, greater "narrowness" of the reproductive age structure led to longer lasting transient cycles at period T (this was Taylor's (1979) result).

For nonlinear, age-structured populations with density-dependent recruitment, the static role of replacement was described by comparing lifetime reproduction to the inverse of early life survival, rather than to 1 as the persistence threshold (Fig. 4.11). The amount of replacement that provided that level could be expressed as a fraction of the natural unfished (or otherwise unfettered) replacement, which was termed the critical replacement threshold (CRT). The dynamic role of replacement was then determined by the sign of the derivative of the egg–recruit function at equilibrium, being one of two different mechanisms depending on its sign. A negative sign indicated over-compensatory density dependence, and a deterministic, unstable population with cycles at a period of twice the cohort scale ($2T$). Examples of this type of behavior were the $2T$ cycles in Dungeness crabs, barnacles, and flour beetles in Chapter 7. A positive slope at equilibrium indicated compensatory density dependence with a deterministically stable equilibrium, but could also exhibit, in the presence of random environmental variability, cohort resonance. This is a stochastic mechanism describing greater sensitivity of populations to two frequency bands in random environmental variability: one at the cohort period (T) and the other at very long periods (low frequencies). In both the case of $2T$ cycles and of cohort resonance, the dynamic role of replacement depended on the shape of the influence function (narrowness again was destabilizing, e.g. Fig. 4.15).

The role of replacement in understanding extinction probabilities is more complex. In Chapter 8, environmental variability was added to a linear, age-structured model to describe probabilities of extinction for populations at low enough densities for density dependence not to be important. The static effect of greater variability, as reflected in the small fluctuations approximation was to reduce λ, indicating that replacement was diminished as variability and covariability among life history parameters increased (Eq. 8.2, Fig. 8.2b). The dynamic role of replacement in this case is described in terms of the distribution of elasticities over age (Figs. 3.6 and 8.2).

When we added consideration of populations being distributed over space in Chapter 9, our understanding of replacement changed in fundamental, unanticipated ways. In the more general, patch-based model used to answer that question (Hastings and Botsford, 2006), we expected, of course that populations could still be self-persistent. Pulliam (1988) had established the view that some populations could have more replacement than needed, and it was possible that individuals from those populations could disperse/migrate to other populations (sources), possibly to benefit persistence of populations without adequate replacement (sinks). Hastings and Botsford (2006) added the understanding that population persistence depended on replacement through loops, not just sources and sinks. A surprising aspect of the Hastings and Botsford (2006) result was that a persistent metapopulation could have no subpopulations with adequate self-replacement, and that lack of replacement could be made up through one or more of a new kind of replacement loop that went through other patches, over several generations (Section 9.5). More generally, in a stochastic environment this phenomenon can be a result of what has been called Parrondo's paradox (Williams and Hastings, 2011). So far the only studies we know of that have attempted a direct measurement how much non-self-persistent migration contributes to persistence are both in coral reef systems: Johnson et al. (2018), who used genetic parentage analysis to estimate replacement in a small fish, and Garavelli et al. (2018) who used simulations from physical circulation models to estimate planktonic larval exchange in spiny lobster.

The most direct application of this understanding of persistence over space has been the design of marine protected areas (MPA). Models that approximate the periodic placement of MPAs along an infinite coastline of actual space (Botsford et al., 2001) found that self-persistent MPAs were possible, and they would protect all species with dispersal distances less than approximately the width of the MPA. However, in addition, once the fraction of coastline in MPAs exceeded the "non-spatial" critical replacement threshold (CRT) for that species, the species could persist regardless of dispersal distance. Again, whether this connection between single population persistence and spatial persistence actually occurs has not yet been studied, but there is interest motivated by a desire to understand whether groups of multiple MPAs actually function as a network (Grorud-Colvert et al., 2014). This case could also initially be studied with physical circulation models to simulate the transport of planktonic larvae.

While the mechanisms underlying replacement in these two spatial models have not been studied or "verified," they have served to remind us what we should be sampling. As mentioned in Section 9.4.1, a number of studies of connectivity through meroplanktonic larval dispersal have been reporting a quantity not involved in replacement or persistence, and are thus not particularly useful for management (Botsford et al., 2009; Burgess et al., 2014), except under certain very restrictive circumstances (Lett et al., 2015).

The recurring theme of replacement as an organizing principle led us, in Chapter 10, to propose that replacement could be used—along with abundance—as a criterion for determining the status and recovery of endangered populations. In Chapter 11, we saw how that approach has been applied, with much success, in modern fisheries management. We revisit this point in Section 12.4.

12.3 Population responses to time scales of environmental variability

Time scales of variability are known to be an important facet of ecology (Levin, 1992; Hastings, 2010). We described recent findings that populations, particularly age-structured populations, are more sensitive to environmental variability at particular time scales. In Chapter 4, we described random time series in terms of their variance per unit frequency at each frequency, where 1/(frequency) is time scale. With that tool we were able to determine how populations transformed the time scales of exogenously driven environmental variability into a different frequency distribution of population variability by filtering. The filtering is described by the population's frequency-selective transfer function, which is basically the Fourier transform of the age-specific influence functions we described earlier. This filtering amplifies variability at cohort time scales and very long scales, via the cohort resonance phenomenon (Bjørnstad et al., 2004; Botsford et al., 2014).

We pointed out in Chapter 8 that cohort resonance sheds light on an old question in ecology: what determines variability in populations, exogenous environmental variability or endogenous population dynamic mechanisms, e.g. Andrewartha and Birch (1954). Our improved understanding of the impact of environmental variability populations through cohort resonance provides an opportunity to see these separate roles of each, with the environmental spectrum representing the exogenous element, and the population transfer function representing the endogenous part.

The consequences of this reshaping of the distributions of time scales could be far reaching. The fact that climate change may be changing time scales of major drivers such as ENSO (Schmidt et al., 2018) elevates the importance of knowing the spectral sensitivity of populations, as does the finding that in one case at least, high levels of environmental variability in the cohort band of sensitivity could increase overall population variability, without increasing the probability of extinction (Kilduff et al., 2018; Section 8.5). From an ecosystem point of view it is notable that this sensitivity can arise in a bottom-up (variability in individual growth rate), as well as a top-down (variability in survival) direction (Worden et al., 2010).

12.4 Applying the lessons of population dynamics

Application of the information in Chapters 1–9 to the practical problems of conservation and management led us to change our approach, focusing more on a closer accounting of what can be known or estimated. That change means that we must inductively determine specific breakpoints where population behavior changes, rather than simply saying, for example, that persistence will be less likely if harvesting is greater. In terms of the value of population dynamics to society, this is the foremost challenge to population dynamics going forward. It needs to be met by continuing to observe and report observations of breakpoints, and determining how they vary across species and environments.

In Chapter 10, trolling the earlier chapters for clues regarding what to monitor to limit extinction turned up two indicators of population state that would be useful metrics of population state: current abundance and current replacement level. Most current approaches to managing to avoid extinction in conservation biology consider only the probability of extinction, but it would not make sense to associate recovery with a specified level of abundance if the replacement were below the replacement threshold. This expands the conventional approach beyond simply monitoring abundance, and it may bring more data to bear (i.e. age data for the estimation of replacement), but our ability to manage extinctions effectively is still limited by lack of knowledge of what are critical lowest possible persistent values of abundance and replacement. Existing analyses of probabilities of extinction do give us some idea of the likelihood of dropping to certain levels of abundance in a specified amount of time. However, it would be useful to establish more general guidelines on what the lower limit should be.

Adopting replacement as a second metric in conservation efforts to avoid extinction would be best if there were a focus on determining whether there was a taxonomic basis for a specified fraction of the natural level, as there is in fisheries (e.g. Fig. 11.5). Reviews of the consistency among the different calculations of probabilities of extinction have not provided great comfort. The study by Fieberg and Ellner (2001; Table 8.1) revealed that most attempts had been stage based, hence were suspect. The one study of the effects of multiple recovery teams doing PVA analyses of the same data sets (for Pacific salmon) led to widely divergent results, which is puzzling, and a concern (Section 10.6).

12.5 What next?

We have attempted in this book to convey our view of the state of the field of population dynamics as it has evolved from the early twentieth century (*pace* Leonardo of Pisa) to the first decades of the twenty-first. In so doing we hope we have imparted our current understanding of how populations function, and why. But one could ask whether we have imparted adequate foresight for readers to make use of this understanding into the future. We think that we have, but we must discuss directly the fact that the environment in which populations of conservation interest exist is changing, due to increasing atmospheric concentrations of greenhouse gases. The effects of climate change and ocean acidification on population dynamics is primarily that any of the important estimates we make to establish the critical thresholds needed for management (e.g. replacement thresholds, Chapters 10 and 11), must be made on the basis of statistics of a *nonstationary process* (i.e. a system whose statistical relationships are changing with time). We hope that our focus on fundamental population behavior with realistic models gives readers a deep enough understanding of population dynamics to predict how the effects of climate change (likely detected at the individual level, as changes to vital rates) could lead to changes in equilibria, generation times, or accompanying dynamics.

Our focus on the time scales associated with population dynamics can help to anticipate problems in detecting the effects of climate change. For example, increasing fishing of a population would, based on our understanding of cohort resonance effects, make the population more sensitive to low-frequency variability than before. Subsequent detection of a new, long-term decline in abundance would raise the question of whether it was due to climate change, lack of adequate replacement, or simply an increase in the sensitivity to low frequencies. This mechanism was identified by Bjørnstad et al. (2004) who termed it a *cloaking effect*. Testing alternative hypotheses in that type of scenario will depend on being able to model realistic population dynamics (Boettiger and Hastings, 2013).

Increasingly, we are facing the need to manage a rising number of species with limited resources. This requires much work to be done in the science of "data-poor" management, where it is difficult to precisely estimate the quantities we have shown are essential for quantifying extinction risk: abundance, population growth rate, and replacement. As we argued in Chapter 10, it would be best to use multiples of these quantities, but in a data-poor context only one might be possible. This problem is attracting growing attention in the realm of global fisheries management (Johannes, 1998; Dowling et al., 2015), and will require ways of predicting those key quantities from other life history attributes of understudied species, e.g. (Patrick et al., 2010; Brooks et al., 2010) and careful approaches to precautionary management, e.g. Hilborn (2009). An outstanding question is what management capabilities are given up when acceding to a data-poor approach?

Finally, one area of particularly active research in population dynamics is in the transient—rather than equilibrium—responses of systems to both environmental and human perturbations (Hastings, 2016; Hastings et al., 2018). The best practices of adaptive management dictate that we make (short-term) predictions about what effect management actions will have, then later evaluate whether reality is meeting expectations. This requires an analytical understanding of what sort of changes should take place in a system that has experienced a disturbance or alteration (Boettiger and Hastings, 2013; White et al., 2013); we hope that our focus on transient and stochastic dynamics in Chapters 4,

8, 10, and 11 prepares readers for this. Additionally, there is a need for efficient estimation of abundance and vital rates at local spatial scales and over short time scales in order to make inferences in noisy systems, (e.g. White et al., 2016; Munch et al., 2017). Our hope is that the understanding developed from the models in this book can help prepare readers to explore these new avenues.

Glossary

Asymptotic An asymptote is a limiting value that a function approaches very closely but never completely reaches. Asymptotic behavior is the behavior of a system that is reached as time gets arbitrarily large. For example, regardless of the initial conditions, a linear population model will eventually approach an asymptotic growth rate.

Autocorrelation Correlation between values in a time series at a specific time lag. The frequency spectrum is the Fourier transform of this function. In the frequency domain, positive autocorrelation appears as low-frequency ("red") variability.

Best available science A phrase used in early (1970s) environmental legislation in the USA to indicate that high uncertainty in an environmental issue was not a viable excuse for putting off a decision, rather the decision needs to made currently using the information at hand. This was later used in lawsuits brought against agencies, claiming that agency science was not the best (National Research Council, 2004).

Boundary conditions A description of the behavior of the state variables in a model at the edge of the model domain. For the types of models in this book, typical boundary conditions include describing what happens at $t = 0$ (because time cannot be negative) or at $N(t) = 0$ (because abundance cannot fall below zero). For partial differential equations, boundary conditions will be used to specify behavior at the edges of the age, size, or spatial domain of interest.

Catastrophic stochasticity/variability Refers to a type of randomness in environmental time series that consists of occasional large, random deviations.

Cohort resonance The response of age-structured populations with density-dependent recruitment to random variability in their environment. Refers to the fact that such populations are generally more sensitive to two frequency bands: 1) frequencies near the inverse of the generation time and 2) very low frequencies. The name refers to the first of these, because it tends to produce cyclic variability that resonates at the time scale it takes for one cohort of new recruits to mature and reproduce.

Cloaking effect A property of cohort resonance in which efforts to detect long-term trends in population vital rates (e.g. increases in harvest mortality or decreases in reproduction) are confounded by a population's heightened sensitivity to low-frequency (long period) environmental variability.

Demographic stochasticity/variability Randomness that arises due to probabilistic demographic processes alone, without influence of environmental variability. Examples include random numbers of births, the sex ratio of offspring in a cohort, or exactly which individuals die if there is a specified probability of mortality in a given year. In large populations this variability is not noticeable due to the law of large numbers, but it can lead to large fluctuations in small populations.

Deterministic skeleton In a stochastic model, this is the underlying deterministic model that would remain if there were no stochasticity.

Diffusivity A measure of the speed with which a substance is able to diffuse in a medium. Units are $\text{length}^2/\text{time}$.

Environmental stochasticity/variability Random variation over time that affects demographic rates (birth, death, growth, etc.) and arises from external, abiotic factors (e.g. weather) rather than due to internal population dynamics or interspecific interactions.

First-order/second-order partial differential equation A first-order partial differential equation (PDE) involves only first derivatives of the dependent variables. A second-order PDE involves the second derivative of at least one of the dependent variables.

Forward problem See inverse problem.

Frequency In the context of a cyclic time series, the frequency is the number of complete cycles (peak to peak) within some time period. It has units of 1/time.

Frequency domain There are two different frames of reference in which to view a time series. In the *time domain*, data are represented with one value for each sequential point in time (the usual way one might plot time series data). In the *frequency domain*, the same data would be represented in terms of the magnitude at different frequencies. A Fourier transformation converts data from the time domain to the frequency domain; an inverse Fourier transformation does the opposite operation.

Generation time Generally speaking, the average age at which an individual's reproductive effort occurs. In Box 4.2 we provide two similar definitions of this quantity, and one other definition.

Geometric growth/Exponential growth This is the basic behavior of any linear (density-independent) system. In simple terms, it is growth caused by the time rate of change of a quantity being proportional to the quantity. Growth can be expressed as growing exponentially at a rate that is given by a constant multiplied by time, or as a sum of such terms.

Index of Imprimitivity If one draws all possible paths through the life cycle graph corresponding to a projection matrix (such as a Leslie matrix), the index of imprimitivity is the smallest common divisor of the lengths of those paths. For an obligate semelparous species, there is only one path and the index is the length of that path. The index will be >1, and the matrix is imprimitive. For (most) iteroparous or non-obligate semelparous species, there will be multiple paths and the index will be 1, indicating a primitive matrix.

Influence functions In an age-structured model, an influence function is the amount of influence at each age that an organism will have on some process (e.g., reproduction, juvenile mortality) . For example the fraction surviving to each age times the fecundity at that age would be an influence function that could affect recruitment.

Invariant A number that is the mathematical combination of two or more parameters; one obtains the same overall results from a model as long as the invariant has the same value, regardless of what the values of the component parameters are.

Inverse problem A modeling exercise in which the parameter values are determined from the output pattern of the data from a population (as opposed to a forward problem, in which input data is from individuals and the model calculates the output data for the population).

Kernel For integro-difference equations, the kernel gives the probability distribution of transfers from one state to other states at each time step. In this book, these appear in spatial equations and in integral projection models.

Master equation In probabilistic models the master equation gives the rates at which states randomly change to other states. For example, in a population model, the probability that there are N individuals goes up with births when there are $N - 1$ individuals, and the probability goes up with deaths when there are $N + 1$ individuals. The probability that there are N individuals goes down when there are births or deaths when starting with N individuals.

Match/mismatch A concept developed primarily in reference to the timing of the larval stage of marine organisms, in which the survival and growth of a cohort is determined in part by whether it is spawned at the right time to match up with a period of suitable environmental conditions or planktonic food availability.

Mean age of reproduction The average age at which an individual's reproductive output occurs. That is, it is the average over all ages at which an individual reproduces, weighted by the total output at each age. This is one of the definitions of generation time in Box. 4.2.

Next generation matrix (or kernel) This is a matrix that projects the state of a population at time t to its state one generation-time later.

Non-negative matrix A matrix in which all elements are greater than or equal to zero.

Phase (1) A developmental stage of an organism's life. (2) The relative position (in space or time) of common points (peaks, troughs, etc.) of two sinusoidal waves.

Population regulation The quality of a population in which one or more demographic rates change with population density (due to either intrinsic or extrinsic factors) so that the population tends towards a nonzero steady state.

Population viability analysis A model analysis that estimates the probability that a given population will fall below a quasi-extinction threshold within some time horizon.

Portfolio effect The phenomenon in which the variance of a sum of multiple varying entities is less than it would be if they varied independently. Originally described for financial assets.

Positive matrix A matrix in which all elements are greater than zero.

Primitive matrix A matrix is primitive if some power of the matrix has entries which are all positive. For a Leslie matrix this represents the idea that at some time in the future an individual now of any age can have an offspring of any age at that future time.

PMM The parametric matrix method for modeling the probability of quasi-extinction in a linear model. If one can describe the multivariate distribution of the elements of a randomly varying projection matrix, then environmental stochasticity can be simulated by letting the projection matrix at each time step be created from random draws from that multivariate distribution.

Quasi-extinction Extinction as defined by a population falling below some nonzero threshold abundance. Most of the population models described in this book do not represent dynamics at very low densities well (e.g. they do not include demographic stochasticity or Allee effects) and, numerically, abundance in these models cannot actually reach zero. Quasi-extinction is meant to suggest that the population has fallen below a level from which it is not likely to recover.

Recruitment The entry of new individuals to a population. The precise meaning varies with context; for example, this could refer to the entry of new individuals to the adult population, or to the observed part of the population, or to the harvested portion of a population.

Reference points (target and limit) Quantities used in resource management, particularly fisheries management. These are pre-agreed-upon values, and in this book we used examples with values of abundance or replacement. If the resource falls below a limit reference point, some pre-agreed management action is triggered (e.g. harvest is reduced substantially). Managers attempt to set harvest rates so that the resource is close to the *target reference point* for harvest.

RTM The random transition matrix approach to modeling the probability of quasi-extinction in a linear model. If one has multiple transition matrices (e.g. Leslie matrices) available, each one corresponding to one observation of transition probabilities (e.g. one year of data), then environmental stochasticity can be simulated by making projections in which the matrix used in each time step is drawn at random from the set of available matrices.

Scorched earth This is an assumption made in some models of marine protected areas (MPAs), that harvest is so intense that no fish survives to reproduce in habitats outside of the protected area. Usually it is a simplifying assumption, meant to represent a limiting case in order to determine whether an MPA or group of MPAs could support an independently self-sustaining population.

SFA The small fluctuations approximation for describing the growth of a randomly varying Leslie matrix.

Spatially implicit/explicit/realistic A spatially implicit model ignores any spatial arrangement and only considers the fraction of habitat in a given state. A spatially explicit model does consider the importance of spatial arrangement (e.g. a reaction–diffusion model), but does not consider the role of realistic features and therefore typically ignores specific underlying habitat variability. A spatially realistic model considers all the details of a specific habitat and would typically only be studied using simulation approaches.

Stationarity The property of a system for which all statistical descriptors (mean, variance, autocorrelation, etc.) remain constant over time. The opposite is *nonstationarity*, in which one of those parameters is changing (e.g. decreasing mean, increasing variance).

Steepness A nondimensional quantity describing the slope at the origin of a stock–recruit curve. It is defined as the percentage of unfished recruitment obtained when the spawning stock is 20% of its unfished size.

Stock assessment A broad term for a type of statistical analysis used in fishery management, in which multiple data sources (e.g. landings, fishery independent survey data) are jointly fitted to a dynamic population model. The outcome is an estimate of the historical and current state of the stock and its level of replacement.

Synchrony The property of two or more time series being highly correlated over time. Sometimes this is due to the time series having cyclic oscillations of the same frequency.

Time domain see *frequency domain.*

Transient Transient behavior is the behavior of a system before it reaches the asymptotic state. This is of interest when the transient behavior differs qualitatively from the asymptotic behavior.

Upwelling An oceanographic phenomenon in which wind blowing parallel to a coastline causes the surface layer of water to move away from the coast, at right angles to the wind direction (due to the Coriolis effect). The offshore movement of the surface layer allows deep, cold, nutrient-rich water to well up to the surface. The nutrients stimulate primary productivity in upwelling zones.

References

Abbott, R. E., D. F. Doak, and M. L. Peterson. (2017). Portfolio effects, climate change, and the persistence of small populations: analyses on the rare plant Saussurea weberi. *Ecology* 98: 1071–81.

Acker, P., A. Robert, R. Bourget, and B. Colas. (2013). Heterogeneity of reproductive age increases the viability of semelparous populations. *Functional Ecology* 28:458–68.

Alexander, S. E. and J. Roughgarden. 1996. Larval transport and population dynamics of intertidal barnacles: a coupled benthic/oceanic model. *Ecological Monographs* 66:259–75.

Alonzo, S. H. and M. Mangel. 2004. The effects of size-selective fisheries on the stock dynamics of and sperm limitation in sex-changing fish. *Fishery Bulletin* 102:1–13.

Amarasekare, P. and R. M. Coutinho. 2013. The intrinsic growth rate as a predictor of population viability under climate warming. *Journal of Animal Ecology* 82:1240–53.

Anderson, J. L. (1996). A method for producing and evaluating probabilistic forecasts from ensemble model integrations. *Journal of Climate* 9:1518–30.

Anderson, S. C., A. B. Cooper, and N. K. Dulvy. 2013. Ecological prophets: quantifying metapopulation portfolio effects. *Methods in Ecology and Evolution* 4:971–81.

Andow, D. A., P. M. Kareiva, S. A. Levin, and A. Okubo. 1990. Spread of invading organisms. *Landscape Ecology* 4:177–88.

Andrewartha, H.G. and L.C. Birch. 1954. *The Distribution and Abundance of Animals*. University of Chicago Press, Chicago, Illinois.

Armbruster, P. and R. Lande. 1993. A population viability analysis for African elephant (*Loxodonta africana*): How big should reserves be? *Conservation Biology* 7:602–10.

Bailey, N. T. J. 1968. Stochastic birth, death, and migration processses for spatially distributed populations. *Biometrika* 55:189–98.

Bakker, V. J., D. F. Doak, G. W. Roemer, D. K. Garcelon, T. J. Connan, S. A. Morrison, et al. 2009. Incorporating ecological drivers and uncertainty into a demographic population viability analysis for the island fox. *Ecological Monographs* 79: 77–108.

Ball, I. R. and H. P. Possingham. 2000. *Marxan (v1.8.2): Marine Reserve Design Using Spatially Explicit Annealing, A Manual*. University of Queensland, Brisbane, Australia.

Banks, H. T., L. W. Botsford, F. Kappel, and C. Wang. 1991. Estimation of growth and survival in size-structured cohort data: an application to larval striped bass (*Morone saxatilis*). *Journal of Mathematical Biology* 30:125–50.

Baranov, F. I. 1918. On the question of the biological basis of fisheries. *Izvestiya Otdela Rybovodstva I Nauchno-promyslovykh Issledovanii* 1:81–128 (in Russian).

Barneche, D. R., D. R. Robertson, C. R. White, and D. J. Marshall. 2018. Fish reproductive-energy output increases disproportionately with body size. *Science* 360:642–5.

Barnett, L. A. K., T. A. Branch, R. A. Ranasinghe, and T. E. Essington. 2017. Old-growth fishes become scarce under fishing. *Current Biology* 27:2843–8.

Barraquand, F., S. Louca, K. C. Abbott, C. A. Cobbold, F. Cordoleani, D. L. DeAngelis, B. D. Elderd, J. W. Fox, P. Greenwood, F. M. Hilker, D. L. Murray, C. R. Stieha, R. A. Taylor, K. Vitense, G. S. K. Wolkowicz, and R. C. Tyson. 2017. Moving forward in circles: challenges and opportunities in modelling population cycles. *Ecology Letters* 20:1074–92.

Basson, M. and J. R. Beddington. 1991. An assessment of the maximum sustainable yield of ivory from African elephant populations. *Mathematical Biosciences* 104:73–95.

Bayoh, M. N. and S. W. Lindsay. 2003. Effect of temperature on the development of the aquatic stages of *Anopheles gambiae* sensu stricto (Diptera: Culicidae). *Bulletin of Entomological Research* 93:375–81.

Beddington, J. R. 1974. Age distribution and the stability of simple discrete time population models. *Journal of Theoretical Biology* 47:65–74.

Bergh, M. O. and W. M. Getz. 1988. Stability of discrete age-structured and aggregated delay-difference population models. *Journal of Mathematical Biology* 26:551–81.

Bergmann, C. 1847. Über die Verhältnisse der Wärmeökonomie der Tiere zu ihrer Grösse. *Göttinger Studien* 3:595–708.

Bernardelli, H. 1941. Population waves. *Journal of the Burma Research Society* 31:1–18.

Bertalanffy, von, L. 1934. Untersuchungen über die Gesetzlichkeit des Wachtums. I. Allgemeine Grundlagen der Theorie: Mathematische und physiologische Geseztlichkeiten des Wachstums bei Wassertieren. *Archiv für Entwicklungsmechanik der Organismen* 131:613–52.

Bertalanffy, von, L. 1951. Metabolic types and growth types. *The American Naturalist* 85:111–7.

Bertalanffy, von, L. 1957. Quantitative laws in metabolism and growth. *The Quarterly Review of Biology* 32:217–31.

Beverton, R. and S. J. Holt. 1957. On the dynamics of exploited fish populations. *Fisheries Investigations, London (series II)* 19:1–533.

Bjørnstad, O. N., J.-M. Fromentin, N.C. Stenseth, and J. Gjøsaeter. 1999. Cycles and trends in cod populations. *Proceedings of the National Academy of Sciences* 96:5066–71.

Bjørnstad, O. N., R.M. Nisbet, and J.-M. Fromentin. 2004. Trends and cohort resonant effects in age-structured populations. *Journal of Animal Ecology* 73:1157–67.

Boettiger, C. and A. Hastings. (2012). Early warning signals and the prosecutor's fallacy. *Proceedings: Biological Sciences* 279: 4734–9.

Bosch, C. A. 1971. Redwoods: A population model. *Science* 172: 345–9.

Botsford, L. W. 1981. The effects of increased individual growth rates on depressed population size. *American Naturalist* 117:38–63.

Botsford, L. W. 1986. Effects of environmental forcing on age-structured populations: Northern California Dungeness crab (*Cancer magister*) as an example. *Canadian Journal of Fisheries and Aquatic Sciences* 43:2345–52.

Botsford, L. W. 1992. Further analysis of Clark's delayed recruitment model. *Bulletin of Mathematical Biology* 54:275–93.

Botsford, L. W. 1997. Dynamics of populations with density-dependent recruitment and age structure. Pages 371–408 in S. Tuljapurkar and H. Caswell, editors. *Structured-Population Models in Marine, Terrestrial, and Freshwater Systems*. Springer Science+Business Media BV, New York.

Botsford, L. W. 2013. *Maximum Sustainable Yield. Oxford Bibliographies in Ecology*. Oxford University Press, New York.

Botsford, L. W. and J. G. Brittnacher. 1998. Viability of Sacramento River winter-run chinook salmon. *Conservation Biology* 12:65–79.

Botsford, L. W., J. Castilla, and C. Peterson. 1997. The management of fisheries and marine ecosystems. *Science* 277:509–15.

Botsford, L. W., A. Hastings, and S. Gaines. 2001. Dependence of sustainability on the configuration of marine reserves and larval dispersal distance. *Ecology Letters* 4:144–50.

Botsford, L. W. and R. C. Hobbs. 1986. Static optimization of yield per recruit with reproduction and fishing costs. *Fisheries Research* 4:181–9.

Botsford, L. W. and R. C. Hobbs. 1995. Recent advances in the understanding of cyclic behavior of Dungeness crab (*Cancer magister*) populations. *ICES Marine Science Symposium* 199: 157–66.

Botsford, L. W., M. D. Holland, J. C. Field, and A. Hastings. 2014b. Cohort resonance: a significant component of fluctuations in recruitment, egg production, and catch of fished populations. *ICES Journal of Marine Science* 71:2158–70.

Botsford, L. W., M. D. Holland, J. F. Samhouri, J. W. White, and A. Hastings. 2011. Importance of age structure in models of the response of upper trophic levels to fishing and climate change. *ICES Journal of Marine Science* 68:1270–83.

Botsford, L. W. and C. A. Lawrence. 2002. Patterns of co-variability among California Current chinook salmon, coho salmon, Dungeness crab, and physical oceanographic conditions. *Progress in Oceanography* 53:283–305.

Botsford, L. W., R. D. Methot Jr., and W. E. Johnston. 1983. Effort dynamics of the Northern California Dungeness Crab (*Cancer magister*) fishery. *Canadian Journal of Fisheries and Aquatic Sciences* 40:337–46.

Botsford, L. W., C. L. Moloney, A. Hastings, and J. L. Largier. 1994. The influence of spatially and temporally varying oceanographic conditions on meroplanktonic metapopulations. *Deep Sea Research II* 41:107–45.

Botsford, L. W., B. D. Smith, and J. F. Quinn. 1994. Bimodality in size distributions: the red sea urchin Strongylocentrotus franciscanus as an example. *Ecological Applications* 4:42–50.

Botsford, L. W., J. W. White, M. H. Carr, and J. E. Caselle. 2014a. Marine Protected Areas in California, USA. Pages 203–49 in Johnson, M.L. and Sandell, J. eds. *Advances in Marine Biology: Marine Managed Areas and Fisheries*, vol. 69. Elsevier, Oxford, UK.

Botsford, L. W., J. W. White, M. H. Carr, and J. E. Caselle. 2014. Marine protected area networks in California, USA. *Advances in Marine Biology* 69:205–51.

Botsford, L. W., J. W. White, M. A. Coffroth, C. B. Paris, S. Planes, T. L. Shearer, S. R. Thorrold, and G. P. Jones. 2009. Connectivity and resilience of coral reef metapopulations in marine protected areas: matching empirical efforts to predictive needs. *Coral Reefs* 28:327–37.

Botsford, L. W. and D. E. Wickham. 1975. Correlation of upwelling index and Dungeness crab catch. *Fishery Bulletin* 73:901–7.

Botsford, L. W. and D. E. Wickham. 1978. Behavior of age-specific, density-dependent models and the Northern California Dungeness crab (*Cancer magister*) fishery. *Journal of the Fisheries Research Board of Canada* 35:833–43.

Boyce, M. S., K. Knopff, J. Northrup, J. Pitt, and L. S. Vors. 2012. Harvest models for changing environments. Pages 293–306 in J. F. Brodie, E. S. Post, and D. F. Doak, editors. *Wildlife Conservation in a Changing Climate*. University of Chicago Press, Chicago.

Brook, B. W., M. A. Burgman, H. R. Akçakaya, J. J. O'Grady, and R. Frankham. 2002. Critiques of PVA ask the wrong questions: throwing the heuristic baby out with the numerical bath water. *Conservation Biology* 16:262–3.

Brown, J. H., J. F. Gillooly, A. P. Allen, V. M. Savage, and G. B. West. 2004. Toward a metabolic theory of ecology. *Ecology* 85:1771–89.

Brussard, P. F., S. A. Levin, L. N. Miller, and R. J. Whittaker. 1971. Redwoods: a population model debunked. *Science* 174:435–6.

Burgess, S. C., K. J. Nickols, C. D. Griesemer, L. A. K. Barnett, A. G. Dedrick, E. V. Satterthwaite, L. Yamane, S. G. Morgan, W. J. W, and L. W. Botsford. 2014. Beyond connectivity: how empirical methods can quantify population persistence to improve marine protected-area design. *Ecological Applications* 24:257–70.

Busch, D. S., D A. Boughton, T. Cooney, P. Lawson, S. T. Lindley, M. McClure, et al. 2013. A practical comparison of viability models used for management of endangered and threatened anadramous Pacific salmonids. *North American Journal of Fisheries Management* 33:1125–41.

Caddy, J. F. and R. Mahon. 1995. *Reference Points for Fisheries Management*. FAO Fisheries Technical Paper no. 347, Rome.

Caddy, J. F. and R. McGarvey. 1996. Targets or limits for management of fisheries? *North American Journal of Fisheries Management* 16:479–87.

Caselle, J. E., B. P. Kinlan, and R. R. Warner. 2010. Temporal and spatial scales of influence on nearshore fish settlement in the Southern California Bight. *Bulletin of Marine Science* 86: 355–85.

Caswell, H. 2001. *Matrix Population Models: Construction, Analysis, and Interpretation*. Second edition. Sinauer Associates, Inc., Sunderland, MA, USA.

Caswell, H. 2007. Sensitivity analysis of transient population dynamics. *Ecology Letters* 10:1–15.

Caswell, H., H. E. Koenig, J. A. Resh, and Q. E. Ross. 1972. An introduction to systems science for ecologists. *Systems Analysis and Simulation in Ecology* 2:3–78.

Caughley, G. 1977. *Analysis of Vertebrate Populations*. Wiley, London.

Caughley, G. and A. Gunn. 1996. *Conservation Biology in Theory and Practice*. Blackwell Science, Cambridge, MA.

Chapman, R. N. 1933. The causes of fluctuations of populations of insects. *Proceedings of the Hawaiian Entomological Society* 8:297–8.

Chapman, R. N. 1928. Quantitative analysis of environmental factors. *Ecology* 9:111–22.

Chaundry, T. W. and E. Phillips. 1936. The convergence of sequences defined by quadratic recurrence-formulae. *Quarterly Journal of Mathematics* (Oxford) 7:74–80.

Chesson, P. 2012. Scale transition theory: its aims, motivations and predictions. *Ecological Complexity* 10:52–68.

Childs, D. Z., M. Rees, K. E. Rose, P. J. Grubb, and S. P. Ellner. 2003. Evolution of complex flowering strategies: an age- and size-structured integral projection model. *Proceedings of the Royal Society B: Biological Sciences* 270:1829–38.

Childs, D. Z., M. Rees, K. E. Rose, P. J. Grubb, and S. P. Ellner. 2004. Evolution of size-dependent flowering in a variable environment: construction and analysis of a stochastic integral projection model. *Proceedings of the Royal Society B: Biological Sciences* 271:425–34.

Chong, Y. C. 1814. Jasaneobo. Pages 1–332 in K. S. Lee (editor). *The History of Korea*. Eul-Yoo Press (in Korean).

Chu, C. and P. B. Adler. 2014. When should plant population models include age structure? *Journal of Ecology* 102:531–43.

Cisneros-Mata, M. A., L. W. Botsford, and J. F. Quinn. 1997. Projecting variability of *Totoaba macdonaldi*, a population with unknown age-dependent variability. *Ecological Applications* 7: 968–80.

Cisneros-Mata, M. A., G. Montemeyor-Lopez, and M. J. Roman-Rodríguez. 1995. Life history and conservation of *Totoaba macdonaldi*. *Conservation Biology* 9:806–14.

Claessen, D., A. M. De Roos, and L. Persson. 2004. Population dynamic theory of size-dependent cannibalism. *Proceedings of the Royal Society B: Biological Sciences* 271:333–40.

Clark, C. 1976. A delayed-recruitment model of population dynamics, with an application to baleen whale populations. *Journal of Mathematical Biology* 3:381–91.

Clark, C. W. 2010. *Mathematical Bioeconomics: The Mathematics of Conservation*. Third edition. Wiley, Hoboken, NJ.

Clark, J. S. (2007). *Models For Ecological Data: An Introduction*. Princeton University Press, Princeton, NJ.

Clark, W. 2002. F 35 percent revisited ten years later. *North American Journal of Fisheries Management* 22:251–7.

Clark, W. G. 1991. Groundfish exploitation rates based on life history parameters. *Canadian Journal of Fisheries and Aquatic Sciences* 48:734–50.

Coale, A.J. 1972. *The Growth and Structure of Human Populations: A Mathematical Approach*. Princeton University Press, Princeton, NJ.

Cobb, K. M., C. D. Charles, H. Cheng, and R. L. Edwards. 2003. El Niño/Southern Oscillation and tropical Pacific climate during the last millennium. *Nature* 424:271–6.

Cohen, J. E. 1979. The cumulative distance from an observed to a stable age structure. *SIAM Journal on Applied Mathematics* 36:169–75.

Cohen, J. E. 1995. Population growth and Earth's human carrying capacity. *Science* 269:341–6.

Collie, J. S., L. W. Botsford, A. Hastings, I. C. Kaplan, J. L. Largier, P. A. Livingston, É. Plagányi, K. A. Rose, B. K. Wells, and F. E. Werner. 2014. Ecosystem models for fisheries management: finding the sweet spot. *Fish and Fisheries* 17:101–25.

Connolly, S. R., B. A. Menge, and J. Roughgarden. 2001. A latitudinal gradient in recruitment of intertidal invertebrates in the northeast Pacific Ocean. *Ecology* 82:1799–813.

Cook, L. M. 1965. Oscillation in the simple logistic growth model. *Nature* 207:316.

Cowen, R. K. and S. Sponaugle. 2009. Larval dispersal and marine population connectivity. *Annual Review of Marine Science* 1:443–66.

Crouse, D. T., L. B. Crowder, and H. Caswell. 1987. A stage-based population model for loggerhead sea turtles and implications for conservation. *Ecology* 68:1412–23.

Crowder, L. B., D. T. Crouse, S. S. Heppell, and T. H. Martin. 1994. Predicting the impact of turtle excluder devices on loggerhead sea turtle populations. *Ecological Applications* 4:437–45.

Cuddington, K., M. J. Fortin, L. R. Gerber, A. Hastings, A. Liebhold, M. O'Connor, and C. Ray. 2013. Process-based models are required to manage ecological systems in a changing world. *Ecosphere* 4:2.

Cushing, D. H. 1990. Plankton production and year-class strength in fish populations: an update of the match/mismatch hypothesis. *Advances in Marine Biology* 26:249–93.

Dayton, P. K. 1973. Two cases of resource partitioning in an intertidal community: making the right prediction for the wrong reason. *American Naturalist* 107:662–70.

DeAngelis, D. L. and L. J. Gross, editors. 1992. *Individual-based Models and Approaches in Ecology: Populations, Communities, and Ecosystems*. Chapman & Hall, New York.

de-Camino-Beck, T. and M. A. Lewis. 2008. On net reproductive rate and the timing of reproductive output. *The American Naturalist* 172:128–39.

de Kroon, H., A. Plaisier, J. van Groenendael, and H. Caswell. 1986. Elasticity: the relative contribution of demographic parameters to population growth rate. *Ecology* 67:1427–31.

De Leo, G. A. and F. Micheli. 2015. The good, the bad and the ugly of marine reserves for fishery yields. *Philosophical Transactions of the Royal Society of London. Series B: Biological Sciences* 370:20140276–12.

de Roos, A. M. 1988. Numerical methods for structured population models: the escalator boxcar train. *Numerical Methods for Partial Differential Equations* 4:173–95.

de Roos, A. M. 1997. A gentle introduction to physiologically structured population models. In S. Tuljapurkar & H. Caswell (Eds.), *Structured-Population Models in Marine, Terrestrial, and Freshwater Systems* (pp. 119–204). Chapman & Hall, New York.

de Roos, A. M., O. Diekmann, and J. A. J. Metz. 1992. Studying the dynamics of structured population models: a versatile technique and its application to *Daphnia*. *The American Naturalist* 139:123–47.

de Roos, A. M. and L. Persson. 2013. *Population and Community Ecology of Ontogenetic Development*. Princeton University Press, Princeton, NJ, USA.

de Valpine, P. 2009. Stochastic development in biologically structured population models. *Ecology* 90:2889–901.

de Valpine, P. and A. Hastings. 2002. Fitting population models incorporating process noise and observation error. *Ecological Monographs* 72:57–76.

de Valpine, P., K. Scranton, J. Knape, K. Ram, and N. J. Mills. 2014. The importance of individual developmental variation in stage-structured population models. *Ecology Letters* 17:1026–38.

den Boer, P. J. 1968. Spreading of risk and stabilization of animal numbers. *Acta Biotheoretica* 18:165–94.

Dennis, B., P. L. Munholland, and J. M. Scott. 1991. Estimation of growth and extinction parameters for endangered species. *Ecological Monographs* 61:115–43.

Denny, M. W., B. Helmuth, G. H. Leonard, C. D. G. Harley, L. J. H. Hunt, and E. K. Nelson. 2004. Quantifying scale in ecology: lessons from a wave-swept shore. *Ecology* 74:513–32.

Devlin, K. 2011. *The Man of Numbers: Fibonacci's Arithmetic Revolution*. Bloomsbury, London.

Diamond, J. M. 1975. The island dilemma: lessons of modern biogeographic studies for the design of natural reserves. *Biological Conservation* 7:129–46.

Diem, J. E. and J. L. McGregor. 1971. Redwoods: a population model debunked. *Science* 174:436.

Doak, D. F., D. Bigger, E. K. Harding, M. A. Marvier, R. E. O'Malley, and D. Thomson. 1998. The statistical inevitability of stability–diversity relationships in community ecology. *The American Naturalist* 151:264–76.

Dorn, M. W. 2002. Advice on West Coast rockfish harvest rates from Bayesian meta-analysis of stock–recruit relationships. *North American Journal of Fisheries Management* 22:280–300.

Dorner, B., R. M. Peterman, and S. L. Haeseker. 2008. Historical trends in productivity of 120 Pacific pink, chum, and sockeye salmon stocks reconstructed by using a Kalman filter. *Canadian Journal of Fisheries and Aquatic Sciences* 65:1842–66.

Drake, P. T., C. A. Edwards, and J. A. Barth. 2011. Dispersion and connectivity estimates along the US west coast from a realistic numerical model. *Journal of Marine Research* 69:1–37.

Duncan, O. D. and B. Duncan. 1955. A methodological analysis of segregation indices. *American Sociological Review* 20:210–17.

Easter, E. E. and J. W. White. 2016. Spatial management for protogynous sex-changing fishes: a general framework for coastal systems. *Marine Ecology Progress Series* 543:223–40.

Easterling, M. R., S. P. Ellner, and P. M. Dixon. 2000. Size-specific sensitivity: applying a new structured population model. *Ecology* 81:694–708.

Edelstein-Keshet, L. 2005. *Mathematical Models in Biology*. Society of Industrial and Applied Mathematics, Philadelphia.

Ellner, S. P., D. Z. Childs, and M. Rees. 2016. *Data-driven Modelling of Structured Populations A Practical Guide to the Integral Projection Model*. Springer, Switzerland.

Ellner, S. P. and M. Rees. 2006. Integral projection models for species with complex demography. *The American Naturalist* 167:410–28.

Elton, C. S. 1958. *The Ecology of Invasions by Animals and Plants*. Butler and Tanner Ltd, London.

Etienne, R., C. J. F. ter Braak, and C. C. Vos. 2004. Application of stochastic patch occupancy models to real metapopulations. Pages 105–32 in *Ecology, Genetics, and Evolution of Metapopulations*. Elsevier Academic Press, Burlington, MA.

Euler, L. 1760. Recherces générales sur la mortalité et la multiplication du genre humaine. *Histoire de l'Académie Royale des Sciences et Belles Lettres* 16:291–9.

Euler, L. 1970. A general investigation into the mortality and multiplication of the human species. *Theoretical Population Biology* 1:307–14.

Evans, M. R., T. G. Benton, V. Grimm, C. M. Lessells, M. A. O'Malley, A. Moustakas, and M. Weisberg. 2014. Data availability and model complexity, generality, and utility: a reply to Lonergan. *Trends in Ecology & Evolution* 29:302–3.

Evans, M. R., M. Bithell, S. J. Cornell, S. R. X. Dall, S. Diaz, S. Emmott, B. Ernande, V. Grimm, D. J. Hodgson, S. L. Lewis, G. M. Mace, M. Morecroft, A. Moustakas, E. Murphy, T. Newbold, K. J. Norris, O. Petchey, M. Smith, J. M. J. Travis, and T. G. Benton. 2013a. Predictive systems ecology. *Proceedings of the Royal Society B: Biological Sciences* 280:20131452.

Evans, M. R., V. Grimm, K. Johst, T. Knuuttila, R. de Langhe, C. M. Lessells, M. Merz, M. A. O'Malley, S. H. Orzack, M. Weisberg, D. J. Wilkinson, O. Wolkenhauer, and T. G. Benton. 2013b. Do simple models lead to generality in ecology? *Trends in Ecology & Evolution* 28:578–83.

Ewens, W. J. 1964. The pseudo-transient distribution and its uses in genetics. *Journal of Applied Probability* 1:141–56.

Ezard, T. H. G., J. M. Bullock, H. J. Dalgleish, A. Millon, F. Pelletier, A. Ozgul, and D. N. Koons. 2010. Matrix models for a changeable world: the importance of transient dynamics in population management. *Journal of Applied Ecology* 47:515–23.

Feller, W. 1939. Die Grundlagen der Volterraschen Theorie des Kampfes ums dasei in Wahrscheinlichkeitstheoretischer Behandlung. *Acta Biotheoretica* 5:11–40.

Feller, W. 1940. On the logistic law of growth and its empirical verifications in biology. *Acta Biotheoretica* 5:51–66.

Ferguson, J. M., F. Carvalho, O. Murillo-García, M. L. Taper, and J. M. Ponciano. 2016. An updated perspective on the role of environmental autocorrelation in animal populations. *Theoretical Ecology* 9:129–48.

Fieberg, J. and S. P. Ellner. 2000. When is it meaningful to estimate an extinction probability? *Ecology* 81:2040–7.

Fieberg, J. and S. P. Ellner. 2001. Stochastic matrix models for conservation and management: a comparative review of methods. *Ecology Letters* 4:244–66.

Field, J. C., S. G. Beyer, and X. He. 2016. *Status of the Chilipepper Rockfish, Sebastes goodei, in the California Current for 2015*. Pacific Marine Fishery Council, Portland, OR, USA.

Figueira, W. F. and L. B. Crowder. 2006. Defining patch contribution in source–sink metapopulations: the importance of including dispersal and its relevance to marine systems. *Population Ecology* 48:215–24.

Finley, C. 2011. *All the Fish in the Sea: Maximum Sustainable Yield and the Failure of Fisheries Management*. University of Chicago Press, Chicago.

Fisher, R. 1937. The wave of advance of advantageous genes. *Annals of Eugenics* 7:355–69.

Foerster, von, H. 1959. Some remarks on changing populations. Pages 382–407 in F. Stohlman, editor. *The Kinetics of Cell Proliferation*. Grune and Stratton, New York.

Fogarty, M., L. Botsford, and F. Werner. 2013. Legacy of the US GLOBEC program: Current and potential contributions to marine ecosystem-based management. *Oceanography* 26:116–27.

Foley, P. 1994. Predicting extinction times from environmental stochasticity and carrying capacity. *Conservation Biology* 8:124–37.

Ford, E. D. 2000. *Scientific Method for Ecological Research*. Cambridge University Press, Cambridge, UK.

Fowler, C. W. 1981. Density dependence as related to life history strategy. *Ecology* 62:602–10.

Fox, G. A. and J. Gurevitch. 2000. Population numbers count: tools for near-term demographic analysis. *The American Naturalist* 156:242–56.

Fox, L. R. 1975. Cannibalism in natural populations. *Annual Review of Ecology and Systematics* 6: 87–106.

Fox, W. W. 1970. An exponential surplus-yield model for optimizing exploited fish populations. *Transactions of the American Fisheries Society* 99:80–8.

Fretwell, S. D. 1969. On territorial behavior and other factors influencing habitat distribution in birds. *Acta Biotheoretica* 19:16–36.

Frobenius, F. G. 1912. Über Matrizen aus nicht negativen Elementen. *Sitzungsberichte der Königlich Preussischen Akademie der Wissenschaften* 23:456–77.

Fulton, E. A. 2010. Approaches to end-to-end ecosystem models. *Journal of Marine Systems* 81: 171–83.

Gaines, S. D., B. Gaylord, and J. L. Largier. 2003. Avoiding current oversights in marine reserve design. *Ecological Applications* 13:32–46.

Gaines, S. and J. Roughgarden. 1985. Larval settlement rate: a leading determinant of structure in an ecological community of the marine intertidal zone. *Proceedings of the National Academy of Sciences of the United States of America* 82:3707–11.

Gaines, S. D., C. White, M. H. Carr, and S. R. Palumbi. 2010. Designing marine reserve networks for both conservation and fisheries management. *Proceedings of the National Academy of Sciences* USA 107:18286–93.

Gantmacher, F. R. 1959. *The Theory of Matrices*. AMS Chelsea Publishing, Providence, Rhode Island, USA.

Garcia, S. M. 1994. The precautionary principle: its implications in capture fisheries management. *Ocean & Coastal Management* 22:99–125.

Gardiner, C. 2009. *Stochastic Methods: A Handbook for the Natural and Social Sciences* (4th edn.). Springer-Verlag, Berlin.

Gardner, R. H., W. G. Cale, and R. V. O'Neill. 1982. Robust analysis of aggregation error. *Ecology* 63:1771–9.

Gaylord, B. and S. D. Gaines. 2000. Temperature or transport? Range limits in marine species mediated solely by flow. *The American Naturalist* 155:769–89.

Gaylord, B., S. D. Gaines, D. A. Siegel, and M. H. Carr. 2005. Marine reserves exploit population structure and life history in potentially improving fisheries yields. *Ecological Applications* 15:2180–91.

Gilbert, C. H. 1916. *Contributions to the Life History of the Sockeye Salmon*. British Columbia Fisheries Department, Vancouver, Canada.

Gilpin, M. E. and F. J. Ayala. 1973. Global models of growth and competition. *Proceedings of the National Academy of Sciences* 70:3590–3.

Goldwyn, E. E. and A. Hastings. 2008. When can dispersal synchronize populations? *Theoretical Population Biology* 73:395–402.

Goldwyn, E. E. and A. Hastings. 2011. The roles of the Moran effect and disperal in synchronizing oscillating populations. *Journal of Theoretical Biology* 289:237–46.

Goodman, D. 1987. Consideration of stochastic demography in the design and management of biological reserves. *Natural Resource Modeling* 1:205–34.

Goodman, L. A. 1969. The analysis of population growth when the birth and death rates depend upon several factors. *Biometrics* 25:659–81.

Goodyear, C. P. 1977. Assessing the impact of power plant mortality on the compensatory reserve of fish populations. Pages 186–95 in *Proceedings of the Conference on Assessing the Effects of Power-Plant-Induced Mortality on Fish Populations*. Pergamon Press, Elmsford, New York.

Goodyear, C. P. 1993. Spawning stock biomass per recruit in fisheries management: foundation and current use. Pages 67–81 in S. J. Smith, J. J. Hunt, and D. Rivard, editors. *Risk Evaluation and Biological Reference Points for Fisheries Management*. Canadian Special Publications in Fisheries and Aquatic Sciences, 120.

Goudriaan, J. 1986. Boxcartrain methods for modelling of ageing, development, delays, and dispersion. Pages 453–73 in J. A. J. Metz and O. Diekmann, editors. *The Dynamics of Physiologically Structured Populations*. Springer Lecture Notes in Biomath, Springer-Verlag, Heidelberg.

Grafton, R. Q. and T. Kompas. 2005. Uncertainty and the active adaptive management of marine reserves. *Marine Policy* 29:471–9.

Graham, M. 1935. Modern theory of exploiting a fishery, and application to North Sea trawling. *ICES Journal of Marine Science* 10:264–74.

Grasman, J. and D. Ludwig. 1983. The accuracy of the diffusion approximation to the expected time to extinction for some discrete stochastic processes. *Journal of Applied Probability* 20:305–21.

Grenfell, B. T., K. Wilson, B.F. Finkenstädt, T.N. Coulson, S. Murray, S.D. Albon, J.M. Pemberton, T.H. Clutton-Brock, and M.J. Crawley. 1998. Noise and determinism in synchronized sheep dynamics. *Nature* 394:674–7.

Griffiths, J. R., D. E. Schindler, J. B. Armstrong, M. D. Scheuerell, D. C. Whited, R. A. Clark, et al. 2014. Performance of salmon fishery portfolios across western North America. *Journal of Applied Ecology* 51:1554–63.

Grimm, V. and S. F. Railsback. 2005. *Individual-based Modeling and Ecology*. Princeton University Press, Princeton, NJ, USA.

Grimm, V., K. Reise, and M. Strasser. 2003. Marine metapopulations: a useful concept? *Helgoland Marine Research* 56:222–8.

Grimm, V. and C. Wissel. 1997. Babel, or the ecological stability discussions: an inventory and analysis. *Oecologia* 109:323–34.

Groot, C. and L. Margolis. 1991. *Pacific Salmon Life Histories*. University of British Columbia Press, Vancouver, British Columbia.

Grüss, A., D. M. Kaplan, S. Guénette, C. M. Roberts, and L. W. Botsford. 2011. Consequences of adult and juvenile movement for marine protected areas. *Biological Conservation* 144:692–702.

Guill, C., B. Drossel, W. Just, and E. Carmack. 2011. A three-species model explaining cyclic dominance of Pacific salmon. *Journal of Theoretical Biology* 276:16–21.

Gulland, J. A. 1983. *Fish Stock Assessment: A Manual of Basic Methods*. FAO/Wiley, Chichester, UK.

Gurney, W. S. C., R. M. Nisbet, and J. H. Lawton. 1983. The systematic formulation of tractable single-species population models incorporating age structure. *Journal of Animal Ecology* 52: 479–95.

Gutiérrez, N. L., R. Hilborn, and O. Defeo. 2011. Leadership, social capital and incentives promote successful fisheries. *Nature* 470:386–9.

Gyllenberg, M. and I. Hanski. 1992. Single-species metapopulation dynamics: a structured model. *Theoretical Population Biology* 42:35–62.

Gyllenberg, M. and I. Hanski. 1997. Habitat deterioration, habitat destruction and metapopulation persistences in a heterogeneous landscape. *Theoretical Population Biology* 52:198–215.

Gyllenberg, M., I. Hanski, and A. Hastings. 1997. Structured metapopulation models. Pages 93–122 in *Metapopulation Biology: Ecology, Genetics, and Evolution*. Academic Press, San Diego, CA, USA.

Halbach, K. 1971. Redwoods: a population model debunked. *Science* 174:436.

Halpern, B. S. 2003. The impact of marine reserves: do reserves work and does reserve size matter? *Ecological Applications* 13:117–37.

Halpern, B. S. and R.R. Warner. 2002. Marine reserves have rapid and lasting effects. *Ecology Letters* 5:361–6.

Hameed, S. O., J. W. White, S. H. Miller, K. J. Nickols, and S. G. Morgan. 2016. Inverse approach to estimating larval dispersal reveals limited population connectivity along 700 km of wave-swept open coast. *Proceedings of the Royal Society B: Biological Sciences* 283:20160370–9.

Hampton, J., J. R. Sibert, P. Kleiber, M. N. Maunder, and S. J. Harley. 2005. Decline of Pacific tuna populations exaggerated? *Nature* 434:E1–2.

Hanski, I. 1991. Single-species metapopulation dynamics: concepts, models and observations. *Biological Journal of the Linnean Society* 42:17–38.

Hanski, I. 1994. A practical model of metapopulation dynamics. *Journal of Animal Ecology* 63: 151–62.

Hanski, I. 1998. Metapopulation dynamics. *Nature* 396:41–9.

Hanski, I. 2001. Spatially realistic theory of metapopulation ecology. *Naturwissenschaften* 88: 372–81.

Hanski, I. A. and M. E. Gilpin. 1997. *Metapopulation Biology: Ecology, Genetics, and Evolution*. Academic Press, San Diego, CA, USA.

Hanski, I. and O. E. Gaggiotti. 2004. *Ecology, Genetics, and Evolution of Metapopulations*. Elsevier Academic Press, Burlington, MA, USA.

Hanski, I. and O. Ovaskainen. 2003. Metapopulation theory for fragmented landscapes. *Theoretical Population Biology* 64:119–27.
Hanski, I. and O. Ovaskainen. 2000. The metapopulation capacity of a fragmented landscape. *Nature* 404:755–8.
Haridas, C. V. and S. Tuljapurkar. 2007. Time, transients and elasticity. *Ecology Letters* 10: 1143–53.
Harris, R. B., L. A. Maguire, and M. L. Shaffer. 1987. Sample sizes for minimum viable population estimation. *Conservation Biology* 1:72–6.
Harrison, G. W. 1979. Stability under environmental stress: resistance, resilience, persistence, and variability. *American Naturalist* 113:659–69.
Hart, D. R. 2006. When do marine reserves increase fishery yield? *Canadian Journal of Fisheries and Aquatic Sciences* 63:1445–9.
Hassell, M. P. 1975. Density-dependence in single-species populations. *Journal of Animal Ecology* 44:283–95.
Hassell, M. P., J. H. Lawton, and R. M. May. 1976. Patterns of dynamical behaviour in single-species populations. *Journal of Animal Ecology* 45:471–86.
Hastings, A. 1977. Spatial heterogeneity and the stability of predator–prey systems. *Theoretical Population Biology* 12:37–48.
Hastings, A. 1987. Cycles in cannibalistic egg–larval interactions. *Journal of Mathematical Biology* 24:651–66.
Hastings, A. 2001. Transient dynamics and persistence of ecological systems. *Ecology Letters* 4: 215–20.
Hastings, A. 2004. Transients: the key to long-term ecological understanding? *Trends in Ecology & Evolution* 19:39–45.
Hastings, A. 2010. Timescales, dynamics, and ecological understanding. *Ecology* 91:3471–89.
Hastings, A. and L. W. Botsford. 1999. Equivalence in yield from marine reserves and traditional fisheries management. *Science* 284:1537.
Hastings, A. and L. Botsford. 2006a. Persistence of spatial populations depends on returning home. *Proceedings of the National Academy of Sciences* 103:6067.
Hastings, A. and L. W. Botsford. 2006b. A simple persistence condition for structured populations. *Ecology Letters* 9:846–52.
Hastings, A. and R. F. Constantino. 1987. Cannibalistic egg–larva interactions in *Tribolium*: an explanation for the oscillations in population numbers. *American Naturalist* 130:36–52.
Hastings, A. and R. F. Constantino. 1991. Oscillations in population numbers: age-dependent cannibalism. *Journal of Animal Ecology* 60:471–82.
Hastings, A. and C. L. Wolin. 1989. Within patch dynamics in a metapopulation. *Ecology* 70: 1261–6.
Hebblewhite, M., M. Percy, and R. Serrouya. 2003. Black bear (*Ursus americanus*) survival and demography in the Bow Valley of Banff National Park, Alberta. *Biological Conservation* 112: 415–25.
Heppell, S. S., C. Pfister, and H. de Kroon, H. 2000. Elasticity analysis in population biology: methods and applications. *Ecology* 81:605–6.
Heppell, S. S., J. R. Walters, and L. B. Crowder. 1994. Evaluating management alternatives for red-cockaded woodpeckers: a modeling approach. *Journal of Wildlife Management* 58:479–87.
Hershey, A. E., J. Pastor, B. J. Peterson, and G. W. Kling. 1993. Stable isotopes resolve the drift paradox for *Baetis* mayflies in an arctic river. *Ecology* 74:2315–25.
Heyde, C. C. and J. E. Cohen. 1985. Confidence intervals for demographic projections based on products of random matrices. *Theoretical Population Biology* 27:120–53.
Higgins, K. A. Hastings, J. N. Sarvela, and L. W. Botsford. 1997. Stochastic dynamics and deterministic skeletons: Population behavior of Dungeness crab. *Science* 276:1431–5.
Hilborn, R. 2010. Pretty good yield and exploited fishes. *Marine Policy* 34:193–6.
Hilborn, R. 2014. Introduction to marine managed areas. *Advances in Marine Biology* 69:1–13.
Hilborn, R., J.-J. Maguire, A. M. Parma, and A. A. Rosenberg. 2001. The precautionary approach and risk management: can they increase the probability of successes in fishery management? *Canadian Journal of Fisheries and Aquatic Sciences* 58:99–107.

Hilborn, R. and M. Mangel. 1997. *The Ecological Detective: Confronting Models with Data*. Princeton University Press, Princeton, NJ, USA.

Hilborn, R., T. P. Quinn, D. E. Schindler, and D. E. Rogers. 2003. Biocomplexity and fisheries sustainability. *Proceedings of the National Academy of Sciences of the United States of America* 100:6564–8.

Hilborn, R. and C. J. Walters. 1992. *Quantitative Fisheries Stock Assessment: Choice, Dynamics and Uncertainty*. Springer Science+Business Media, Dordrecht.

Hill, M. F., A. Hastings, and L. W. Botsford. 2002. The effects of small dispersal rates on extinction times in structured metapopulation models. *The American Naturalist* 160:389–402.

Hixon, M. A., S. W. Pacala, and S. A. Sandin. 2002. Population regulation: Historical context and contemporary challenges of open vs. closed systems. *Ecology* 83:1490–508.

Hjort, J. 1914. Fluctuations in the great fisheries of northern Europe. *Rapport et Proces-Verbaux des Réunions du Conseil Permanent International pour L'Exploitation de la Mer* XX:1–228.

Hjort, J., G. Jahn, and P. Ottestad. 1933. The optimum catch. *Hvalradets Skrifter* 7:92–107.

Hjort, J. and E. Lea. 1914. The age of herring. *Nature* 94:60–1.

Hobbs, R. C. and L. W. Botsford. 1989. Dynamics of an age-structured prey with density- and predation-dependent recruitment: the Dungeness crab and a nemertean egg predator worm. *Theoretical Population Biology* 36:1–22.

Holland, D. S. and R. J. Brazee. 1996. Marine reserves for fisheries management. *Marine Resource Economics* 11:157–72.

Holling, C. S. 1968. Development of models of ecological processes. *SIAM Review* 10:476.

Holling, C. S. 1973. Resilience and stability of ecological systems. *Annual Review of Ecology and Systematics* 4:1–23.

Holmes, E. E. (2001). Estimating risks in declining populations with poor data. *Proceedings of the National Academy of Sciences of the United States of America* 98:5072–7.

Hopf, J. K., G. P. Jones, D. H. Willamson, and S. R. Connolly. 2015. Fishery consequences of marine reserves: short-term pain for longer-term gain. *Ecological Applications* 26:818–29.

Houde, E. D. 1989. Comparative growth, mortality, and energetics of marine fish larvae: temperature and implied latitudinal effects. *Fishery Bulletin* 87:471–95.

Hsieh, C.-H., C. Anderson, and G. Sugihara. 2008. Extending nonlinear analysis to short ecological time series. *American Naturalist* 171:71–80.

Hsieh, C.-H., C. S. Reiss, J. R. Hunter, J. R. Beddington, R. M. May, and G. Sugihara. 2006. Fishing elevates variability in the abundance of exploited species. *Nature* 443:859–62.

Huffaker, C. B. 1958. Experimental studies on predation: dispersion factors and predator–prey oscillations. *Hilgardia* 27:343–83.

Hunsicker, M. E., L. Ciannelli, K. M. Bailey, J. A. Buckel, J. Wilson White, J. S. Link, T. E. Essington, S. Gaichas, T. W. Anderson, R. D. Brodeur, K.-S. Chan, K. Chen, G. Englund, K. T. Frank, V. Freitas, M. A. Hixon, T. Hurst, D. W. Johnson, J. F. Kitchell, D. Reese, G. A. Rose, H. Sjodin, W. J. Sydeman, H. W. van der Veer, K. Vollset, and S. Zador. 2011. Functional responses and scaling in predator–prey interactions of marine fishes: contemporary issues and emerging concepts. *Ecology Letters* 14:1288–99.

Huston, M., D. DeAngelis, and W. Post. 1988. New computer models unify ecological theory. *BioScience* 38:682–91.

Hutchinson, G. E. 1978. *An Introduction to Population Ecology*. Yale University Press, New Haven, CT, USA.

International Council for Exploration of the Seas (ICES). 1983. Report of the working group on methods of fish stock assessments. *ICES C.M. 1983*/Assess:17.

International Council for Exploration of the Seas (ICES). 1984. Report of the working group on methods of fish stock assessments. *ICES C.M. 1984*/Assess:19.

Ims, R., J. Henden, and S. Killengreen. 2008. Collapsing population cycles. *Trends in Ecology & Evolution* 23:79–86.

Iooss, G. and D. Joseph. 1990. *Elementary Stability and Bifurcation Theory*. Second edition. Springer-Verlag, New York.

Ives, A. R. and S. R. Carpenter. 2007. Stability and diversity of ecosystems. *Science* 317:58–62.

Jacobi, M. N. and P. R. Jonsson. 2011. Optimal networks of nature reserves can be found through eigenvalue perturbation theory of the connectivity matrix. *Ecological Applications* 21:1861–70.

Jensen, J. L. W. V. 1906. Sur les fonctions convexes et les inégalités entre les valeurs moyennes. *Acta Mathematica* 30:175–93.

Johannes, R. E. 1978. Traditional marine conservation areas in Oceania and their demise. *Annual Review of Ecology and Systematics* 9:349–64.

Judson, O. P. 1994. The rise of the individual-based model in ecology. *Trends in Ecology & Evolution* 9:9–14.

Kaplan, D., L. Botsford, and S. Jorgensen. 2006. Dispersal per recruit: an efficient method for assessing sustainability in marine reserve networks. *Ecological Applications* 16:2248–63.

Kaplan, D. M. 2006. Alongshore advection and marine reserves: consequences for modeling and management. *Marine Ecology - Progress Series* 309:11–24.

Kaplan, D. M., L. W. Botsford, M. R. O'Farrell, S. D. Gaines, and S. Jorgensen. 2009. Model-based assessment of persistence in proposed marine protected area designs. *Ecological Applications* 19:433–48.

Kaplan, K. A., L. Yamane, L. W. Botsford, M. L. Baskett, A. Hastings, S. Worden, and J. W. White. 2019. Setting expected timelines of fished population recovery for the adaptive management of a marine protected area network. *Ecological Applications*, in press.

Kendall, B. E., C. J. Briggs, W. W. Murdoch, P. Turchin, S. P. Ellner, E. McCauley, R. M. Nisbet, and S. N. Wood. 1999. Why do populations cycle? A synthesis of statistical and mechanistic modeling approaches. *Ecology* 80:1789–805.

Kendall, B. E., S. P. Ellner, E. McCauley, S. N. Wood, C. J. Briggs, W. W. Murdoch, and P. Turchin. 2018. Population cycles in the pine looper moth: dynamical tests of mechanistic hypotheses. *Ecological Monographs* 75:259–76.

Keyfitz, N. 1968. *Introduction to the Mathematics of Population*. Addison-Wesley, Reading, Massachusetts.

Kierstead, H. and L. B. Slobodkin. 1953. The size of water masses containing plankton blooms. *Journal of Marine Research* 12:141–7.

Kilduff, D. P., L. W. Botsford, and L. C. Thompson. 2018. Frequency content of environmental variability and extinction risk of age-structured populations: Chinook salmon (*Oncorhynchus tschawytscha*) as an example. *Theoretical Ecology* https://doi.org/10.1007/s12080-018-0401-4

Kilduff, D. P., E. Di Lorenzo, L. W. Botsford, and S. L. H. Teo. 2015. Changing central Pacific El Niños reduce stability of North American salmon survival rates. *Proceedings of the National Academy of Sciences of the United States of America* 112:10962–6.

Kinzig, A. P., S. W. Pacala, and D. Tilman, editors. 2001. *The Functional Consequences of Biodiversity: Empirical Progress and Theoretical Extensions*. Princeton University Press, Princeton, NJ, USA.

Koellner, T. and O. J. Schmitz. 2006. Biodiversity, ecosystem function, and investment risk. *BioScience* 56:977–85.

Kot, M. 2001. *Elements of Mathematical Ecology*. Cambridge University Press, Cambridge, UK.

Kot, M., M. A. Lewis, and P. van den Driessche. 1996. Dispersal data and the spread of invading organisms. *Ecology* 77:2027–42.

Krebs, C. J. 1972. *Ecology: The Experimental Analysis of Distribution and Abundance*. Harper & Row, New York.

Krebs, C. J., R. Boonstra, and S. Boutin. 2018. Using experimentation to understand the 10-year snowshoe hare cycle in the boreal forest of North America. *Journal of Animal Ecology* 87: 87–100.

Krebs, C. J., K. Kielland, J. Bryant, M. O'Donoghue, F. Doyle, C. McIntyre, D. DiFolco, N. Berg, S. Carriere, R. Boonstra, S. Boutin, A. J. Kenney, D. G. Reid, K. Bodony, J. Putera, H. K. Timm, and T. Burke. 2013. Synchrony in the snowshoe hare (Lepus americanus) cycle in northwestern North America, 1970–2012. *Canadian Journal of Zoology* 91:562–72.

Kritzer, J. P. and P. F. Sale. 2004. Metapopulation ecology in the sea: from Levins' model to marine ecology and fisheries science. *Fish and Fisheries* 5:131–40.

Kritzer, J. P. and P. F. Sale. 2006. *Marine Metapopulations*. Academic Press, San Diego, CA, USA.

Krkosek, M., R. Hilborn, R. M. Peterman, and T. P. Quinn. 2011. Cycles, stochasticity and density dependence in pink salmon population dynamics. *Proceedings of the Royal Society B: Biological Sciences* 278:2060–8.

Kroeker, K. J., R. L. Kordas, R. N. Crim, and G. G. Singh. 2010. Meta-analysis reveals negative yet variable effects of ocean acidification on marine organisms. *Ecology Letters* 13:1419–34.

Lande, R., S. Engen, and B.-E. Sæther. 2003. *Stochastic Population Dynamics in Ecology and Conservation*. Oxford University Press, Oxford, UK.

Lande, R. and S. H. Orzack. 1988. Extinction dynamics of age-structured populations in a fluctuating environment. *Proceedings of the National Academy of Sciences* 85:7418–21.

Largier, J. L. 2003. Considerations in estimating larval dispersal distances from oceanographic data. *Ecological Applications* 13:71–89.

Larkin, P. A. 1977. An epitaph for the concept of maximum sustainable yield. *Transactions of the American Fisheries Society* 106:1–11.

Law, R. (1983). A model for the dynamics of a plant population containing individuals classified by age and size. *Ecology* 64:224–30.

Le Cren, E. D. 1958. Observations on the growth of perch (*Perca fluviatilis* L.) over twenty-two years with special reference to the effects of temperature and changes in population density. *Journal of Animal Ecology* 27:287–334.

Le Cren, E. D., C. Kipling, and J. C. McCormack. 1977. A study of the number, biomass and year-class strengths of perch (*Perca fluviatilis* L.) in Windermere from 1941 to 1966. *Journal of Animal Ecology* 46:281–307.

Le Quesne, W. J. F. and J. K. Pinnegar. 2012. The potential impacts of ocean acidification: scaling from physiology to fisheries. *Fish and Fisheries* 13:333–44.

Lee, R. 1974. The formal dynamics of controlled populations and the echo, the boom, and the bust. *Demography* 11:563–85.

Lefkovitch L. P. 1965. The study of population growth in organisms grouped by stages. *Biometrics*, 21, 1–18.

Leggett, W. and E. DeBlois. 1994. Recruitment in marine fishes: Is it regulated by starvation and predation in the egg and larval stages? *Netherlands Journal of Sea Research* 32:119–34.

Leigh, E. 1981. The average lifetime of a population in a varying environment. *Journal of Theoretical Biology* 90: 213–39.

Leopold, A., L. K. Sowls, and D. L. Spencer. 1947. A survey of over-populated deer ranges in the United States. *The Journal of Wildlife Management* 11:162–77.

Leslie, P. H. 1945. On the use of matrices in certain population mathematics. *Biometrika* 33: 183–212.

Leslie, P. H. 1957. An analysis of the data for some experiments carried out by Gause with populations of the protozoa, *Paramecium aurelia* and *Paramecium caudatum*. *Biometrika* 44: 314–27.

Leslie, P. H. 1966. The intrinsic rate of increase and the overlap of successive generations in a population of guillemots (*Uria aalge* Pont.). *Journal of Animal Ecology* 35:291–301.

Lester, S. E., B.S. Halpern, K. Grorud-Colvert, J. Lubchenco, B.I. Ruttenberg, S.D. Gaines, S. Airame, and R.R. Warner. 2009. Biological effects within no-take marine reserves: a global synthesis. *Marine Ecology – Progress Series* 384:33–46.

Letcher, B. H., J. A. Priddy, J. R. Walters, and L. B. Crowder. 1998. An individual-based, spatially-explicit simulation model of the population dynamics of the endangered red-cockaded woodpecker, Picoides borealis. *Biological Conservation* 86:1–14.

Lett, P. F., R. K. Mohn, and D. F. Gray. 1981. Density-dependent processes and management strategy for the Northwest Atlantic harp seal population. Pages 135–57 in C. W. Fowler and T. D. Smith, editors. *Dynamics of Large Mammal Populations*. John Wiley & Sons Ltd, New York.

Levin, S. A. 1992. The problem of pattern and scale in ecology. *Ecology* 73:1943–67.

Levin, S. A. and C. P. Goodyear. 1980. Analysis of an age-structured fishery model. *Journal of Mathematical Biology* 9:245–74.

Levin, S. A. and R. T. Paine. 1974. Disturbance, patch formation, and community structure. *Proceedings of the National Academy of Sciences of the United States of America* 71: 2744–7.

Levins, R. 1966. The strategy of model building in population biology. *American Scientist* 54: 421–31.

Levins, R. 1969. Some demographic and genetic consequences of environmental heterogeneity for biological control. *Bulletin of the Entomological Society of America*, 15, 237–40.

Levins, R. 1970. Extinction. *Lecture Notes in Mathematics* 2:75–107.

Lewis, E. G. 1942. On the generation and growth of a population. *Sankhya*: *The Indian Journal of Statistics* 6:93–6.

Lewis, M. A. 1997. Variability, patchiness, and jump dispersal in the spread of an invading population. Pages 46–74 in D. Tilman and P. Kareiva, editors. *Spatial Ecology*. Princeton University Press, Princeton, NJ, USA.

Lewontin, R. C. and D. Cohen. 1969. On population growth in a randomly varying environment. *Proceedings of the National Academy of Sciences* 62:1056–60.

Li, J., L. Cai, and Y. Li. 2017. Stage-structured wild and sterile mosquito population models and their dynamics. *Journal of Biological Dynamics* 11:79–101.

Li, T.-Y. and J. A. Yorke. 1975. Period three implies chaos. *American Mathematical Monthly* 82: 985–92.

Liebhold, A., V. Sork, M. Peltonen, W. Koenig, O. N. Bjørnstad, R. Westfall, J. Elkinton, and J. M. H. Knops. 2004b. Within-population spatial synchrony in mast seeding of North American oaks. *Oikos* 104:156–64.

Liebhold, A., D. K. Walter, and O. N. Bjørnstad. 2004a. Spatial synchrony in population dynamics. *Annual Review of Ecology and Systematics* 35:467–90.

Lockwood, D. R., A. Hastings, and L. W. Botsford. 2002. The effects of dispersal patterns on marine reserves: Does the tail wag the dog? *Theoretical Population Biology* 61:297–309.

Lonergan, M. 2014. Data availability constrains model complexity, generality, and utility: a response to Evans et al. *Trends in Ecology & Evolution* 29:301–2.

Loreau, M. 2010. *From Populations to Ecosystems: Theoretical Foundations for a New Ecological Synthesis*. Princeton University Press, Princeton, NJ, USA.

Loreau, M. and C. de Mazancourt. 2013. Biodiversity and ecosystem stability: a synthesis of underlying mechanisms. *Ecology Letters* 16:106–15.

Lotka, A. J. 1907. Relation between birth rates and death rates. *Science* 26:21–2.

Lotka, A. J. 1922. The stability of the normal age distribution. *Proceedings of the National Academy of Sciences* 8:339–45.

Lotka, A. J. 1939. Contact points of population study with related branches of science. *Proceedings of the American Philosophical Society* 80:601–26.

Lowe, C. G., D. T. Topping, D. P. Cartamil, and Y. P. Papastamatiou. 2003. Movement patterns, home range, and habitat utilization of adult kelp bass *Paralabrax clathratus* in a temperate no-take marine reserve. *Marine Ecology Progress Series* 256:205–16.

Lubchenco, J. and B. A. Menge. 1978. Community development and persistence in a low rocky intertidal zone. *Ecological Monographs* 48:67–94.

Lubchenco, J., S. R. Palumbi, S. D. Gaines, and S. Andelman. 2003. Plugging a hole in the ocean: the emerging science of marine reserves. *Ecological Applications* 13:S3–7.

Lubina, J. A. and S. A. Levin. 1988. The spread of a reinvading species: range expansion in the California sea otter. *American Naturalist* 131:526–43.

Luck, G. W., G. C. Daily, and P. R. Ehrlich. 2003. Population diversity and ecosystem services. *Trends in Ecology & Evolution* 18:331–6.

Ludwig, D. 1999. Is it meaningful to estimate a probability of extinction? *Ecology* 80: 298–310.

M'Kendrick, A. G. 1926. Applications of mathematics to medical problems. *Proceedings of the Edinburgh Mathematical Society* 44:98–130.

MacArthur, R. H. and E. O. Wilson. 1967. *The Theory of Island Biogeography*. Princeton University Press. Princeton, NJ, USA.

Mace, P. M. 1994. Relationships between common biological reference points used as thresholds and targets of fisheries management strategies. *Canadian Journal of Fisheries and Aquatic Sciences* 51:110–22.

Mace, P. M. 2001. A new role for MSY in single-species and ecosystem approaches to fisheries stock assessment and management. *Fish and Fisheries* 2:2–32.

Mace, P. M. and M. Sissenwine. 1993. How much spawning per recruit is enough? *Canadian Special Publications in Fisheries and Aquatic Sciences* 120:101–18.

Macfayden, A. 1963. *Animal Ecology: Aims and Methods*. Second edition. Pitman, London.

Maciel, G. A. and F. Lutscher. 2013. How individual movement response to habitat edges affects population persistence and spatial spread. *American Naturalist* 182:42–52.

Maguire, L. A., G. F. Wilhere, and Q. Dong. 1995. Population viability analysis for red-cockaded woodpeckers in the Georgia Piedmont. *Journal of Wildlife Management* 59:533–42.

Malthus, T. R. 1960. *On Population*. Random House, New York.

Mangel, M. 1998. No-take areas for sustainability of harvested species and a conservation invariant for marine reserves. *Ecology Letters* 1:87–90.

Markowitz, H. 1952. Portfolio selection. *The Journal of Finance* 7:77–91.

Marquet, P. A., A. P. Allen, J. H. Brown, J. A. Dunne, B. J. Enquist, J. F. Gillooly, P. A. Gowaty, J. L. Green, J. Harte, S. P. Hubbell, J. O'Dwyer, J. G. Okie, A. Ostling, M. Ritchie, D. Storch, and G. B. West. 2014. On theory in ecology. *BioScience* 64:701–10.

Márquez-Farías, J. F. and F. J. Rosales-Juárez. 2013. Intrinsic rebound potential of the endangered (Totoaba macdonaldi) population, endemic to the Gulf of California, México. *Fisheries Research* 147:150–3.

Maunder, M. N. and A. E. Punt. 2013. A review of integrated analysis in fisheries stock assessment. *Fisheries Research* 142:61–74.

Maunder, M., J. Sibert, A. Fonteneau, J. Hampton, P. Kleiber, and S. Harley. 2006. Interpreting catch per unit effort data to assess the status of individual stocks and communities. *ICES Journal of Marine Science* 63:1373–85.

May, R. M. 1972. On relationships among various types of population models. *American Naturalist* 107:46–57.

May, R. M. 1973. *Stability and Complexity in Model Ecosystems*. Princeton University Press, Princeton, NJ, USA.

May, R. M. 1974. Biological populations with nonoverlapping generations: stable points, stable cycles, and chaos. *Science* 186:645–7.

May, R. M. and G. F. Oster. 1976. Bifurcations and dynamic complexity in simple ecological models. *American Naturalist* 110:573–99.

Maynard Smith, J. 1968. *Mathematical Ideas in Biology*. Cambridge University Press, Cambridge, UK.

Maynard Smith, J. 1974. *Models in Ecology*. Cambridge University Press, Cambridge, UK.

McClenachan, L. and J. N. Kittinger. 2013. Multicentury trends and the sustainability of coral reef fisheries in Hawai'i and Florida. *Fish and Fisheries* 14:239–55.

McClure, M. M., E. E. Holmes, B. L. Sanderson, and C. E. Jordan. 2003. A large-scale, multispecies status assessment: anadromous salmonids in the Columbia River basin. *Ecological Applications* 13:964–89.

McCullough, D. R. 1981. Population dynamics of the Yellowstone grizzly bear. Pages 173–96 in C. W. Fowler and T. D. Smith, editors. *Dynamics of Large Mammal Populations*. John Wiley & Sons Ltd, New York.

McElhany, P., C. Busack, M. Chilcote, S. Kolmes, B. McIntosh, J. Myers, et al. 2006. *Revised Viability Criteria for Salmon and Steelhead in the Willamette and Lower Columbia Basins* (pages 1–178). Willamette/Lower Columbia Technical Recovery Team and Oregon Department of Fish and Wildlife, Salem, OR, USA.

McKelvey, R., D. Hankin, K. Yanosko, and C. Snygg. 1980. Stable cycles in multistage recruitment models: an application to the northern California Dungeness crab (Cancer magister) fishery. *Canadian Journal of Fisheries and Aquatic Sciences* 37:2323–45.

Meiri, S. and T. Dayan. 2003. On the validity of Bergmann's rule. *Journal of Biogeography* 30:331–51.

Mendelssohn, R. 1976. Optimization problems associated with a Leslie matrix. *American Naturalist* 110:339–49.

Menge, B. A. and D. N. L. Menge. 2018. Testing the intermittent upwelling hypothesis: comment. *Ecology* 100: e02476.

Mesterton-Gibbons, M. 1993. Why demographic elasticities sum to one: a postscript to de Kroon et al. *Ecology* 74:2467–8.

Methot, R.D., Jr. and C. R. Wetzel. 2013. Stock synthesis: a biological and statistical framework for fish stock assessment and fishery management. *Fisheries Research* 142: 86–99.

Metz, J. A. J. and O. Diekmann, editors. 1986. *The Dynamics of Physiologically Structured Populations*. Springer-Verlag, Berlin.

Metz, J. A. J. and O. Diekmann. 2014. Formulating models for structured populations. Pages 78–116 in J. A. J. Metz and O. Diekmann, editors. *The Dynamics of Physiologically Structured Populations*. Springer-Verlag, Berlin.

Miller, T. J., L. B. Crowder, J. A. Rice, and E. Marschall. 1988. Larval size and recruitment mechanisms in fishes: toward a conceptual framework. *Canadian Journal of Fisheries and Aquatic Sciences* 45:1657–70.

Mitarai, S., D. A. Siegel, J. R. Watson, C. Dong, and J. C. McWilliams. 2009. Quantifying connectivity in the coastal ocean with application to the Southern California Bight. *Journal of Geophysical Research* 114:C02010–21.

Moffitt, E. A., L. W. Botsford, D. M. Kaplan, and M. R. O'Farrell. 2009. Marine reserve networks for species that move within a home range. *Ecological Applications* 19:1835–47.

Moffitt, E. A., J. W. White, and L. W. Botsford. 2011. The utility and limitations of size and spacing guidelines for designing marine protected area (MPA) networks. *Biological Conservation* 144: 306–18.

Moilanen, A., K. A. Wilson, and H. P. Possingham. 2008. *Spatial Conservation Prioritization: Quantitative and Computational Tools*. Oxford University Press, Oxford, UK.

Moloney, C. L., L. W. Botsford, and J. L. Largier. 1994. Development, survival and timing of metamorphosis of planktonic larvae in a variable environment: the Dungeness crab as an example. *Marine Ecology – Progress Series* 113:61–79.

Moore, J. W., M. McClure, L. A. Rogers, and D. E. Schindler. 2010. Synchronization and portfolio performance of threatened salmon. *Conservation Letters* 3:340–8.

Moran, P. A. P. 1950. Some remarks on animal population dynamics. *Biometrics* 6:250–8.

Moran, P. A. P. 1953. The statistical analysis of the Canadian lynx cycle. *Australian Journal of Zoology* 1:163–73.

Morgan, S. G., J. L. Fisher, S. H. Miller, S. T. McAfee, and J. L. Largier. 2009. Nearshore larval retention in a region of strong upwelling and recruitment limitation. *Ecology* 90:3489–502.

Morris, W. F. and D. F. Doak. 2002. *Quantitative Conservation Biology Theory and Practice of Population Viability Analysis*. Sinauer Associates Inc. Sunderland, MA, USA.

Müller, K. 1954. *Investigations on the Organic Drift in North Swedish Streams*. Pages 202–7. Institute of Freshwater Research, Drottingholm, Sweden.

Murdoch, W., B. Kendall, R. Nisbet, and C. Briggs. 2002. Single-species models for many-species food webs. *Nature* 417:541–3.

Murphy, G. I. 1977. Clupeoids. Page 372 in *Fish Population Dynamics*, J. A. Gulland, editor. Wiley, New York.

Murray, S. N., R. F. Ambrose, J. A. Bohnsack, and L. W. Botsford. 1999. No-take reserve networks: sustaining fishery populations and marine ecosystems. *Fisheries* 24:11–25.

Myers, R. A. and B. Worm. 2003. Rapid worldwide depletion of predatory fish communities. *Nature* 423:280–3.

Nahin, P.J. 2006. *Dr. Euler's Fabulous Formula Cures Many Mathematical Ills*. Princeton University Press, Princeton NJ, USA.

National Research Council. 2004. *Improving the Use of the "Best Scientific Information Available" Standard in Fisheries Management*. National Academies Press, Washington, D.C. http://doi.org/10.17226/11045

Neave, F. 1953. Principles affecting the size of pink and chum salmon populations in British Columbia. *Canadian Journal of Fisheries and Aquatic Sciences* 9:450–91.

Neubert, M. G. and H. Caswell. 1997. Alternatives to resilience for measuring the responses of ecological systems to perturbations. *Ecology* 78:653–65.

Nicholson, A. 1933. The balance of animal populations. *Journal of Animal Ecology* 2:131–78.

Nicholson, A. and V. A. Bailey. 1935. The balance of animal populations—part I. *Journal of Zoology* 105:551–98.

Nickols, K. J., J. W. White, J. L. Largier, and B. Gaylord. 2015. Marine population connectivity: Reconciling large-scale dispersal and high self-retention. *American Naturalist* 185:196–211.

Nickols, K. J., J. W. White, D. Malone, M. H. Carr, R. M. Starr, M. L. Baskett, A. Hastings, and L. W. Botsford. 2019. Setting expectations for adaptive management of marine protected areas. *Journal of Applied Ecology*, in press.

Nisbet, R. M. 1997. Delay-differential equations for structured populations. In S. Tuljapurkar & H. Caswell, editors, *Structured-Population Models in Marine, Terrestrial, and Freshwater Systems* (pp. 89–118). Chapman & Hall, New York.

Nisbet, R. M. and J. R. Bence. 1989. Alternative dynamic regimes for canopy-forming kelp: a variant on density-vague population regulation. *The American Naturalist* 134:377–408.

Nisbet, R.M. and W.S.C. Gurney. 1982. *Modelling Fluctuating Populations*. John Wiley & Sons Ltd., Chicester, UK.

O'Farrell, M. R. and L. W. Botsford. 2005. Estimation of change in lifetime egg production from length frequency data. *Canadian Journal of Fisheries and Aquatic Science* 62:1626–39.

O'Leary, B. C., M. Winther-Janson, J. M. Bainbridge, J. Aitken, J. P. Hawkins, and C. M. Roberts. 2016. Effective coverage targets for ocean protection. *Conservation Letters* 9:398–404.

Okubo, A. 1978. Horizontal dispersion and critical scales for phytoplankton patches. Pages 21–42 in J. H. Steele, editor. *Spatial Pattern in Plankton Communities*. NATO Conference Series (IV Marine Sciences). Boston, USA.

Okubo, A. and S. A. Levin. 2001. *Diffusion and Ecological Problems: Modern Perspectives*. Second edition. Springer Science+Business Media, New York.

Oster, G. F., D. M. Auslander, and T. T. Allen. 1976. Deterministic and stochastic effects in population dynamics. *Journal of Dynamic Systems, Measurement, and Control* 98:44–8.

Ovaskainen, O. 2003. Habitat destruction, habitat restoration and eigenvector–eigenvalue relations. *Mathematical Biosciences* 181:165–76.

Ovaskainen, O. 2017. The interplay between immigration and local population dynamics in metapopulations. *Annales Zoologici Fennici* 54:113–21.

Ovaskainen, O. and I. Hanski. 2001. Spatially structured metapopulation models: Global and local assessment of metapopulation capacity. *Theoretical Population Biology* 60:281–302.

Ovaskainen, O. and I. Hanski. 2003. How much does an individual habitat fragment contribute to metapopulation dynamics and persistence? *Theoretical Population Biology* 64:481–95.

Ovaskainen, O., H. J. de Knegt, and M. D. M. Delgado. 2016. *Quantitative Ecology and Evolutionary Biology: Integrating Models with Data*. Oxford University Press, Oxford, UK.

Ovaskainen, O. and B. Meerson. 2010. Stochastic models of population extinction. *Trends in Ecology & Evolution* 25:643–52.

Paine, R. T. and S. A. Levin. 1981. Intertidal landscapes: disturbance and the dynamics of pattern. *Ecological Monographs* 51:145–78.

Parlett, B. 1970. Ergodic properties of populations I: The one sex model. *Theoretical Population Biology* 1:191–207.

Pearcy, W. G. 1992. *Ocean Ecology of North Pacific Salmonids*. University of Washington Press, Seattle, USA.

Pelc, R. A., R. R. Warner, S. D. Gaines, and C. B. Paris. 2010. Detecting larval export from marine reserves. *Proceedings of the National Academy of Sciences* USA 107:18266–71.

Pella, J. J. and P. K. Tomlinson. 1969. A generalized stock production model. *Inter-American Tropical Tuna Commission Bulletin* 13:421–96.

Pelletier, D., J. Claudet, J. Ferraris, L. Benedetti-Cecchi, and J. A. Garcìa-Charton. 2008. Models and indicators for assessing conservation and fisheries-related effects of marine protected areas. *Canadian Journal of Fisheries and Aquatic Sciences* 65:765–79.

Pelletier, D. and S. Mahevas. 2005. Spatially explicit fisheries simulation models for policy evaluation. *Fish and Fisheries* 6:307–49.

Pennycuick, C. J., R. M. Compton, and L. Beckingham. 1968. A computer model for simulating the growth of a population, or of two interacting populations. *Journal of Theoretical Biology* 18: 316–29.

Perron, O. 1907. Zur theorie der matrices. *Mathematische Annalen* 64:248–63.

Pew Oceans Commission. 2003. America's living oceans: charting a course for sea change. https://www.pewtrusts.org/-/media/assets/2003/06/02/full_report.pdf

Pikitch, E., E. A. Santora, A. Babcock, A. Bakun, R. Bonfil, D. O. Conover, P. A. O. Dayton, P. Doukakis, D. Fluharty, and B. Heheman. 2004. Ecosystem-based fishery management. *Science* 305:346–7.

Pikitch, E. K. 1987. Use of a mixed-species yield-per-recruit model to explore the consequences of various management policies for the Oregon flatfish fishery. *Canadian Journal of Fisheries and Aquatic Sciences* 44:349–59.

Pimm, S. L. 1984. The complexity and stability of ecosystems. *Nature* 307:321–6.

Plagányi, É. E., A. E. Punt, R. Hillary, E. B. Morello, O. Thébaud, T. Hutton, R. D. Pillans, J. T. Thorson, E. A. Fulton, A. D. M. Smith, F. Smith, P. Bayliss, M. Haywood, V. Lyne, and P. C. Rothlisberg. 2014. Multispecies fisheries management and conservation: tactical applications using models of intermediate complexity. *Fish and Fisheries* 15:1–22.

Platt, J. R. 1964. Strong inference. *Science* 146:347–53.

Polis, G. A. 1981. The evolution and dynamics of intraspecific predation. *Annual Review of Ecology and Systematics* 12:225–51.

Pollock, D. E. 1979. Predator–prey relationships between the rock lobster *Jasus lalandi* and the mussel *Aulacomya ater* at Robben Island on the Cape West Coast of Africa. *Marine Biology* 52: 347–56.

Popper, K. R. 1959. *The Logic of Scientific Discovery*. Hutchinson, London.

Possingham, H. P. and J. Roughgarden. 1990. Spatial population dynamics of a marine organism with a complex life cycle. *Ecology* 71:973–85.

Possingham, H. P., S. Tuljapurkar, J. Roughgarden, and M. Wilks. 1994. Population cycling in space-limited organisms subject to density-dependent predation. *The American Naturalist*, 143:563–82.

Pulliam, H. 1988. Sources, sinks, and population regulation. *American Naturalist* 132:652.

Purves, D., J. Scharlemann, M. Harfoot, and T. Newbold. 2013. Time to model all life on earth. *Nature* 493:295–7.

Railsback, S. F. and V. Grimm. 2011. *Agent-based and Individual-based Modeling: A Practical Introduction*. Princeton University Press, Princeton, NJ, USA.

Rankin, D. J. and H. Kokko. 2007. Do males matter? The role of males in population dynamics. *Oikos* 116:335–48.

Rasmuson, L. K. 2013. The biology, ecology and fishery of the Dungeness crab, *Cancer magister*. In *Advances in Marine Biology*, vol. 65. (Lesser, M. ed.), Elsevier, Oxford, UK. pp. 95–148.

Rassweiler, A., C. Costello, and D. A. Siegel. 2012. Marine protected areas and the value of spatially optimized fishery management. *Proceedings of the National Academy of Sciences* 109:11884–9.

Reid, C. 1899. *The Origin of the British Flora*. Dulau, London.

Restrepo, V. R. and J. E. Powers. 1999. Precautionary control rules in US fisheries management: specification and performance. *ICES Journal of Marine Science: Journal du Conseil* 56:846–52.

Ricard, D., C. Minto, O. P. Jensen, and J. K. Baum. 2011 Evaluating the knowledge base and status of commercially exploited marine species with the RAM Legacy Stock Assessment Database. *Fish and Fisheries* 13:380–98.

Ricker, W. E. 1945. A method of estimating minimum size limits for obtaining maximum yield. *Copeia* 1945:84–94.

Ricker, W. E. 1950. Cycle dominance among the Fraser sockeye. *Ecology* 31:6–26.

Ricker, W. E. 1954. Stock and recruitment. *Journal of the Fisheries Research Board of Canada* 11: 559–623.

Roberts, C. M. and N. V. Polunin. 1991. Are marine reserves effective in management of reef fisheries? *Reviews in Fish Biology and Fisheries* 1:65–91.

Robertson, P. A. and A. A. Rosenberg. 1988. Harvesting gamebirds. Pages 177–201 in P. J. Hudson and M. R. W. Rands, editors. *Ecology and Management of Gamebirds*. Blackwell Science Ltd, Boston, USA.

Robson, D. S. and D. G. Chapman. 1961. Catch curves and mortality rates. *Transactions of the American Fisheries Society* 90:181–9.

Roughgarden, J., S. Gaines, and H. Possingham. 1988. Recruitment dynamics in complex life cycles. *Science* 241:1460–6.

Roughgarden, J. and Y. Iwasa. 1986. Dynamics of a metapopulation with space-limited subpopulations. *Theoretical Population Biology* 29:235–61.

Roughgarden, J., Y. Iwasa, and C. Baxter. 1985. Demographic theory for an open marine population with space-limited recruitment. *Ecology* 66:54–67.

Rosenberg, A. A., S. Swartz, and G. C. Darcy. 1993. *Overview Paper: The Scientific Basis for Definitions of Overfishing in the United States*. Pages 6–18 in A. A. Rosenberg, editor. National Oceanic and Atmospheric Administration, Silver Spring, MD.

Rothschild, B. J. and A. Suda. 1977. Population dynamics of tuna. Pages 309–34 in J. A. Gulland, editor. *Fish Population Dynamics*. John Wiley & Sons Ltd., New York.

Ruel, J. J. and M. P. Ayres. 1999. Jensen's inequality predicts effects of environmental variation. *Trends in Ecology & Evolution* 14:361–6.

Sale, P. F., I. Hanski, and J. P. Kritzer. 2006. The merging of metapopulation theory and marine ecology: establishing the historical context. Pages 3–28 in *Marine Metapopulations*. Academic Press, San Diego, CA, USA.

Salmon, W. C. 1973. *Logic*. Prentice Hall, Englewood Cliffs, NJ, USA.

Sauer, J. R. and N. Slade. 1985. Mass-based demography of a hispid cotton rat (Sigmodon hispidus) population. *Journal of Mammology* 66:316–328.

Sauer, J. R. and N. Slade. 1986a. Size-dependent population dynamics of *Microtus ochrogaster*. *American Naturalist* 127:902–908.

Sauer, J. R. and N. Slade. 1986b. Size-dependent population dynamics of *Microtus ochrogaster*. *The American Naturalist* 127:902–8.

Sauer, J. R. and N. Slade. 1987. Uinta ground squirrel demography: is body mass a better categorical variable than age? *Ecology* 68:642–50.

Schaefer, M. B. 1954. Some aspects of the dynamics of populations important to the management of the commercial marine fisheries. *Inter-American Tropical Tuna Commission Bulletin* 1:25–56.

Schaefer, M. B. 1967a. Fishery dynamics and present status of the yellowfin tuna population of the Eastern Pacific Ocean. *Inter-American Tropical Tuna Commission Bulletin* 12:89–136.

Schaefer, M. B. 1967b. Some considerations of population dynamics and economics in relation to the management of the commercial marine fisheries. *Journal of the Fisheries Research Board of Canada* 14:669–81.

Schele, L. and D. Freidel. 1990. *A Forest of Kings: The Untold Story of the Ancient Maya*. Morrow, New York.

Schindler, D. E. and R. Hilborn. 2015. Prediction, precaution, and policy under global change. *Science* 347:953–4.

Schindler, D. E., R. Hilborn, B. Chasco, C. P. Boatright, T. P. Quinn, L. A. Rogers, and M. S. Webster. 2010. Population diversity and the portfolio effect in an exploited species. *Nature* 465:609–12.

Schmidt, A. E., L. W. Botsford, D. P. Kilduff, R. W. Bradley, J. Jahncke, and J. M. Eadie. 2018. Changing environmental spectra influence age-structured populations: increasing ENSO frequency could diminish variance and extinction risk in long-lived seabirds. *Theoretical Ecology* 11:367–77.

Schnute, J. 1977. Improved estimates from the Schaefer production model: theoretical considerations. *Journal of the Fisheries Research Board of Canada* 34:583–603.

Schoener, T. W. 1973. Population growth regulated by intraspecific competition for energy or time: some simple representations. *Theoretical Population Biology* 4:56–84.

Schwager, M., K. Johst, and F. Jeltsch. 2006. Does red noise increase or decrease extinction risk? Single extreme events versus series of unfavorable conditions. *American Naturalist* 167:879–88.

Secor, D. H. 2014. The unit stock concept: bounded fish and fisheries. Pages 7–28, Chapter 2 in S. X. Cadrin, L. A. Kerr, and S. Mariani, editors. *Stock Identification Methods: Applications in Fishery Science*. Academic Press, London.

Secor, D. H. 2015. *Migration Ecology of Marine Fishes*. Johns Hopkins University Press, Baltimore, MD, USA.

Sethi, S. P. 2019. *Optimal Control Theory*. Third edition. Springer US, Cham, Switzerland.

Shanks, A. L. 2013. Atmospheric forcing drives recruitment variation in the Dungeness crab (*Cancer magister*), revisited. *Fisheries Oceanography* 22:263–72.

Shanks, A. L. and S. G. Morgan. 2018. Testing the intermittent upwelling hypothesis: upwelling, downwelling, and subsidies to the intertidal zone. *Ecological Monographs* 88:22–35.

Shanks, A. L. and G. C. Roegner. 2007. Recruitment limitation in Dungeness crab populations is driven by variation in atmospheric forcing. *Ecology* 88:1726–37.

Sharpe, F. R. and A. J. Lotka. 1911. A problem in age-distribution. *Philosophical Magazine* 21:435–8.

Sibert, J., J. Hampton, P. Kleiber, and M. Maunder. 2006. Biomass, size, and trophic status of top predators in the Pacific Ocean. *Science* 314:1773–6.

Sigler, L. 2002. *A Translation into Modern English of Leonardo Pisano's Book of Calculations*. Springer-Verlag, New York.

Simberloff, D. S. and E. O. Wilson. 1969. Experimental zoogeography of islands: the colonization of empty islands. *Ecology* 50:278–96.

Simberloff, D. S., and E. O. Wilson. 1970. Experimental zoogeography of islands. A two-year record of colonization. *Ecology* 51:934–7.

Sissenwine, M. and J.G. Shepherd. 1987. An alternative perspective on recruitment overfishing and biological reference points. *Canadian Journal of Fisheries and Aquatic Sciences* 44: 913–18.

Skellam, J. G. 1951. Random dispersal in theoretical populations. *Biometrika* 38:196–218.

Slade, N. and D. F. Balph. 1974. Population ecology of Uinta ground squirrels. *Ecology* 55:989–1003.

Slobodkin, L. B. 1953. A possible initial condition for red tides on the coast of Florida. *Journal of Marine Research* 12:148–55.

Smith, A. T. and M. E. Gilpin. 1997. Spatially correlated dynamics in a pika metapopulation. Pages 407–28 in *Metapopulation Biology: Ecology, Genetics, and Evolution*. Academic Press, San Diego, CA, USA.

Smith, B. D., L. W. Botsford, and S. R. Wing. 1998. Estimation of growth and mortality parameters from size frequency distributions lacking age patterns: the red sea urchin (*Strongylocentrotus franciscanus*) as an example. *Canadian Journal of Fisheries and Aquatic Sciences* 55:1236–47.

Smith, D. and N. Keyfitz, editors. 1977. *Mathematical Demography: Selected Papers*. Springer-Verlag, Berlin.

Smith, H. 2011. *An Introduction to Delay Differential Equations with Applications to the Life Sciences*. Springer-Verlag, New York.

Smith, H. S. 1935. The role of biotic factors in the determination of population densities. *Journal of Economic Entomology* 28:873–98.

Smith, T. and A. E. Punt. 2001. The gospel of maximum sustainable yield in fisheries management: birth, crucifixion, and reincarnation. Pages 41–66 in J. D. Reynolds, G. M. Mace, K. H. Redford, and J. G. Robinson, editors. *Conservation of Exploited Species*. Cambridge University Press, Cambridge, UK.

Smith, T. D. 1994. *Scaling Fisheries: The Science of Measuring the Effects of Fishing*. Cambridge University Press, Cambridge, UK.

Speirs, D. C. and W. S. C. Gurney. 2001. Population persistence in rivers and estuaries. *Ecology* 82:1219–37.

Starr, R. M., D.E. Wendt, C.L. Barnes, C.I. Marks, D. Malone, G. Waltz, K.T. Schmidt, J. Chiu, A.L. Launer, N.C. Hall, and N. Yochum. 2015. Variation in responses of fishes across multiple reserves within a network of Marine Protected Areas in temperate waters. *PLoS ONE* 10:e0118502.

Steiner, U. K., S. Tuljapurkar, and T. Coulson. 2014. Generation time, net reproductive rate, and growth in stage-age-structured populations. *The American Naturalist* 183:771–83.

Stott, I., S. Townley, and D.J. Hodgson. 2011. A framework for studying transient dynamics of population projection matrix models. *Ecology Letters* 14:959–70.

Strong, D. R. 1986. Density-vague population change. *Trends in Ecology & Evolution* 1:39–42.

Sykes, Z. M. 1969. On discrete stable population theory. *Biometrics* 25:285–93.

Taubert, F., M. W. Jahn, H.-J. Dobner, T. Wiegand, and A. Huth. 2015. The structure of tropical forests and sphere packings. *Proceedings of the National Academy of Sciences USA* 112:15125–9.

Taylor, B. L. 1995. The reliability of using population viability analysis for risk classification of species. *Conservation Biology* 9:551–8.

Taylor, F. 1979. Convergence to the stable age distribution in populations of insects. *The American Naturalist* 113:511–30.

Tegner, M. J. and P. K. Dayton. 1981. Population structure, recruitment and mortality of two sea urchins (*Strongylocentrotus franciscanus* and *S. purpuratus*) in a kelp forest. *Marine Ecology – Progress Series* 5:255–68.

Thibaut, L. M., S. R. Connolly, and H. P. A. Sweatman. 2012. Diversity and stability of herbivorous fishes on coral reefs. *Ecology* 93:891–901.

Thomas, C. D. and I. Hanski. 1997. Butterfly metapopulations. Pages 359–86 in *Metapopulation Biology: Ecology, Genetics, and Evolution*. Academic Press, San Diego, CA, USA.

Thomas, C. J., J. Lambrechts, E. Wolanski, V. A. Traag, V. D. Blondel, E. Deleersnijder, and E. Hanert. 2014. Numerical modelling and graph theory tools to study ecological connectivity in the Great Barrier Reef. *Ecological Modelling* 272:160–74.

Thomson, D. M. 2005. Matrix models as a tool for understanding invasive plant and native plant interactions. *Conservation Biology* 19:917–28.

Tilman, D. 1999. The ecological consequences of changes in biodiversity: a search for general principles. *Ecology* 80:1455–74.

Tilman, D. and P. Kareiva. 1997. *Spatial Ecology: The Role of Space in Population Dynamics and Interspecific Interactions*. Princeton University Press, Princeton, NJ, USA.

Tilman, D. and C. Lehman. 2002. Biodiversity, composition, and ecosystem processes: theory and concepts. in A. P. Kinzig, S. W. Pacala, and D. Tilman, editors. *The Functional Consequences of Biodiversity: Empirical Progress and Theoretical Extensions*. Princeton University Press, Princeton, NJ, USA.

Timmermann, A., J. Oberhuber, A. Bacher, and M. Esch. 1999. Increased El Niño frequency in a climate model forced by future greenhouse warming. *Nature* 398:694–7.

Tinker, M. T., D. F. Doak, and J. A. Estes. 2008. Using demography and movement behavior to predict range expansion of the southern sea otter. *Ecological Applications* 18:1781–94.

Torrence, C. and G.P. Compo. 1998. A practical guide to wavelet analysis. *Bulletin of the American Meteorological Society* 79:61–78.

Treml, E. A., P. N. Halpin, D. L. Urban, and L. F. Pratson. 2007. Modeling population connectivity by ocean currents, a graph-theoretic approach for marine conservation. *Landscape Ecology* 23: 19–36.

Trucco, E. 1965a. Mathematical models for cellular systems. The von Foerster equation. Part I. *Bulletin of Mathematical Biophysics* 27:285–304.

Trucco, E. 1965b. Mathematical models for cellular systems. The von Foerster equation. Part II. *Bulletin of Mathematical Biophysics* 27:449–71.

Tuljapurkar, S. 1989. An uncertain life: demography in random environments. *Theoretical Population Biology* 35:227–94.

Tuljapurkar, S. 1990. *Population Dynamics in Variable Environements*. Springer-Verlag, Berlin.

Tuljapurkar, S. D. 1982. Population dynamics in variable environments. II. Correlated environments, sensitivity analysis and dynamics. *Theoretical Population Biology* 21:114–40.

Tuljapurkar, S. D. and S. H. Orzack. 1980. Population dynamics in variable environments. I. Long-run growth rates and extinction. *Theoretical Population Biology* 18:314–42.

Turchin, P. 2003. *Complex Population Dynamics*. Princeton University Press, Princeton, NJ, USA.

US Commission on Ocean Policy. 2004. *An Ocean Blueprint for the 21st Century*. Washington, DC, USA.

Usher, M. B. 1966. A matrix approach to the management of renewable resources, with special reference to selection forests. *Journal of Applied Ecology* 3:355–67.

Usher, M. B. 1972. Developments in the Leslie matrix model. Pages 29–60 in J. N. R. Jeffers, editor. *Mathematical Models in Ecology*. Blackwell, Oxford, UK.

Utida, S. 1957. Population fluctuation, an experimental and theoretical approach. *Cold Spring Harbor Symposium in Quantitative Biology* 22:139–51.

Utida, S. 1967. Damped oscillation of population density at equilibrium. *Researches in Population Ecology* 9:1–9.

Valenzuela-Quiñonez, F., F. Arreguín-Sánchez, S. Salas-Márquez, F. J. García-De León, J. C. Garza, M. J. Roman-Rodríguez, and J. A. De-Anda-Montañez. 2015. Critically endangered totoaba Totoaba macdonaldi: signs of recovery and potential threats after a population collapse. *Endangered Species Research* 29:1–11.

van de Pol, M., Y. Vindenes, B. E. Sæther, S. Engen, B J. Ens, K. Oosterbeek, and J. M. Tinbergen. 2011. Poor environmental tracking can make extinction risk insensitive to the colour of environmental noise. *Proceedings of the Royal Society B: Biological Sciences* 278:3713–22.

van Groenendael, J., H. de Kroon, and H. Caswell. 1988. Projection matrices in population biology. *Trends in Ecology & Evolution* 3:264–9.

Vandermeer, J. H. 1975. 377: On the construction of the population projection matrix for a population grouped in unequal stages. *Biometrics* 31:239–42.

Vandermeer, J. H. 1978. Choosing category size in a stage projection matrix. *Oecologia* 32: 79–84.

Varley, G. C., G. R. Gradwell, and M. P. Hassell. 1973. *Insect Population Ecology*. Blackwell, Oxford, UK.

Vasseur, D. A. 2007. Populations embedded in trophic communities respond differently to coloured environmental noise. *Theoretical Population Biology* 72:186–96.

Verhulst, P.-F. 1838. Notice sur la loi que la population suit dans son accroissement. Correspondance mathématique et physique publiée par a. *Quetelet* 10:113–21.

Volterra, V. 1926. Fluctuations in the abundance of a species considered mathematically. *Nature* 118:558–60.

Walters, C. 1986. *Adaptive Management of Renewable Resources*. McMillan, New York.

Walters, C. J., R. Hilborn, and R. Parrish. 2007. An equilibrium model for predicting the efficacy of marine protected areas in coastal environments. *Canadian Journal of Fisheries and Aquatic Sciences* 64:1009–18.

Walters, C. J. and S. J. D. Martell. 2004. *Fisheries Ecology and Management*. Princeton University Press, Princeton, NJ, USA.

Walters, J. R., L. B. Crowder, and J. A. Priddy. 2002. Population viability analysis for red-cockaded woodpeckers using an individual-based model. *Ecological Applications* 12:249–60.

Walters, J. R., P. D. Doerr, and J. H. Carter. 1988. The cooperative breeding system of the red-cockaded woodpecker. *Ethology* 78:275–305.

Wang, B. and Y. Wang. 1996. Temporal structure of the Southern Oscillation as revealed by waveform and wavelet analysis. *Journal of Climate* 9:1586–98.

Waples, R. S. 1991. Pacific Salmon, *Oncorhynchus* spp., and the definition of "species" under the Endangered Species Act. *Marine Fisheries Review* 53:11–22.

Waters, T. F. 1972. The drift of stream insects. *Annual Review of Entomology* 17:253–72.

Watson, J. R., D. A. Siegel, B. E. Kendall, S. Mitarai, A. Rassweiler, and S. D. Gaines. 2011. Identifying critical regions in small-world marine metapopulations. *Proceedings of the National Academy of Sciences* 108:E907–13.

Webster, J. R., J. B. Waide, and B. C. Patten. 1974. Nutrient recycling and the stability of ecosystems. Pages 1–27 in F. G. Horwell, J. B. Gentry, and M. H. Smith, editors. *Mineral Cycling in Southeastern Ecosystems*. National Technical Information Service, Springfield, VA, USA.

Weller, D. E. 1987. A reevaluation of the −3/2 power rule of plant self-thinning. *Ecological Monographs* 57:23–43.

Werner, P. A. 1975. Predictions of fate from rosette size in teasel (*Dipsacus fullonum* L.). *Oecologia* 20:197–201.

Werner, P. A. and H. Caswell. 1977. Population growth rates and age versus stage-distribution models for teasel (*Dipsacus sylvestris* Huds.). *Ecology*, 58(5), 1103–11.

White, C. and B. Kendall. 2007. A reassessment of equivalence in yield from marine reserves and traditional fisheries management. *Oikos* 116:2039–43.

White, G. C. and K. P. Burnham. 1999. Program MARK: survival estimation from populations of marked animals. *Bird Study* 46:S120–39.

White, J. W., L. W. Botsford, M. L. Baskett, L. A. Barnett, R. J. Barr, and A. Hastings. 2011. Linking models with monitoring data for assessing performance of no-take marine reserves. *Frontiers in Ecology and the Environment* 9:390–9.

White, J. W., L.W. Botsford, A. Hastings, M. L. Baskett, D.M. Kaplan, and L. A. K. Barnett. 2013a. Transient responses of fished populations to marine reserve establishment. *Conservation Letters* 6:180–91.

White, J. W., L. W. Botsford, A. Hastings, and M. D. Holland. 2014a. Stochastic models reveal conditions for cyclic dominance in sockeye salmon populations. *Ecological Monographs* 84: 69–90.

White, J. W., L. W. Botsford, A. Hastings, and J. L. Largier. 2010a. Population persistence in marine reserve networks: incorporating spatial heterogeneities in larval dispersal. *Marine Ecology - Progress Series* 398:49–67.

White, J. W., L. W. Botsford, and E. A. Moffitt. 2010c. Decision analysis for designing marine protected areas for multiple species with uncertain fishery status. *Ecological Applications* 20: 1523–41.

White, J. W., L. W. Botsford, E. A. Moffitt, and D. T. Fischer. 2010b. Decision analysis for designing marine protected areas for multiple species with uncertain fishery status. *Ecological Applications* 20:1523–41.

White, J. W., B. J. Cole, G. N. Cherr, R. E. Connon, and S. M. Brander. 2017. Scaling up endocrine disruption effects from individuals to populations: outcomes depend on how many males a population needs. *Environmental Science & Technology* 51:1802–10.

White, J. W., K. J. Nickols, D. Malone, M. H. Carr, R. M. Starr, F. Cordoleani, M. L. Baskett, A. Hastings, and L. W. Botsford. 2016. Fitting state-space integral projection models to size-structured time series data to estimate unknown parameters. *Ecological Applications* 26:2675–92.

White, J. W., A. J. Scholz, A. Rassweiler, C. Steinback, L. W. Botsford, S. Kruse, C. Costello, S. Mitarai, D. A. Siegel, P. T. Drake, and C. A. Edwards. 2013b. A comparison of approaches used for economic analysis in marine protected area network planning in California. *Ocean and Coastal Management* 74:77–89.

White, J. W., J. Schroeger, P. T. Drake, and C. A. Edwards. 2014b. The value of larval connectivity information in the static optimization of marine reserve design. *Conservation Letters* 7:533–44.

White, S. M., P. Rohani, and S. M. Sait. 2010. Modelling pulsed releases for sterile insect techniques: fitness costs of sterile and transgenic males and the effects on mosquito dynamics. *Journal of Applied Ecology* 47:1329–39.

Wielgus, R. B., F. Sarrazin, R. Ferriere, and J. Clobert. 2001. Estimating effects of adult male mortality on grizzly bear population growth and persistence using matrix models. *Biological Conservation* 98:293–303.

Williams, J. L., T. E. X. Miller, and S. P. Ellner. 2012. Avoiding unintentional eviction from integral projection models. *Ecology* 93:2008–14.

Williams, K. S. and C. Simon. 1995. The ecology, behavior, and evolution of periodical cicadas. *Annual Review of Ecology and Systematics* 40:269–95.

Williams, P. D. and A. Hastings. 2013. Stochastic dispersal and population persistence in marine organisms. *American Naturalist* 182:271–82.

Williamson, M. 1972. *The Analysis of Biological Populations*. Arnold, London.

Wilson, E. O. and D. S. Simberloff. 1969. Experimental zoogeography of islands: defaunation and monitoring techniques. *Ecology* 50:267–78.

Wolkovich, E. M., B. I. Cook, K. K. McLauchlan, and T. J. Davies. 2014. Temporal ecology in the Anthropocene. *Ecology Letters* 17:1365–79.

Wood, L. J., L. Fish, J. Laughren, and D. Pauly. 2008. Assessing progress towards global marine protection targets: shortfalls in information and action. *Oryx* 42:340–51.

Woodley, S., B. Bertzky, N. Crawhall, N. Dudley, J. M. Londoño, K. MacKinnon, K. Redford, and T. Sandwith. 2012. Meeting Aichi Target 11: What does success look like for protected area systems? *Parks* 18:23–36.

Worden, L., L.W. Botsford, A. Hastings, and M.D. Holland. 2010. Theoretical population biology. *Theoretical Population Biology* 78:239–49.

Worm, B., E. B. Barbier, N. Beaumont, J. E. Duffy, C. Folke, B. S. Halpern, J. B. C. Jackson, H. K. Lotze, F. Micheli, S. R. Palumbi, E. Sala, K. A. Selkoe, J. J. Stachowicz, and R. Watson. 2006. Impacts of biodiversity loss on ocean ecosystem services. *Science* 314:787–90.

Worm, B., R. Hilborn, J. K. Baum, T. A. Branch, J. S. Collie, C. Costello, M. J. Fogarty, E. A. Fulton, J. A. Hutchings, S. Jennings, O. P. Jensen, H. K. Lotze, P. M. Mace, T. R. McClanahan, C. Minto, S. R. Palumbi, A. M. Parma, D. Ricard, A. A. Rosenberg, R. Watson, and D. Zeller. 2009. Rebuilding global fisheries. *Science* 325:578–85.

Yamane, L., L. W. Botsford, and D. P. Kilduff. 2017. Tracking restoration of population diversity via the portfolio effect. *Journal of Applied Ecology* 55:472–81.

Yoda, K., T. Kira, H. Ogawa, and K. Hozumi. 1963. Self-thinning in overcrowded pure stands under cultivated and natural conditions (intraspecific competition among higher plants XI). *Journal of the Institute of Polytechnics, Osaka City University, Series D* 14:107–29.

Young, T. P. and C.K. Augspurger. 1991. Ecology and evolution of long-lived semelparous plants. *Trends in Ecology & Evolution* 6:285–9.

Zadeh, L. A. 1973. Outline of new approach to the analysis of complex structures and decision processes. *IEEE Transactions on Systems, Man, and Cybernetics* 3:28–44.

Index

1T cycles 118
2*T* cycles 99, 166–176, 179, 276

Accuracy 3
African elephant (*Loxodonta africana*) 252–253
Age-lumping 82–84
Antarctic fin whale (*Balaenoptera physalus*) 177–179

Barnacle (*Balanus glandula)* 177, 183–187, 226
Basin of attraction 38
Beetle, flour (*Tribolium spp.)* 177, 187–189
Beverton-Holt model 35, 107, 115, 169–171, 292
Bifurcation 171–172
Binomial distribution 13
Birth-death process 195
Black bear (*Ursus americanus*) 82–84
Boundary conditions 23, 63, 127, 135, 217, 225

California coastal redwoods (*Sequoia sempervirens*) 148–149, 156
Cannibalism 177, 181–183, 187–189
Catastrophic variability 195
Catch per unit effort 52, 269
Central limit theorem 47–48
Characteristic equation 41, 69, 74, 76, 170, 177
Characteristics, method of 61–63, 126–127
Cobwebbing 36–37, 43–45
Cod, Atlantic (*Gadus morhua*) 112–113, 116–118,
Cohort distribution 123, 134–138
Cohort replacement rate 254–255
Cohort resonance 88, 112–119, 166, 120
Compensation ratio 276–277
Complex numbers 88–93, 97–98, 175, 178
Conditional arguments 7
Continuous time 24
CPUE *see Catch per unit effort*
CRR *see Cohort replacement rate*
CRT *see Critical replacement threshold*
Critical replacement threshold 238–242, 279, 287, 290

Deductive logic 10
Demographic stochasticity 12–14, 194–195
Diffusion 198–199, 217–222
Discrete time 24
Dungeness crab (*Metacarcinus* (formerly *Cancer*) *magister*) 8, 135–138, 169, 172, 177, 180–183, 189–191
Dynamic pool model *see single cohort model*

Echo effect 96–98, 114
Eigenvalue 40–43, 73–85, 88–104, 148, 159, 196
  Dominant 116–119
  2nd and 3rd 88–104, 116, 119
Eigenvector 73–77, 90
Elasticity 77–81, 83, 153, 155, 160, 177, 181, 188, 195, 199–201
Environmental variability 194
Equilibrium 38, 40–42, 37, 44, 51–52, 106–107, 168–170, 177, 184, 188
Escalator–boxcar train 138–139
Exponential growth 24–29, 40, 69
Extinction (quasi-extinction) 45–49, 195, 197–199, 207–212, 231, 249–254, 257–259

Fibonacci series 23, 41, 74
Fibonacci 22–23
'Filling in' of age distributions 104, 297–300
FLEP *see Fraction of lifetime egg production*
Fourier transform 109–112, 116–118
Fraction of lifetime egg production 290–292

Generality (as a property of models) 3
Generation time 91, 96–98, 114, 142, 167
Goemetric growth *see exponential growth*

Harvest 189, 267–286
Holism 3, 4

IBM *see Individual based model*
IBT *see Island biogeography theory*
  Individual based model 14
Inductive logic 9–10
Influence function 65, 169, 172–174, 185–191
Integral projection models 139–142, 155, 159, 223
Inverse problem 55
IPM *see Integral projection models*
Island fox (*Urocyon littoralis*) 259–261
i-state 12, 14–16, 60, 157
Incidence function model (of metapopulations) 231–235
Integro-difference equation 187, 223
Island biogeography theory 230–231
Iteroparous 94, 116–117

J *see Jacobian matrix*
Jacobian matrix 114–115
Jensen's inequality 50, 197

Kernel 139–142, 159, 223–224

L *see Leslie matrix*
Larval settlement 137, 183
Lefkovitch 147–149, 151–155, 160
Leonardo of Pisa *see Fibonacci*
LEP *see Lifetime egg production*
Leslie matrix 54, 66–84, 90–105, 114–115, 139, 141, 145, 147, 149, 159–160, 197, 200, 232–233
Levins model (of metapopulations) 230–233
Lifetime egg production 276–278, 282–285
Lifetime reproductive rate 70, 107–108, 142, 160, 162, 249
Linear models 25–30, 40–42

Logistic models 30–37, 51–52, 267–270
Logistic models, discrete time 34–37

M'Kendrick/von Foerster model 54–64, 68, 82, 122–128, 135, 138, 142, 145, 227–229
Marine protected areas 104, 286–300
Marine reserves *see also Marine protected areas* 5, 239–242
Markovian 11, 155–156
Master equation 203
Maximum sustainable yield 52, 269–270, 281–284, 289, 295
Measurement error 15–16
Metapopulation 215, 230–242
MPA *see Marine protected areas*
MSY *see Maximum sustainable yield*

Nemertean worm (*Carcinonemertes errantia*) 177, 181–182
Network persistence 235–242, 236
Next generation kernel 142

Obligate semelparous 93
Observation error *see Measurement error*
Ocean acidification 17
Optimal harvest 29, 275, 280–289
Overcompensatory 26, 166
Overfishing 282–285, 290

Perron-Frobenius 77, 82, 93–94, 171
Persistence 39, 235–242, 249, 278
Perturbation 40
Plaice, North Sea (*Pleuronectes platessa*) 273–275
Population, closed 172, 184–185
Population, open 168, 172, 184–185
Population spread 216–225
Population viability analysis 207–212, 247, 253–261
Portfolio effect 243, 261–264
Power spectrum 110–112, 116–117
Precautionary approach 281–286
Precision 3
Process error 15–16
p-state 12, 14–15
PVA *see Population viability analysis*

Quasi-extinction *see Extinction*

R_0 *see Lifetime reproductive rate*
Realism 1, 3
Red sea urchins (*Mesocentrotus* (formerly *Strongylocentrotus*) *franciscanus*) 128–134, 142
Red-cockaded woodpecker (*Picoides borealis*) 14–15
Reaction-diffusion model 217–219
Renewal equation 54, 64–66, 68, 82, 98, 128, 168, 178
Reproductive rate 65
Reproductive value 76, 79
Resilience 39
Ricker model 35, 43–45, 107, 169, 252
Ring-necked pheasant (*Phasianus colchicus*) 28–29
Rockfishes (*Sebastes* spp.) 104, 279–285, 293–294, 297

SAD *see Stable age distribution*
Salmon, Pacific (*Oncorhynchus* spp) 93–95, 112–119, 247, 255–258
  Chinook salmon (*O. tshawytscha*) 29, 204–205, 254–255, 263
  Pink salmon (*O. gorbuscha*) 95
  Sockeye salmon (*O. nerka*) 116–119, 262
Sea otter (*Enhydra lutris*) 221–222
Sea turtle, loggerhead (*Caretta caretta*) 152–154, 156–158, 210–212
Self persistence 238–243
Semelparous 93, 95, 116–117
Sensitivity 77–81, 83, 196–197, 199
Separation of variables 26, 30
SFA *see Small fluctuation approximation*
Single cohort model 270–275
Small fluctuation approximation 195–212
Space, explicit 215, 231–235
Space, implicit 215
Space, realistic 216
Spawning potential ratio 277–285, 290
Spawning stock biomass per recruit 277–278
SPOM *see Stochastic patch occupancy model*
SPR *see Spawning potential ratio*
SSBR *see Spawning stock biomass per recruit*
Stability 37–45
Stability, global 39
Stability, local 38–39, 166–176
Stable age distribution 71–73, 77, 79, 95, 98–105, 153, 155
Stand distribution 123, 128–134
Stochastic patch occupancy model 231–235
Stock assessment 10, 284
Stock 35
Strategic models 1, 3–4
Structural error 15–16
Surplus yield model 51, 267
Synchrony, spatial 243–244
System state 1, 11, 157

Tactical models 1, 3–4, 292–295
Taylor series 43, 113
Teasel (*Dipsacus sylvestris*) 149–151
TED *see Turtle exclusion device*
Totoaba (*Totoaba macdonaldi*) 77–81, 198–203, 208–209
Tuna, yellowfin (*Thunnus albacares*) 268–269
Turtle exclusion device 152–154, 210–212

Uinta ground squirrel (*Spermophilus armatus*) 151–152

von Bertalanffy model 78, 129–130, 139, 271, 275, 292

Wavelet 118–120

Yield per recruit 271–275, 278, 280–281
Yield per recruit model *see Single cohort model*
YPR *see Yield per recruit*